RESONANT POWER CONVERTERS

RESONANT POWER CONVERTERS

MARIAN K. KAZIMIERCZUK
Wright State University

DARIUSZ CZARKOWSKI
University of Florida

A Wiley-Interscience Publication
JOHN WILEY & SONS, INC.
New York / Chichester / Brisbane / Toronto / Singapore

This text is printed on acid-free paper.

Library of Congress Cataloging in Publication Data:

Kazimierczuk, Marian.
Resonant power converters / Marian K. Kazimierczuk, Dariusz Czarkowski.
p. cm.
"A Wiley-Interscience publication."
Includes bibliographical references and index.
ISBN 0-471-04706-6 (acid-free paper)
1. Electric current converters. 2. Electric resonators. 3. Power electronics. I. Czarkowski, Dariusz. II. Title.
TK7872.C8K39 1995
621.3815'322--dc20 94-31255

Printed in the United States of America

10 9 8 7 6 5 4 3 2 1

To our families

CONTENTS

PREFACE

This book is about the analysis and design of dc-ac resonant inverters, high-frequency rectifiers, and dc-dc resonant converters that are basic building blocks of various high-frequency, high-efficiency energy processors. The past decade has initiated a revolution in and unprecedented growth of power electronics. Continuing advances in this area have resulted in dc and ac energy sources that are smaller, more efficient, lighter, less expensive, and more reliable than ever before. Power processors are widely used in computer, telecommunication, instrumentation, automotive, aerospace, defense, and consumer industries. Dc-dc converters are being used in power supplies to power practically all electronic circuits that contain active devices. The increasing complexity of modern electronic systems is imposing challenging demands on the capabilities of circuit designers.

Many design problems encountered in a great diversity of products can be solved using the unique capabilities of resonant technology. Information on resonant power processors is scattered throughout many different technical journals and application notes. This volume brings the principles of resonant technology to students and practicing design engineers. The state-of-the-art technology of high-frequency resonant power processors is covered in a systematic manner for the first time. The reader will be introduced to the topologies, characteristics, terminology, and mathematics of resonant converters. The book provides students and engineers with a sound understanding of existing high-frequency inverters, rectifiers, and dc-dc resonant converters and presents a general and easy-to-use tool of analysis and design of resonant power circuits.

The text provides rigorous in-depth analysis to help the reader understand how and why the power converters are built as they are. The fundamental-frequency component method is used throughout the entire book. This approach leads to relatively simple closed-form analytic expressions for converter characteristics, which provides good insight into circuit operation and greatly simplifies the design process. Graphical representations of various characteristics are emphasized throughout the text because they provide a visual picture of circuit operation and often yield insights not readily obtained from purely algebraic treatments.

This book is intended as a textbook for senior-level and graduate students in electrical engineering and as a reference for practicing design engineers, researchers, and consultants in industry. The objective of the book is to develop in the reader the ability to analyze and design high-frequency power electronic circuits. A knowledge of network analysis, electronic circuits and devices, complex algebra, Fourier series, and Laplace transforms is required to handle the mathematics in this book. Numerous analysis and design examples are included throughout the textbook. An extensive list of references is provided in each chapter. Problems are placed at the end of each chapter. Selected problem answers are given at the end of the book. Complete solutions for all problems are included in the *Solutions Manual*, which is available from the publisher for those instructors who adopt the book for their courses.

The book is divided into three parts: Part I. "Rectifiers," Part II. "Inverters," and Part III. "Converters."

High-frequency rectifiers are covered in Chapters 2 through 5. Chapter 2 deals with Class D current-driven rectifiers, and Chapter 3 is devoted to the study of Class D voltage-driven rectifiers. Each of these chapters contains analyses of three types of rectifiers, namely, the half-wave, transformer center-tapped, and bridge rectifier. Chapter 4 presents two Class E low dv/dt rectifiers, whereas Chapter 5 deals with two Class E low di/dt rectifiers.

High-frequency resonant inverters are discussed in Chapters 6 through 14. The Class D series resonant converter is thoroughly covered in Chapter 6. Many topics discussed in this chapter apply also to other resonant inverters presented in the following chapters. The Class D parallel resonant inverter is the topic of Chapter 7. Chapters 8 and 9 discuss Class D series-parallel and Class D CLL resonant inverters, respectively. Chapter 10 discusses zero-voltage-switching techniques in resonant inverters. The Class D current-source inverter is covered in Chapter 11. An example of a constant-frequency phase-controlled Class D resonant inverter, namely, the single-capacitor phase-controlled resonant inverter, is given in Chapter 12. The Class E resonant inverters are analyzed in Chapters 13 and 14. Chapter 13 deals with a zero-voltage-switching Class E inverter, and Chapter 14 presents a zero-current-switching Class E inverter.

Converters are studied in Part III, which ties together the material of Parts I and II. Resonant dc-dc converters that are a result of cascading resonant inverters with high-frequency rectifiers are presented in Chapters 15 through 21. Chapters 15 through 19 discuss converters, with inverters presented in

Chapters 6 through 10. Hence, Chapter 15 covers a Class D series resonant converter, Chapter 16 presents a Class D parallel resonant converter, Chapter 17 deals with a Class D series-parallel resonant converter, Chapter 18 gives an analysis of a Class D CLL resonant converter, and Chapter 19 discusses a Class D current-source converter. An example of matching a Class D inverter with a Class E rectifier that leads to a Class D-E resonant converter is presented in Chapter 20. Chapter 21 gives an analysis of a single-capacitor phase-controlled resonant converter that belongs to a broad family of phase-controlled converters.

We are pleased to express our gratitude to many individuals for their help during the preparation of this book. We wish to thank two anonymous reviewers for many helpful comments. The first author had the privilege to teach numerous superb students at the Technical University of Warsaw, Warsaw, Poland, and at Wright State University, Dayton, Ohio. He would like to express his deepest appreciation to them for their research contributions, ideas, suggestions, and critical evaluations of the original manuscript. Special thanks go to our students Becky Roman–Amador, Ronald L. Bobb, and Joseph P. Harrington for proofreading and editorial improvements. Recent support provided under National Science Foundation research grant ECS-8922694 is gratefully acknowledged. We wish to acknowledge the assistance provided by the College of Engineering and Computer Science and the Department of Electrical Engineering of Wright State University as well as the Florida Power Affiliates and Power Electronics Consortium at the Department of Electrical Engineering, University of Florida.

Throughout the entire course of this project, the support provided by John Wiley & Sons, Wiley-Interscience Division, was excellent. We wish to express our sincere thanks to George J. Telecki, Senior Editor; Rose Leo Kish, Editorial Assistant; Lisa Van Horn, Managing Editor; and Angioline Loredo, Associate Managing Editor. It has been a real pleasure working with them. Last but not least, we wish to thank our families for their support.

The authors invite the readers to contact them directly or through the publisher with comments and suggestions about this book.

Marian K. Kazimierczuk
Dariusz Czarkowski

LIST OF SYMBOLS

c_{pR} Power-output capability of rectifier
C Resonant capacitance
C_c Coupling capacitance
C_{ds} Drain-source capacitance of MOSFET
$C_{ds(25V)}$ Drain-source capacitance of MOSFET at $V_{DS} = 25$ V
C_f Filter capacitance
C_{fmin} Minimum value of C_f
C_{gd} Gate-drain capacitance of MOSFET
C_{gs} Gate-source capacitance of MOSFET
C_{iss} MOSFET input capacitance at $V_{DS} = 0$, $C_{iss} = C_{gs} + C_{gd}$
C_{oss} MOSFET output capacitance at $V_{GD} = 0$, $C_{oss} = C_{gs} + C_{ds}$
C_{out} Transistor output capacitance
C_{rss} MOSFET transfer capacitance, $C_{rss} = C_{gd}$
C_s Equivalent series-resonant capacitance
D_k k-th diode
f Switching frequency
f_o Resonant frequency
f_p Frequency of pole of transfer function
f_p Corner frequency of output filter
f_r Resonant frequency of L-C_s-R_s circuit
f_s Switching frequency
f_z Frequency of zero of transfer function
f_H Upper 3-dB frequency
i Current through resonant circuit
i_{cr} Ac component of i_{CR}
i_i Ac current source

i_o Ac current load
i_{Cf} Current through filter capacitance
i_{CR} Current through the C_f-R_L circuit
i_{Dk} Current through k-th diode
i_R Input current of rectifier
i_S Switch current
i_{Sk} Current through k-th switch
I_l Capacitor dc leakage current
I_m Amplitude of i
I_n n-th harmonic of the current to R_L-C_f-r_C circuit
I_{pk} Magnitude of cross-conduction current
I_{rms} rms value of i
$I_{Cf(rms)}$ rms value of i_{Cf}
I_{DM} Peak current of diode
I_{Drms} rms current of diode
I_D Average current through diode
I_O Dc output current
I_{OFF} Current at which the transistor turns off
I_{Omax} Maximum value of I_O
I_{SM} Peak current of switch
k Ratio R_L/r_C
K_I Current transfer function of rectifier
L Resonant inductance
L_e Inductance of electrodes
L_f Filter inductance
L_{fmin} Minimum value of L_f
L_t Inductance of terminations
L_{ESL} Equivalent series inductance
M Dc-dc voltage function of converter
$\mathbf{M_{VI}}$ Voltage transfer function of inverter
M_{VI} Amplitude of the voltage transfer function of inverter
M_{Vs} Voltage transfer function of switches
M_{Vr} Voltage transfer function of resonant circuit
M_{VR} Voltage transfer function of rectifier
n Transformer turns ratio
P_i Input power of rectifier
P_{lc} Ac conduction loss in filter inductor and capacitor
P_r Conduction loss in r
P_{rC} Conduction loss in filter capacitor
P_{tf} Average value of power loss associated with current fall time t_f
P_{tr} Average value of power loss associated with voltage rise time t_r
P_{toff} Turn-off switching losses
P_{ton} Turn-on switching losses
P_C Overall conduction loss of rectifier
P_D Total diode conduction loss

P_M Overall power dissipation in MOSFET (excluding gate drive power)
P_I Dc input power of converter
P_O Dc output power
P_{RF} Conduction loss in R_F
P_{VF} Conduction loss in V_F
P_T Overall power dissipation in inverter
PF Power factor
Q_g Gate charge
Q_k k-th transistor
Q_o Unloaded quality factor at f_o
Q_{oL} Quality factor of inductor
Q_L Loaded quality factor at f_o
r Total parasitic resistance
r_d Resistance representing dielectric losses
r_e Resistance of electrodes
r_t Resistance of terminations
r_C ESR of filter capacitor
r_{DS} On-resistance of MOSFET
R Overall resitance of series-resonant circuit
R Overall resitance of inverter
R Real part of **Z**
R_i Input resistance of rectifier
R_I Insulation resistance
R_F Diode forward resistance
R_G Gate resistance
R_L Dc load resistance
R_{Lmin} Minimum value of R_L
S_k k-th switch
t_f Fall time
t_r Rise time
T_a Ambient temperature
T_o Operating temperature
THD Total harmonic distortion
v Input voltage of series resonant circuit
v_c Voltage across C_f
v_i Ac voltage source
v_{i1} Fundamental component of v_{DS2}
v_{i1} Fundamental component of v
v_o Ac component of v_O
v_{Dk} Voltage across k-th diode
v_{DS} Drain-source voltage
v_{ESR} Voltage across r_C
v_{GSk} Drive voltage of k-th MOSFET
v_O Output voltage
v_R Input voltage of rectifier

v_{Ri} Voltage across R_i
v_1 Fundamental component of v
V_c Peak-to-peak value of v_c
$V_{c(max)}$ Maximum value of v_c
$V_{c(min)}$ Minimum value of v_c
V_m Amplitude of v_{i1}
V_n N-th harmonic of the output voltage of the rectifier
V_r Peak-to-peak value of ripple voltage
V_r Ripple voltage
V_{rms} rms value of v and v_{i1}
V_{rESR} Peak-to-peak value of v_{ESR}
V_{Cm} Amplitude of the voltage across capacitance
$\mathbf{V_{CR}}$ Voltage across a series combination of capacitance and resistance
V_{DS} Drain-source dc voltage
V_{DM} Reverse peak voltage of diode
V_F Diode forward voltage
V_{GSpp} Peak-to-peak gate-to-source voltage
V_I Dc input voltage of converter
V_{Lm} Amplitude of the voltage across inductance
V_O Dc output voltage
V_{Ri} rms value of v_{Ri}
V_{SM} Peak voltage of switch
V_{1m} Amplitude of v_{i1}
V_{1rms} rms value of fundamental component
V_{1rms} rms value of v_{i1}
W Energy stored in capacitance
X Imaginary part of $\mathbf{Z}$
$\mathbf{Z}$ Input impedance of resonant circuit
Z Magnitude of $\mathbf{Z}$
Z_o Characteristic impedance of resonant circuit
η Efficiency of converter
η_I Efficiency of inverter
η_{rc} Efficiency of resonant circuit
η_R Efficiency of rectifier
η_{tr} Efficiency of transformer
μ_n Mobility of electrons
μ_p Mobility of holes
ψ Angle between v_{i1} and i
ψ Phase of Z
ω Operating angular frequency
ω_o Resonant angular frequency
θ Phase of M_{Vr}

CHAPTER 1

INTRODUCTION

Modern electronic systems demand high-quality, small, lightweight, reliable, and efficient power processors. Linear power regulators can handle only low power levels (typically below 20 W), have a very low efficiency, and have a low power density because they require low-frequency (50 or 60 Hz) line transformers and filters. The higher the operating frequency, the smaller and lighter the transformers, filter inductors, and capacitors. In addition, dynamic characteristics of converters improve with increasing operating frequencies. The bandwidth of a control loop is usually determined by the corner frequency of the output filter. Therefore, high operating frequencies allow for achieving a faster dynamic response to rapid changes in the load current and/or the input voltage. As a result, high-frequency power technology, which employs semiconductor power switches, has developed rapidly in recent years.

High-frequency power processors can be classified into three categories:

- inverters (dc-ac converters)
- rectifiers (ac-dc converters)
- dc-dc converters

Dc-ac inverters, whose block diagram is depicted in Fig. 1.1(a), convert dc energy into ac energy. The input power source is either a dc voltage source or a dc current source. Inverters deliver ac power to a load impedance. In many applications, a sinusoidal output voltage or current is required. To generate a sinusoidal voltage and/or current waveforms, dc-ac inverters contain a resonant circuit; therefore, they are called *resonant dc-ac inverters*.

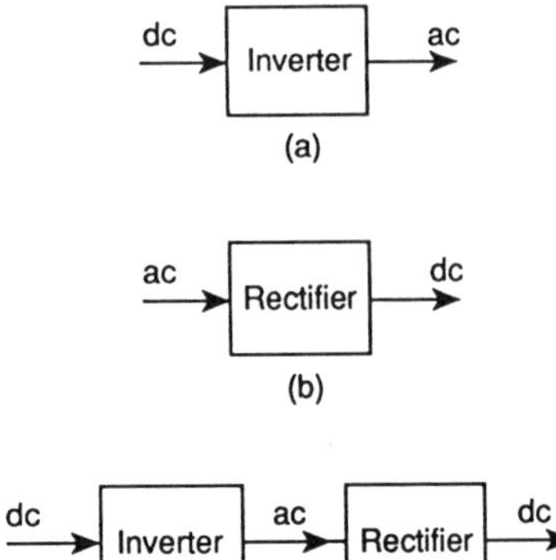

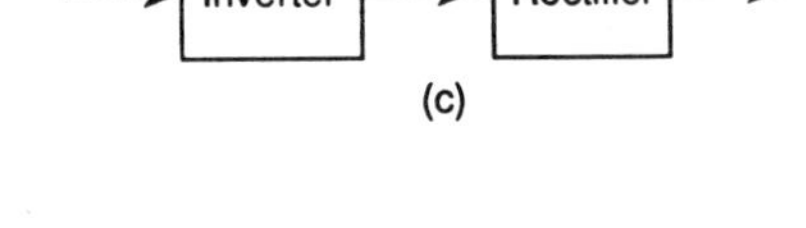

Figure 1.1: Block diagrams of high-frequency power processors. (a) Inverters (dc-ac converters). (b) Rectifiers (ac-dc converters). (c) dc-dc converters.

Power MOSFETs are usually used as switching devices in resonant inverters at high frequencies and isolated-gate bipolar transistors (IGBTs) and MOS-controlled thyristors (MCTs) at low frequencies.

A block diagram of an ac-dc rectifier is depicted in Fig. 1.1(b). Rectifiers convert an ac voltage or current into a dc voltage. At low frequencies of 50, 60, and 400 Hz, peak rectifiers are widely used; however, the ratio of the diode peak current to the diode average current is very high in these rectifiers and the diode current waveforms contain a large amount of harmonics. Therefore, peak rectifiers are not used at high frequencies. In this book, rectifiers suitable for high-frequency applications are given and analyzed.

High-frequency rectifiers can be divided into unregulated diode rectifiers, unregulated synchronous rectifiers, and regulated synchronous rectifiers. Both *pn* junction diodes and Schottky diodes are used in the first group of circuits. Schottky diodes are used only in low output voltage applications because their breakdown voltage is relatively low, typically less than 100 V. They have low forward voltage drops of the order of 0.3 to 0.4 V and do not suffer from reverse recovery, resulting in high rectifier efficiency. The leakage current in Schottky diodes is much higher than that in junction diodes. When the peak value of the diode voltage exceeds 100 V, *pn* junction diodes must be used. Power *pn* junction diodes have a forward voltage drop of about 1 V and a reverse recovery effect that limits the operating frequency of rectifiers.

In both unregulated and regulated synchronous rectifiers, power MOSFETs are used. Unlike diodes, power MOSFETs do not have an offset voltage. If their on-resistance is low, the forward voltage drops are low, yielding high efficiency.

High-frequency power processors are used in dc-dc power conversion. A block diagram of a dc-dc converter is shown in Fig. 1.1(c). The functions of dc-dc converters are as follows:

- to convert a dc input voltage V_I into a dc output voltage V_O;
- to regulate the dc output voltage against load and line variations;
- to reduce the ac voltage ripple on the dc output voltage below the required level;
- to provide isolation between the input source and the load (isolation is not always required);
- to protect the supplied system from electromagnetic interference (EMI);
- to satisfy various international and national safety standards.

Pulse-width modulated (PWM) converters [1]–[5] are well described in literature and are still widely used in low and medium power applications. However, PWM rectangular voltage and current waveforms cause turn-on and turn-off losses that limit the operating frequency. Rectangular waveforms generate broad-band electromagnetic energy and thus increase the potential for electromagnetic interference (EMI). The inability of PWM converters to operate efficiently at very high frequencies imposes a limit on the size of reactive components of the converter and, thereby, on power density. In search of converters capable of operating at higher frequencies, power electronics engineers started to develop converter topologies that shape either the sinusoidal current or the sinusoidal voltage waveform, significantly reducing switching losses. The key idea is to use a resonant circuit with a sufficiently high quality factor. Such converters are called *resonant dc-dc converters*.

A resonant dc-dc converter is obtained by cascading a resonant dc-ac inverter and a high-frequency rectifier, as shown in Fig. 1.1(c). The dc input power is first converted into ac power by the inverter, and then the ac power is converted back to dc power by the rectifier. If isolation is required, a high-frequency transformer, which is much smaller than a low-frequency transformer, can be inserted between the inverter and the rectifier.

The cascaded representation of a resonant dc-dc converter is convenient from analytical point of view. If the input current or the input voltage of the rectifier is sinusoidal, only the power of the fundamental component is converted from ac to dc power. In this case, the rectifier can be replaced by the input impedance, defined as the ratio of the fundamental components of the input voltage to the input current. In turn, the input impedance of the rectifier can be used as an ac load of the inverter. Thus, the inverter can be analyzed and designed as a separate stage, independently of the rectifier. If the loaded quality factor of a resonant circuit is high enough and the switching frequency is close enough to the resonant frequency, a resonant inverter usually operates in continuous conduction mode and forces either a sinusoidal output current or a sinusoidal output voltage, depending on the resonant circuit topology. Therefore, the entire inverter can be replaced by a sinusoidal current source or a sinusoidal voltage source that drives the rectifier. As a result, the analysis and design of the rectifier can be carried out independent of the inverter. Finally, the two stages—the inverter and the rectifier—can be cascaded, in a manner similar to other cells in electronic systems.

The cascaded inverter and rectifier should be compatible. A rectifier that requires an input voltage source (called a voltage-driven rectifier) should be connected to an inverter whose output behaves like a voltage source. This takes place in inverters that contain a parallel-resonant circuit. Similarly, a rectifier that requires an input current source should be connected to an inverter whose output behaves like a current source. Inverters that contain a series-resonant circuit force a sinusoidal output current. Characteristics of a dc-dc converter, for example, efficiency or voltage transfer function, can be obtained simply as a product of characteristics of an inverter and a rectifier. For example, nine converters can be built using three types of inverters and three types of rectifiers, assuming that the inverters and rectifiers are compatible. To obtain characteristics of all converters using the state-space approach, a tedious analysis of nine complex circuits is required, and the results are given in the form of graphs rather than equations. In addition, the entire analysis must be repeated with every change of the converter topology. In contrast, the cascaded representation allows one to obtain characteristics of nine converters from the analysis of only six simple blocks (three inverters and three rectifiers). Moreover, the results are given as closed-form expressions, which makes it easier to investigate effects of various parameters on the converter performance. Because of its advantages, the fundamental-frequency approach outlined earlier shall be used throughout this book. If the loaded quality factor of the resonant circuit is very low and/or the switching frequency is much lower or much higher than the resonant frequency, the current and voltage waveforms may significantly differ from sine waves. The converter may even enter a discontinuous conduction mode. In such cases, the state-space analysis should be used.

1.1 REFERENCES

1. R. P. Severns and G. Bloom, *Modern DC-to-DC Switchmode Power Converter Circuits*, New York: Van Nostrand Reinhold Company, 1985.
2. R. G. Hoft, *Semiconductor Power Electronics*, New York: Van Nostrand Reinhold Company, 1986.
3. M. H. Rashid, *Power Electronics*, Englewood Cliffs, NJ: Prentice Hall, 1988.
4. N. Mohan, T. M. Undeland, and W. P. Robbins, *Power Electronics: Converters, Applications and Design*, New York: John Wiley & Sons, 1989.
5. J. G. Kassakian, M. S. Schlecht, and G. C. Verghese, *Principles of Power Electronics*, Reading, MA: Addison-Wesley, 1991.

PART I

RECTIFIERS

CHAPTER 2

CLASS D CURRENT-DRIVEN RECTIFIERS

2.1 INTRODUCTION

A resonant dc-dc converter consists of a high-frequency resonant dc-ac inverter and a high-frequency rectifier. A high-frequency rectifier is an ac-dc converter that is driven by a high-frequency ac energy source. The input source may be either a high-frequency current source or a high-frequency voltage source. Rectifiers that are driven by a current source are called *current-driven rectifiers* [1]–[3]. Some dc-ac inverters contain a series-resonant circuit at the output, for example, Class D or Class E inverters. A series-resonant circuit with a high loaded quality factor Q_L (i.e., $Q_L \geq 3$) behaves approximately like a sinusoidal current source. For this reason, the current-driven rectifiers are compatible with the aforementioned resonant inverters. In some of these rectifiers, the diode current and voltage waveforms are similar to the corresponding transistor waveforms in Class D voltage-switching inverters (studied in Part II of this book). Specifically, the diode current waveform is a half sine wave and the diode voltage waveform is a square wave. The on-duty cycle of each diode is 50%. Therefore, these rectifiers are referred to as *Class D rectifiers* [1]–[3].

This chapter presents three topologies of Class D current-driven rectifiers: the half-wave rectifier, the center-tapped rectifier, and the bridge rectifier. Analyses of these rectifiers are given, taking into account the diode threshold voltage, the diode forward resistance, and the equivalent series resistance (ESR) of the filter capacitor.

2.2 ASSUMPTIONS

The analysis of the Class D current-driven rectifiers is carried out under the following assumptions:

1) The diode in the ON state is modeled by a series combination of a constant-voltage battery V_F and a constant resistance R_F, where V_F represents the diode threshold voltage (or the diode forward offset voltage) and R_F represents the diode forward resistance, as shown in Fig. 2.1.
2) The diode in the OFF state is modeled by an infinite resistance, which means that its junction capacitance and leakage current are neglected.
3) The charge-carrier lifetime is zero for *pn* junction diodes and the diode junction capacitance and lead inductance are zero.
4) The rectifier is driven by an ideal sinusoidal current source.
5) The ripple voltage V_r on the dc output voltage V_O is low, for example, $V_r/V_O \leq 1\%$.

2.3 CLASS D HALF-WAVE RECTIFIER

2.3.1 Circuit Operation

A circuit of a Class D half-wave rectifier is shown in Fig. 2.2(a). It consists of two diodes D_1 and D_2 and a large filter capacitor C_f. Resistor R_L represents a dc load. The rectifier is driven by a sinusoidal current source i_R. The rectifier may be coupled to the current source by a transformer with a turns ratio n and a coupling capacitor C_c. In the transformerless half-wave rectifier, both

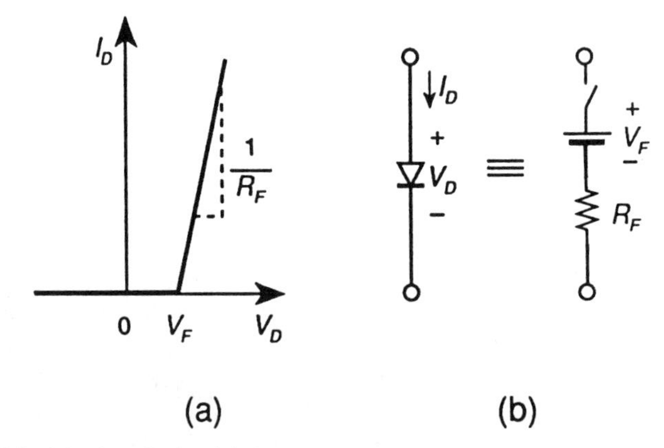

Figure 2.1: Model of a diode. (a) Piecewise-linear *I*-*V* characteristic of a diode. (b) Battery-resistance large-signal model of a diode.

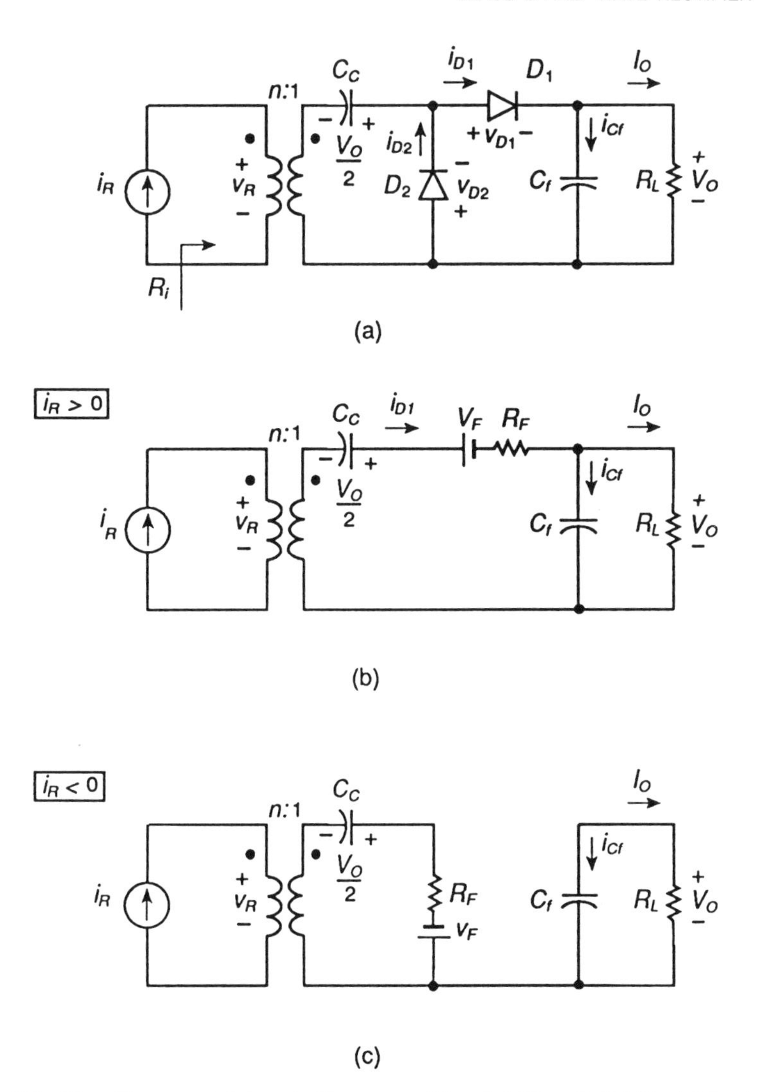

Figure 2.2: Class D current-driven half-wave rectifier. (a) Circuit. (b) Model for $i_R > 0$. (c) Model for $i_R < 0$.

the source and the load can be connected to the same ground, as opposed to the transformerless bridge rectifier. Models of the rectifier are shown in Fig. 2.2(b) and (c) for $i_R > 0$ and $i_R < 0$, respectively. Figure 2.3 depicts the current and voltage waveforms in the transformer version of the rectifier. When $i_R > 0$, diode D_2 is OFF and diode D_1 is ON. The current through diode D_1 charges the filter capacitor C_f. When $i_R < 0$, diode D_1 is OFF and diode D_2 is ON. Diode D_2 acts as a freewheeling diode (i.e., the diode that closes the current path) and provides the path for the current i_R. The on-duty cycle

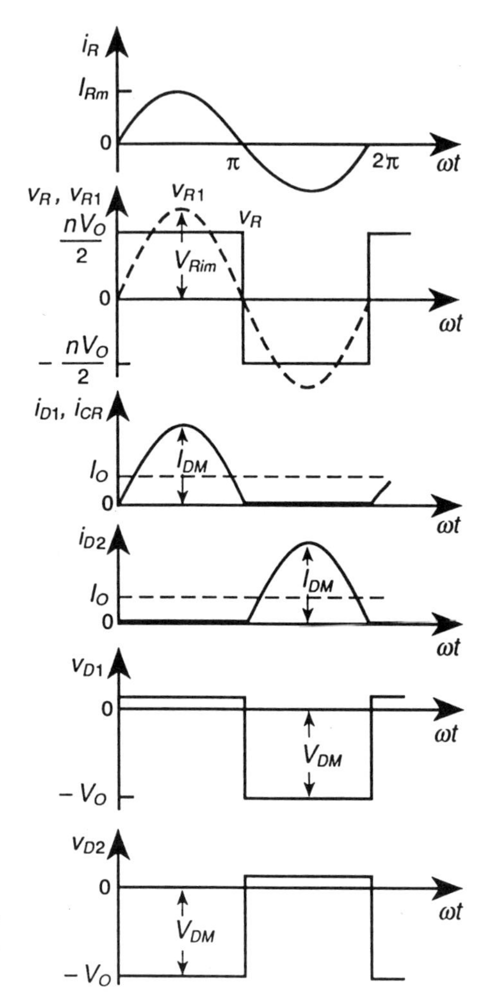

Figure 2.3: Current and voltage waveforms in Class D transformer current-driven half-wave rectifier.

of each diode is 50%. The capacitor C_f is discharged through resistor R_L, maintaining a nearly constant output voltage V_O. The diode currents i_{D1} and i_{D2} are half sine waves, and the diode voltages v_{D1} and v_{D2} are square waves. The input voltage is a square wave, whose high level is approximately $nV_O/2$ and whose low level is approximately $-nV_O/2$ for the transformer version of the rectifier. For the transformerless version, the high level is nearly V_O and the low level is nearly zero. A negative dc output voltage V_O may be obtained by reversing both diodes. Since the diode currents are half sine waves, the diodes turn off at low di/dt, reducing the reverse-recovery current for *pn* junction diodes and associated switching loss and noise. On the other hand, the diodes turn on and off at high dv/dt, causing a current to flow through the diode junction capacitances. This results in switching losses and noise.

2.3.2 Currents and Voltages

According to assumption 4) and Fig. 2.3, the rectifier is excited by a sinusoidal input current

$$i_R = I_{Rm} sin\omega t, \tag{2.1}$$

where I_{Rm} is the amplitude of i_R. The current through the diode D_1 is

$$i_{D1} = \begin{cases} nI_{Rm} sin\omega t, & \text{for} \quad 0 < \omega t \le \pi \\ 0, & \text{for} \quad \pi < \omega t \le 2\pi, \end{cases} \tag{2.2}$$

where n is the transformer turns ratio. Hence, one can find the dc component of the output current

$$I_O = \frac{1}{2\pi} \int_0^{2\pi} i_{D1} d(\omega t) = \frac{nI_{Rm}}{2\pi} \int_0^{\pi} sin\omega t d(\omega t) = \frac{nI_{Rm}}{\pi}. \tag{2.3}$$

Thus, the dc output current I_O is directly proportional to the amplitude of the input current I_{Rm}. Equation (2.3) leads to the ac-to-dc current transfer function

$$K_I \equiv \frac{I_O}{I_{Rrms}} = \frac{\sqrt{2} I_O}{I_{Rm}} = \frac{\sqrt{2} n}{\pi} \approx 0.45n \tag{2.4}$$

and the dc component of the output voltage

$$V_O = I_O R_L = \frac{nI_{Rm} R_L}{\pi}, \tag{2.5}$$

where $I_{Rrms} = I_{Rm}/\sqrt{2}$ is the rms value of the input current. It follows from (2.5) that the dc output voltage V_O is directly proportional to I_{Rm} and therefore can be regulated against load and line variations by varying I_{Rm} in such a way that the product $I_{Rm} R_L$ is held constant.

Since the input current is sinusoidal, the input power contains only the power of the fundamental component. The fundamental component v_{R1} of the input voltage v_R is in phase with the input current i_R, as shown in Fig. 2.3. Therefore, using (2.3) the input power can be expressed as

$$P_i = \frac{I_{Rm}^2 R_i}{2} = \frac{\pi^2 I_O^2 R_i}{2n^2}, \tag{2.6}$$

where R_i is the input resistance of the rectifier at the fundamental frequency f. The dc output power is

$$P_O = I_O^2 R_L. \tag{2.7}$$

2.3.3 Power Factor

Neglecting the voltage drops across the diodes, the input voltage v_R is a square wave given by

$$v_R = \begin{cases} \frac{nV_O}{2}, & \text{for} \quad 0 < \omega t \leq \pi \\ -\frac{nV_O}{2}, & \text{for} \quad \pi < \omega t \leq 2\pi. \end{cases} \tag{2.8}$$

Hence, one can determine the rms value of v_R

$$V_{Rrms} = \sqrt{\frac{1}{2\pi}\int_0^{2\pi} v_R^2 d(\omega t)} = \sqrt{\frac{n^2 V_O^2}{4\pi}\int_0^{\pi} d(\omega t)} = \frac{nV_O}{2}, \tag{2.9}$$

the amplitude of the fundamental component of the input voltage

$$V_{R1m} = \frac{1}{\pi}\int_0^{2\pi} v_R sin\omega t d(\omega t) = \frac{2}{\pi}\int_0^{\pi} \frac{nV_O}{2} sin\omega t d(\omega t) = \frac{2nV_O}{\pi}, \tag{2.10}$$

the rms value of the fundamental component of the input voltage

$$V_{R1rms} = \frac{V_{R1m}}{\sqrt{2}} = \frac{2nV_O}{\sqrt{2}\pi} = \frac{\sqrt{2}nV_O}{\pi}, \tag{2.11}$$

and the *power factor*

$$\begin{aligned} PF &= \frac{\text{Real Power}}{\text{Apparent Power}} = \frac{P_i}{I_{Rrms}V_{Rrms}} = \frac{I_{Rrms}V_{R1rms}}{I_{Rrms}V_{Rrms}} = \frac{V_{R1rms}}{V_{Rrms}} \\ &= \frac{V_{R1rms}}{\sqrt{V_{R1rms}^2 + V_{R2rms}^2 + V_{R3rms}^2 + \ldots}} = \frac{2\sqrt{2}}{\pi} \approx 0.9, \end{aligned} \tag{2.12}$$

where V_{R1rms}, V_{R2rms}, V_{R3rms}, ... are the rms values of the harmonics of the rectifier input voltage. The *total harmonic distortion* of the rectifier input voltage is

$$\begin{aligned} THD &= \sqrt{\frac{V_{R2rms}^2 + V_{R3rms}^2 + V_{R3rms}^2 + \ldots}{V_{R1rms}^2}} = \sqrt{\frac{1}{PF^2} - 1} = \sqrt{\frac{\pi^2}{8} - 1} \\ &\approx 0.4834. \end{aligned} \tag{2.13}$$

2.3.4 Power-Output Capability

The peak forward current through each of the diodes D_1 and D_2 is

$$I_{DM} = nI_{Rm} = \pi I_O, \tag{2.14}$$

and the peak reverse voltage across each of these diodes is

$$V_{DM} = V_O. \tag{2.15}$$

Hence, one obtains the power-output capability of the rectifier

$$c_{pR} = \frac{P_O}{I_{DM} V_{DM}} = \frac{I_O V_O}{I_{DM} V_{DM}} = \frac{1}{\pi} \approx 0.318. \tag{2.16}$$

The dc output power that can be achieved at given peak values of the diode current I_{DM} and voltage V_{DM} is

$$P_O = c_{pR} I_{DM} V_{DM}. \tag{2.17}$$

2.3.5 Efficiency

Consider now power losses and efficiency of the rectifier. The average current I_D of the diode D_1 is equal to the dc output current I_O and is given by (2.3). It follows from Fig. 2.3 that the average current of diode D_2 is also I_O. Hence, power loss in one diode due to V_F is

$$P_{VF} = V_F I_D = V_F I_O = \frac{V_F}{V_O} P_O. \tag{2.18}$$

The rms value of the current through the diode is obtained from (2.2) and (2.3)

$$\begin{aligned} I_{Drms} &= \sqrt{\frac{1}{2\pi} \int_0^{2\pi} i_{D1}^2 d(\omega t)} = \sqrt{\frac{n^2 I_{Rm}^2}{2\pi} \int_0^{\pi} sin^2 \omega t d(\omega t)} \\ &= \frac{n I_{Rm}}{2} = \frac{\pi I_O}{2}, \end{aligned} \tag{2.19}$$

which gives the power loss in one diode due to R_F

$$P_{RF} = R_F I_{Drms}^2 = \frac{\pi^2 I_O^2 R_F}{4} = \frac{\pi^2 R_F}{4 R_L} P_O. \tag{2.20}$$

From (2.18) and (2.20), the overall conduction loss in one diode is

$$\begin{aligned} P_D &= P_{VF} + P_{RF} = I_O V_F + R_F I_{Drms}^2 \\ &= I_O V_F + \frac{\pi^2 I_O^2 R_F}{4} = P_O \left(\frac{V_F}{V_O} + \frac{\pi^2 R_F}{4 R_L} \right). \end{aligned} \tag{2.21}$$

If $1/\omega C_f << R_L$, the current through filter capacitor C_f is

$$i_{Cf} \approx i_{D1} - I_O = \begin{cases} I_O(\pi sin\omega t - 1), & \text{for} \quad 0 < \omega t \leq \pi \\ -I_O, & \text{for} \quad \pi < \omega t \leq 2\pi, \end{cases} \tag{2.22}$$

its rms value is

$$I_{Cf(rms)} = \sqrt{\frac{1}{2\pi}\int_0^{2\pi} i_{Cf}^2 d(\omega t)} = I_O\sqrt{\frac{\pi^2}{4} - 1}, \tag{2.23}$$

and the power dissipated in the ESR of the filter capacitor r_C is

$$P_{rC} = r_C I_{Cf(rms)}^2 = r_C I_O^2 \left(\frac{\pi^2}{4} - 1\right) = P_O \frac{r_C}{R_L}\left(\frac{\pi^2}{4} - 1\right). \tag{2.24}$$

Thus, one obtains the overall conduction loss

$$\begin{aligned} P_C &= 2P_D + P_{rC} = 2I_O V_F + \frac{\pi^2 I_O^2 R_F}{2} + r_C I_O^2\left(\frac{\pi^2}{4} - 1\right) \\ &= P_O\left[\frac{2V_F}{V_O} + \frac{\pi^2 R_F}{2R_L} + \frac{r_C}{R_L}\left(\frac{\pi^2}{4} - 1\right)\right]. \end{aligned} \tag{2.25}$$

Neglecting switching losses, the input power is found as

$$\begin{aligned} P_i &= \frac{P_O + P_C}{\eta_{tr}} = \frac{P_O + 2P_D + P_{rC}}{\eta_{tr}} \\ &= \frac{P_O}{\eta_{tr}}\left[1 + \frac{2V_F}{V_O} + \frac{\pi^2 R_F}{2R_L} + \frac{r_C}{R_L}\left(\frac{\pi^2}{4} - 1\right)\right], \end{aligned} \tag{2.26}$$

where η_{tr} is the efficiency of the transformer. The efficiency of the rectifier is obtained from (2.7) and (2.25)

$$\eta_R = \frac{P_O}{P_i} = \frac{P_O\eta_{tr}}{P_O + P_C} = \frac{\eta_{tr}}{1 + \frac{P_C}{P_O}} = \frac{\eta_{tr}}{1 + \frac{2V_F}{V_O} + \frac{\pi^2 R_F}{2R_L} + \frac{r_C}{R_L}(\frac{\pi^2}{4} - 1)}. \tag{2.27}$$

2.3.6 Input Resistance

Substitution of (2.6) and (2.7) into (2.26) yields the input resistance

$$R_i \equiv \frac{V_{R1m}}{I_{Rm}} = \frac{2n^2 R_L}{\pi^2 \eta_{tr}}\left[1 + \frac{2V_F}{V_O} + \frac{\pi^2 R_F}{2R_L} + \frac{r_C}{R_L}\left(\frac{\pi^2}{4} - 1\right)\right] = \frac{2n^2 R_L}{\pi^2 \eta_R}, \tag{2.28}$$

where V_{R1m} is the amplitude of the fundamental component v_{R1} of the rectifier input voltage v_R.

2.3.7 Voltage Transfer Function

The input power of the rectifier can be expressed as

$$P_i = \frac{V_{R1rms}^2}{R_i}, \tag{2.29}$$

where $V_{R1rms} = V_{R1m}/\sqrt{2}$ is the rms value of the fundamental component of the input voltage. From (2.7) and (2.29), the efficiency of the rectifier is

$$\eta_R = \frac{P_O}{P_i} = \left(\frac{V_O}{V_{R1rms}}\right)^2 \frac{R_i}{R_L}. \tag{2.30}$$

Hence, using (2.27) and (2.28), one obtains the ac-to-dc voltage transfer function

$$M_{VR} \equiv \frac{V_O}{V_{R1rms}} = \sqrt{\frac{\eta_R R_L}{R_i}} = \frac{\pi \eta_{tr}}{n\sqrt{2}[1 + \frac{2V_F}{V_O} + \frac{\pi^2 R_F}{2R_L} + \frac{r_C}{R_L}(\frac{\pi^2}{4} - 1)]}$$
$$= \frac{\pi \eta_R}{n\sqrt{2}}. \tag{2.31}$$

Example 2.1

Find the efficiency η_R, the voltage transfer function M_{VR}, and the input resistance R_i for a Class D half-wave rectifier of Fig. 2.2(a) at $V_O = 5$ V and $I_O = 20$ A. The rectifier employs Schottky diodes with $V_F = 0.5$ V and $R_F = 0.025\ \Omega$ and a filter capacitor with $r_C = 20$ mΩ. The transformer turns ratio is $n = 5$. Assume the transformer efficiency $\eta_{tr} = 0.96$.

Solution: The load resistance of the rectifier is $R_L = V_O/I_O = 0.25\ \Omega$, and the output power is $P_O = I_O V_O = 100$ W. Substituting (2.18) and (2.20) into (2.21), one obtains the power loss in the diode as

$$P_D = P_{VF} + P_{RF} = V_F I_O + \frac{\pi^2 I_O^2 R_F}{4}$$
$$= 0.5 \times 20 + \frac{\pi^2 \times 20^2 \times 0.025}{4} = 34.67 \text{ W}. \tag{2.32}$$

The power loss in the filter capacitor is obtained from (2.24)

$$P_{rC} = r_C I_O^2 \left(\frac{\pi^2}{4} - 1\right) = 11.74 \text{ W}. \tag{2.33}$$

Hence, using (2.25),

$$P_C = 2P_D + P_{rC} = 69.34 + 11.74 = 81.08 \text{ W}. \tag{2.34}$$

From (2.27),

$$\eta_R = \frac{P_O \eta_{tr}}{P_O + P_C} = \frac{100 \times 0.96}{100 + 81.08} = 53\%. \tag{2.35}$$

The input resistance of the rectifier can be obtained using (2.28)

$$R_i = \frac{2n^2 R_L}{\pi^2 \eta_R} = \frac{2 \times 5^2 \times 0.25}{\pi^2 0.53} = 2.39\ \Omega. \tag{2.36}$$

The voltage transfer function is calculated from (2.31)

$$M_{VR} = \frac{\pi \eta_R}{n\sqrt{2}} = \frac{\pi \times 0.53}{5\sqrt{2}} = 0.235. \tag{2.37}$$

It can be seen that all parameters of the rectifier are considerably altered by nonzero values of V_F, R_F, and r_C at a low value of V_O. In particular, the efficiency is very low.

2.3.8 Ripple Voltage

An equivalent circuit of the rectifier's output filter is shown in Fig. 2.4, where the capacitor is modeled by a capacitance C_f and an ESR r_C. Using (2.22), one obtains the voltage across the filter capacitor ESR

$$v_{ESR} = r_C i_{Cf} = \begin{cases} r_C I_O(\pi sin\omega t - 1), & \text{for} \quad 0 < \omega t \le \pi \\ -r_C I_O, & \text{for} \quad \pi < \omega t \le 2\pi \end{cases} \tag{2.38}$$

and the ac component of the voltage across the filter capacitance C_f

$$\begin{aligned} v_c &= \frac{1}{\omega C_f} \int_0^{\omega t} i_{Cf} d(\omega t) + v_c(0) \\ &= \begin{cases} \frac{\pi I_O}{\omega C_f}(1 - \frac{\omega t}{\pi} - cos\omega t) + v_c(0), & \text{for} \quad 0 < \omega t \le \pi \\ -\frac{I_O t}{C_f} + \frac{2\pi I_O}{\omega C_f} + v_c(0), & \text{for} \quad \pi < \omega t \le 2\pi, \end{cases} \end{aligned} \tag{2.39}$$

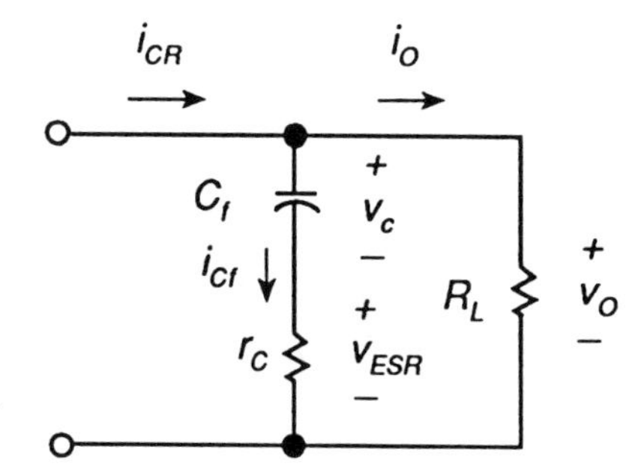

Figure 2.4: An equivalent circuit of the rectifier's output filter.

where $v_c(0) = -\pi I_O/(2\omega C_f)$ is the initial value of v_c at $\omega t = 0$. From (2.38) and (2.39), the ac component of the output voltage is

$$v_r = v_{ESR} + v_c = \begin{cases} r_C I_O(\pi sin\omega t - 1) + \frac{\pi I_O}{\omega C_f}(\frac{1}{2} - \frac{\omega t}{\pi} - cos\omega t), & \text{for } 0 < \omega t \le \pi \\ -r_C I_O - \frac{I_O t}{C_f} + \frac{3\pi I_O}{2\omega C_f}, & \text{for } \pi < \omega t \le 2\pi. \end{cases} \tag{2.40}$$

From (2.39), the minimum value of the ac component of the voltage across the filter capacitance $V_{c(min)}$ occurs at $\omega t_{min} = arcsin(1/\pi) = 18.56°$, and the maximum value of the ac component of the voltage across the filter capacitance $V_{c(max)}$ occurs at $\omega t_{max} = \pi - arcsin(1/\pi) = 161.43°$. Substitution of these values into (2.39) gives the peak-to-peak value of the ac component of the voltage across the filter capacitance

$$V_c \equiv V_{c(max)} - V_{c(min)} = \frac{2I_O}{\omega C_f}\left[\sqrt{\pi^2 - 1} - arccos\left(\frac{1}{\pi}\right)\right]$$
$$\approx \frac{3.46 I_O}{\omega C_f} = \frac{0.55 I_O}{f C_f} = \frac{0.55 V_O}{f C_f R_L} = \frac{3.46 f_H V_O}{f}, \tag{2.41}$$

where $f_H = 1/(2\pi R_L C_f)$ is the upper 3-dB frequency of the C_f-R_L low-pass output filter. The maximum value of V_c occurs at the full-load resistance $R_L = R_{Lmin}$ at which $f_H = f_{Hmax} = 1/(2\pi C_f R_{Lmin})$. The maximum ripple voltage on the ESR of the filter capacitor is

$$V_{rESR} = \pi r_C I_{Omax} = \frac{\pi r_C V_O}{R_{Lmin}}. \tag{2.42}$$

It is possible to find analytically the peak-to-peak value of the voltage v_r in terms of ω, I_O, C_f, and r_C. However, the resulting expression is too complicated to be useful in designing the value of the filter capacitor. Figure 2.5 shows the waveforms illustrating the ripple voltage for $f = 1$ MHz, $I_O = 0.4$ A, $C_f = 6.6$ μF, and $r_C = 0.03$ Ω. It can be seen that the peak-to-peak value of the ac component of the output voltage is always less than $V_c + V_{rESR}$. As a rule of thumb, it can be assumed that the peak-to-peak value of the output voltage ripple is

$$V_r = \begin{cases} V_c, & \text{for } V_c \gg V_{rESR} \\ V_c + V_{rESR}, & \text{for } V_c \approx V_{rESR} \\ V_{rESR}, & \text{for } V_c \ll V_{rESR}. \end{cases} \tag{2.43}$$

Condition $V_c = V_{rESR}$ is equivalent to $\pi^2 f C_f r_C = 1.73$.

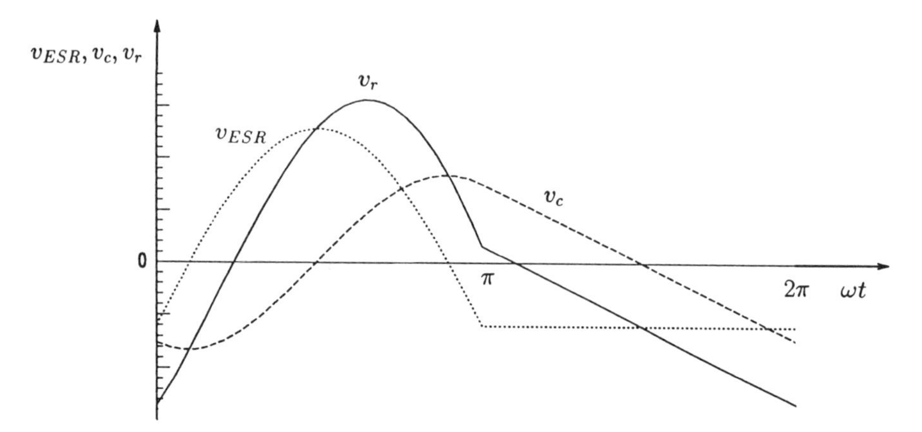

Figure 2.5: Current and voltage waveforms that illustrate the ripple voltage V_r for the Class D current-driven half-wave rectifier at $f = 1$ MHz, $I_O = 0.4$ A, $C_f = 6.6$ μF, and $r_C = 0.03$ Ω.

Example 2.2

Design a filter capacitor for a Class D half-wave rectifier operating at a switching frequency $f = 1$ MHz. The output voltage of the rectifier is $V_O = 14$ V, and the minimum load resistance is $R_{Lmin} = 35$ Ω. It is specified that the ripple voltage cannot be greater than 0.5% of V_O. The ESR of the filter capacitor is $r_C = 0.03$ Ω at 1 MHz.

Solution: The maximum ripple voltage is

$$V_r = 0.005 V_O = 0.07 \text{ V}. \tag{2.44}$$

The maximum output current is

$$I_{Omax} = \frac{V_O}{R_{Lmin}} = \frac{14}{35} = 0.4 \text{ A}. \tag{2.45}$$

From (2.42),

$$V_{rESR} = \pi r_C I_{Omax} = \pi \times 0.03 \times 0.4 = 0.038 \text{ V}. \tag{2.46}$$

Thus, using (2.43),

$$V_c = V_r - V_{rESR} = 0.07 - 0.038 = 0.032 \text{ V}. \tag{2.47}$$

From (2.41),

$$C_{fmin} = \frac{0.55 I_{Omax}}{f V_c} = \frac{0.55 \times 0.4}{10^6 \times 0.032} = 6.9 \text{ } \mu\text{F}. \tag{2.48}$$

2.4 CLASS D TRANSFORMER CENTER-TAPPED RECTIFIER

2.4.1 Currents and Voltages

Figure 2.6 shows a Class D current-driven transformer center-tapped rectifier and its models. The current and voltage waveforms are shown in Fig. 2.7. The input current i_R is sinusoidal and given by (2.1). For $i_R > 0$, D_2 is OFF and D_1 is ON. For $i_R < 0$, D_1 is OFF and D_2 is ON. Therefore, the current to the C_f-R_L circuit is

$$i_{CR} = i_{D1} + i_{D2} = n \mid i \mid = nI_{Rm} \mid sin\omega t \mid . \tag{2.49}$$

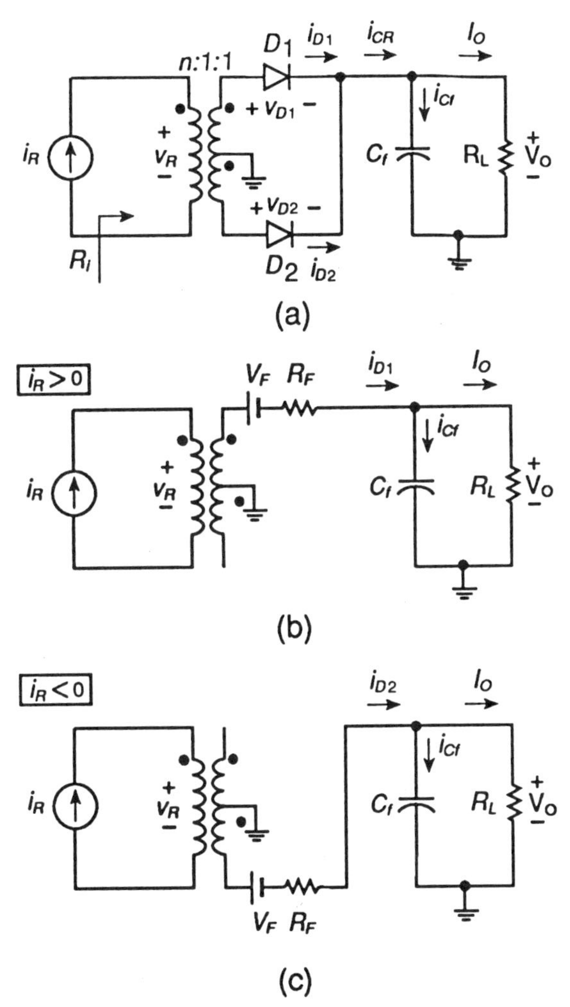

Figure 2.6: Class D current-driven transformer center-tapped rectifier. (a) Circuit. (b) Model for i_R >0. (c) Model for i_R <0.

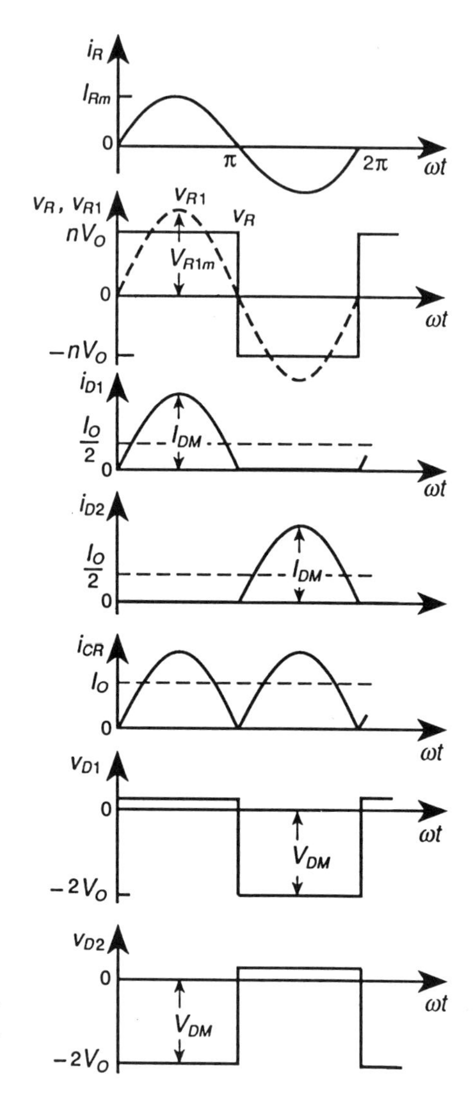

Figure 2.7: Current and voltage waveforms in Class D transformer current-driven center-tapped rectifier.

This yields the dc component of the output current

$$I_O = \frac{1}{2\pi}\int_0^{2\pi}(i_{D1} + i_{D2})d(\omega t) = \frac{nI_{Rm}}{\pi}\int_0^{\pi} sin\omega t d(\omega t) = \frac{2nI_{Rm}}{\pi}, \qquad (2.50)$$

which is directly proportional to I_{Rm}. Note that I_O doubled over the half-wave rectifier for the same value of I_{Rm}. The input power is

$$P_i = \frac{I_{Rm}^2 R_i}{2} = \frac{\pi^2 I_O^2 R_i}{8n^2}. \qquad (2.51)$$

The output power P_O is given by (2.7).

2.4.2 Power Factor

Assuming that $V_F = 0$ and $R_F = 0$, the input voltage v_R is a square wave expressed by

$$v_R = \begin{cases} nV_O, & \text{for} \quad 0 < \omega t \le \pi \\ -nV_O, & \text{for} \quad \pi < \omega t \le 2\pi. \end{cases} \tag{2.52}$$

The rms value of v_R is

$$V_{Rrms} = \sqrt{\frac{1}{2\pi}\int_0^{2\pi} v_R^2 d(\omega t)} = \sqrt{\frac{n^2 V_O^2}{\pi}\int_0^{\pi} d(\omega t)} = nV_O, \tag{2.53}$$

the amplitude of the fundamental component of the input voltage is

$$V_{R1m} = \frac{1}{\pi}\int_0^{2\pi} v_R sin\omega t d(\omega t) = \frac{2}{\pi}\int_0^{\pi} nV_O sin\omega t d(\omega t) = \frac{4nV_O}{\pi}, \tag{2.54}$$

the rms value of the fundamental component of the input voltage is

$$V_{R1rms} = \frac{V_{R1m}}{\sqrt{2}} = \frac{2\sqrt{2}nV_O}{\pi}, \tag{2.55}$$

and the power factor is

$$PF = \frac{P_i}{I_{Rrms}V_{Rrms}} = \frac{I_{Rrms}V_{R1rms}}{I_{Rrms}V_{Rrms}} = \frac{V_{R1rms}}{V_{Rrms}} = \frac{2\sqrt{2}}{\pi} \approx 0.9. \tag{2.56}$$

Hence, the total harmonic distortion of the rectifier input voltage is

$$THD = \sqrt{\frac{1}{PF^2} - 1} = \sqrt{\frac{\pi^2}{8} - 1} \approx 04834. \tag{2.57}$$

2.4.3 Power-Output Capability

The peak forward currents of the rectifier diodes are

$$I_{DM} = nI_{Rm} = \frac{\pi I_O}{2}, \tag{2.58}$$

the peak reverse voltages are

$$V_{DM} = 2V_O, \tag{2.59}$$

and the power-output capability is

$$c_{pR} = \frac{P_O}{I_{DM}V_{DM}} = \frac{I_O V_O}{I_{DM}V_{DM}} = \frac{1}{\pi} \approx 0.318. \tag{2.60}$$

Note that c_{pR} is the same as for the half-wave rectifier.

2.4.4 Efficiency

The rms value of the current through each diode is

$$I_{Drms} = \sqrt{\frac{1}{2\pi}\int_0^{2\pi} i_{D1}^2 d(\omega t)} = \sqrt{\frac{n^2 I_{Rm}^2}{2\pi}\int_0^{\pi} sin^2\omega t d(\omega t)}$$
$$= \frac{nI_{Rm}}{2} = \frac{\pi I_O}{4} \quad (2.61)$$

and the power loss in the diode forward resistance R_F is

$$P_{RF} = R_F I_{Drms}^2 = \frac{\pi^2 I_O^2 R_F}{16} = \frac{\pi^2 R_F}{16R_L} P_O. \quad (2.62)$$

Using (2.50), the diode average current is

$$I_D = \frac{1}{2\pi}\int_0^{2\pi} i_{D1} d(\omega t) = \frac{1}{2\pi}\int_0^{\pi} nI_{Rm} sin\omega t d(\omega t) = \frac{nI_{Rm}}{\pi} = \frac{I_O}{2}, \quad (2.63)$$

and the power dissipated in each diode due to V_F is

$$P_{VF} = V_F I_D = \frac{V_F I_O}{2} = \frac{V_F}{2V_O} P_O. \quad (2.64)$$

Thus, the total conduction loss per diode is

$$P_D = P_{VF} + P_{RF} = \frac{V_F I_O}{2} + \frac{\pi^2 I_O^2 R_F}{16} = P_O\left(\frac{V_F}{2V_O} + \frac{\pi^2 R_F}{16R_L}\right). \quad (2.65)$$

The power P_{VF} is two times lower and the power P_{RF} is four times lower for the transformer center-tapped rectifier than for the half-wave rectifier at the same value of I_O. This is because I_D and I_{Rrms} are reduced two times, P_{VF} is proportional to I_D, and P_{RF} is proportional to I_{Rrms}^2.

Using (2.49) and assuming that ESR and ESL are zero and $1/\omega C_f \ll R_L$, the current through the filter capacitor C_f can be approximated by

$$i_{Cf} \approx i_{CR} - I_O = nI_{Rm} \mid sin\omega t \mid - I_O = I_O\left(\frac{\pi}{2} \mid sin\omega t \mid -1\right), \quad (2.66)$$

which leads to the rms value of the current through the filter capacitor

$$I_{Cf(rms)} = \sqrt{\frac{1}{2\pi}\int_0^{2\pi} i_{Cf}^2 d(\omega t)} = I_O\sqrt{\frac{\pi^2}{8} - 1} \approx 0.4834 I_O \quad (2.67)$$

and the power dissipated in the ESR

$$P_{rC} = r_C I_{Cf(rms)}^2 = r_C I_O^2\left(\frac{\pi^2}{8} - 1\right) = \frac{r_C}{R_L}\left(\frac{\pi^2}{8} - 1\right) P_O. \quad (2.68)$$

The total conduction loss is then

$$P_C = 2P_D + P_{rC} = V_F I_O + \frac{\pi^2 I_O^2 R_F}{8} + r_C I_O^2 \left(\frac{\pi^2}{8} - 1\right)$$
$$= P_O \left[\frac{V_F}{V_O} + \frac{\pi^2 R_F}{8R_L} + \frac{r_C}{R_L}\left(\frac{\pi^2}{8} - 1\right)\right]. \tag{2.69}$$

The input power can be expressed as

$$P_i = \frac{P_O + P_C}{\eta_{tr}} = \frac{P_O}{\eta_{tr}}\left[1 + \frac{V_F}{V_O} + \frac{\pi^2 R_F}{8R_L} + \frac{r_C}{R_L}\left(\frac{\pi^2}{8} - 1\right)\right]. \tag{2.70}$$

The efficiency is

$$\eta_R = \frac{P_O}{P_i} = \frac{P_O \eta_{tr}}{P_O + 2P_D + P_{rC}} = \frac{\eta_{tr}}{1 + \frac{V_F}{V_O} + \frac{\pi^2 R_F}{8R_L} + \frac{r_C}{R_L}(\frac{\pi^2}{8} - 1)}. \tag{2.71}$$

2.4.5 Input Resistance

Substitution of (2.7) and (2.51) into (2.70) gives the input resistance

$$R_i = \frac{8n^2 R_L}{\pi^2 \eta_{tr}}\left[1 + \frac{V_F}{V_O} + \frac{\pi^2 R_F}{8R_L} + \frac{r_C}{R_L}\left(\frac{\pi^2}{8} - 1\right)\right] = \frac{8n^2 R_L}{\pi^2 \eta_R}. \tag{2.72}$$

2.4.6 Voltage Transfer Function

Using (2.30), (2.72), and (2.71), one obtains the ac-to-dc voltage transfer function of the rectifier as

$$M_{VR} \equiv \frac{V_O}{V_{Rrms}} = \sqrt{\frac{\eta_R R_L}{R_i}} = \frac{\pi \eta_{tr}}{2\sqrt{2}n[1 + \frac{V_F}{V_O} + \frac{\pi^2 R_F}{8R_L} + \frac{r_C}{R_L}(\frac{\pi^2}{8} - 1)]}$$
$$= \frac{\pi \eta_R}{2\sqrt{2}n}. \tag{2.73}$$

Example 2.3

A Class D center-tapped rectifier of Fig. 2.6 is built using Schottky diodes (e.g., Motorola MBR20100CT) with $V_F = 0.5\ V$ and $R_F = 0.025\ \Omega$ and a filter capacitor C_f with $r_C = 20\ m\Omega$. The rectifier is operated at $V_O = 5\ V$ and $I_O = 20\ A$. Find the efficiency η_R, the voltage transfer function M_{VR}, and the input resistance R_i for this rectifier. The transformer turns ratio is $n = 5$. Assume the transformer efficiency $\eta_{tr} = 0.96$.

Solution: The load resistance of the rectifier is $R_L = V_O/I_O = 0.25\ \Omega$, and the output power is $P_O = I_O V_O = 100\ W$. Substitution of (2.62) and (2.64) into (2.65) yields the diode conduction loss

$$P_D = P_{VF} + P_{RF} = \frac{V_F I_O}{2} + \frac{\pi^2 I_O^2 R_F}{16}$$
$$= \frac{0.5 \times 20}{2} + \frac{\pi^2 \times 20^2 \times 0.025}{16} = 11.2 \text{ W}. \tag{2.74}$$

Using (2.68), one obtains the power loss in the filter capacitor

$$P_{rC} = r_C I_O^2 \left(\frac{\pi^2}{8} - 1\right) = 0.02 \times 20^2 \left(\frac{\pi^2}{8} - 1\right) = 1.87 \text{ W}. \tag{2.75}$$

From (2.69),

$$P_C = 2P_D + P_{rC} = 22.4 + 1.87 = 24.27 \text{ W}. \tag{2.76}$$

Using (2.71),

$$\eta_R = \frac{P_O \eta_{tr}}{P_O + P_C} = \frac{100 \times 0.96}{100 + 24.27} = 77.25\%. \tag{2.77}$$

The input resistance of the rectifier can be obtained from (2.72)

$$R_i = \frac{8n^2 R_L}{\pi^2 \eta_R} = \frac{8 \times 5^2 \times 0.25}{\pi^2 \times 0.7725} = 6.56 \ \Omega. \tag{2.78}$$

The voltage transfer function is calculated using (2.73)

$$M_{VR} = \frac{\pi \eta_R}{2\sqrt{2}n} = \frac{\pi \times 0.7725}{2\sqrt{2} \times 5} = 0.172. \tag{2.79}$$

Note that the efficiency of the center-tapped rectifier is much higher than the efficiency of the half-wave rectifier at the same parameters (see Example 2.1).

2.4.7 Ripple Voltage

The frequency of the output ripple for the center-tapped rectifier is twice the operating frequency. From (2.66), one obtains the voltage drop across the filter capacitor ESR

$$v_{ESR} = r_C i_{Cf} = r_C I_O \left(\frac{\pi}{2} sin\omega t - 1\right), \quad \text{for} \quad 0 < \omega t \leq \pi. \tag{2.80}$$

Hence, from (2.58) the maximum peak-to-peak ripple voltage on the ESR of the filter capacitor is

$$V_{rESR} = r_C I_{DM} = \frac{\pi r_C I_{Omax}}{2} = \frac{\pi r_C V_O}{2R_{Lmin}}. \tag{2.81}$$

The ac component of the voltage across the output capacitance is from (2.66)

$$v_c = \frac{1}{\omega C_f}\int_0^{\omega t} i_{Cf} d(\omega t) + v_c(0) = \frac{I_O}{\omega C_f}\left(\frac{\pi}{2} - \omega t - \frac{\pi}{2} cos \omega t\right) + v_c(0), \quad (2.82)$$

where $v_c(0) = 0$. Thus, the ac component of the output voltage is expressed as

$$v_r = v_{ESR} + v_c = r_C I_O\left(\frac{\pi}{2} sin\omega t - 1\right) + \frac{I_O}{\omega C_f}\left(\frac{\pi}{2} - \omega t - \frac{\pi}{2} cos\omega t\right) + v_c(0), \quad \text{for} \quad 0 < \omega t \le \pi. \quad (2.83)$$

The minimum value of the ac component of the voltage across the filter capacitance $V_{c(min)}$ occurs at $\omega t_{min} = arcsin(2/\pi) = 39.54°$, and the maximum value of the ac component of the voltage across the filter capacitance $V_{c(max)}$ occurs at $\omega t_{max} = \pi - arcsin(2/\pi) = 140.46°$. Thus,

$$V_c = V_{c(max)} - V_{c(min)} = \frac{I_O}{\omega C_f}\left[\sqrt{\pi^2 - 4} - 2arc\ cos\left(\frac{2}{\pi}\right)\right] \approx \frac{0.66 I_O}{\omega C_f} = \frac{0.105 I_O}{f C_f} = \frac{0.105 V_O}{f C_f R_L} = \frac{0.66 f_H V_O}{f}. \quad (2.84)$$

Figure 2.8 depicts the waveforms that illustrate the ripple voltage for $f =$ 1 MHz, $I_O = 0.4$ A, $C_f = 3.3$ μF, and $r_C = 0.03$ Ω. The peak-to-peak value of the ac component of the output voltage is always less than V_c +

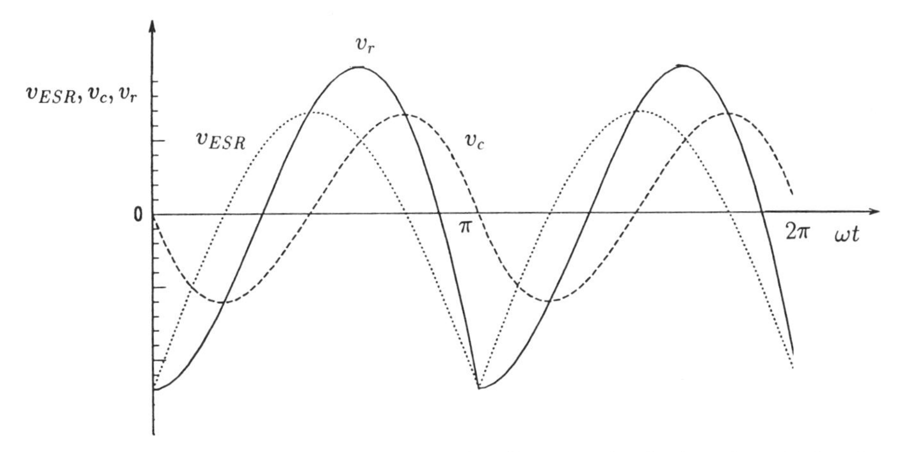

Figure 2.8: Current and voltage waveforms that illustrate the ripple voltage V_r for the Class D current-driven transformer center-tapped and bridge rectifiers at $f = 1$ MHz, $I_O = 0.4$ A, $C_f = 3.3$ μF, and $r_C = 0.03$ Ω.

V_{rESR}. Therefore, it can be assumed that the peak-to-peak value of the output voltage ripple is

$$V_r = \begin{cases} V_c, & \text{for} \quad V_c \gg V_{rESR} \\ V_c + V_{rESR}, & \text{for} \quad V_c \approx V_{rESR} \\ V_{rESR}, & \text{for} \quad V_c \ll V_{rESR}. \end{cases} \tag{2.85}$$

Condition $V_c = V_{rESR}$ is equivalent to $\pi^2 f C_f r_C = 0.66$. In low-voltage high-current applications, filter capacitors with a low value of the ESR are required in the current-driven rectifiers because the peak-to-peak capacitor current is equal to the peak value of the diode current I_{DM}.

Example 2.4

Design a filter capacitor for a Class D center-tapped rectifier operating with a switching frequency $f = 1$ MHz. The output voltage of the rectifier is $V_O = 14$ V, and the minimum load resistance is $R_{Lmin} = 35\ \Omega$. It is specified that the ripple voltage cannot be greater than 0.2% V_O. The ESR of the filter capacitor is $r_C = 0.03\ \Omega$ at 1 MHz.

Solution: The maximum ripple voltage is

$$V_r = 0.002 V_O = 0.028 \text{ V}. \tag{2.86}$$

The maximum output current is

$$I_{Omax} = \frac{V_O}{R_{Lmin}} = \frac{14}{35} = 0.4 \text{ A}. \tag{2.87}$$

From (2.81),

$$V_{rESR} = \frac{\pi}{2} r_C I_{Omax} = \frac{\pi}{2} \times 0.03 \times 0.4 = 0.019 \text{ V}. \tag{2.88}$$

Thus, using (2.85),

$$V_c = V_r - V_{rESR} = 0.028 - 0.019 = 0.009 \text{ V}. \tag{2.89}$$

From (2.84),

$$C_{fmin} = \frac{0.105 I_{Omin}}{f V_c} = \frac{0.105 \times 0.4}{10^6 \times 0.009} = 4.7\ \mu\text{F}. \tag{2.90}$$

2.5 CLASS D BRIDGE RECTIFIER

Figure 2.9 shows a circuit of a Class D bridge rectifier, along with its models. The waveforms are depicted in Fig. 2.10. For the transformerless version of the bridge rectifier, either source or load can be grounded. The transformer version of the rectifier allows for connecting both the source and the load to either the same ground or different grounds. When $i_R > 0$, D_1 and D_3 are

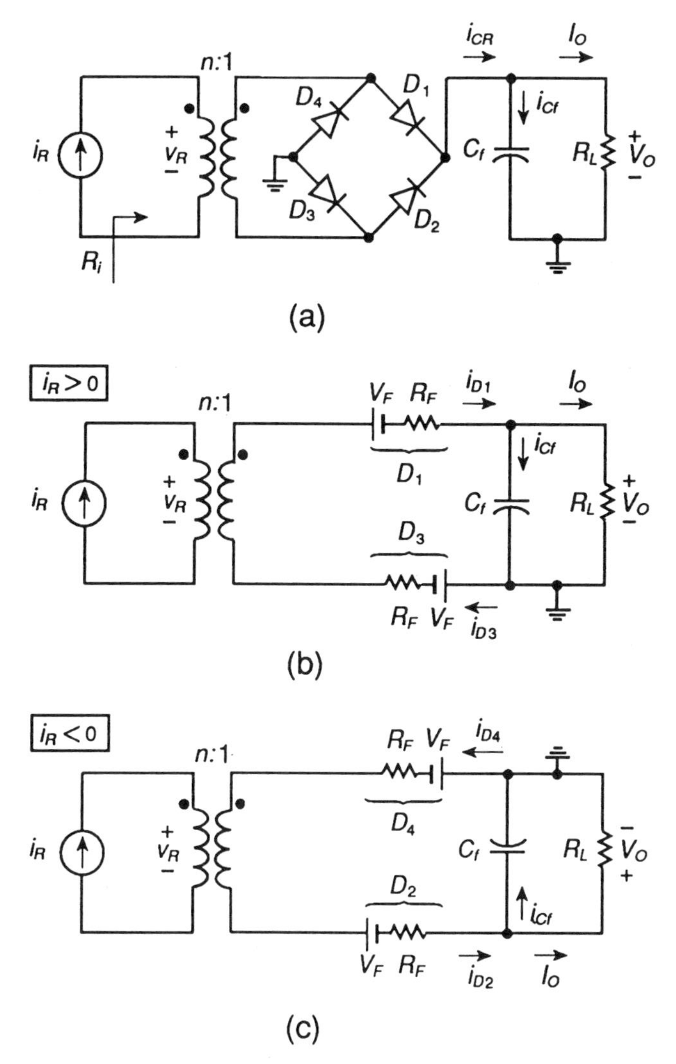

Figure 2.9: Class D current-driven bridge rectifier. (a) Circuit. (b) Model for $i_R > 0$. (c) Model for $i_R < 0$.

ON, while D_2 and D_4 are OFF. When $i_R < 0$, D_2 and D_4 are ON, while D_1 and D_3 are OFF. Expressions (2.49) to (2.57) and (2.61) to (2.68) remain the same.

2.5.1 Power-Output Capability

The peak forward currents of the rectifier diodes are

$$I_{DM} = nI_{Rm} = \frac{\pi I_O}{2}, \tag{2.91}$$

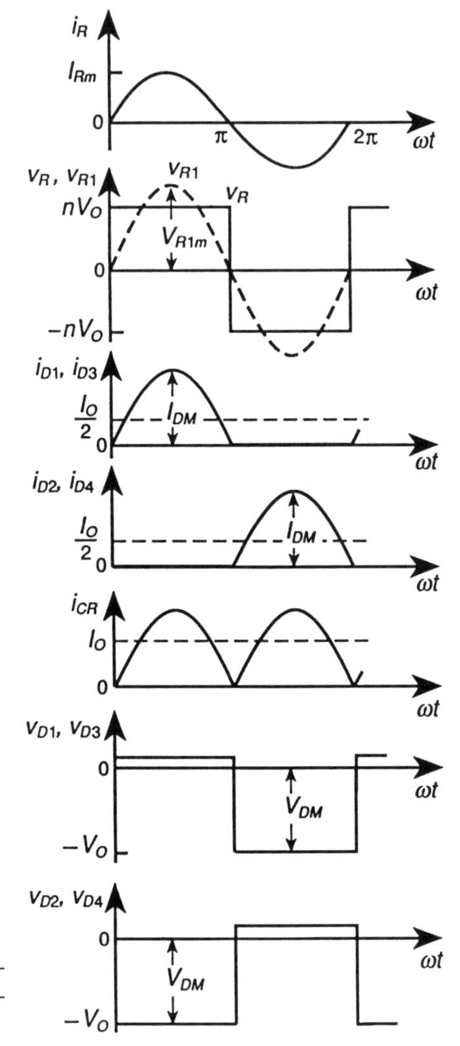

Figure 2.10: Current and voltage waveforms in Class D current-driven bridge rectifier.

the peak reverse voltages of the diodes are

$$V_{DM} = V_O, \tag{2.92}$$

and the power-output capability is

$$c_{pR} = \frac{P_O}{I_{DM}V_{DM}} = \frac{I_O V_O}{I_{DM}V_{DM}} = \frac{2}{\pi} \approx 0.637. \tag{2.93}$$

2.5.2 Efficiency

Using (2.65) and (2.68), the overall conduction loss of the rectifier is found as

$$P_C = 4P_D + P_{rC} = 2V_F I_O + \frac{\pi^2 I_O^2 R_F}{4} + r_C I_O^2 \left(\frac{\pi^2}{8} - 1\right)$$
$$= P_O \left[\frac{2V_F}{V_O} + \frac{\pi^2 R_F}{4R_L} + \frac{r_C}{R_L}\left(\frac{\pi^2}{8} - 1\right)\right]. \quad (2.94)$$

The input power is

$$P_i = \frac{P_O + P_C}{\eta_{tr}} = \frac{P_O + 4P_D + P_{rC}}{\eta_{tr}}$$
$$= \frac{P_O}{\eta_{tr}}\left[1 + \frac{2V_F}{V_O} + \frac{\pi^2 R_F}{4R_L} + \frac{r_C}{R_L}\left(\frac{\pi^2}{8} - 1\right)\right]. \quad (2.95)$$

From (2.94) and (2.95), the efficiency is expressed as

$$\eta_R = \frac{P_O}{P_i} = \frac{P_O \eta_{tr}}{P_O + P_C} = \frac{\eta_{tr}}{1 + \frac{2V_F}{V_O} + \frac{\pi^2 R_F}{4R_L} + \frac{r_C}{R_L}(\frac{\pi^2}{8} - 1)}. \quad (2.96)$$

2.5.3 Input Resistance

From (2.7), (2.51), and (2.95), one obtains the input resistance

$$R_i = \frac{8n^2 R_L}{\pi^2 \eta_{tr}}\left[1 + \frac{2V_F}{V_O} + \frac{\pi^2 R_F}{4R_L} + \frac{r_C}{R_L}\left(\frac{\pi^2}{8} - 1\right)\right] = \frac{8n^2 R_L}{\pi^2 \eta_R}. \quad (2.97)$$

2.5.4 Voltage Transfer Function

Using (2.30), (2.97), and (2.96), the ac-to-dc voltage transfer function is obtained as

$$M_{VR} \equiv \frac{V_O}{V_{Rrms}} = \sqrt{\frac{\eta_R R_L}{R_i}} = \frac{\pi \eta_{tr}}{2\sqrt{2}n[1 + \frac{2V_F}{V_O} + \frac{\pi^2 R_F}{4R_L} + \frac{r_C}{R_L}(\frac{\pi^2}{8} - 1)]}$$
$$= \frac{\pi \eta_R}{2\sqrt{2}n}. \quad (2.98)$$

The parameters of the three current-driven rectifiers are given in Table 2.1, assuming that the losses are zero.

Example 2.5

Find the efficiency η_R, the voltage transfer function M_{VR}, and the input resistance R_i for a Class D bridge rectifier of Fig. 2.9 at $V_O = 100$ V and $I_O = 1$ A. The rectifier employs *pn* junction diodes with $V_F = 0.9$ V and $R_F = 0.04\ \Omega$, and a filter capacitor C_f with $r_C = 50$ mΩ. The transformer turns ratio is $n = 2$. Assume the transformer efficiency $\eta_{tr} = 0.97$. Calculate also current and voltage stresses for the diodes.

Table 2.1. Parameters of Lossless Current-Driven Rectifiers

Parameter	Half-Wave Rectifier	Transformer Center-Tapped Rectifier	Bridge Rectifier
M_{VR}	$\frac{\pi}{n\sqrt{2}}$	$\frac{\pi}{2\sqrt{2}n}$	$\frac{\pi}{2\sqrt{2}n}$
R_i	$\frac{2n^2 R_L}{\pi^2}$	$\frac{8n^2 R_L}{\pi^2}$	$\frac{8n^2 R_L}{\pi^2}$
I_{DM}	πI_O	$\frac{\pi I_O}{2}$	$\frac{\pi I_O}{2}$
V_{DM}	V_O	$2V_O$	V_O

Solution: The load resistance of the rectifier is $R_L = V_O/I_O = 100\ \Omega$, and the output power is $P_O = I_O V_O = 100$ W.

Using (2.94),

$$P_C = 4P_D + P_{rC} = 1.9 + 0.01 = 1.91 \text{ W}. \tag{2.99}$$

From (2.96),

$$\eta_R = \frac{P_O \eta_{tr}}{P_O + P_C} = \frac{100 \times 0.97}{100 + 1.91} = 95.18\%. \tag{2.100}$$

The input resistance of the rectifier is obtained from (2.97)

$$R_i = \frac{8n^2 R_L}{\pi^2 \eta_R} = \frac{8 \times 2^2 \times 100}{\pi^2 \times 0.9518} = 340.65\ \Omega. \tag{2.101}$$

From (2.98), the voltage transfer function is

$$M_{VR} = \frac{\pi \times 0.9518}{2\sqrt{2} \times 2} = 0.529. \tag{2.102}$$

The stresses for the diodes can be computed from (2.91) and (2.92)

$$I_{DM} = \frac{\pi I_O}{2} = \frac{\pi \times 1}{2} = 1.57 \text{ A} \tag{2.103}$$

$$V_{DM} = V_O = 100 \text{ V}. \tag{2.104}$$

The efficiency of this rectifier is high because of the high output voltage and low output current. The bridge rectifier is not suitable for low-voltage and high-current applications. Two diodes conducting at the same time cause poor efficiency (see Problem 2.3). This rectifier is intended for high output voltage applications, because the voltage stresses for the diodes are low.

Example 2.6

Plot the efficiencies η_R versus load resistance R_L for Class D current-driven half-wave, center-tapped, and full-bridge rectifiers at $V_O = 5$ V, $V_F = 0.5$ V, $R_F = 0.025\ \Omega$, $r_C = 0.02\ \Omega$, and $\eta_{tr} = 96\%$. Discuss the results.

Solution: The expressions (2.27), (2.71), and (2.96) were used to compute the plots of the efficiencies for the rectifiers. The results are shown in Fig. 2.11. It can be seen that the center-tapped rectifier has the highest efficiency and the half-wave rectifier has the lowest efficiency. At light loads, the efficiences of full-bridge and half-wave rectifiers are nearly the same. The efficiencies decrease dramatically at low load resistances for all the rectifiers. The ripple voltage for the bridge rectifier is the same as that for the center-tapped rectifier.

2.6 EFFECT OF EQUIVALENT SERIES RESISTANCE AND EQUIVALENT SERIES INDUCTANCE

An equivalent circuit of a capacitor is depicted in Fig. 2.12. A distributed resistance of the capacitor is modeled by lumped resistance r_C, called an *equivalent series resistance* (ESR) and given by

$$r_C = r_d + r_e + r_t, \tag{2.105}$$

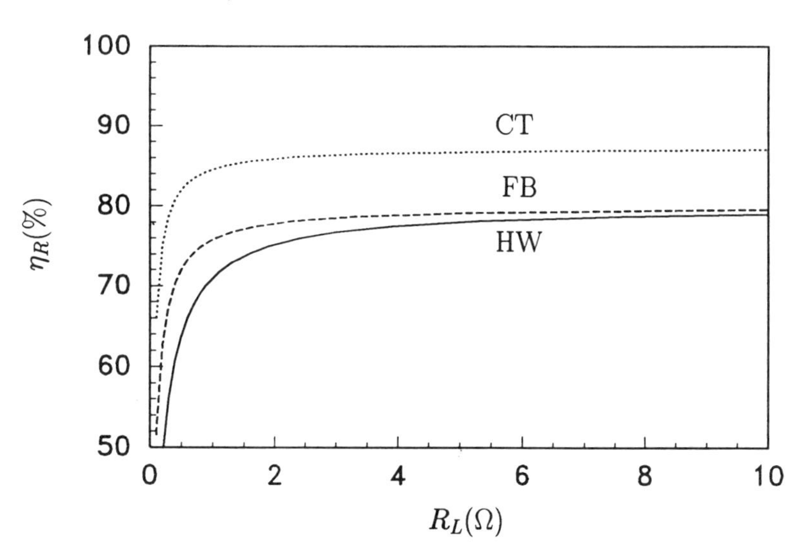

Figure 2.11: Plots of efficiencies for Class D current-driven half-wave (HW), center-tapped (CT), and full-bridge (FB) rectifiers of Example 2.6.

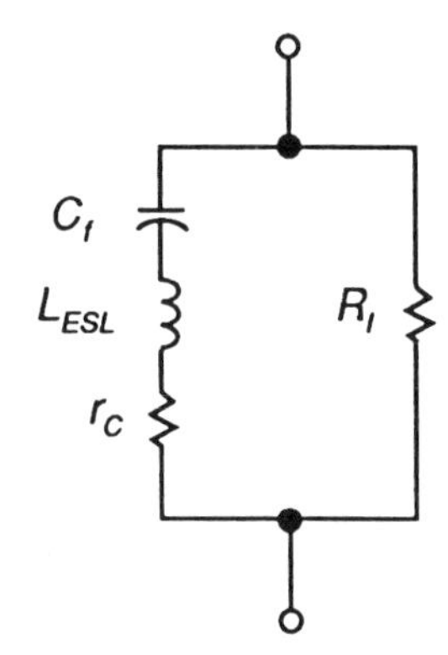

Figure 2.12: Equivalent circuit of a capacitor.

where r_d represents the dielectric losses, r_e is the resistance of electrodes (plates), and r_t is the resistance of terminations.

A distributed inductance of the capacitor is modeled by lumped inductance L_{ESL}, termed an *equivalent series inductance* (ESL), and given by

$$L_{ESL} = L_e + L_t, \tag{2.106}$$

where L_e is the inductance of electrodes and L_t is the inductance of terminations. The resistance R_l represents the insulation resistance whose typical values range from 10 kΩ to 1 GΩ. A dc leakage current I_l flowing through R_l can be as high as 5 mA for electrolytic capacitors. Heating caused by the leakage current I_l is usually negligible for low dc output voltages ($V_O \leq 200\ V$). Two types of filter capacitors have been used in dc-dc converters: 1) aluminum or tantalum electrolytic capacitors and 2) multilayer ceramic capacitors. The parasitic elements ESR and ESL of filter capacitors have an adverse effect on the level of the output ripple. In addition, ESR has a detrimental impact on the efficiency. The typical values of ESR and ESL are as follows: r_C = 40 mΩ to 2 Ω and L_{ESL} = 10 to 25 nH for electrolytic capacitors, and r_C = 2 to 50 mΩ and L_{ESL} = 5 to 10 nH for ceramic capacitors. Capacitors with higher capacitances usually have lower resistances r_C and higher inductances L_{ESL}. The ESR initially decreases and then increases with frequency. The resistance r_C depends on the type of the capacitor and decreases with increasing temperature T, as shown in Fig. 2.13. As capacitors age, r_C may increase (e.g., by 40 %).

Figure 2.14(a) shows a small-signal model of Class D rectifiers, in which r_C represents ESR of C_f, ESL is assumed to be negligible, and R_l is assumed to be infinity. The output-to-input transfer function is equal to the input impedance of the R_L-C_f-r_C circuit in the s-domain and is given by

$$Z(s) = \frac{v_r}{i_{cr}} = \frac{R_L(r_C + \frac{1}{sC_f})}{R_L + r_C + \frac{1}{sC_f}} = \frac{r_C R_L}{r_C + R_L} \frac{1 + \frac{1}{sC_f r_C}}{1 + \frac{1}{sC_f(R_L + r_C)}} = \frac{R_L}{k+1} \frac{1 + \frac{\omega_z}{s}}{1 + \frac{\omega_p}{s}}, \tag{2.107}$$

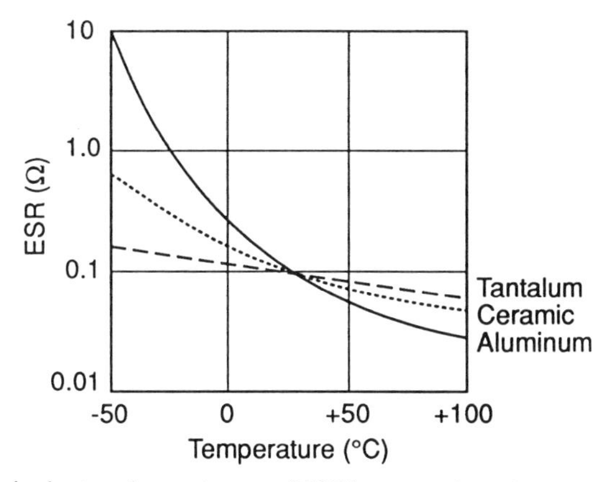

Figure 2.13: Equivalent series resistance (ESR) r_C as a function of temperature T for various types of capacitors.

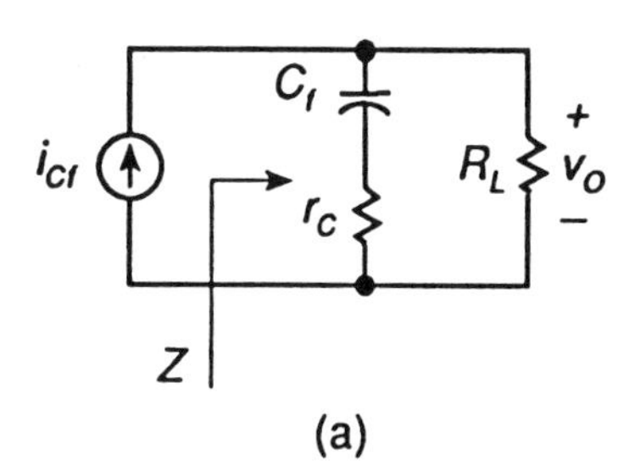

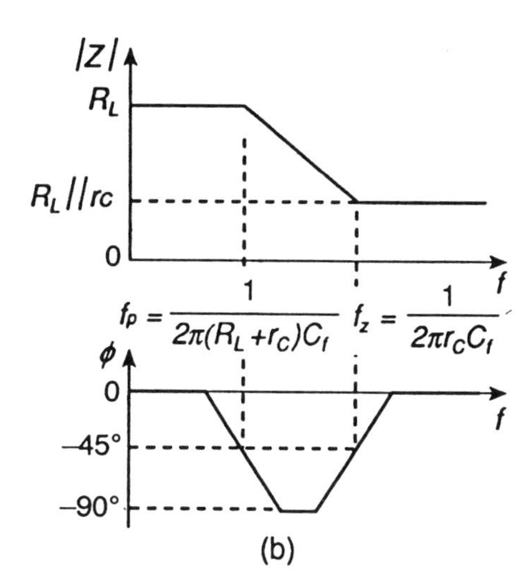

Figure 2.14: (a) Small-signal model of Class D rectifiers. (b) Bode plots of the impedance Z for the R_L-C_f-r_C circuit.

where $k = R_L/r_C$, $f_z = \omega_z/2\pi = 1/(2\pi r_C C_f)$ is the frequency of the zero, and $f_p = \omega_p/2\pi = 1/[2\pi C_f(R_L + r_C)]$ is the frequency of the pole. Note that $f_z/f_p = R_L/r_C + 1 = k + 1$ and $R_L \parallel r_C = R_L/(k+1)$; for $r_C \ll R_L$, $R_L \parallel r_C \approx r_C$. From (2.107),

$$Z(jf) = \frac{R_L}{k+1}\frac{1 - j\frac{f_z}{f}}{1 - j\frac{f_p}{f}} = |Z| e^{j\phi}, \tag{2.108}$$

where

$$|Z| = \frac{R_L}{k+1}\frac{\sqrt{1 + (\frac{f_z}{f})^2}}{\sqrt{1 + (\frac{f_p}{f})^2}} \tag{2.109}$$

$$\phi = arctan\left(\frac{f_p}{f}\right) - arctan\left(\frac{f_z}{f}\right). \tag{2.110}$$

The Bode plots of the impedance Z are displayed in Fig. 2.14(b). For $f \geq f_z$, $|Z| \approx R_L \parallel r_C = R_L/(k+1)$ is independent of the frequency. Therefore, the attenuation of harmonics higher than f_z does not increase with frequency. The ac component of i_{D1} consists of the fundamental and higher harmonics. The amplitudes of the n-th harmonic of the current I_n to the R_L-C_f-r_C circuit and the output voltage V_n are related by

$$V_n = I_n |Z|, \tag{2.111}$$

which becomes

$$V_n \approx I_n(R_L \parallel r_C) = \frac{I_n R_L}{k+1}, \quad \text{for} \quad f \geq f_z. \tag{2.112}$$

Thus, V_n is independent of C_f and depends only on r_C and R_L for $f \geq f_z$. If the operating frequency is greater than f_z, the ac components of the current $i_{cr} = i_{D1} - I_O$ to the R_L-C_f-r_C circuit and the output voltage v_r are related by

$$v_r = (R_L \parallel r_C) i_{cr}, \tag{2.113}$$

where $i_{cr} = i_{D1} - I_O$. The the peak-to-peak ripple voltage is

$$V_r = r_C I_{pp} = r_C I_{DM}. \tag{2.114}$$

To obtain a low value of V_r, the capacitor C_f should be selected so that $r_C \ll R_L$. This condition may be difficult to satisfy at a low V_O because R_L is very low and becomes comparable with r_C.

The frequency f_z is a performance parameter of filter capacitors. For electrolytic capacitors, $\tau_z = r_C C \approx 65$ μs and therefore $f_z = 1/(2\pi r_C C) \approx$

2.5 kHz over a wide range of capacitances and breakdown voltages. A tantalum capacitor 550D has $C = 330\ \mu\text{F}/6$ V and $r_C = 40\ \text{m}\Omega$ for $f < 100$ kHz. Hence, $f_z = 1/2\pi r_C C = 12$ kHz.

At high frequencies, the ESL may significantly affect the impedance of the filter capacitor. For example, if a capacitor has $C = 330\ \mu\text{F}$ and $L_{ESL} = 10$ nH, its self-resonant frequency is $f_{sr} = 1/(2\pi\sqrt{LC}) = 87.6$ kHz. For $f > f_{sr}$, the reactance of the capacitor increases with frequency. The self-resonant frequency f_{sr} increases when the lengths of the capacitor leads decrease. Figure 2.15 shows a small-signal model of Class D rectifiers, taking into account ESL of the filter capacitor. The impedance Z is given by

$$Z(s) = R_L \frac{s^2 L_{ESL} C_f + s r_C C_f + 1}{s^2 L_{ESL} C_f + s(r_C + R_L) C_f + 1}. \tag{2.115}$$

Figure 2.15(b) depicts the Bode plots of the impedance (or transfer function) Z for the case of real zeros f_{z1} and f_{z2} and real poles f_{p1} and f_{p2}. Note that in

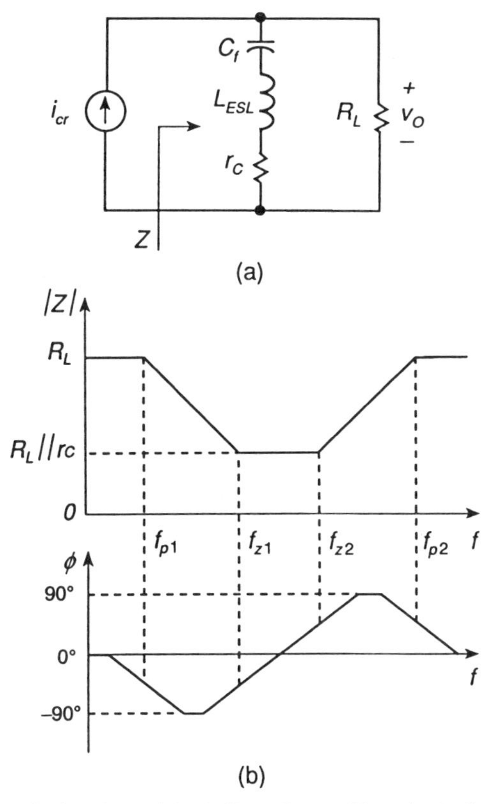

Figure 2.15: (a) Small-signal model of Class D rectifiers including ESL of the filter capacitor. (b) Bode plots of the impedance Z for the R_L-C_f-r_C-L_{ESL} circuit.

the high-frequency range ESL causes $| Z |$ to increase back to the R_L value, therefore reducing the filtering effect.

The ac ripple current flowing through the capacitor's ESR causes power dissipation in the form of heat, reducing the efficiency. Moreover, this heat causes the internal operating temperature T_o to rise, which can severely limit the lifetime of the capacitor. The capacitor operating temperature T_o depends on the ambient temperature T_a and the temperature rise ΔT caused by the power dissipated in r_C, that is, $T_o = T_a + \Delta T$. For example, the increased operating temperature of an electrolytic capacitor accelerates vaporization of the electrolyte. As the amount of electrolyte decreases, r_C increases, which, in turn, raises the operating temperature even higher; it is a positive feedback mechanism. This succession of events may eventually lead to *thermal runaway* and destruction of the capacitor. To lower the ambient temperature and thereby the operating temperature, the filter capacitor should be placed far from other hot devices such as power diodes. However, this will increase the lead length, increasing both ESR and ESL. To minimize r_C and L_{ESL}, the connections should be as wide as possible.

2.7 SYNCHRONOUS RECTIFIERS

A disadvantage of diode rectifiers is a low efficiency at a low dc output voltage such as 3.3 V. This is because the diode forward voltage drop is relatively high due to the diode offset voltage V_F. When the diode forward voltage is comparable with the dc output voltage, the diode conduction loss becomes very high compared with the output power. Therefore, the rectifier's efficiency is significantly reduced, the operating temperature is increased, the diode leakage current is increased, large heat sinks are needed, the rectifier's size is increased, and the reliability is reduced. The aforementioned problems can be alleviated by the application of synchronous rectifiers [2] and [4]–[24].

Figure 2.16 shows an unregulated synchronous Class D current-driven transformer center-tapped rectifier. The synchronous rectifier is obtained by replacing diodes in a conventional rectifier shown in Fig. 2.6(a) with low on-resistance power MOSFETs. The MOSFETs are driven in synchronism with an ac current (or voltage) source in such a manner that each transistor alternately conducts and is cut off on successive half-cycles, so that a unidirectional current flows in the circuit, providing rectifying action. Power MOSFETs do not have the offset voltage. Therefore, if the MOSFET on-resistance r_{DS} is low, a low device forward voltage may be achieved. Power MOSFETs with on-resistances r_{DS} on the order of 10 to 18 mΩ are available. Lower on-resistances may be obtained by paralleling power MOSFETs. As a result, the conduction losses of the rectifying devices can be significantly reduced, yielding high efficiency. If $v_{DS} = r_{DS} i_S < 0.7$ V, the antiparallel MOSFET body diode is off and all of the switch current flows through the MOSFET channel. Since the diode does not conduct, the power loss asso-

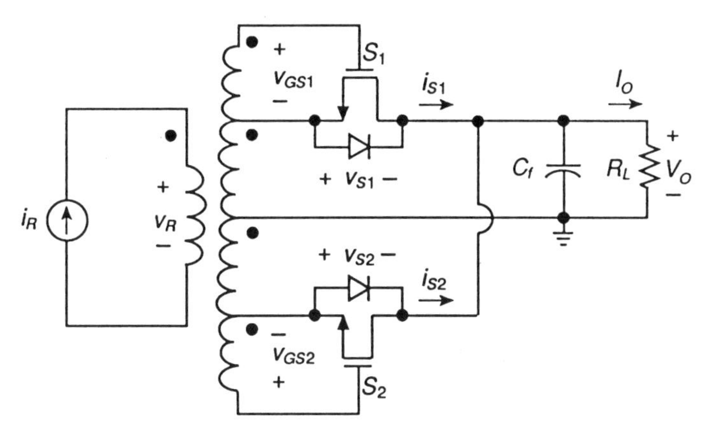

Figure 2.16: Unregulated synchronous Class D current-driven transformer center-tapped rectifier.

ciated with its reverse recovery is zero. In addition, the leakage current of power MOSFETs is much lower than that of rectifier diodes. However, power MOSFETs require a driver that not only adds to the circuit complexity but consumes power. The gate-drive power at low frequencies is low, but at high frequencies may become a dominant component of the total power loss. Since power MOSFETs are controllable devices, synchronous rectifiers can be used as regulated rectifiers [2], [13], [16], [19], [22]–[24].

The rectifier's input current i_R is sinusoidal and given by (2.1). For $i_R > 0$, S_2 is OFF and S_1 is ON. For $i_R < 0$, S_1 is OFF and S_2 is ON. The currents, voltages, input and output power, power factor, peak current and voltage of the MOSFETs, and power-output capability are given by (2.49) through (2.60).

2.7.1 Gate-Drive Power

The input capacitance of a power MOSFET C_{iss} is highly nonlinear. For this reason, the estimation of the gate-drive power based on the input capacitance is difficult. A simpler method relies on the concept of the gate charge. The energy required to charge and discharge the MOSFET input capacitance is

$$W_G = Q_g V_{GSpp}, \tag{2.116}$$

where Q_g is the gate charge given in data sheets and V_{GSpp} is the peak-to-peak gate-to-source voltage. Hence, the drive power of each MOSFET associated with turning the device on and off is

$$P_G = f Q_g V_{GSpp}. \tag{2.117}$$

Note that the drive power increases proportionally with frequency and becomes comparable with the conduction power loss at high frequencies (usually above 0.5 MHz). Therefore, synchronous rectifiers are not recommended for operation at very high frequencies.

2.7.2 Efficiency

The rms value of the current through each switch is derived in (2.61) and is

$$I_{Srms} = \frac{\pi I_O}{4}. \tag{2.118}$$

Thus, the power loss in the MOSFET forward resistance r_{DS} is

$$P_{rDS} = r_{DS} I_{Srms}^2 = \frac{\pi^2 I_O^2 r_{DS}}{16} = \frac{\pi^2 r_{DS}}{16 R_L} P_O. \tag{2.119}$$

Since there is no parasitic voltage source as opposed to a diode, the power loss P_{rDS} is the total conduction loss per switch. The power loss in the filter capacitor is given by (2.68). Hence, the total conduction loss is

$$P_C = 2P_{rDS} + P_{rC} = \frac{\pi^2 I_O^2 r_{DS}}{8} + r_C I_O^2 \left(\frac{\pi^2}{8} - 1\right)$$
$$= P_O \left[\frac{\pi^2 r_{DS}}{8R_L} + \frac{r_C}{R_L}\left(\frac{\pi^2}{8} - 1\right)\right]. \tag{2.120}$$

The overall power dissipation in the synchronous rectifier is

$$P_T = P_C + 2P_G = P_O \left[\frac{\pi^2 r_{DS}}{8R_L} + \frac{r_C}{R_L}\left(\frac{\pi^2}{8} - 1\right) + \frac{2fQ_g V_{GSpp}}{P_O}\right]. \tag{2.121}$$

Note that the switching losses in MOSFETs have been neglected in (2.121) because synchronous rectifiers usually have low output voltage, which makes the switching losses very small. The turn-on and turn-off switching losses in power MOSFETs are discussed in detail in Sections 6.7.2 and 6.7.3. The efficiency of the synchronous rectifier is

$$\eta_R = \frac{P_O}{P_i} = \frac{P_O \eta_{tr}}{P_C + P_{rC} + 2P_G} = \frac{\eta_{tr}}{1 + \frac{\pi^2 r_{DS}}{8R_L} + \frac{r_C}{R_L}\left(\frac{\pi^2}{8} - 1\right) + \frac{2fQ_g V_{GSpp}}{P_O}}. \tag{2.122}$$

2.7.3 Input Resistance

Substitution of (2.7), (2.51), and (2.120) into (2.26) gives the input resistance

$$R_i = \frac{8n^2 R_L}{\pi^2 \eta_{tr}} \left[1 + \frac{\pi^2 r_{DS}}{8R_L} + \frac{r_C}{R_L}\left(\frac{\pi^2}{8} - 1\right) + \frac{2fQ_g V_{GSpp}}{P_O}\right] = \frac{8n^2 R_L}{\pi^2 \eta_R}. \tag{2.123}$$

2.7.4 Voltage Transfer Function

Using (2.30), (2.122), and (2.123), one obtains the voltage transfer function of the rectifier as

$$M_{VR} \equiv \frac{V_O}{V_{Rrms}} = \sqrt{\frac{\eta_R R_L}{R_i}} = \frac{\pi \eta_{tr}}{2\sqrt{2}n[1 + \frac{\pi^2 r_{DS}}{8R_L} + \frac{r_C}{R_L}(\frac{\pi^2}{8} - 1) + \frac{2fQ_g V_{GSpp}}{P_O}]}$$
$$= \frac{\pi \eta_R}{2\sqrt{2}n}. \quad (2.124)$$

Example 2.7

An unregulated synchronous Class D center-tapped rectifier of Fig. 2.16 is built using SMP60N06-18 MOSFETs (Siliconix) with $r_{DS} = 0.018\ \Omega$ and $Q_g = 100$ nC and a filter capacitor C_f with $r_C = 20$ mΩ. The rectifier is operated at $V_O = 5$ V, $I_O = 20$ A, and $f = 100$ kHz. Find the efficiency η_R, the voltage transfer function M_{VR}, and the input resistance R_i for this rectifier. The transformer turns ratio is $n = 5$. Assume the transformer efficiency $\eta_{tr} = 0.96$ and the peak-to-peak gate-to-source voltage $V_{GSpp} = 10$ V.

Solution: The load resistance of the rectifier is $R_L = V_O/I_O = 0.25\ \Omega$, and the output power is $P_O = I_O V_O = 100$ W. Equation (2.119) yields the MOSFET conduction loss

$$P_{rDS} = \frac{\pi^2 I_O^2 r_{DS}}{16} = \frac{\pi^2 \times 20^2 \times 0.018}{16} = 4.44 \text{ W}. \quad (2.125)$$

Using (2.68), one obtains the power loss in the filter capacitor

$$P_{rC} = r_C I_O^2 \left(\frac{\pi^2}{8} - 1\right) = 0.02 \times 20^2 \left(\frac{\pi^2}{8} - 1\right) = 1.87 \text{ W}. \quad (2.126)$$

From (2.120),

$$P_C = 2P_{rDS} + P_{rC} = 2 \times 4.44 + 1.87 = 10.75 \text{ W}. \quad (2.127)$$

From (2.117), the gate drive power of each transistor is

$$P_G = fQ_g V_{GSpp} = 100 \times 10^3 \times 100 \times 10^{-9} \times 10 = 0.1 \text{ W}. \quad (2.128)$$

Using (2.122),

$$\eta_R = \frac{P_O \eta_{tr}}{P_O + P_C + 2P_G} = \frac{100 \times 0.96}{100 + 10.75 + 0.2} = 86.53\%. \quad (2.129)$$

The input resistance of the rectifier can be obtained from (2.123)

$$R_i = \frac{8n^2 R_L}{\pi^2 \eta_R} = \frac{8 \times 5^2 \times 0.25}{\pi^2 \times 0.8653} = 5.85\ \Omega. \tag{2.130}$$

The voltage transfer function is calculated using (2.124)

$$M_{VR} = \frac{\pi \eta_R}{2\sqrt{2}n} = \frac{\pi \times 0.8653}{2\sqrt{2} \times 5} = 0.192. \tag{2.131}$$

Note that the efficiency of the synchronous center-tapped rectifier is much higher than the efficiency of the conventional center-tapped rectifier at the same parameters (see Example 2.3).

2.8 SUMMARY

- Class D current-driven rectifiers have simple topologies. They contain two diodes and a capacitive first-order output filter.
- The dc output current I_O is directly proportional to the amplitude of the input current I_{Rm}.
- The on-duty cycle of each diode is 50%.
- The diode threshold voltage V_F, the diode forward resistance R_F, and the filter capacitor ESR reduce the efficiency of the rectifiers, especially at low output voltages V_O and high output currents I_O.
- The efficiency increases with an increasing load resistance R_L and is higher at a higher dc output voltage.
- The center-tapped rectifier has the highest efficiency, while the half-wave rectifier has the lowest. In the half-wave rectifier the average current through each diode is I_O, whereas in the center-tapped and bridge rectifier the average current through each diode is $I_O/2$. Therefore, the rms value of the diode current in the center-tapped and bridge rectifier is two times less than that in the half-wave rectifier, resulting in four times lower power loss in R_F per diode.
- The half-wave rectifier and the bridge rectifier are suitable for high-voltage applications because the diode peak reverse voltage is $V_{DM} = -V_O$, while the center-tapped rectifier is suitable for low-voltage applications because $V_{DM} = -2V_O$.
- An advantage of the half-wave rectifier is that both the source and the load can be connected to the same ground without a transformer.
- A disadvantage of the bridge rectifier is that the source and the load cannot be connected to the same ground without a transformer.
- The secondary winding of the transformer in the center-tapped rectifier requires twice as many turns as that in the bridge and half-wave rectifiers.
- In all three rectifiers, a negative output voltage can be obtained by reversing all the diodes.

- The rms current of the filter capacitor is very high, and therefore the capacitor must be rated accordingly.
- For the C_f-r_C model of the filter capacitor, the corner frequency of the output filter $f_p = 1/[2\pi C_f(R_L + r_C)]$ depends on the load resistance R_L. Above the frequency of the zero $f_z = 1/(2\pi r_C C_f)$, the slope of the magnitude of the output filter is zero.
- The ESL of the filter capacitor may destroy the filtering effect at very high frequencies because the capacitor behaves like an inductor.
- The diodes turn off at low di/dt because their currents are half-wave sinusoids. This results in low reverse-recovery peak current and low noise level.
- Synchronous Class D current-driven rectifiers can be obtained by replacing diodes in conventional Class D current-driven rectifiers with low on-resistance power MOSFETs to achieve high efficiency and/or controllability of the output voltage.
- The efficiency of current-driven synchronous rectifiers is low at light loads because the current flows through the switch in both directions, increasing conduction loss.
- Synchronous rectifiers are not suitable for applications at very high frequencies because the gate-drive power becomes very large at high frequencies, reducing efficiency.
- The Class D current-driven rectifiers can be used in resonant dc-dc converters, in which the inverter contains a series-resonant circuit that forces a sinusoidal current.

2.9 REFERENCES

1. M. K. Kazimierczuk, "Class D current-driven rectifiers for resonant dc/dc converter applications," *IEEE Trans. Ind. Electron.*, vol. IE-38, pp. 344–354, Oct. 1991.
2. M. K. Kazimierczuk and J. Jóźwik, "Class E zero-voltage-switching and zero-current-switching rectifiers," *IEEE Trans. Circuits Syst.*, vol. CAS-37, pp. 436–444, Mar. 1990.
3. M. K. Kazimierczuk and X. T. Bui, "Class E dc/dc resonant converters," *Power Conversion Conf. (PCI'88)*, Dearborn, MI, Oct. 1988, pp. 69-93.
4. M. W. Smith and K. Owyang, "Improving the efficiency of low output power voltage switched-mode converters with synchronous rectification," *Proc. Powercon 7*, H-4, 1980, pp. 1–13.
5. R. S. Kagan and M. H. Chi, "Improving power supply efficiency with synchronous rectifiers," *Proc. Powercon 9*, D-4, 1982, pp. 1–5.
6. R. A. Blanchard and R. Severns, "Designing switched-mode power converter for very low temperature operation," *Proc. Powercon 10*, D-2, 1983, pp. 1–11.
7. M. Alexander, R. Blanchard, and R. Severns, "MOSFETs move in on low volt-

age rectification," *MOSPOWER Application Handbook*, Siliconix, 1984, pp. 5.69–5.87.

8. E. Oxner, "Using power MOSFETs as high-frequency synchronous and bridge rectifiers in switch-mode power supplies," *MOSPOWER Application Handbook*, Siliconix, 1984, pp. 5.87–5.94.
9. R. A. Blanchard and P. E. Thibodeau, "The design of a high efficiency, low voltage power supply using MOSFET synchronous rectification and current mode control," *IEEE Power Electronics Specialists Conference Record*, 1985, pp. 355–361.
10. P. L. Hower, G. M. Kepler, and R. Patel, "Design, performance and application of a new bipolar synchronous rectifier," *IEEE Power Electronics Specialists Conference Record*, 1985, pp. 247–256.
11. R. A. Blanchard and P. E. Thibodeau, "Use of depletion mode MOSFET devices in synchronous rectification," *IEEE Power Electronics Specialists Conference Record*, 1986, pp. 81–81.
12. J. Blanc, R. A. Blanchard, and P. E. Thibodeau, "Use of enhencement-and depletion-mode MOSFETs in synchronous rectification," *Power Electronics Conference*, San Jose, CA 1987, pp. 1–8.
13. M. K. Kazimierczuk and J. Jóźwik, "Resonant dc/dc converter with Class-E inverter and Class-E rectifier," *IEEE Trans. Ind. Electron.*, vol. IE-36, pp. 568–576, Nov. 1989.
14. D. A. Grant and J. Grower, *Power MOSFETs: Theory and Applications*, John Wiley & Sons New York: 1989, pp. 250–256.
15. J. Blanc, "Bidirectional dc-to-dc converter using synchronous rectifiers," *Power Conversion and Intelligent Motion Conference*, Long Beach, Ca, 1989, pp. 15–20.
16. J. Jóźwik and M. K. Kazimierczuk "Analysis and design of Class E^2 dc/dc resonant converter," *IEEE Trans. Ind. Electron.*, vol. IE-37, pp. 173–183, Apr. 1990.
17. W. A. Tabisz, F. C. Lee, and D. Y. Chen, "A MOSFET resonant synchronous rectifier for high-frequency dc/dc converter," *IEEE Power Electronics Specialists Conference Record*, 1990, pp. 769–779.
18. B. Mohandes, "MOSFET synchronous rectifiers achieve 90% efficiency," *Power Conversion and Intelligent Motion*, Part I, June 1991 pp. 10–13; and Part II, July 1991, pp. 55–61.
19. M. K. Kazimierczuk and K. Puczko, "Class E low dv/dt synchronous rectifier with controlled duty cycle and output voltage," *IEEE Trans. Circuits Syst.*, vol. 37, pp. 1165–1172, Oct. 1991.
20. F. Goodenough, "Synchronous rectifier ups PC battery life," *Electronic Design*, pp. 47–52, Apr. 16, 1992.
21. J. A. Cobos, O. Garcia, J. Uceda, and A. de Hoz, "Self driven synchronous rectification in resonant topologies: Forward ZVS-MRC, forward ZCS-QRC and LCC-PRC," *IEEE Industrial Electronics Conf.*, San Diego, CA, Nov. 1992, pp. 185–190.
22. M. Mikolajewski and M. K. Kazimierczuk, "Zero-voltage-ripple rectifiers and dc/dc resonant converters," *IEEE Trans. Power Electronics*, vol. PE-6, pp. 12–17, Jan. 1993.

23. M. K. Kazimierczuk and M. J. Mescher, "Series resonant converter with phase-controlled synchronous rectifier," *IEEE Int. Conf. on Industrial Electronics, Control, Instrumentation, and Automation (IECON'93)*, Lahaina, Hawaii, Nov. 15–19, 1993, pp. 852–856.

24. M. K. Kazimierczuk and M. J. Mescher, "Class D converter with regulated synchronous half-wave rectifier," *IEEE Applied Power Electronics Conf.*, Orlando, FL, Feb. 13–17, 1994, vol. 2, pp. 1005–1011.

2.10 REVIEW QUESTIONS

2.1 What is a current-driven rectifier?

2.2 Sketch the waveforms for the Class D current-driven half-wave rectifier.

2.3 What is the average current through each diode in the Class D current-driven half-wave, transformer center-tapped, and bridge rectifiers?

2.4 Sketch the waveforms for the Class D current-driven transformer center-tapped rectifier.

2.5 Compare the current and voltage stresses of the diodes in the half-wave, transformer center-tapped, and bridge rectifiers at the same output voltage and output power.

2.6 Compare the efficiencies of the half-wave, transformer center-tapped, and bridge rectifiers at the same output voltage and output power.

2.7 Can both the ac input source and the load be grounded in a transformerless bridge rectifier?

2.8 What is a synchronous rectifier?

2.9 What is the motivation behind using synchronous rectifiers?

2.10 What is the difference between unregulated and regulated synchronous rectifiers?

2.11 How does the gate-drive power of power MOSFETs depend on frequency?

2.12 Is the efficiency of the current-driven synchronous rectifiers high at light loads?

2.11 PROBLEMS

2.1 Compare the efficiency of the Class D current-driven center-tapped rectifier of Fig. 2.6 with *pn* junction diodes and Schottky diodes for $V_O = 5$ V and $I_O = 10$ A. The *pn* junction diode has $V_F = 0.8$ V and $R_F = 75$ mΩ. The Schottky diode has $V_F = 0.4$ V and $R_F = 25$ mΩ. Assume that in both cases the ESR of the filter capacitor is $r_C = 20$ mΩ and the transformer efficiency is $\eta_{tr} = 96\%$.

2.2 Derive an equation for the power factor of the transformerless version of the Class D current-driven half-wave rectifier. Compare the result with the expression for the transformer version of the rectifier.

2.3 Calculate the efficiency η_R, the voltage transfer function M_{VR}, and the input resistance R_i for a Class D bridge rectifier of Fig. 2.9 at $V_O = 5$ V and $I_O = 20$ A. The rectifier employs Schottky diodes with $V_F = 0.4$ V and $R_F = 0.025\ \Omega$ and a filter capacitor C_f with $r_C = 20$ mΩ. The transformer turns ratio is $n = 5$. Assume the transformer efficiency $\eta_{tr} = 96\%$.

2.4 Prove that $v_c(0)$ in (2.39) is equal to $-\pi I_O/(2\omega C_f)$.

2.5 A half-wave rectifier, in which the maximum output current I_{Omax} is 5 A, operates at a switching frequency $f = 200$ kHz. The filter capacitance is $C_f = 100\ \mu$F. What value of the ESR of the filter capacitor cannot be exceeded if it is specified that the ripple voltage V_r must be less than 0.3 V.?

2.6 Repeat Problem 2.5 for a current-driven transformer center-tapped rectifier.

2.7 The output filter of a Class D current-driven rectifier has the following parameters: $R_L = 2\ \Omega$, $r_C = 0.05\ \Omega$, $C_f = 47\ \mu$F, and $L_{ESL} = 20$ nH. Calculate the frequency at which the effect of ESL becomes significant. Is this frequency dependent upon R_L in a properly designed rectifier?

2.8 A current-driven transformer center-tapped rectifier with output voltage $V_O = 3.3$ V and maximum output current $I_{Omax} = 10$ A employs Schottky diodes with $V_F = 0.3$ V and $R_F = 20$ mΩ, a filter capacitor with $C = 100\ \mu$F and $r_C = 5$ mΩ, and a transformer with $n = 5$ and $\eta_{tr} = 96\%$. The operating frequency is $f = 150$ kHz. Find η_R, M_{VR}, R_i, and V_r.

2.9 Replace the diodes in the circuit given in Problem 2.8 with SMP60N03-10L power MOSFETs (Siliconix) to obtain an unregulated synchronous rectifier. The on-resistance of the power MOSFETs is $r_{DS} = 10$ mΩ, and the gate charge is $Q_g = 100$ nC. Calculate the drive power and the rectifier efficiency. Assume the peak-to-peak gate-to-source voltage $V_{GSpp} = 10$ V. Compare the efficiencies of the diode rectifier and the synchronous rectifier. What is the operating frequency at which the drive power is equal to the conduction loss?

CHAPTER 3

CLASS D VOLTAGE-DRIVEN RECTIFIERS

3.1 INTRODUCTION

Rectifiers that are driven by a voltage source are referred to as *voltage-driven rectifiers* [1], [2]. If a diode current waveform is a square wave and a diode voltage waveform is a half-wave sinusoid, the rectifiers are called *Class D rectifiers*. This is because these waveforms are similar to the corresponding waveforms of the Class D current-source resonant inverter (discussed in Part II). In this chapter, three Class D voltage-driven rectifiers are given and analyzed: the half-wave rectifier, the transformer center-tapped rectifier, and the bridge rectifier.

3.2 ASSUMPTIONS

The analysis of the Class D voltage-driven rectifiers is carried out under the following assumptions:

1) The diode in the ON state is modeled by a series combination of a constant-voltage battery V_F and a constant resistance R_F, where V_F represents the diode threshold voltage and R_F represents the diode forward resistance.
2) The diode in the OFF state is modeled by an open switch.
3) The minority charge-carrier lifetime is zero for *pn* junction diodes, and the diode junction capacitance and lead inductance are zero.

4) The rectifiers are driven by an ideal sinusoidal voltage source, and the amplitude of the input voltage V_{Rm} is much higher than V_F.
5) The filter inductance L_f is large enough that its ac current ripple is negligible.

3.3 CLASS D HALF-WAVE RECTIFIER

3.3.1 Currents and Voltages

A circuit of the voltage-driven half-wave rectifier is depicted in Fig. 3.1(a). It consists of a transformer, if needed, diodes D_1 and D_2, and a second order

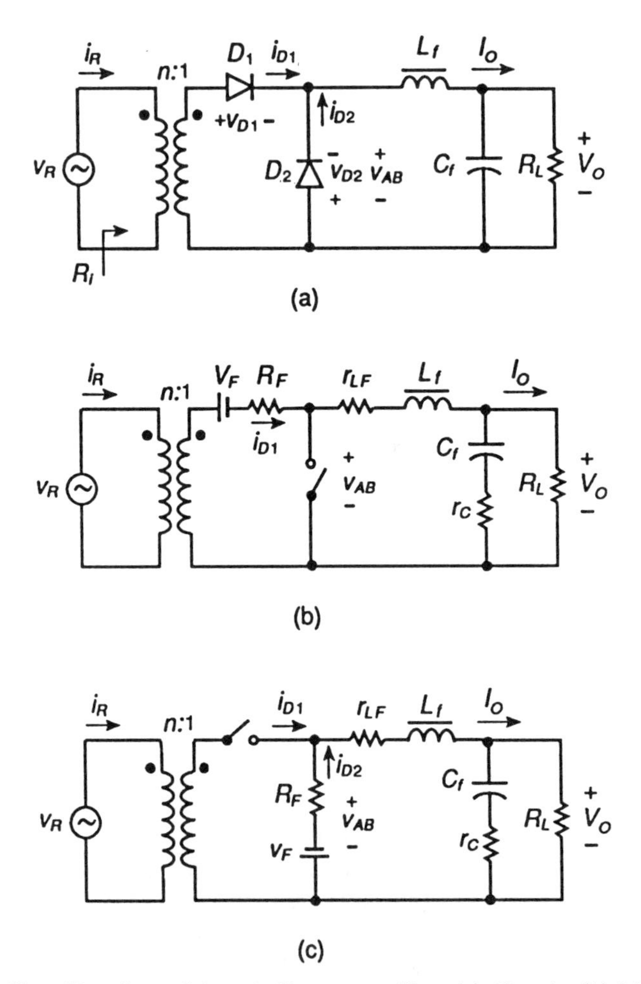

Figure 3.1: Class D voltage-driven half-wave rectifier. (a) Circuit. (b) Model when diode D_1 is on and diode D_2 is off. (c) Model when diode D_1 is off and diode D_2 is on.

L_f-C_f low-pass filter. Resistance R_L is a dc load. The rectifier is driven by a sinusoidal voltage source v_R. If the filter inductance is large enough, its ripple current is small and the inductor current is approximately equal to the dc output current I_O. In this case, the output filter and the load resistance can be replaced by a current sink. Assuming the ideal transformer, the input voltage source v_R can be reflected from the primary to the secondary side of the transformer to become v_R/n. When $v_R > 0$, diode D_1 is on and diode D_2 is off. An equivalent circuit is shown in Fig. 3.1(b). When $v_R < 0$, diode D_1 is off and diode D_2 is on. An equivalent circuit is depicted in Fig. 3.1(c).

Figure 3.2 shows current and voltage waveforms that explain the principle of operation of the half-wave rectifier. According to assumption 4), the input voltage of the rectifier is sinusoidal and given by

$$v_R = V_{Rm} sin\omega t, \tag{3.1}$$

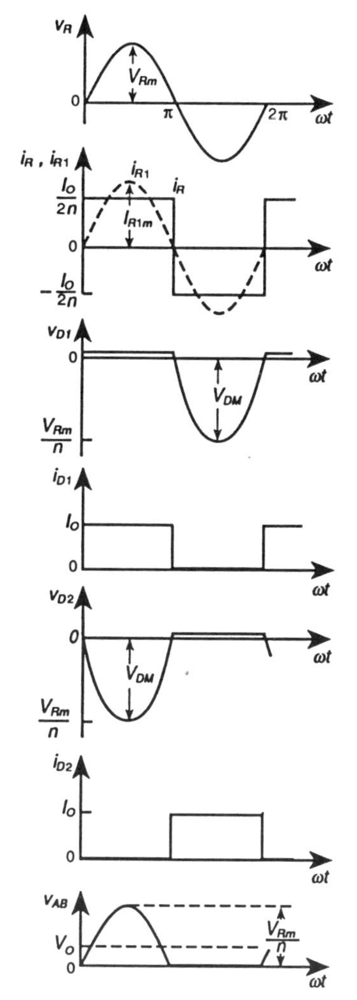

Figure 3.2: Current and voltage waveforms in Class D voltage-driven half-wave rectifier.

where V_{Rm} is the amplitude of v_R. The voltage at the input of the output filter is

$$v_{AB} = -v_{D2} = \begin{cases} \frac{v_R}{n} = \frac{V_{Rm}}{n} sin\omega t, & \text{for } 0 < \omega t \leq \pi \\ 0, & \text{for } \pi < \omega t \leq 2\pi, \end{cases} \quad (3.2)$$

where n is the transformer turns ratio. The average value of the voltage across the filter inductor L_f is zero. For this reason, the average value of the voltage v_{AB} is equal to the dc output voltage

$$V_O = \frac{1}{2\pi}\int_0^{2\pi} v_{AB} d(\omega t) = \frac{1}{2\pi}\int_0^{\pi} V_{Rm} sin\omega t d(\omega t) = \frac{V_{Rm}}{\pi n}. \quad (3.3)$$

Thus, the dc output voltage V_O is directly proportional to the amplitude of the input voltage V_{Rm}. Therefore, V_O may be regulated by controlling V_{Rm}.

The input current of the rectifier is a square wave and is given by

$$i_R = \begin{cases} I_O/2n, & \text{for } 0 < \omega t \leq \pi \\ -I_O/2n, & \text{for } \pi < \omega t \leq 2\pi. \end{cases} \quad (3.4)$$

This waveform exhibits an odd symmetry with respect to ωt, that is, $i_R(-\omega t) = -i_R(\omega t)$. In such a case, the amplitude of the fundamental component of i_R can be found as

$$I_{R1m} = \frac{2}{\pi}\int_0^{\pi} i_R sin\omega t d(\omega t) = \frac{I_O}{\pi n}\int_0^{\pi} sin\omega t d(\omega t) = \frac{2I_O}{\pi n}. \quad (3.5)$$

The fundamental component of the input current i_R is

$$i_{R1} = I_{R1m} sin\omega t = \frac{2I_O}{\pi n} sin\omega t. \quad (3.6)$$

Since the rectifier's input voltage v_R is sinusoidal, the input power P_i contains only the power of the fundamental component

$$P_i = \frac{I_{R1m}^2 R_i}{2} = \frac{2I_O^2 R_i}{\pi^2 n^2}, \quad (3.7)$$

where $R_i = V_{Rim}/I_{R1m}$ is the input resistance of the rectifier at the fundamental frequency f. The dc output power is

$$P_O = I_O^2 R_L. \quad (3.8)$$

3.3.2 Power Factor

From (3.4) and (3.5), one can find the rms value of the input current i_R

$$I_{Rrms} = \sqrt{\frac{1}{2\pi}\int_0^{2\pi} i_R^2 d(\omega t)} = \sqrt{\frac{I_O^2}{8\pi n^2}\int_0^{2\pi} d(\omega t)} = \frac{I_O}{2n} \tag{3.9}$$

and the rms value of the fundamental component i_{R1} of the input current i_R

$$I_{R1rms} = \frac{I_{R1m}}{\sqrt{2}} = \frac{2I_O}{\sqrt{2}\pi n} = \frac{\sqrt{2}I_O}{\pi n}. \tag{3.10}$$

Hence, the power factor is

$$\begin{aligned} PF &\equiv \frac{P_i}{I_{Rrms}V_{Ri}} = \frac{I_{R1rms}V_{Ri}}{I_{Rrms}V_{Ri}} = \frac{I_{R1rms}}{I_{Rrms}} \\ &= \frac{I_{R1rms}}{\sqrt{I_{R1rms}^2 + I_{R2rms}^2 + I_{R3rms}^2 + \ldots}} = \frac{2\sqrt{2}}{\pi} \approx 0.9. \end{aligned} \tag{3.11}$$

where I_{R1rms}, I_{R2rms}, I_{R3rms}, ... are the rms values of the harmonics of the rectifier input current. The total harmonic distortion of the rectifier input current is

$$\begin{aligned} THD &= \sqrt{\frac{I_{R2rms}^2 + I_{R3rms}^2 + I_{R4rms}^2 + \ldots}{I_{R1rms}^2}} = \sqrt{\frac{1}{PF^2} - 1} \\ &= \sqrt{\frac{\pi^2}{8} - 1} \approx 0.4834. \end{aligned} \tag{3.12}$$

3.3.3 Current and Voltage Stresses

The peak values of the diode forward current and the diode reverse voltage are

$$I_{DM} = I_O \tag{3.13}$$

$$V_{DM} = \frac{V_{Rm}}{n} = \pi V_O, \tag{3.14}$$

which leads to the rectifier's power-output capability

$$c_{pR} \equiv \frac{P_O}{I_{DM}V_{DM}} = \frac{I_O V_O}{I_{DM}V_{DM}} = \frac{V_O}{V_{DM}} = \frac{1}{\pi} \approx 0.318. \tag{3.15}$$

3.3.4 Efficiency

Since the current of each diode is $i_D = I_O$ during one-half of the cycle, one can find the average value of the diode current

$$I_D = \frac{1}{2\pi}\int_0^{2\pi} i_D d(\omega t) = \frac{1}{2\pi}\int_0^{\pi} I_O d(\omega t) = \frac{I_O}{2}, \tag{3.16}$$

the power loss associated with V_F

$$P_{VF} = V_F I_D = \frac{V_F I_O}{2} = \frac{V_F}{2V_O} P_O, \tag{3.17}$$

the rms value of the diode current

$$I_{Drms} = \sqrt{\frac{1}{2\pi}\int_0^{2\pi} i_D^2 d(\omega t)} = \sqrt{\frac{I_O^2}{2\pi}\int_0^{\pi} d(\omega t)} = \frac{I_O}{\sqrt{2}}, \tag{3.18}$$

the power loss in R_F

$$P_{RF} = R_F I_{Drms}^2 = \frac{I_O^2 R_F}{2} = \frac{R_F}{2R_L} P_O, \tag{3.19}$$

and the total conduction loss in each diode

$$P_D = P_{VF} + P_{RF} = \frac{I_O V_F}{2} + \frac{I_O^2 R_F}{2} = \frac{P_O}{2}\left(\frac{V_F}{V_O} + \frac{R_F}{R_L}\right). \tag{3.20}$$

The dc conduction loss in the filter inductor is

$$P_{rL} = r_{LF} I_O^2 = \frac{r_{LF} P_O}{R_L}, \tag{3.21}$$

where r_{LF} is the dc ESR of the filter inductor.

To calculate the ac conduction loss in the inductor and the loss in the ESR of the filter capacitor, it is assumed that the ripple current is no longer negligible and it flows entirely through the filter capacitor. Using (3.1) and (3.3), the voltage across the filter inductance can be approximated by

$$v_L = \begin{cases} \frac{v_R}{n} - V_O = V_O(\pi sin\omega t - 1), & \text{for} \quad 0 < \omega t \le \pi \\ -V_O, & \text{for} \quad \pi < \omega t \le 2\pi. \end{cases} \tag{3.22}$$

The filter inductor and capacitor ac current i_C is given by

$$\begin{aligned} i_C &= \frac{1}{\omega L_f}\int_0^{\omega t} v_L d(\omega t) \\ &= \begin{cases} \frac{V_O}{\omega L_f}[\pi(1 - cos\omega t) - \omega t] + i_C(0), & \text{for} \quad 0 < \omega t \le \pi \\ -\frac{V_O t}{L_f} + \frac{2\pi V_O}{\omega L_f} + i_C(0), & \text{for} \quad \pi < \omega t \le 2\pi, \end{cases} \end{aligned} \tag{3.23}$$

where $i_C(0) = -V_O/(4fL_f)$ satisfies the condition

$$\frac{1}{2\pi}\int_0^{2\pi} i_C d(\omega t) = 0, \tag{3.24}$$

which states that the average current through the capacitor is zero for steady state. The rms value of the capacitor current is given by

$$I_{Crms} = \sqrt{\frac{1}{2\pi}\int_0^{2\pi} i_C^2 d(\omega t)} = a_{hw}\frac{I_O R_L}{fL_f}, \tag{3.25}$$

where $a_{hw} = (\sqrt{1/3 - 2/\pi^2})/2 = 0.1808$. Hence, the losses in the ac ESR of the filter inductor r_{Lf} and in the ESR of the filter capacitor r_C are

$$P_{lc} = a_{hw}^2 \frac{r_{ac} I_O^2 R_L^2}{f^2 L_f^2} = a_{hw}^2 \frac{r_{ac} R_L}{f^2 L_f^2} P_O, \tag{3.26}$$

where the ac resistance of the filter inductor and the filter capacitor at the operating frequency is

$$r_{ac} = r_{Lf} + r_C. \tag{3.27}$$

The total conduction loss in the half-wave rectifier (excluding the transformer) is obtained from (3.20), (3.21), and (3.26)

$$P_C = 2P_D + P_{rL} + P_{lc} = P_O\left(\frac{V_F}{V_O} + \frac{R_F + r_{LF}}{R_L} + \frac{a_{hw}^2 r_{ac} R_L}{f^2 L_f^2}\right). \tag{3.28}$$

Assuming that the switching losses in the diodes are zero, the rectifier input power is

$$P_i = \frac{P_O + 2P_D + P_{rL} + P_{lc}}{\eta_{tr}} = \frac{P_O}{\eta_{tr}}\left(1 + \frac{V_F}{V_O} + \frac{R_F + r_{LF}}{R_L} + a_{hw}^2 \frac{r_{ac} R_L}{f^2 L_f^2}\right), \tag{3.29}$$

where η_{tr} is the efficiency of the transformer. The efficiency of the rectifier is obtained from (3.8) and (3.29)

$$\eta_R \equiv \frac{P_O}{P_i} = \frac{P_O \eta_{tr}}{P_O + 2P_D + P_{rL} + P_{lc}} = \frac{\eta_{tr}}{1 + \frac{V_F}{V_O} + \frac{R_F + r_{LF}}{R_L} + a_{hw}^2 \frac{r_{ac} R_L}{f^2 L_f^2}}. \tag{3.30}$$

3.3.5 Input Resistance

Substitution of (3.7), (3.8), (3.20), (3.21), and (3.26) into (3.29) yields the input resistance of the rectifier

$$R_i \equiv \frac{V_{Rm}}{I_{R1m}} = \frac{\pi^2 n^2 R_L}{2\eta_{tr}} \left(1 + \frac{V_F}{V_O} + \frac{R_F + r_{LF}}{R_L} + a_{hw}^2 \frac{r_{ac} R_L}{f^2 L_f^2}\right) = \frac{\pi^2 n^2 R_L}{2\eta_R}. \quad (3.31)$$

3.3.6 Voltage Transfer Function

The input power of the rectifier can be also described as

$$P_i = \frac{V_{Ri}^2}{R_i}, \quad (3.32)$$

where $V_{Ri} = V_{Rm}/\sqrt{2}$ is the rms value of the input voltage v_R. Using (3.8) and (3.32),

$$\eta_R = \frac{P_O}{P_i} = \left(\frac{V_O}{V_{Ri}}\right)^2 \frac{R_i}{R_L}. \quad (3.33)$$

Hence, from (3.31) and (3.30), the ac-to-dc voltage transfer function of the rectifier can be expressed as

$$M_{VR} \equiv \frac{V_O}{V_{Ri}} = \sqrt{\frac{\eta_R R_L}{R_i}} = \frac{\sqrt{2}\eta_{tr}}{\pi n \left(1 + \frac{V_F}{V_O} + \frac{R_F + r_{LF}}{R_L} + a_{hw}^2 \frac{r_{ac} R_L}{f^2 L_f^2}\right)}. \quad (3.34)$$

Example 3.1

Find the efficiency η_R, the voltage transfer function M_{VR}, and the input resistance R_i for a Class D half-wave rectifier of Fig. 3.1(a) at $V_O = 5$ V and $I_O = 20$ A. The operating frequency of the rectifier is 100 kHz. The rectifier is built using Schottky diodes (e.g., Motorola MBR20100CT) with $V_F = 0.5$ V and $R_F = 0.025\ \Omega$. The value of the filter inductance is $L_f = 1$ mH. The dc ESR of the inductor is $r_{LF} = 0.1\ \Omega$ and the ac ESR of the inductor is $r_{Lf} = 1.85\ \Omega$. A filter capacitor with $r_C = 50$ mΩ is employed. The transformer turns ratio is $n = 5$. Assume the transformer efficiency $\eta_{tr} = 0.96$.

Solution: The load resistance of the rectifier is $R_L = V_O/I_O = 0.25\ \Omega$, and the output power is $P_O = I_O V_O = 100$ W. From (3.20), the total conduction loss in each diode is

$$P_D = \frac{I_O V_F}{2} + \frac{I_O^2 R_F}{2} = \frac{20 \times 0.5}{2} + \frac{20^2 \times 0.025}{2} = 10 \text{ W}. \quad (3.35)$$

The dc conduction loss in the filter inductor is obtained using (3.21)

$$P_{rL} = I_O^2 r_{LF} = 20^2 \times 0.1 = 40 \text{ W}. \quad (3.36)$$

The ac conduction loss in the filter inductor and the filter capacitor is calculated from (3.26) and (3.27) as

$$P_{lc} = a_{hw}^2 \frac{(r_{Lf} + r_C)R_L}{f^2 L_f^2} P_O$$
$$= 0.1808^2 \frac{(1.85 + 0.05) \times 0.25}{(100 \times 10^3 \times 0.001)^2} \times 100 = 0.16 \text{ mW}. \quad (3.37)$$

It can be seen that relatively high values of the filter inductance and switching frequency result in negligible ac losses. From (3.30), the efficiency of the rectifier is

$$\eta_R = \frac{P_O \eta_{tr}}{P_O + 2P_D + P_{rL} + P_{lc}} = \frac{100 \times 0.96}{100 + 2 \times 10 + 40 + 0.00016} = 60\%. \quad (3.38)$$

The input resistance of the rectifier is given by (3.31)

$$R_i = \frac{\pi^2 n^2 R_L}{2\eta_R} = \frac{\pi^2 \times 5^2 \times 0.25}{2 \times 0.6} = 51.4\ \Omega. \quad (3.39)$$

From (3.34), the ac-to-dc voltage transfer function of the rectifier is obtained as

$$M_{VR} = \sqrt{\frac{\eta_R R_L}{R_i}} = \sqrt{\frac{0.6 \times 0.25}{51.4}} = 0.054. \quad (3.40)$$

3.3.7 Ripple Voltage

An equivalent circuit of the parallel connection of the filter capacitor and the load resistance is shown in Fig. 2.4. The capacitor is modeled by a capacitance C_f and an ESR r_C. The ac current through the capacitor i_C, given by (3.23), causes the output ripple voltage

$$v_o = v_c + v_{ESR} = \frac{1}{\omega C_f} \int_0^{\omega t} i_C d(\omega t) + r_C i_C, \quad (3.41)$$

where $v_c(0) = 0$. Substitution of (3.23) into the above equation leads to quite complicated expressions. Therefore, it is assumed that the current through the inductor is a cosinusoid $i_{C1} = I_{C1} cos\omega t$, with an amplitude equal to half of the peak-to-peak value I_{Cpp} of the current i_C. Under this assumption, the output ripple voltage becomes

$$v_o \approx v_{c1} + v_{ESR1} = \frac{I_{C1}}{\omega C_f} sin\omega t + r_C I_{C1} cos\omega t, \quad (3.42)$$

where v_{c1} and v_{ESR1} are voltages across the filter capacitance and the ESR of the filter capacitor, respectively, caused by current i_{C1}. Using (3.23), the value of I_{C1} is found to be

$$I_{C1} = \frac{I_{Cpp}}{2} = \frac{0.28V_O}{fL_f}. \tag{3.43}$$

When the components v_{c1} and v_{ESR1} have equal amplitudes, the peak-to-peak value of the output ripple voltage is $V_c = \sqrt{2}I_{C1}/(\omega C_f) = \sqrt{2}r_C I_{C1}$. Thus, it can be stated that

$$V_c \le \sqrt{2}I_{C1}\max\left(\frac{1}{\omega C_f}, r_C\right) = \frac{0.28\sqrt{2}V_O}{fL_f}\max\left(\frac{1}{\omega C_f}, r_C\right), \tag{3.44}$$

which gives the minimum filter inductance

$$L_{fmin} = \frac{0.28\sqrt{2}V_O}{fV_{cmax}}\max\left(\frac{1}{\omega C_f}, r_C\right). \tag{3.45}$$

Example 3.2

Design the reactive component values for the output filter of the Class D voltage-driven half-wave rectifier operating at the switching frequency $f =$ 150 kHz. It is specified that the peak-to-peak ripple voltage cannot be higher than 0.1% of the dc output voltage. The ESR of the filter capacitor is $r_C =$ 20 mΩ.

Solution: In (3.42), we want to keep the amplitude of the voltage across the filter capacitance v_{c1} less or equal to the amplitude of the voltage across the ESR of the filter capacitor v_{ESR1}. It follows from (3.44) that the minimum value of the filter capacitor is

$$C_{fmin} = \frac{1}{\omega r_C} = \frac{1}{2\pi f r_C} = \frac{1}{2\pi \times 150 \times 10^3 \times 0.02} = 53\ \mu\text{F}. \tag{3.46}$$

Let $C = 68\ \mu$F.

Substituting 0.1% of V_O for V_{cmax} into (3.45), one can calculate the minimum value of the filter inductance

$$L_{fmin} = \frac{0.28\sqrt{2}r_C V_O}{fV_{cmax}} = \frac{0.28 \times \sqrt{2} \times 0.02}{150 \times 10^3 \times 0.001} = 53\ \mu\text{H}. \tag{3.47}$$

3.4 CLASS D TRANSFORMER CENTER-TAPPED RECTIFIER

3.4.1 Currents and Voltages

Figure 3.3(a) depicts a circuit of a Class D voltage-driven transformer center-tapped rectifier. Its simplified circuit and models are shown in Fig. 3.3(b) through (d). Figure 3.4 shows current and voltage waveforms of the trans-

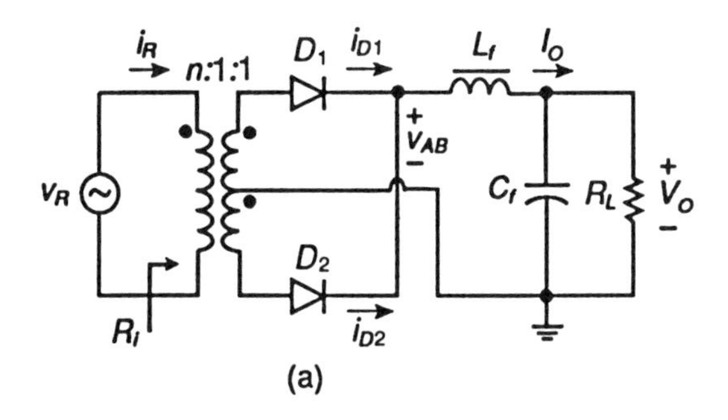

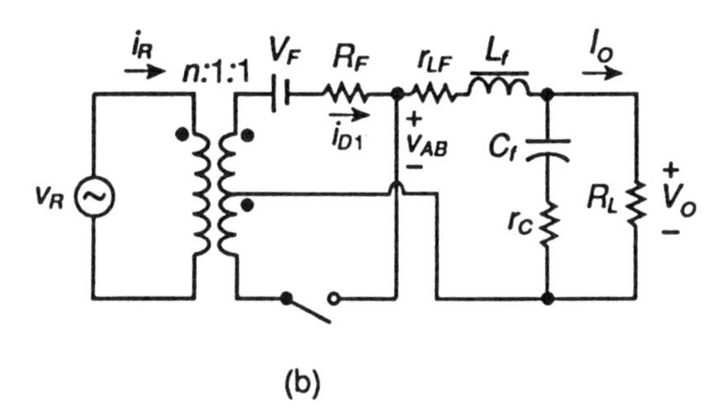

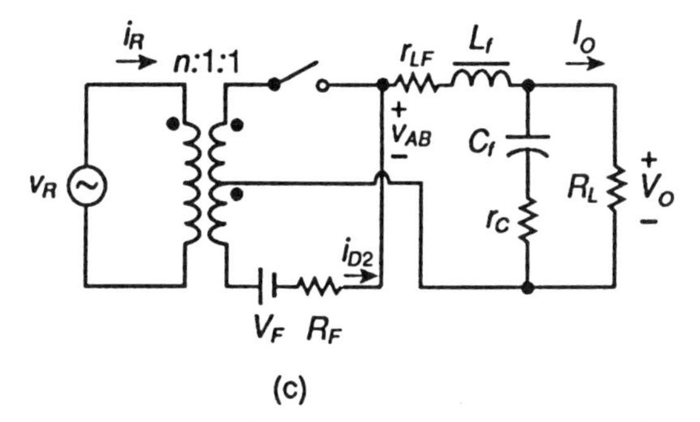

Figure 3.3: Class D transformer voltage-driven center-tapped rectifier. (a) Circuit. (b) Model when diode D_1 is on and diode D_2 is off. (c) Model when diode D_1 is off and diode D_2 is on.

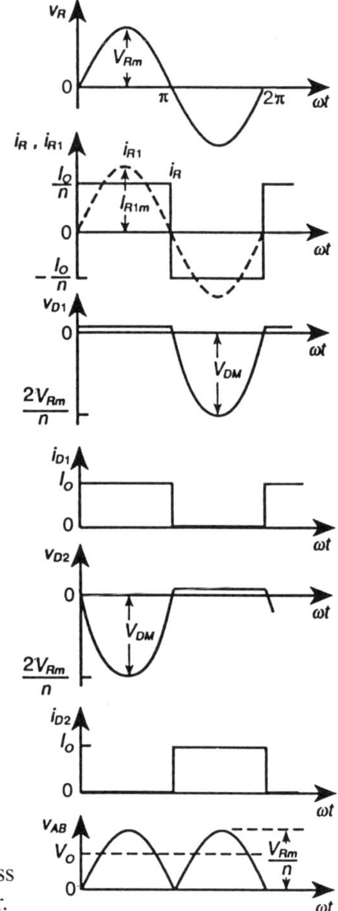

Figure 3.4: Current and voltage waveforms in Class D voltage-driven transformer center-tapped rectifier.

former center-tapped rectifier. The input voltage is sinusoidal and given by (3.1). The input voltage of the output filter is

$$v_{AB} = \frac{|\, v_R \,|}{n} = \frac{V_{Rm}}{n} \,|\, sin\omega t \,| \,. \tag{3.48}$$

Because the average voltage across the filter inductor is zero,

$$V_O = \frac{1}{2\pi}\int_0^{2\pi} v_{AB} d(\omega t) = \frac{1}{\pi}\int_0^{\pi} v_{AB} d(\omega t) = \frac{1}{\pi}\int_0^{\pi} \frac{V_{Rm}}{n} sin\omega t d(\omega t) = \frac{2V_{Rm}}{\pi n}. \tag{3.49}$$

It follows from this equation that the dc output voltage V_O is directly proportional to the amplitude of the input voltage V_{Rm}.

The input current of the rectifier is a square wave and is given by

$$i_R = \begin{cases} I_O/n, & \text{for} \quad 0 < \omega t \le \pi \\ -I_O/n, & \text{for} \quad \pi < \omega t \le 2\pi. \end{cases} \tag{3.50}$$

Because of the odd symmetry of i_R with respect to ωt, the amplitude of fundamental component of i_R is found as

$$I_{R1m} = \frac{2}{\pi} \int_0^{\pi} i_R sin\omega t d(\omega t) = \frac{2I_O}{\pi n} \int_0^{\pi} sin\omega t d(\omega t) = \frac{4I_O}{\pi n}, \tag{3.51}$$

from which the waveform of the fundamental component of the input current is

$$i_{R1} = I_{R1m} sin\omega t = \frac{4I_O}{\pi n} sin\omega t. \tag{3.52}$$

The input power is

$$P_i = \frac{I_{R1m}^2 R_i}{2} = \frac{8I_O^2 R_i}{\pi^2 n^2}. \tag{3.53}$$

3.4.2 Power Factor

From (3.51), the rms value of the fundamental component of the input current is

$$I_{R1rms} = \frac{I_{R1m}}{\sqrt{2}} = \frac{4I_O}{\sqrt{2}\pi n} = \frac{2\sqrt{2}I_O}{\pi n}. \tag{3.54}$$

Using (3.50), one can obtain the rms value of the input current

$$I_{Rrms} = \sqrt{\frac{1}{2\pi} \int_0^{2\pi} i_R^2 d(\omega t)} = \sqrt{\frac{I_O^2}{\pi n^2} \int_0^{\pi} d(\omega t)} = \frac{I_O}{n}. \tag{3.55}$$

Hence, the power factor is found as

$$PF \equiv \frac{P_i}{I_{Rrms} V_{Ri}} = \frac{I_{R1rms} V_{Ri}}{I_{Rrms} V_{Ri}} = \frac{I_{R1rms}}{I_{Rrms}} = \frac{2\sqrt{2}}{\pi} \approx 0.9. \tag{3.56}$$

Thus, the total harmonic distortion of the rectifier current is

$$THD = \sqrt{\frac{1}{PF^2} - 1} = \sqrt{\frac{\pi^2}{8} - 1} \approx 0.4834. \tag{3.57}$$

The peak value of the diode current is given by (3.14). The peak value of the diode voltage is

$$V_{DM} = \frac{2V_{Rm}}{n} = \pi V_O \tag{3.58}$$

and is the same as the peak value of the diode voltage for the half-wave rectifier at the same value of V_O. Hence, the power-output capability is given by (3.15).

3.4.3 Efficiency

The output power is given by (3.8), the power dissipated in one diode is given by (3.20), and the conduction loss in the inductor is given by (3.21). The voltage across the filter inductor is

$$v_L = v_{AB} - V_O = V_O \left(\frac{\pi}{2} \mid sin\omega t \mid -1 \right), \tag{3.59}$$

resulting in the inductor current

$$\begin{aligned} i_L &= \frac{1}{\omega L} \int_0^{\omega t} v_L d(\omega t) + I_O \\ &= \frac{V_O}{\omega L_f} \left(-\frac{\pi}{2} cos\omega t - \omega t + \frac{\pi}{2} \right) + I_O, \quad \text{for} \quad 0 < \omega t \leq \pi \end{aligned} \tag{3.60}$$

and the current through the filter capacitor

$$i_C = i_L - I_O = \frac{V_O}{\omega L_f} \left(-\frac{\pi}{2} cos\omega t - \omega t + \frac{\pi}{2} \right), \quad \text{for} \quad 0 < \omega t \leq \pi. \tag{3.61}$$

The rms value of the capacitor current is given by

$$I_{Crms} = \sqrt{\frac{1}{\pi} \int_0^{\pi} i_C^2 d(\omega t)} = a_{ct} \frac{I_O R_L}{f L_f}, \tag{3.62}$$

where $a_{ct} = (\sqrt{5/24 - 2/\pi^2})/2 = 0.0377$. Hence, the losses in the ac ESR of the filter inductor r_{Lf} and in the ESR of the filter capacitor r_C are

$$P_{lc} = a_{ct}^2 \frac{r_{ac} I_O^2 R_L^2}{f^2 L_f^2} = \frac{a_{ct}^2 r_{ac} R_L}{f^2 L_f^2} P_O. \tag{3.63}$$

From (3.35), (3.21), and (3.63), one arrives at the overall conduction loss of the center-tapped rectifier (excluding the power loss in the transformer)

$$P_C = 2P_D + P_{rL} + P_{lc} = P_O \left(\frac{V_F}{V_O} + \frac{R_F + r_{LF}}{R_L} + \frac{a_{ct}^2 r_{ac} R_L}{f^2 L_f^2} \right). \tag{3.64}$$

From (3.8), (3.20), (3.21), and (3.63), the efficiency η_R is obtained as

$$\eta_R \equiv \frac{P_O}{P_i} = \frac{P_O \eta_{tr}}{P_O + 2P_D + P_{rL} + P_{lc}} = \frac{\eta_{tr}}{1 + \frac{V_F}{V_O} + \frac{R_F + r_{LF}}{R_L} + a_{ct}^2 \frac{r_{ac} R_L}{f^2 L_f^2}}. \tag{3.65}$$

3.4.4 Input Resistance

Substitution of (3.8), (3.20), (3.53), (3.21), and (3.63) into (3.29) yields the input resistance of the rectifier

$$R_i \equiv \frac{V_{Rm}}{I_{R1m}} = \frac{\pi^2 n^2 R_L}{8\eta_{tr}}\left(1 + \frac{V_F}{V_O} + \frac{R_F + r_{LF}}{R_L} + a_{ct}^2 \frac{r_{ac} R_L}{f^2 L_f^2}\right) = \frac{\pi^2 n^2 R_L}{8\eta_R}. \quad (3.66)$$

3.4.5 Voltage Transfer Function

From (3.34) and (3.66),

$$M_{VR} \equiv \frac{V_O}{V_{Ri}} = \sqrt{\frac{\eta_R R_L}{R_i}} = \frac{2\sqrt{2}\eta_{tr}}{\pi n\left(1 + \frac{V_F}{V_O} + \frac{R_F + r_{LF}}{R_L} + a_{ct}^2 \frac{r_{ac} R_L}{f^2 L_f^2}\right)}. \quad (3.67)$$

Example 3.3

A Class D center-tapped rectifier of Fig. 3.3(a) is operated at $V_O = 5$ V, $I_O = 20$ A, and $f = 100$ kHz. The rectifier employs Schottky diodes (e.g., Motorola MBR20100CT) with $V_F = 0.5$ V and $R_F = 0.025\ \Omega$. The filter inductance is $L_f = 1$ mH. The dc ESR of the filter inductor is $r_{LF} = 0.1\ \Omega$ and the ac ESR of the inductor is $r_{Lf} = 1.85\ \Omega$. A filter capacitor has ESR $r_C = 50$ mΩ. The transformer turns ratio is $n = 5$. Assume the transformer efficiency $\eta_{tr} = 0.96$. Find the efficiency η_R, the voltage transfer function M_{VR}, and the input resistance R_i for this rectifier.

Solution: The load resistance of the rectifier is $R_L = V_O/I_O = 0.25\ \Omega$, and the output power is $P_O = I_O V_O = 100$ W. The total conduction loss in each diode and the dc conduction loss in the filter inductor are the same as those in Example 3.1 for the half-wave rectifier, that is, $P_D = 10$ W and $P_{rL} = 40$ W, respectively. The ac conduction loss in the filter inductor and the filter capacitor is obtained from (3.63)

$$P_{lc} = a_{ct}^2 \frac{r_{ac} I_O^2 R_L^2}{f^2 L_f^2} = 0.0377^2 \frac{(1.85 + 0.05) \times 20^2 \times 0.25^2}{(100 \times 10^3 \times 0.001)^2} = 6.75\ \mu\text{W}. \quad (3.68)$$

Because the operating conditions are the same and ac conduction losses are negligible, the efficiency of the center-tapped rectifier equals the efficiency of the half-wave rectifier of Example 3.1 and is

$$\eta_R = \frac{P_O \eta_{tr}}{P_O + 2P_D + P_{rL} + P_C} = \frac{100 \times 0.96}{100 + 2 \times 10 + 40 + 6.75 \times 10^{-6}} = 60\%. \quad (3.69)$$

From (3.66), one obtains the input resistance of the rectifier

$$R_i = \frac{\pi^2 n^2 R_L}{8\eta_R} = \frac{\pi^2 \times 5^2 \times 0.25}{8 \times 0.6} = 12.85\ \Omega. \tag{3.70}$$

The ac-to-dc voltage transfer function of the rectifier is calculated using (3.67)

$$M_{VR} = \sqrt{\frac{\eta_R R_L}{R_i}} = \sqrt{\frac{0.6 \times 0.25}{12.85}} = 0.108. \tag{3.71}$$

3.4.6 Ripple Voltage

Let us assume for simplicity, similarly to the discussion in Section 3.3.7, that the current through the inductor is a cosinusoid $i_{C1} = I_{C1} cos\omega t$ with an amplitude equal to half of the peak-to-peak value I_{Cpp} of the current i_C. Using (3.61), the value of I_{C1} is found to be

$$I_{C1} = \frac{I_{Cpp}}{2} = \frac{0.05 V_O}{f L_f}. \tag{3.72}$$

The output ripple voltage can be approximated by

$$v_o \approx v_{c1} + v_{ESR1} = \frac{I_{C1}}{\omega C_f} sin\omega t + r_C I_{C1} cos\omega t, \tag{3.73}$$

and the boundary of the maximum peak-to-peak value of the output ripple voltage can be given as

$$V_{cmax} = \sqrt{2} I_{C1} \max\left(\frac{1}{\omega C_f}, r_C\right) = \frac{0.05\sqrt{2} V_O}{f L_f} \max\left(\frac{1}{\omega C_f}, r_C\right), \tag{3.74}$$

which yields the minimum filter inductance

$$L_{min} = \frac{0.05\sqrt{2} V_O}{f V_{cmax}} \max\left(\frac{1}{\omega C_f}, r_C\right). \tag{3.75}$$

Example 3.4

Design the reactive component values for the output filter of the Class D voltage-driven transformer center-tapped rectifier that operates at the switching frequency $f = 150$ kHz. The peak-to-peak ripple voltage is required to be no more than 0.1% of the dc output voltage. Assume that the ESR of the filter capacitor is $r_C = 20$ mΩ.

Solution: The design procedure of the filter capacitor is the same as that in Example 3.2. The result is

$$C_{fmin} = \frac{1}{2\pi f r_C} = 53\ \mu\text{F}. \tag{3.76}$$

The minimum value of the filter inductance is obtained by substitution of 0.1% of V_O for V_{cmax} into (3.75)

$$L_{fmin} = \frac{0.05\sqrt{2} r_C V_O}{f V_{cmax}} = \frac{0.05 \times \sqrt{2} \times 0.02}{150 \times 10^3 \times 0.001} = 9.4\ \mu\text{H}. \tag{3.77}$$

Comparing this result with that of Example 3.2, it can be seen that the transformer center-tapped rectifier requires more than 5 times less filter inductance than the half-wave rectifier for the same operating conditions.

3.5 CLASS D BRIDGE RECTIFIER

Figure 3.5(a) shows a circuit of a Class D bridge rectifier. The waveforms are shown in Fig. 3.6. The output power P_O, the diode conduction loss P_D, the inductor conduction loss, the losses in the filter capacitor, and the input power P_i are given by (3.8), (3.20), (3.21), (3.63), and (3.53), respectively. The total conduction loss in the bridge rectifier without the transformer is

$$P_C = 4P_D + P_{rL} + P_{lc} = P_O \left(\frac{2V_F}{V_O} + \frac{2R_F + r_{LF}}{R_L} + \frac{a_b^2 r_{ac} R_L}{f^2 L_f^2} \right). \tag{3.78}$$

Because

$$P_i = \frac{P_O + 4P_D + P_{rL} + P_{lc}}{\eta_{tr}} = \frac{P_O}{\eta_{tr}} \left(1 + \frac{2V_F}{V_O} + \frac{2R_F + r_{LF}}{R_L} + a_b^2 \frac{r_{ac} R_L}{f^2 L_f^2} \right), \tag{3.79}$$

the efficiency can be found as

$$\eta_R \equiv \frac{P_O}{P_i} = \frac{P_O \eta_{tr}}{P_O + 4P_D + P_{rL} + P_{lc}} = \frac{\eta_{tr}}{1 + \frac{2V_F}{V_O} + \frac{2R_F + r_{LF}}{R_L} + a_b^2 \frac{r_{ac} R_L}{f^2 L_f^2}}, \tag{3.80}$$

where $a_b = a_{ct} = (\sqrt{5/24 - 2/\pi^2})/2 = 0.0377$. The input resistance is

$$R_i = \frac{V_{Rm}}{I_{R1m}} = \frac{\pi^2 n^2 R_L}{8\eta_{tr}} \left(1 + \frac{2V_F}{V_O} + \frac{2R_F + r_{LF}}{R_L} + a_b^2 \frac{r_{ac} R_L}{f^2 L_f^2} \right) = \frac{\pi^2 n^2 R_L}{8\eta_R}. \tag{3.81}$$

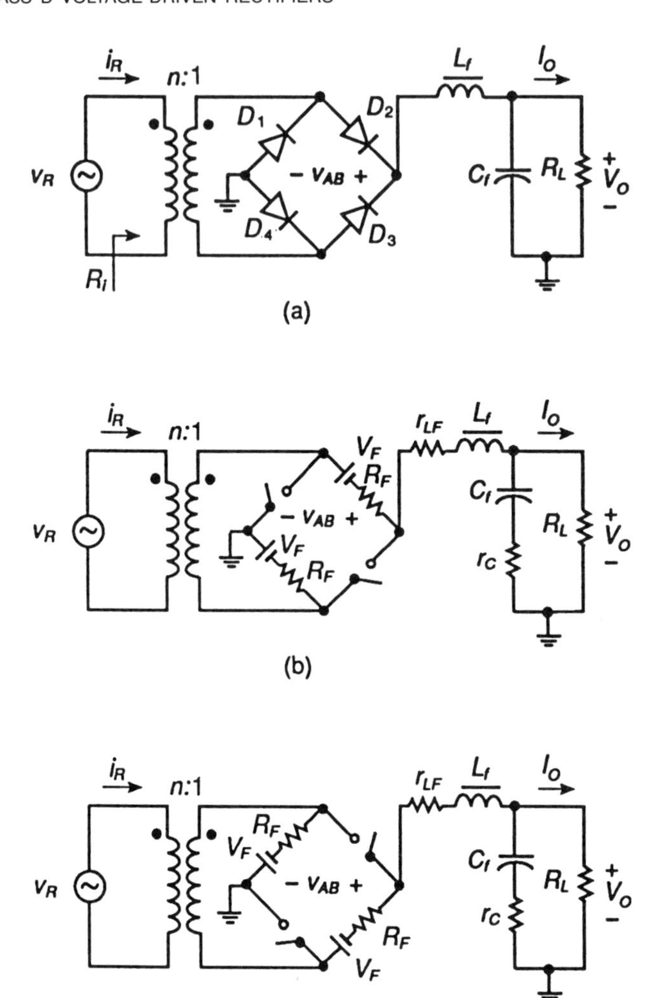

Figure 3.5: Class D voltage-driven bridge rectifier. (a) Circuit. (b) Model when diodes D_1 and D_3 are on and diodes D_2 and D_4 are off. (c) Model when diodes D_1 and D_3 are off and diodes D_2 and D_4 are on.

From (3.34), (3.81), and (3.80),

$$M_{VR} \equiv \frac{V_O}{V_{Rrms}} = \sqrt{\frac{\eta_R R_L}{R_i}} = \frac{2\sqrt{2}\eta_{tr}}{\pi n\left(1 + \frac{2V_F}{V_O} + \frac{2R_F + r_{LF}}{R_L} + a_b^2 \frac{r_{ac} R_L}{f^2 L_f^2}\right)}. \tag{3.82}$$

The peak value of the diode current is given by (3.14). The peak value of the diode voltage is

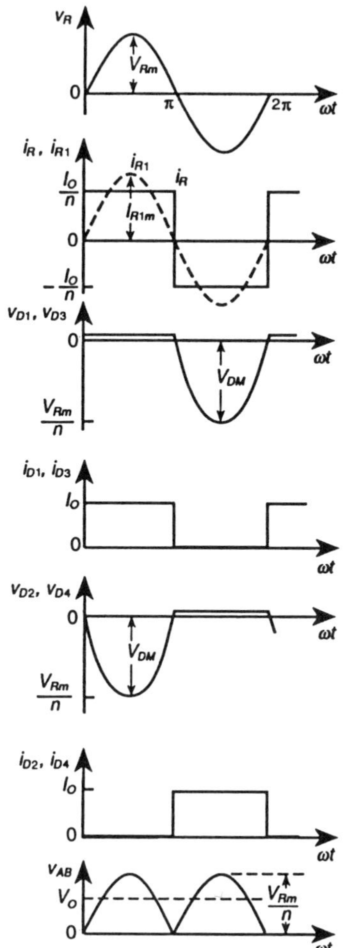

Figure 3.6: Current and voltage waveforms in Class D voltage-driven bridge rectifier.

$$V_{DM} = \frac{V_{Rm}}{n} = \frac{\pi}{2} V_O \tag{3.83}$$

The power output capability of the bridge rectifier is

$$c_{pR} \equiv \frac{P_O}{I_{DM} V_{DM}} = \frac{I_O V_O}{I_{DM} V_{DM}} = \frac{2}{\pi} \approx 0.637. \tag{3.84}$$

The power factor and the total harmonic distortion are given by (3.56) and (3.57).

Various parameters of the voltage-driven rectifiers are given in Table 3.1 for the lossless case.

Table 3.1. Parameters of Lossless Voltage-Driven Rectifiers

Parameter	Half-Wave Rectifier	Transformer Center-Tapped Rectifier	Bridge Rectifier
M_{VR}	$\frac{\sqrt{2}}{\pi n}$	$\frac{2\sqrt{2}}{\pi n}$	$\frac{2\sqrt{2}}{\pi n}$
R_i	$\frac{\pi^2 n^2 R_L}{2}$	$\frac{\pi^2 n^2 R_L}{8}$	$\frac{\pi^2 n^2 R_L}{8}$
I_{DM}	I_O	I_O	I_O
V_{DM}	πV_O	πV_O	$\frac{\pi}{2} V_O$

Example 3.5

Calculate the efficiency η_R, the voltage transfer function M_{VR}, and the input resistance R_i for a Class D bridge rectifier of Fig. 3.5(a) at $V_O = 100$ V and $I_O = 1$ A. The operating frequency of the rectifier is 100 kHz. The rectifier employs *p-n* junction diodes with $V_F = 0.9$ V and $R_F = 0.04\ \Omega$. The value of the filter inductance is $L_f = 1$ mH. The dc ESR of the inductor is $r_{LF} = 0.1\ \Omega$, and the ac ESR of the inductor is $r_{Lf} = 1.85\ \Omega$. A filter capacitor with $r_C = 50$ mΩ is employed. The transformer turns ratio is $n = 2$. Assume the transformer efficiency $\eta_{tr} = 0.97$.

Solution: The load resistance of the rectifier is $R_L = V_O/I_O = 100\ \Omega$, and the output power is $P_O = I_O V_O = 100$ W. The total conduction loss in each diode is found using (3.20) as

$$P_D = \frac{I_O V_F}{2} + \frac{I_O^2 R_F}{2} = \frac{1 \times 0.9}{2} + \frac{1^2 \times 0.04}{2} = 0.47 \text{ W}. \tag{3.85}$$

From (3.21), the dc conduction loss in the filter inductor is

$$P_{rL} = I_O^2 r_{LF} = 1^2 \times 0.1 = 0.1 \text{ W}. \tag{3.86}$$

The ac conduction loss in the filter inductor and the filter capacitor is obtained from (3.63)

$$P_{lc} = a_b^2 \frac{r_{ac} I_O^2 R_L^2}{f^2 L_f^2} = 0.0377^2 \frac{(1.85 + 0.05) \times 1^2 \times 100^2}{(100 \times 10^3 \times 0.001)^2} = 2.7 \text{ mW}. \tag{3.87}$$

The efficiency of the bridge rectifier is calculated from (3.80)

$$\eta_R \equiv \frac{P_O}{P_i} = \frac{P_O \eta_{tr}}{P_O + 4P_D + P_{rL} + P_{lc}}$$
$$= \frac{100 \times 0.97}{100 + 4 \times 0.47 + 0.1 + 0.0027} = 95.1\%. \quad (3.88)$$

Using (3.81), one obtains the input resistance

$$R_i = \frac{\pi^2 n^2 R_L}{8\eta_R} \frac{\pi^2 \times 2^2 \times 100}{8 \times 0.951} = 518.9\ \Omega. \quad (3.89)$$

The ac-to-dc voltage transfer function of the rectifier can be calculated using (3.82)

$$M_{VR} = \sqrt{\frac{\eta_R R_L}{R_i}} = \sqrt{\frac{0.951 \times 100}{518.9}} = 0.428. \quad (3.90)$$

3.6 SYNCHRONOUS RECTIFIERS

Synchronous rectifiers [2]–[24] can be used to reduce the forward voltage of the rectifying devices and achieve high efficiencies at low output voltages. Figure 3.7 depicts a circuit of an unregulated Class D voltage-driven transformer center-tapped rectifier. It is obtained by replacing diodes in the conventional rectifier shown in Fig. 3.3(a) by power MOSFETs. The expressions for currents, voltages, input power, and power factor of the rectifier are given by (3.48) through (3.56). The peak values of the diode current and voltage, and the power-output capability are given by (3.13), (3.14), and (3.15), respectively.

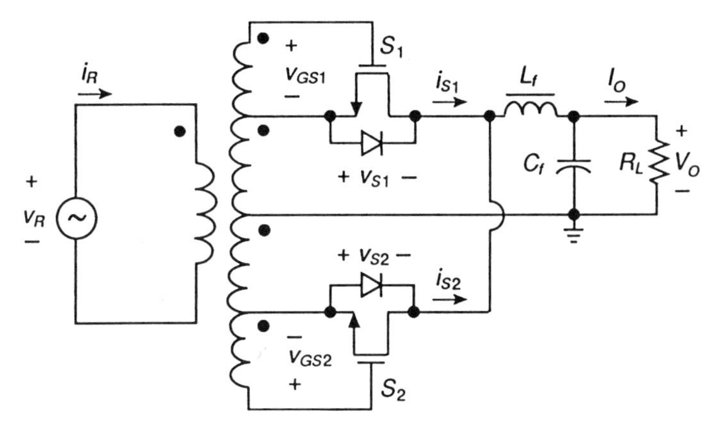

Figure 3.7: Unregulated synchronous Class D voltage-driven transformer center-tapped rectifier.

3.6.1 Efficiency

The output power of the rectifier is given by (3.8). The rms value of the current through each switch I_{Srms} is the same as the rms value of the current through each diode in a conventional transformer center-tapped rectifier and is derived in (3.18) as

$$I_{Srms} = \frac{I_O}{\sqrt{2}}. \tag{3.91}$$

Thus, the conduction loss per switch is

$$P_{rDS} = r_{DS} I_{Srms}^2 = \frac{I_O^2 r_{DS}}{2} = \frac{r_{DS}}{2R_L} P_O. \tag{3.92}$$

The dc conduction loss in the inductor is given by (3.21). The ac losses in the ac ESR of the filter inductor r_{Lf} and in the ESR of the filter capacitor r_C are expressed by (3.63). The drive-power loss is given by (2.117).

From (3.8), (3.21), (3.63), (3.92), and (2.117), the efficiency η_R is obtained as

$$\begin{aligned}\eta_R &\equiv \frac{P_O}{P_i} = \frac{P_O \eta_{tr}}{P_O + 2P_{rDS} + P_{rL} + P_{lc} + 2P_G} \\ &= \frac{\eta_{tr}}{1 + \frac{r_{DS}+r_{LF}}{R_L} + a_{ct}^2 \frac{r_{ac} R_L}{f^2 L_f^2} + \frac{2fQ_g V_{GSpp}}{P_O}}.\end{aligned} \tag{3.93}$$

3.6.2 Input Resistance

Substitution of (3.8), (3.20), (3.53), (3.63), and (3.92) instead of (3.21) into (3.29) and taking into account the drive-power loss yields the input resistance of the rectifier

$$\begin{aligned}R_i &\equiv \frac{V_{Rm}}{I_{R1m}} = \frac{\pi^2 n^2 R_L}{8\eta_{tr}} \left(1 + \frac{r_{DS} + r_{LF}}{R_L} + a_{ct}^2 \frac{r_{ac} R_L}{f^2 L_f^2} + \frac{2fQ_g V_{GSpp}}{P_O}\right) \\ &= \frac{\pi^2 n^2 R_L}{8\eta_R}.\end{aligned} \tag{3.94}$$

3.6.3 Voltage Transfer Function

From (3.34) and (3.94),

$$M_{VR} \equiv \frac{V_O}{V_{Ri}} = \sqrt{\frac{\eta_R R_L}{R_i}} = \frac{2\sqrt{2}\eta_{tr}}{\pi n \left(1 + \frac{r_{DS}+r_{LF}}{R_L} + a_{ct}^2 \frac{r_{ac} R_L}{f^2 L_f^2} + \frac{2fQ_g V_{GSpp}}{P_O}\right)}. \tag{3.95}$$

Example 3.6

An unregulated synchronous Class D voltage-driven center-tapped rectifier of Fig. 3.7 is operated at $V_O = 5$ V, $I_O = 20$ A, and $f = 100$ kHz. The rectifier employs SMP60N06-18 MOSFETs (Siliconix) with $r_{DS} = 0.018\ \Omega$ and $Q_g = 100\ nC$. The filter inductance is $L_f = 1$ mH. The dc ESR of the filter inductor is $r_{LF} = 0.1\ \Omega$ and the ac ESR of the inductor is $r_{Lf} = 1.85\ \Omega$. A filter capacitor has ESR $r_C = 50$ mΩ. The transformer turns ratio is $n = 5$. Assume the transformer efficiency $\eta_{tr} = 0.96$ and the peak-to-peak gate-to-source voltage $V_{GSpp} = 10$ V. Find the efficiency η_R, the voltage transfer function M_{VR}, and the input resistance R_i for this rectifier. Compare results to those of Example 3.3.

Solution: The load resistance of the rectifier is $R_L = V_O/I_O = 0.25\ \Omega$, and the output power is $P_O = I_O V_O = 100$ W. From (3.92) the total conduction loss in each transistor is

$$P_{rDS} = \frac{I_O^2 r_{DS}}{2} = \frac{20^2 \times 0.018}{2} = 3.6 \text{ W}. \tag{3.96}$$

The dc conduction loss in the filter inductor is obtained using (3.21)

$$P_{rL} = I_O^2 r_{LF} = 20^2 \times 0.1 = 40 \text{ W}. \tag{3.97}$$

The ac conduction loss in the filter inductor and the filter capacitor is calculated from (3.63)

$$P_{lc} = a_{ct}^2 \frac{r_{ac} I_O^2 R_L^2}{f^2 L_f^2} = 0.0377^2 \frac{(1.85 + 0.05) \times 20^2 \times 0.25^2}{(100 \times 10^3 \times 0.001)^2} = 6.75\ \mu\text{W}. \tag{3.98}$$

The gate-drive power of each transistor can be calculated, using (2.117), as

$$P_G = f Q_g V_{GSpp} = 100 \times 10^3 \times 100 \times 10^{-9} \times 10 = 0.1 \text{ W}. \tag{3.99}$$

Using (3.6.1), the efficiency of the rectifier is

$$\begin{aligned} \eta_R &= \frac{P_O \eta_{tr}}{P_O + 2P_{rDS} + P_{rL} + P_C + 2P_G} \\ &= \frac{100 \times 0.96}{100 + 2 \times 3.6 + 40 + 6.75 \times 10^{-6} + 0.2} = 65.1\%. \end{aligned} \tag{3.100}$$

The efficiency of the unregulated synchronous transformer center-tapped rectifier is 5% higher than that of the conventional rectifier of Example 3.3 because of smaller losses in the switches. However, the dc losses in the filter inductor still do not allow for a significant increase in the efficiency. From (3.94), one obtains the input resistance of the rectifier

$$R_i = \frac{\pi^2 n^2 R_L}{8\eta_R} = \frac{\pi^2 \times 5^2 \times 0.25}{8 \times 0.651} = 11.84\ \Omega. \tag{3.101}$$

The ac-to-dc voltage transfer function of the rectifier is calculated using (3.95) and (3.6.1)

$$M_{VR} = \frac{2\sqrt{2}\eta_R}{n\pi} = \frac{2\sqrt{2} \times 0.651}{5 \times \pi} = 0.117. \tag{3.102}$$

3.7 SUMMARY

- Class D voltage-driven rectifiers have a second-order L_f-C_f output filter.
- The dc output voltage V_O is directly proportional to the amplitude of the input voltage V_{Rm}.
- The on-duty cycle of each diode is 50%.
- The peak-to-peak and rms current through the filter capacitor C_f is relatively low.
- The conduction loss in the ESR of the filter capacitor is low.
- The corner frequency of the output filter $f_0 = 1/(2\pi\sqrt{L_f C_f})$ is independent of the load resistance R_L.
- The diodes turn off at a high di/dt, causing a high peak reverse-recovery current and generating noise.
- Synchronous Class D voltage-driven rectifiers are obtained by replacing diodes in the corresponding conventional rectifiers with low on-resistance power MOSFETs.

3.8 REFERENCES

1. M. K. Kazimierczuk, W. Szaraniec, and S. Wang, "Analysis and design of parallel resonant converter with high Q_L," *IEEE Trans. Aerospace Electron. Syst.*, vol. AES-27, pp. 35–50, Jan. 1992.
2. M. K. Kazimierczuk and J. Jóźwik, "Class E zero-voltage-switching and zero-current-switching rectifiers," *IEEE Trans. Circuits Syst.*, vol. CAS-37, pp. 436–444, Mar. 1990.
3. M. W. Smith and K. Owyang, "Improving the efficiency of low output power voltage switched-mode converters with synchronous rectification," *Proc. Powercon 7*, H-4, 1980, pp. 1–13.
4. R. S. Kagan and M. H. Chi, "Improving power supply efficiency with synchronous rectifiers," *Proc. Powercon 9*, D-4, 1982, pp. 1–5.
5. R. A. Blanchard and R. Severns, "Designing switched-mode power converter for very low temperature operation," *Proc. Powercon 10*, D-2, 1983, pp. 1–11.

6. M. Alexander, R. Blanchard, and R. Severns, "MOSFETs move in on low voltage rectification," *MOSPOWER Application Handbook*, Siliconix, 1984, pp. 5.69–5.87.
7. E. Oxner, "Using power MOSFETs as high-frequency synchronous and bridge rectifiers in switch-mode power supplies," *MOSPOWER Application Handbook*, Siliconix, 1984, pp. 5.87–5.94.
8. R. A. Blanchard and P. E. Thibodeau, "The design of a high efficiency, low voltage power supply using MOSFET synchronous rectification and current mode control," *IEEE Power Electronics Specialists Conference Record*, 1985, pp. 355–361.
9. P. L. Hower, G. M. Kepler, and R. Patel, "Design, performance and application of a new bipolar synchronous rectifier," *IEEE Power Electronics Specialists Conference Record*, 1985, pp. 247–256.
10. R. A. Blanchard and P. E. Thibodeau, "Use of depletion mode MOSFET devices in synchronous rectification," *IEEE Power Electronics Specialists Conference Record*, 1986, pp. 81–81.
11. J. Blanc, R. A. Blanchard, and P. E. Thibodeau, "Use of enhancement- and depletion-mode MOSFETs in synchronous rectification," *Power Electronics Conference*, San Jose, CA, 1987, pp. 1–8.
12. M. K. Kazimierczuk and J. Jóźwik, "Class E zero-voltage switching and zero-current switching rectifiers," *IEEE Trans. Circuits Syst.*, vol. CAS-37, pp. 436–444, Mar. 1990.
13. M. K. Kazimierczuk and J. Jóźwik, "Resonant dc/dc converter w ith Class-E inverter and Class-E rectifier," *IEEE Trans. Ind. Electron.*, vol. IE-36, pp. 568–576, Nov. 1989.
14. D. A. Grant and J. Grower, *Power MOSFETs: Theory and Applications*, John Wiley & Sons New York: 1989, pp. 250–256.
15. J. Blanc, "Bidirectional dc-to-dc converter using synchronous rectifiers," *Power Conversion and Intelligent Motion Conference*, Long Beach, CA, 1989, pp. 15–20.
16. J. Jóźwik and M. K. Kazimierczuk "Analysis and design of Class E^2 dc/dc resonant converter," *IEEE Trans. Ind. Electron.*, vol. IE-37, pp. 173–183, Apr. 1990.
17. W. A. Tabisz, F. C. Lee, and D. Y. Chen, "A MOSFET resonant synchronous rectifier for high-frequency dc/dc converter," *IEEE Power Electronics Specialists Conference Record*, 1990, pp. 769–779.
18. B. Mohandes, "MOSFET synchronous rectifiers achieve 90% efficiency," *Power Conversion and Intelligent Motion*, Part I, June 1991 pp. 10–13; and Part II, July 1991, pp. 55–61.
19. M. K. Kazimierczuk and K. Puczko, "Class E low dv/dt synchronous rectifier with controlled duty cycle and output voltage," *IEEE Trans. Circuits Syst.*, vol. 37, pp. 1165–1172, Oct. 1991.
20. F. Goodenough, "Synchronous rectifier ups PC battery life," *Electronic Design*, pp. 47–52, Apr. 16, 1992.
21. J. A. Cobos, O. Garcia, J. Uceda, and A. de Hoz, "Self driven synchronous rectification in resonant topologies: Forward ZVS-MRC, forward ZCS-QRC and LCC-PRC," *IEEE Industrial Electronics Conf.*, San Diego, CA, Nov. 1992, pp. 185–190.
22. M. Mikolajewski and M. K. Kazimierczuk, "Zero-voltage-ripple rectifiers and

dc/dc resonant converters," *IEEE Trans. Power Electronics*, vol. PE-6, pp. 12–17, Jan. 1993.

23. M. K. Kazimierczuk and M. J. Mescher, "Series resonant converter with phase-controlled synchronous rectifier," *IEEE Int. Conf. on Industrial Electronics, Control, Instrumentation, and Automation (IECON '93)*, Lahaina, Hawaii, Nov. 15–19, 1993, pp. 852–856.
24. M. K. Kazimierczuk and M. J. Mescher, "Class D converter with regulated synchronous half-wave rectifier," *IEEE Applied Power Electronics Conf.*, Orlando, FL, Feb. 13–17, 1994, vol. 2, pp. 1005–1011.

3.9 REVIEW QUESTIONS

3.1 What is the relationship between the amplitude of the input voltage V_{Rm} and the output voltage V_O in the half-wave voltage-driven rectifier?

3.2 What is the relationship between the amplitude of the input voltage V_{Rm} and the output voltage V_O in the transformer center-tapped and bridge voltage-driven rectifiers?

3.3 Compare the diode peak currents I_{DM} in terms of I_O for the Class D voltage-driven half-wave, transformer center-tapped, and bridge rectifiers.

3.4 Compare the diode peak reverse voltages V_{DM} in terms of V_O for the Class D voltage-driven half-wave, transformer center-tapped, and bridge rectifiers.

3.5 Is the corner frequency of the output filter of the Class D voltage-driven rectifiers dependent of the load?

3.6 Explain how the synchronous Class D voltage-driven rectifiers are obtained.

3.10 PROBLEMS

3.1 Calculate the efficiency η_R, the voltage transfer function M_{VR}, and the input resistance R_i for a Class D half-wave rectifier of Fig. 3.1(a) at $V_O = 100$ V and $I_O = 1$ A. The operating frequency of the rectifier is $f = 100$ kHz. The rectifier employs *p-n* junction diodes with $V_F = 0.9$ V and $R_F = 0.04$ Ω. The value of the filter inductance is $L_f = 1$ mH. The dc ESR of the inductor is $r_{LF} = 0.1$ Ω, and the ac ESR of the inductor is $r_{Lf} = 1.85$ Ω. The ESR of the filter capacitor is $r_C = 50$ mΩ. The transformer turns ratio is $n = 2$. Assume the transformer efficiency $\eta_{tr} = 97\%$.

3.2 Repeat Problem 3.1 for the transformer center-tapped rectifier of Fig. 3.3(a).

3.3 Derive an expression for the power factor of the Class D transformerless voltage-driven half-wave rectifier. Explain why it is different than that for the transformer version of the half-wave rectifier.

3.4 Calculate the efficiency η_R, the voltage transfer function M_{VR}, and the input resistance R_i for a Class D bridge rectifier of Fig. 3.5(a) at $V_O = 5$ V and $I_O = 20$ A. The operating frequency of the rectifier is 100 kHz. The rectifier employs Schottky diodes with $V_F = 0.4$ V and $R_F = 0.025\ \Omega$. The value of the filter inductance is $L_f = 1$ mH. The dc ESR of the inductor is $r_{LF} = 0.1\ \Omega$ and the ac ESR of the inductor is $r_{Lf} = 1.85\ \Omega$. A filter capacitor with $r_C = 50$ mΩ is employed. The transformer turns ratio is $n = 5$. Assume the transformer efficiency $\eta_{tr} = 0.96$.

3.5 Derive Equation (3.43).

3.6 Derive Equation (3.72).

3.7 Show that $a_{hw} = (\sqrt{1/3 - 2/\pi^2})/2 = 0.1808$.

3.8 Show that $a_{ct} = (\sqrt{5/24 - 2/\pi^2})/2 = 0.0377$.

CHAPTER 4

CLASS E LOW *dv/dt* RECTIFIERS

4.1 INTRODUCTION

Class E rectifiers [1]–[14] offer a new means of high-frequency, high-efficiency, low-noise rectification. There are two groups of these rectifiers: Class E low dv/dt rectifiers and Class E low di/dt rectifiers. They contain either a single reactive component (in addition to the output filter) or a resonant circuit. Class E rectifiers are counterparts of Class E zero-voltage-switching (ZVS) inverters, studied in Chapter 13. This is because the diode current and voltage waveforms in the Class E low dv/dt rectifiers are time-reversed images of the corresponding transistor current and voltage waveforms in the Class E ZVS inverters. Class E rectifiers may be applied in resonant dc/dc power converters. The primary advantage of Class E low dv/dt rectifiers is that the diode turns on and off at low dv/dt and turns off at low di/dt, thus resulting in lower switching losses and better EMI performance. This chapter presents the principle of operation, analysis, and design procedure for Class E low dv/dt and Class E resonant low dv/dt rectifiers. Normalized parameters of the rectifiers are given in tables and sample designs are presented for both rectifiers.

4.2 LOW *dv/dt* RECTIFIER WITH A PARALLEL CAPACITOR

4.2.1 Principle of Operation

A Class E low dv/dt rectifier with a parallel capacitor [8] is shown in Fig. 4.1(a). The rectifier consists of a diode, a shunt capacitor C, and a second-order low-pass output filter L_f-C_f. Dc power is delivered to the load resistance R_L. The rectifier is driven by a sine-wave current source i_R and may be coupled to the ac source with a transformer. The shunt capacitor C shapes the voltage across the diode in such a way that the diode turns on and off at low dv/dt, reducing the current through the diode junction capacitance at both transitions. An important advantage of the rectifier is that the diode junction capacitance, the winding output capacitance of the transformer (if any), and the winding capacitance of the filter inductor L_f are absorbed into the shunt capacitance C. The winding capacitance of the filter inductor L_f is included in C because the filter capacitor C_f is almost an ideal short circuit for the ac component. The L_f-C_f output filter ensures that the ripple in the output voltage is below a specified level. The corner frequency of the filter, $f_o = 1/(2\pi\sqrt{L_f C_f})$, is independent of the load resistance R_L; that is a desirable feature of the filter. Assuming that the current through the filter inductor L_f is approximately constant and equal to the dc output current I_O, the L_f-C_f output filter and the load resistor R_L can be replaced by a current sink I_O as shown in Fig. 4.1(b). Figures 4.1(c) and (d) show the rectifier models when the diode is OFF and ON, respectively. If a negative dc output voltage is required, the diode polarity should be simply reversed.

The idealized current and voltage waveforms in the rectifier are depicted in Fig. 4.1(e) for the diode ON duty ratio $D = 0.5$, that is, for the conduction angle of the diode equal to 180°. The input current i_R is a sine wave and the output current I_O is constant. The diode and capacitor C connected in parallel are driven by a current source $I_O - i_R$. When the diode is OFF, the current $I_O - i_R$ flows through the capacitor C. When the diode is ON, the current $I_O - i_R$ flows through the diode. The diode turns on when its voltage increases to the diode threshold voltage, for example, 0.4 V for a Schottky diode. The diode turns off when its forward current decreases to zero. The current flowing through the shunt capacitor C shapes the voltage across the diode and the capacitor when the diode is OFF. The general equation is $i_C = C dv_D/dt$. The capacitor current i_C at turn-off is zero; therefore, the derivative of the diode voltage dv_D/dt is also zero at turn-off. The diode voltage v_D gradually decreases when the capacitor current i_C is negative, it reaches its minimum value when i_C is again zero, and it slowly rises to zero when i_C is positive. Thus, the diode turns on and off at low dv/dt, reducing switching losses and switching noise. Moreover, the absolute value of the diode current derivative at turn-off is quite small; thus, the effect of the reverse-recovery charge is alleviated. During the reverse recovery of a *pn* junction diode, the diode voltage remains low while reverse-recovery current is flowing through the diode, until the completion of the storage time. Therefore, the power dissipation in the diode is low. The shunt capacitor keeps the diode reverse voltage low during the remainder of the diode turn-off transient, after the completion of the storage time, and thus reduces the diode power dissipation during that part of the reverse-recovery transient. Schottky diodes have essentially no storage charge. The current through the junction capacitance is reduced be-

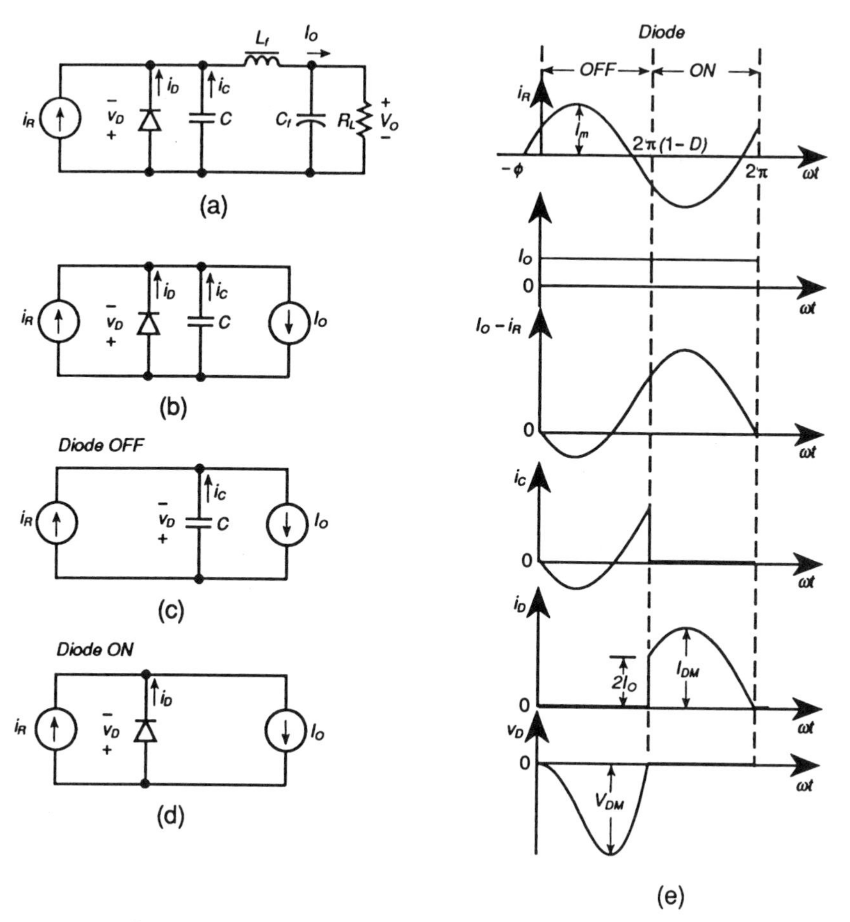

Figure 4.1: Class E low dv/dt rectifier with a parallel capacitor. (a) Circuit. (b) Model. (c) Model for the diode off-state. (d) Model for the diode on-state. (e) Current and voltage waveforms.

cause Schottky diodes turn on and off at low dv/dt in the Class E low dv/dt rectifier.

4.2.2 Assumptions

The analysis of the rectifier of Fig. 4.1(a) is based on the equivalent circuits of Fig. 4.1(b)–(d) and the following assumptions:

1) The diode is ideal, that is, its threshold voltage and ON resistance are zero, its OFF resistance is infinity, its junction capacitance is independent of voltage and is absorbed into the shunt capacitance C, and its charge-carrier lifetime is zero.

2) The parasitic shunt capacitance of the filter inductor L_f is included in C because the filter capacitor C_f is essentially a short circuit for ac.
3) The filter inductance L_f is large enough so that its current is approximately constant and equal to the dc output current I_O.
4) The rectifier is driven by an ideal sine-wave current source.

4.2.3 Characterization of the Rectifier at Any *D*

According to assumption 4), the input current is sinusoidal

$$i_R = I_m sin(\omega t + \phi), \tag{4.1}$$

where I_m is the amplitude and ϕ is the phase angle indicated in Fig. 4.1(e). The basic equation for the rectifier model of Fig. 4.1(b) is

$$I_O - i_R = i_D + i_C. \tag{4.2}$$

Figure 4.1(c) shows the model of the rectifier when the diode is OFF, that is, for $0 < \omega t \leq 2\pi(1 - D)$. The capacitor C is driven by two current sources, i_R and I_O. Using (4.1) and (4.2), the current through capacitor C is

$$i_C = I_O - i_R = I_O - I_m sin(\omega t + \phi). \tag{4.3}$$

According to Fig. 4.1(e), $i_D(0) = 0$ and therefore $i_C(0) = 0$. Hence, from (4.3),

$$I_O = I_m sin\phi. \tag{4.4}$$

Thus, (4.3) becomes

$$\begin{aligned} i_C &= I_O \left[1 - \frac{sin(\omega t + \phi)}{sin\phi}\right] \\ &= I_O \left(1 - cos\omega t - \frac{sin\omega t}{tan\phi}\right), \quad \text{for} \quad 0 < \omega t \leq 2\pi(1 - D). \end{aligned} \tag{4.5}$$

Since $v_D(0) = 0$, the voltage across the diode and capacitor C is found as

$$v_D = v_C = \frac{1}{\omega C}\int_0^{\omega t} i_C d(\omega t) = \frac{V_O}{\omega C R_L}\left(\omega t - sin\omega t + \frac{cos\omega t - 1}{tan\phi}\right). \tag{4.6}$$

The ideal diode turns on when its voltage reaches zero, that is, $v_D(2\pi(1 - D)) = 0$. Substituting this condition into (4.6), one obtains the relationship between the diode ON duty ratio D and the phase angle ϕ

$$tan\phi = \frac{1 - cos2\pi D}{2\pi(1 - D) + sin2\pi D}. \tag{4.7}$$

This expression is illustrated in Fig. 4.2.

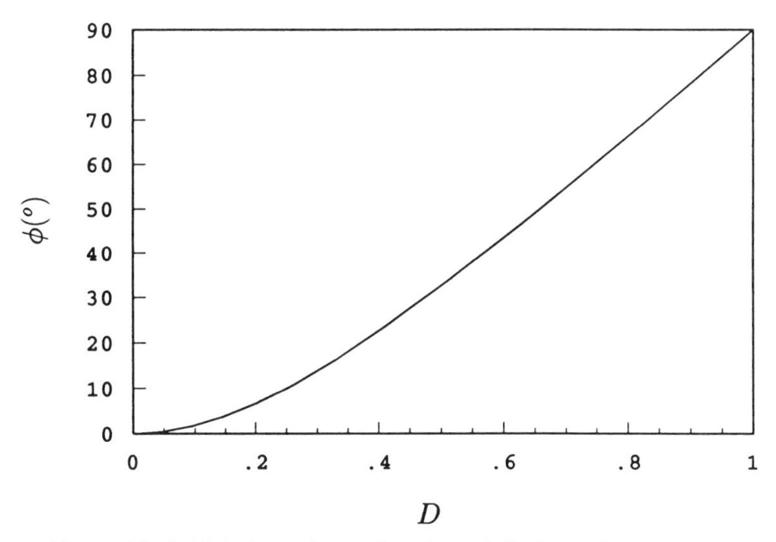

Figure 4.2: Initial phase ϕ as a function of diode ON duty cycle D.

The average value of the diode voltage waveform is $V_D = -V_O$ because the dc component of the voltage across L_f is zero. Thus, using (4.6),

$$V_D = -V_O = \frac{1}{2\pi}\int_0^{2\pi(1-D)} v_D d(\omega t)$$
$$= \frac{V_O}{2\pi\omega CR_L}\left[2\pi^2(1-D)^2 + cos2\pi D - 1 - \frac{2\pi(1-D) + sin2\pi D)}{tan\phi}\right]. \tag{4.8}$$

Rearranging this equation and using (4.7), one obtains the relationship between the diode ON duty cycle D and ωCR_L

$$\omega CR_L = \frac{1}{2\pi}\left\{1 - 2\pi^2(1-D)^2 - cos2\pi D + \frac{[2\pi(1-D) + sin2\pi D]^2}{1 - cos2\pi D}\right\}. \tag{4.9}$$

This relation is plotted in Fig. 4.3, which shows that D decreases from 1 to zero as ωCR_L increases from zero to infinity.

Substitution of (4.7) into (4.6) yields the diode voltage waveform normalized with respect to the dc output voltage V_O

$$\frac{v_D}{V_O} = \begin{cases} \frac{1}{\omega CR_L}\{\omega t - sin\omega t + \frac{[2\pi(1-D)+sin2\pi D](cos\omega t-1)}{1-cos2\pi D}\}, & \text{for } 0 < \omega t \leq 2\pi(1-D) \\ 0, & \text{for } 2\pi(1-D) < \omega t \leq 2\pi, \end{cases} \tag{4.10}$$

where ωCR_L is given by (4.9).

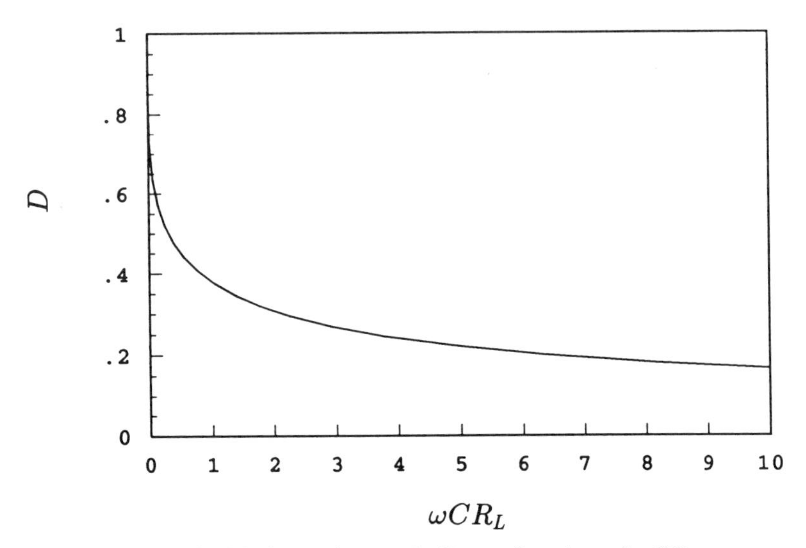

Figure 4.3: Diode ON duty cycle D as a function of ωCR_L.

Figure 4.1(d) shows the rectifier model when the diode is ON, that is, for $2\pi(1-D) < \omega t \leq 2\pi$. The diode is driven by two current sources, i_R and I_O. From (4.3) and (4.4),

$$i_D = I_O - i_R = I_O - I_m sin(\omega t + \phi) = I_O \left[1 - \frac{sin(\omega t + \phi)}{sin\phi}\right]$$
$$= I_O \left(1 - cos\omega t - \frac{sin\omega t}{tan\phi}\right), \quad \text{for} \quad 2\pi(1-D) < \omega t \leq 2\pi. \quad (4.11)$$

Substitution of (4.7) into (4.11) gives the diode current waveform normalized with respect to the dc output current I_O

$$\frac{i_D}{I_O} = \begin{cases} 0, & \text{for} \quad 0 < \omega t \leq 2\pi(1-D) \\ 1 - cos\omega t - \frac{[2\pi(1-D)+sin2\pi D]sin\omega t}{1-cos2\pi D}, & \text{for} \quad 2\pi(1-D) < \omega t \leq 2\pi. \end{cases} \quad (4.12)$$

The waveforms of the diode current and voltage are depicted in Fig. 4.4 for $D = 0.75$, 0.5, and 0.25. As the load resistance R_L increases at a constant value of ωC, the diode ON duty cycle D decreases, V_{DM}/V_O decreases, and I_{DM}/I_O increases.

From (4.12), the peak value of the diode current I_{DM} occurs at $\omega t_{im} = 3\pi/2 - \phi$ for $D > 0.28$ and at $\omega t_{im} = 2\pi(1-D)$ for $D \leq 0.28$. The result is

$$\frac{I_{DM}}{I_O} = \begin{cases} 1 - cos2\pi D + \frac{sin2\pi D}{tan\phi}, & \text{for} \quad 0 < D \leq 0.28 \\ 1 + \frac{1}{sin\phi}, & \text{for} \quad 0.28 < D \leq 1. \end{cases} \quad (4.13)$$

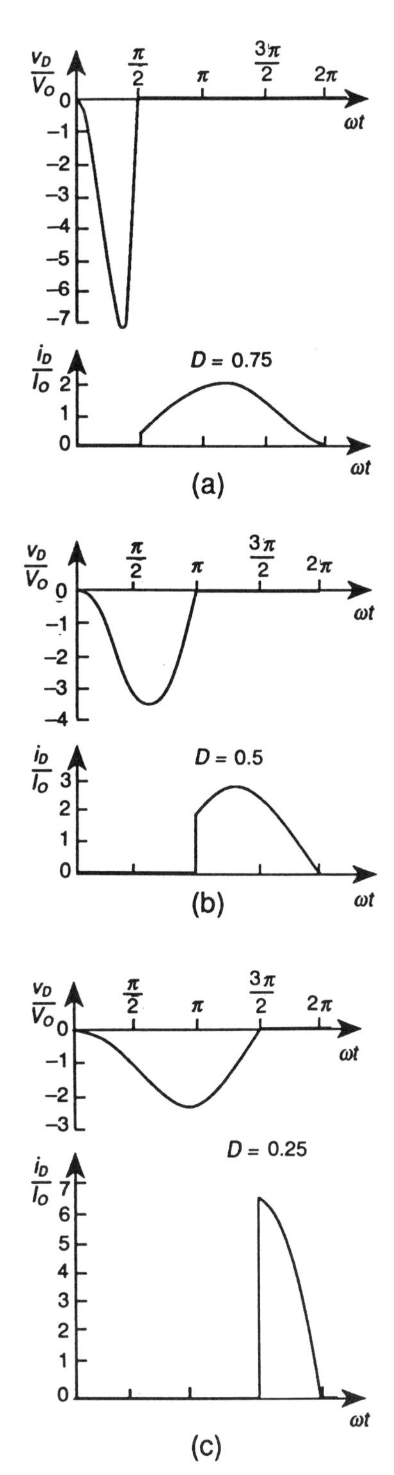

Figure 4.4: Diode voltage and current waveforms normalized with respect to dc output voltage V_O and current I_O, respectively. (a) For $D = 0.75$. (b) For $D = 0.5$. (c) For $D = 0.25$.

Figure 4.5(a) illustrates I_{DM}/I_O as a function of D, and Fig. 4.5(b) illustrates it as a function of ωCR_L. The ratio I_{DM}/I_O increases from 2 to ∞ as D decreases from 1 to 0, or as ωCR_L increases from 0 to ∞. However, the absolute value of I_{DM} decreases with ωCR_L at a fixed value of V_O because $I_O = V_O/R_L$ decreases with R_L. Therefore, the maximum value of I_{DM} occurs at the minimum load resistance R_{Lmin}.

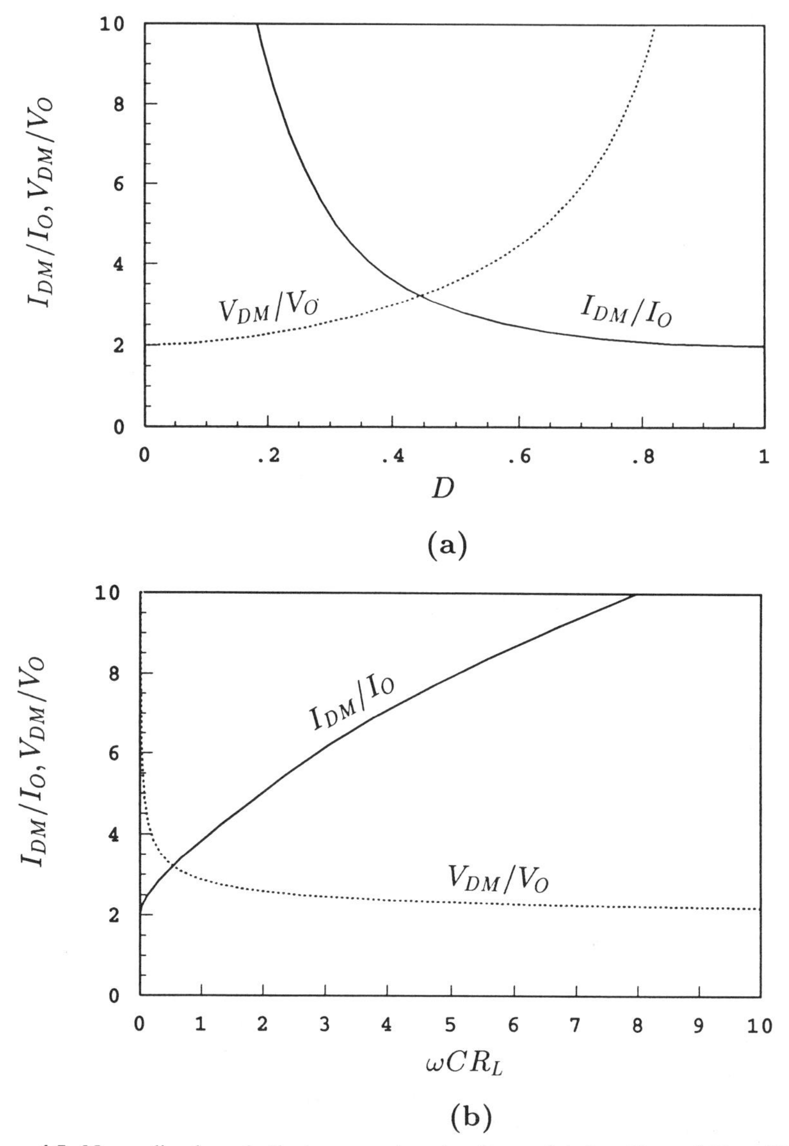

Figure 4.5: Normalized peak diode current and voltage. (a) I_{DM}/I_O and V_{DM}/V_O as functions of D. (b) I_{DM}/I_O and V_{DM}/V_O as functions of ωCR_L.

The peak values of the diode reverse voltage V_{DM} can be found by differentiating (4.10) and setting the result equal to zero. The minimum value of v_D occurs at $\omega t_{vm} = \pi - 2\phi$, and the peak value of v_D is given by

$$\frac{V_{DM}}{V_O} = -\frac{v_D(\omega t_{vm})}{V_O} = \frac{1}{\omega C R_L}\left(2\phi - \pi + \frac{2}{tan\phi}\right), \tag{4.14}$$

where $\omega C R_L$ is given by (4.9) and $tan\phi$ is given by (4.7). Notice that ϕ should be expressed in radians. Figure 4.5(a) depicts V_{DM}/V_O as a function of D, and Fig. 4.5(b) depicts it as a function of $\omega C R_L$. The ratio V_{DM}/V_O decreases from ∞ to 2 as D decreases from 1 to 0, or as $\omega C R_L$ increases from 0 to infinity. Thus, the maximum value of V_{DM} occurs at the minimum load resistance R_{Lmin} if the dc output voltage V_O is held constant.

The normalized power-output capability of the rectifier is defined as

$$c_p = \frac{P_O}{I_{DM}V_{DM}} = \frac{I_O V_O}{I_{DM}V_{DM}}. \tag{4.15}$$

Factor c_p was calculated using (4.13) and (4.14) and plotted as a function of D in Fig. 4.6(a) and as a function of $\omega C R_L$ in Fig. 4.6(b). The maximum value of c_p occurs at $D = 0.5$ and equals 0.0981. Factor c_p first increases with $\omega C R_L$, reaches its maximum value at $\omega C R_L = 1/\pi$, and then decreases with $\omega C R_L$.

The ac-to-dc current transfer function is defined as

$$M_{IR} = \frac{I_O}{I_R} = \frac{I_O}{I_m/\sqrt{2}} = \sqrt{2}sin\phi = \frac{V_O}{I_R R_L}, \tag{4.16}$$

where I_R is the rms value of the input current. For $D = 0.5$, $M_{IR} = \sqrt{8/(\pi^2+4)}$ $= 0.7595$. Figure 4.7(a) and (b) shows M_{IR} as a function of D and as a function of $\omega C R_L$, respectively. As seen, M_{IR} decreases from $\sqrt{2}$ to 0 as D decreases from 1 to 0, or as $\omega C R_L$ increases from 0 to ∞. From (4.16), the amplitude of the input current is

$$I_m = \frac{\sqrt{2}I_O}{M_{IR}} = \frac{\sqrt{2}V_O}{M_{IR}R_L}. \tag{4.17}$$

In practice, it is important to know the input impedance of the rectifier. It is sufficient to determine the input impedance of the rectifier at only the operating frequency f because the input current is a sine wave at that frequency. This impedance may be represented as the series combination of the input resistance R_i and the input capacitance C_i, as shown in Fig. 4.8. The input voltage of the rectifier is $v_R = -v_D$. The fundamental component of this voltage can be expressed as

$$v_{R1} = v_{Ri} + v_{Ci} = V_{Rim}sin(\omega t + \phi) - V_{Cim}cos(\omega t + \phi), \tag{4.18}$$

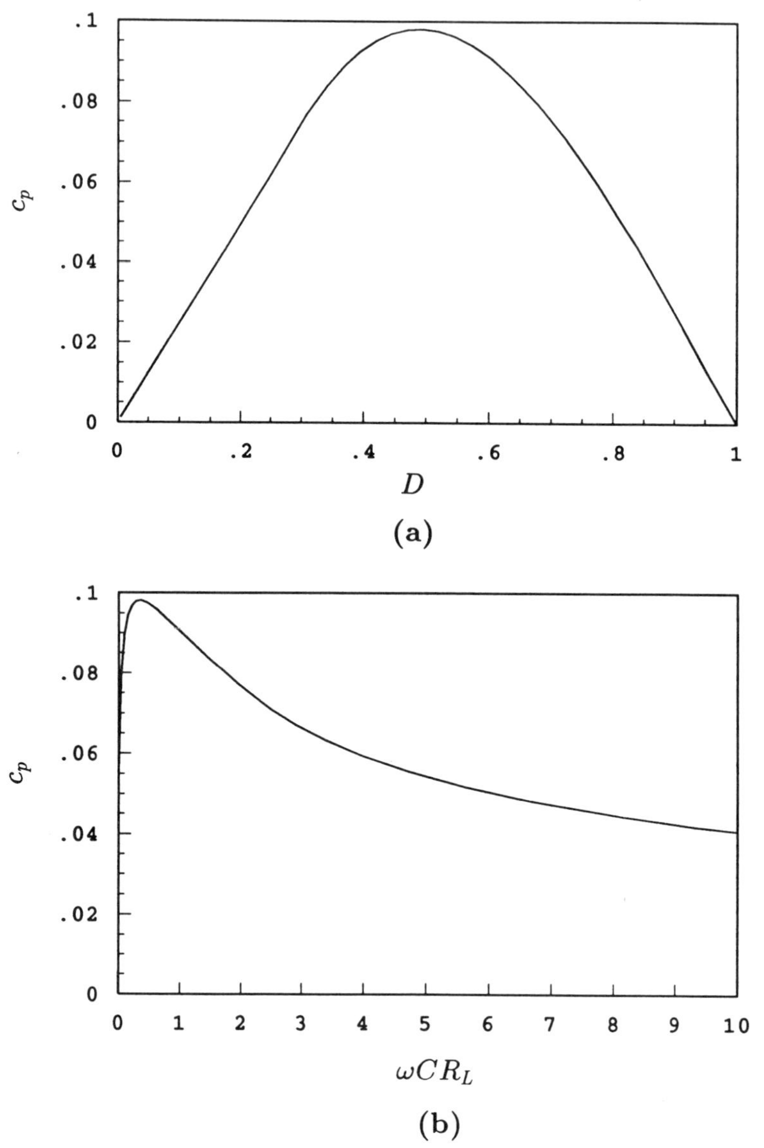

Figure 4.6: Power-output capability c_p. (a) c_p as a function of D. (b) c_p as a function ωCR_L.

where V_{Rim} and V_{Cim} are the amplitudes of the fundamental components of the voltages v_{Ri} and v_{Ci} across R_i and C_i, respectively. Neglecting power losses, the output power $P_O = I_O^2 R_L$ equals the input power $P_i = I_m^2 R_i/2$, that is, $I_O^2 R_L = I_m^2 R_i/2$. Hence, from (4.16) one obtains the input resistance R_i at the operating frequency f

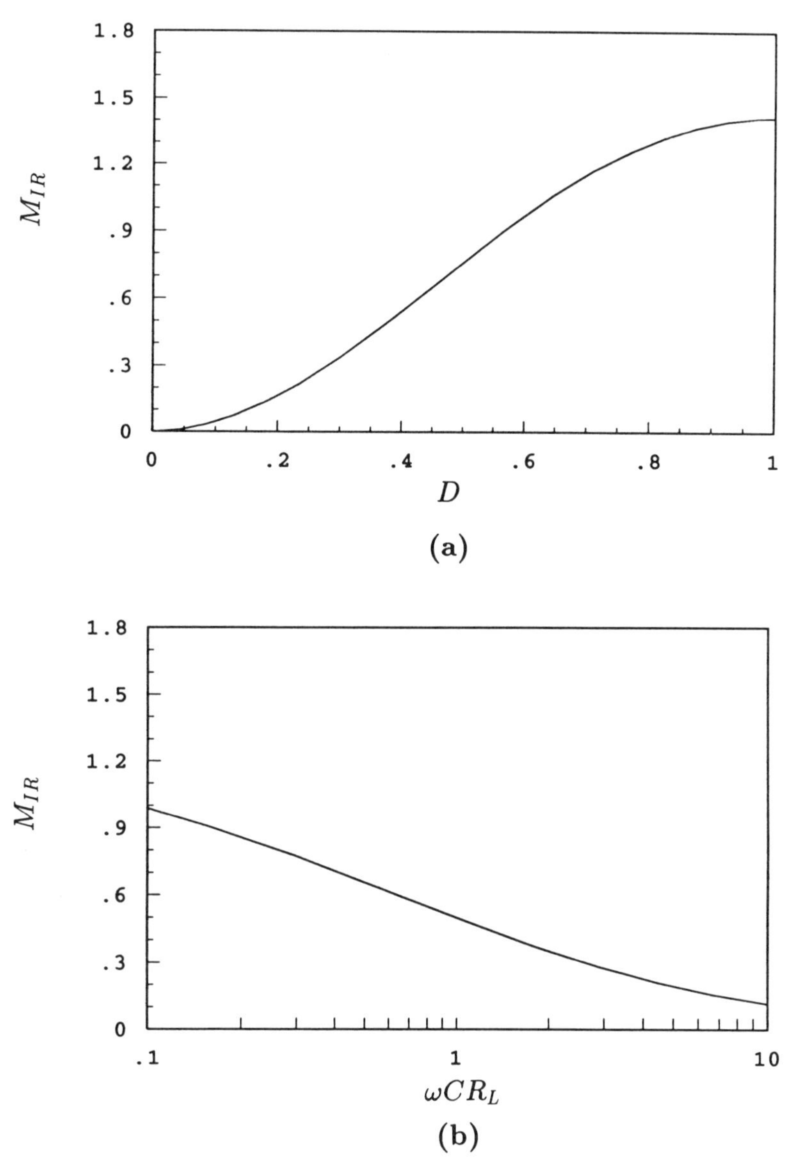

Figure 4.7: Ac-to-dc current transfer function M_{IR}. (a) M_{IR} as a function of D. (b) M_{IR} as a function of ωCR_L.

$$\frac{R_i}{R_L} = 2sin^2\phi. \tag{4.19}$$

The plots of R_i/R_L versus D and versus ωCR_L are shown in Fig. 4.9(a) and (b), respectively. As seen, R_i/R_L decreases from 2 to 0 as D decreases from 1 to 0, or as ωCR_L increases from 0 to ∞.

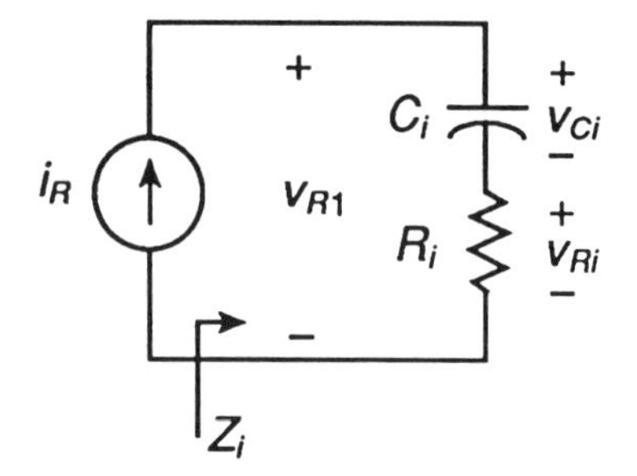

Figure 4.8: Model of the input impedance of the rectifier at the operating frequency f.

From (4.9) and (4.19), the input resistance R_i normalized with respect to $1/\omega C$ can be expressed as

$$\omega C R_i = (\omega C R_L)\left(\frac{R_i}{R_L}\right) = \frac{sin^2\phi}{\pi}\left[1 - 2\pi^2(1-D)^2 - cos2\pi D + \frac{2\pi(1-D) + sin2\pi D}{tan\phi}\right]. \quad (4.20)$$

Figure 4.10 plots $\omega C R_i$ against D and $\omega C R_L$. The maximum value of $\omega C R_i$ occurs at $D = 0.35$ or $\omega C R_L = 1.3253$ and equals 0.2525. For $D \geq 0.35$ or $\omega C R_L \geq 1.3253$, $\omega C R_i$ decreases as $\omega C R_L$ increases. Thus, the rectifier acts as an impedance inverter and is, therefore, compatible with Class E ZVS inverters [10]–[14] studied in Chapter 13.

Substitution of $v_R = -v_D$ given by (4.6) into the Fourier formula gives

$$V_{Cim} = -\frac{1}{\pi}\int_0^{2\pi} v_R cos(\omega t + \phi) d(\omega t) = \frac{1}{\pi}\int_0^{2\pi} v_D cos(\omega t + \phi) d(\omega t)$$
$$= \frac{V_O}{\pi \omega C R_L}\left\{\frac{\left[\pi(1-D) + sin2\pi D - \frac{1}{4}sin4\pi D\right] cos\phi}{tan\phi} + \left[\pi(1-D) + sin2\pi D + \frac{1}{4}sin4\pi D\right] sin\phi - cos\phi sin^2 2\pi D + 2\pi(1-D) sin(2\pi D - \phi)\right\}. \quad (4.21)$$

Hence, from (4.4) the input reactance of the rectifier at the operating frequency f is

$$X_i = \frac{1}{\omega C_i} \equiv \frac{V_{Cim}}{I_m} = \frac{1}{\pi \omega C}\left\{\pi(1-D) + sin2\pi D - \frac{1}{4}sin4\pi D cos2\phi - \frac{1}{2}sin2\phi sin^2 2\pi D - 2\pi(1-D) sin\phi sin(2\pi D - \phi)\right\}. \quad (4.22)$$

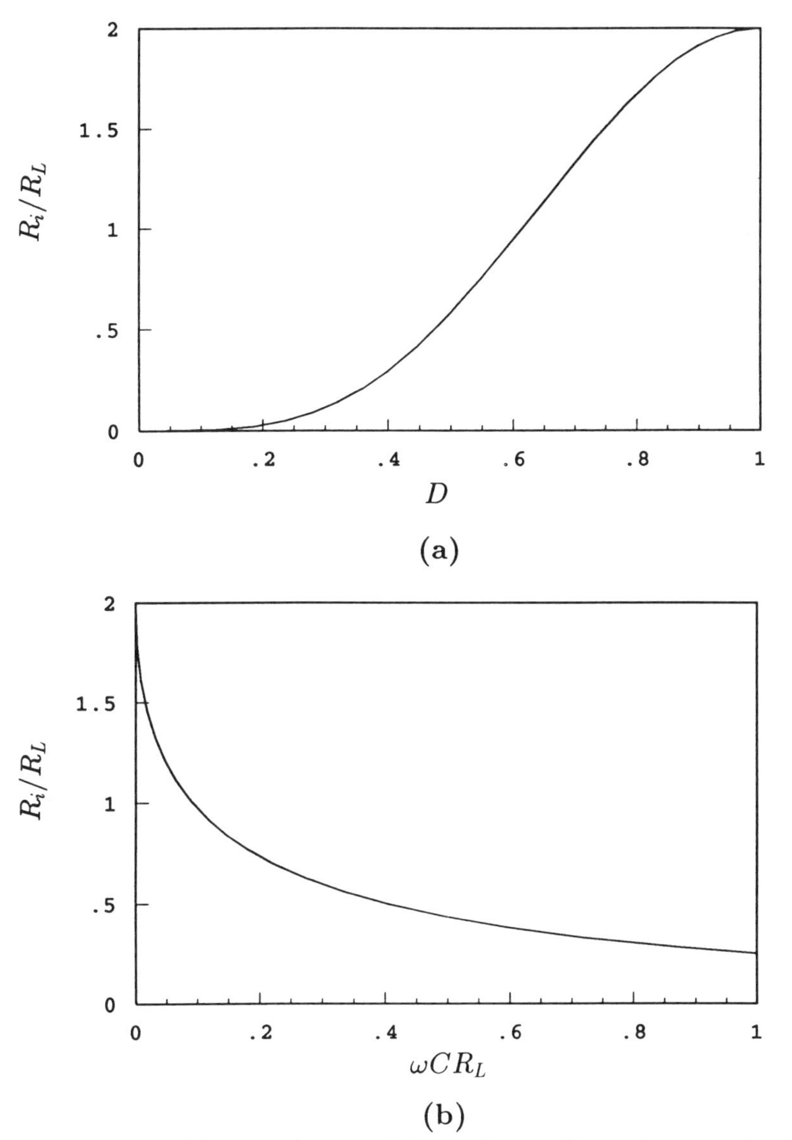

Figure 4.9: Normalized input resistance of the rectifier R_i/R_L at the operating frequency f. (a) R_i/R_L as a function of D. (b) R_i/R_L as a function of $\omega C R_L$.

Simplifying this expression, one arrives at the input capacitance of the rectifier at the fundamental frequency f

$$\frac{C_i}{C} = \pi \Big[\pi(1-D) + sin 2\pi D - \frac{1}{4} cos 2\phi \, sin 4\pi D - \frac{1}{2} sin 2\phi \, sin^2 2\pi D \\ -2\,\pi(1-D) sin\phi \, sin(2\pi D - \phi) \Big]^{-1}. \quad (4.23)$$

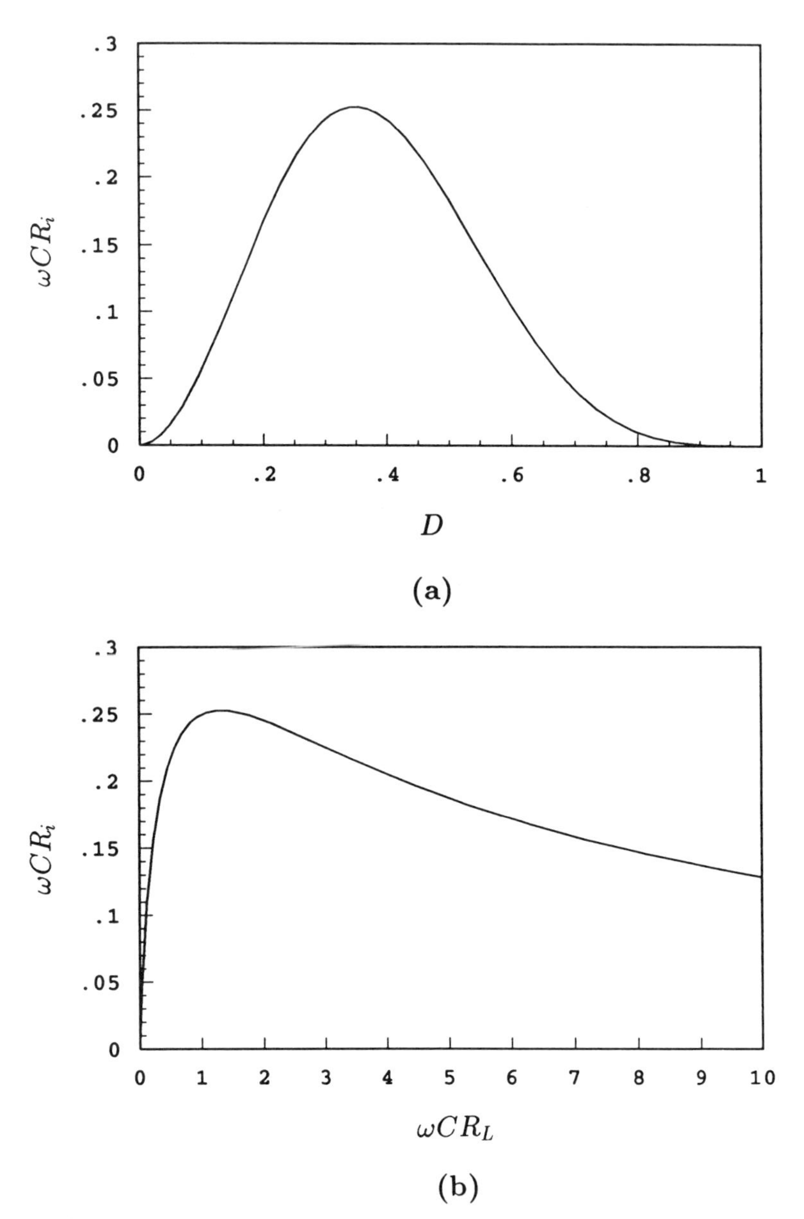

Figure 4.10: Normalized input resistance $\omega C R_i$. (a) $\omega C R_i$ versus D. (b) $\omega C R_i$ versus $\omega C R_L$.

Figure 4.11(a) and (b) shows plots of C_i/C as a function of D and as a function of $\omega C R_L$. It can be seen that C_i/C decreases from infinity to 1 as D decreases from 1 to 0, or as $\omega C R_L$ increases from 0 to ∞.

Neglecting power losses, the output power $P_O = V_O^2/R_L = I_O^2 R_L$ equals the input power $P_i = V_{R1}^2/R_i = I_{rms}^2 R_i$, where V_{R1} is the rms value of v_{R1}.

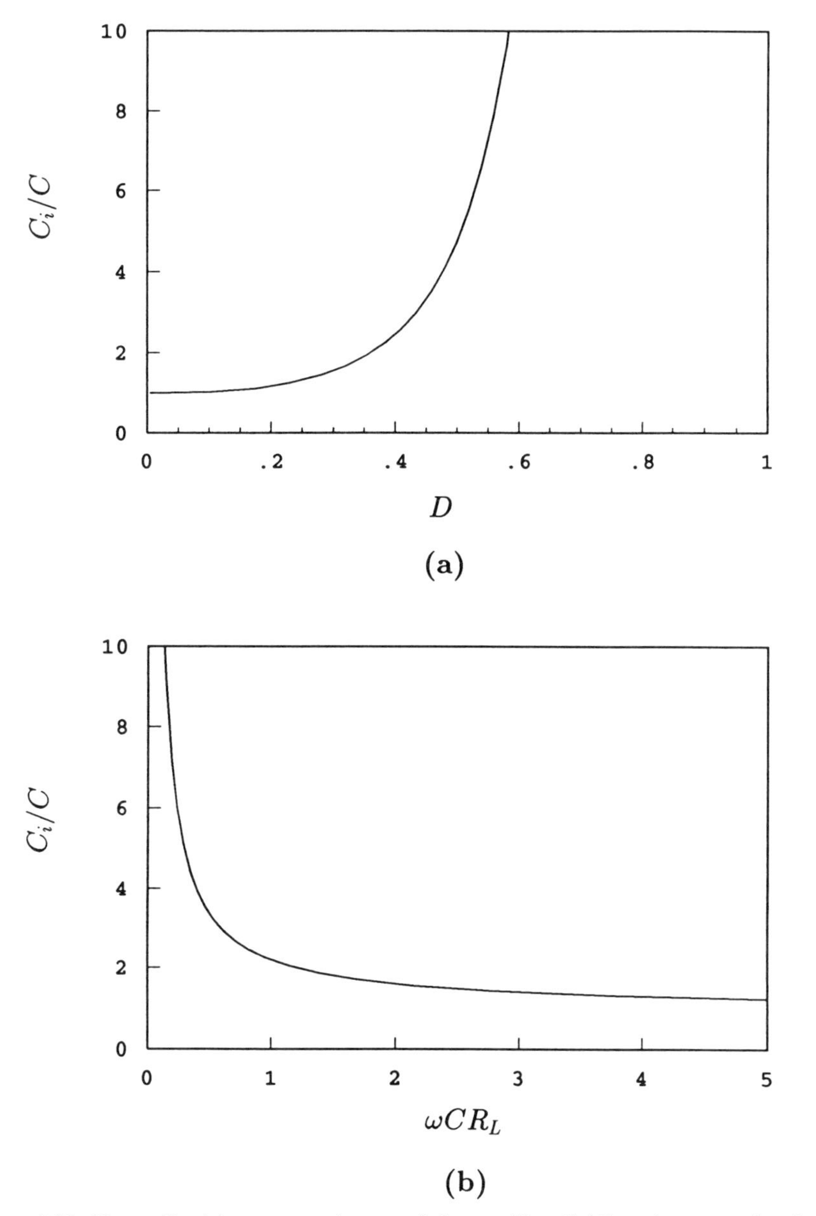

Figure 4.11: Normalized input capacitance of the rectifier C_i/C at the operating frequency f. (a) C_i/C as a function of D. (b) C_i/C as a function of $\omega C R_L$.

Hence, from (4.16) the voltage transfer function is

$$M_{VR} = \frac{V_O}{V_{R1}} = \frac{I_{rms}}{I_O} = \frac{1}{M_{IR}} = \sqrt{\frac{R_L}{R_i}} = \frac{1}{\sqrt{2} sin\phi}. \tag{4.24}$$

The plots of M_{VR} as a function of D and as a function of ωCR_L are shown in Fig. 4.12(a) and (b), respectively. M_{VR} increases from $1/\sqrt{2}$ to ∞ as D decreases from 1 to 0, or as ωCR_L increases from 0 to ∞.

The numerical values of the rectifier parameters for various values of D are given in Table 4.1.

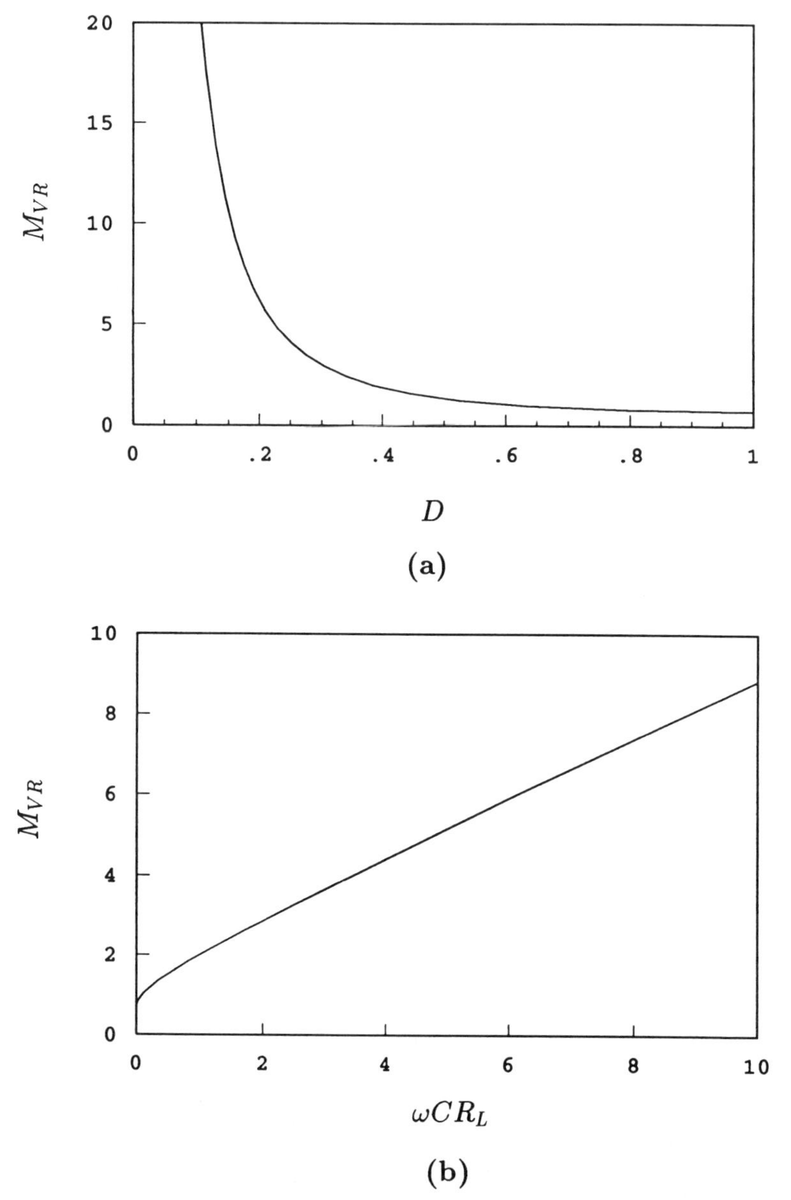

Figure 4.12: Voltage transfer function of the rectifier M_{VR}. (a) M_{VR} as a function of D. (b) M_{VR} as a function of ωCR_L.

Table 4.1. Parameters of Class E Low *dv/dt* Rectifier with a Parallel Capacitor

D	$\phi(^\circ)$	$\omega C R_L$	I_{DM}/I_O	V_{DM}/V_O	R_i/R_L	$\omega C R_i$	C_i/C	M_{VR}
0	0	∞	∞	2	0	0	1	∞
0.05	0.45	125.34	39.69	2.022	0.0001	0.0152	1.0031	90.704
0.1	1.75	29.962	19.40	2.079	0.0019	0.0560	1.0233	23.124
0.15	3.83	12.398	12.48	2.164	0.0089	0.1109	1.0738	10.572
0.2	6.59	6.3295	8.918	2.274	0.0264	0.1669	1.1665	6.1578
0.25	9.93	3.5855	6.712	2.407	0.0595	0.2132	1.3167	4.1007
0.3	13.75	2.1481	5.207	2.566	0.1130	0.2427	1.5483	2.9749
0.4	22.54	0.8276	3.608	2.976	0.2940	0.2433	2.4443	1.8443
0.5	32.48	0.3183	2.862	3.562	0.5768	0.1836	4.7259	1.3167
0.6	43.21	0.1114	2.460	4.447	0.9377	0.1045	11.831	1.0327
0.7	54.49	0.0316	2.228	5.927	1.3254	0.0419	42.944	0.8686
0.75	60.28	0.0147	2.151	7.112	1.5085	0.0221	100.88	0.8142
0.8	66.14	0.0058	2.093	8.889	1.6728	0.0097	293.83	0.7732
0.9	78.02	0.0035	2.022	17.78	1.9138	0.0004	8837.6	0.7229
1	90	0	2	∞	2	0	∞	0.7071

When the diode is ON, the voltage drop across the filter inductor L_f is V_O. Hence, the steady-state peak-to-peak value of the ripple current in the filter inductor is

$$I_r = \frac{V_O(1 - D_{max})T}{L_f}. \tag{4.25}$$

If it is required that the peak-to-peak value of the ripple current cannot exceed $I_{r(max)}$, the minimum value of the filter inductance can be found as

$$L_{f(min)} = \frac{(1 - D_{max})V_O}{f I_{r(max)}}. \tag{4.26}$$

Placing, as a rule of thumb, the corner frequency of the output filter $f_o = 1/(2\pi\sqrt{L_f C_f})$ at least a decade below the operating frequency of the rectifier f, the minimum value of the filter capacitor is given by

$$C_{f(min)} = \frac{25}{\pi^2 f^2 L_{f(min)}}. \tag{4.27}$$

4.2.4 Parameters for D = 0.5

The rectifier has the following parameters for $D = 0.5$:

$$tan\phi = \frac{2}{\pi} \tag{4.28}$$

$$sin\phi = \frac{2}{\sqrt{\pi^2 + 4}} \tag{4.29}$$

$$cos\phi = \frac{\pi}{\sqrt{\pi^2 + 4}} \tag{4.30}$$

$$\phi = 0.5669 \text{ rad} = 32.48^\circ \tag{4.31}$$

$$\omega C R_L = \frac{1}{\pi} \tag{4.32}$$

$$\frac{I_{DM}}{I_O} = \frac{1}{2}\sqrt{\pi^2 + 4} + 1 = 2.862 \tag{4.33}$$

$$\frac{V_{DM}}{V_O} = 2\pi\phi = 3.562 \tag{4.34}$$

$$c_p = \frac{I_O}{I_{IM}}\frac{V_O}{V_{DM}} = 0.0981 \tag{4.35}$$

$$\frac{R_i}{R_L} = \frac{8}{\pi^2 + 4} = 0.5768 \tag{4.36}$$

$$\omega C R_i = \frac{8}{\pi(\pi^2 + 4)} = 0.1836 \tag{4.37}$$

$$\frac{C_i}{C} = \frac{2(\pi^2 + 4)}{\pi^2 - 4} = 4.726 \tag{4.38}$$

$$M_{IR} = \sqrt{2} sin\phi = \sqrt{\frac{8}{\pi^2 + 4}} = 0.7595 \tag{4.39}$$

$$M_{VR} = \frac{1}{2}\sqrt{\frac{\pi^2}{2} + 2} = 1.3167. \tag{4.40}$$

4.2.5 Design Example

Example 4.1

Design a Class E rectifier with the following specifications: $V_O = 5$ V, $I_O = 0$ to 0.2 A, and $f = 1$ MHz.

Solution: The minimum load resistance is

$$R_{Lmin} = \frac{V_O}{I_{Omax}} = \frac{5}{0.2} = 25\ \Omega, \tag{4.41}$$

and the maximum load resistance is infinity. The maximum output power is

$$P_{Omax} = V_O I_{Omax} = 5 \times 0.2 = 1 \text{ W}. \tag{4.42}$$

Assume that $D = 0.5$ at $R_L = R_{Lmin}$. From (4.33), the maximum value of the diode peak current is

$$I_{DMmax} = 2.862 I_{Omax} = 2.862 \times 0.2 = 0.57 \text{ A}. \tag{4.43}$$

From (4.34), the maximum value of the diode reverse peak voltage is

$$V_{DMmax} = 3.562 V_O = 3.562 \times 5 = 17.8 \text{ V}. \tag{4.44}$$

The value of the capacitance C can be calculated using (4.32)

$$C = \frac{1}{\pi \omega R_{Lmin}} = \frac{1}{2\pi^2 f R_{Lmin}} = \frac{1}{2\pi^2 \times 10^6 \times 25} = 2 \text{ nF}. \tag{4.45}$$

The voltage rating of the capacitor should be at least 25 V. From (4.16), (4.17), and (4.29), the maximum value of the amplitude of the input current is

$$I_{m(max)} = \frac{\sqrt{2} I_{Omax}}{M_{IR}} = \frac{\sqrt{2} \times 0.2}{0.7595} = 0.37 \text{ A}. \tag{4.46}$$

Let us assume that the maximum allowable peak-to-peak current ripple in the filter inductor is 10% of the output current. From (4.26), the minimum value of the filter inductance is

$$L_{f(min)} = \frac{(1-D)V_O}{0.1 f I_O} = \frac{(1-0.5) \times 5}{0.1 \times 10^6 \times 0.2} = 125 \ \mu\text{H}. \tag{4.47}$$

The minimum value of the filter capacitance can be obtained from (4.27) as

$$C_{f(min)} = \frac{25}{\pi^2 f^2 L_{f(min)}} = \frac{25}{\pi^2 \times 10^{12} \times 125 \times 10^{-6}} = 20.3 \text{ nF}. \tag{4.48}$$

4.3 RESONANT LOW *dv/dt* RECTIFIER

4.3.1 Circuit Description

A circuit diagram of a Class E resonant voltage-driven low dv/dt rectifier [11], is shown in Fig. 4.13(a). It consists of a rectifying diode, a resonant capacitor C connected in parallel with the diode, a resonant inductor L connected in series with the parallel combination of the diode and the capacitor, and a first-order low-pass output filter C_f-R_L. Resistor R_L represents a load to which a dc power is to be delivered. The rectifier is driven by a sinusoidal voltage source v_R. If the diode is replaced by a controllable switch (e.g., a power MOSFET), a Class E synchronous regulated rectifier is obtained. The ac source v_R and the rectifier can be coupled by an isolation transformer to provide the desired ac-to-dc voltage transfer function and/or to satisfy strict safety requirements governed by international and national regulations, such as IEC (International Electrotechnical Commission), UL (Underwriters Laboratories), VDE (Verband Deutscher Elektrotechniker), and CSA (Canadian Standards Association). The advantages of the rectifier topology are that the diode junction capacitance is absorbed into the resonant capacitance C and

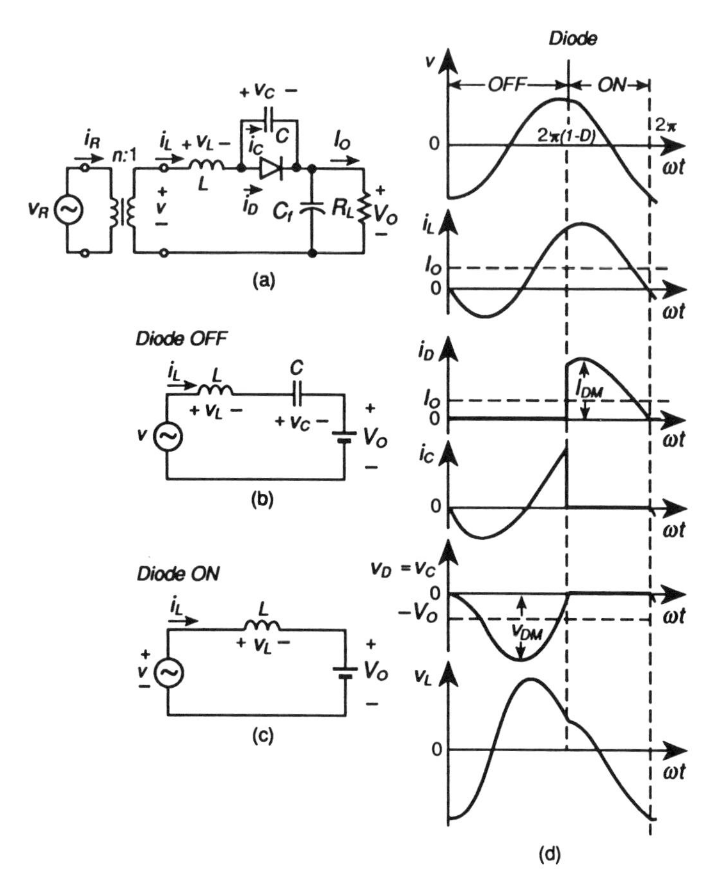

Figure 4.13: Class E resonant voltage-driven low dv/dt rectifier. (a) Circuit. (b) Model of the rectifier when the diode is OFF. (c) Model of the rectifier when the diode is ON. (d) Current and voltage waveforms.

the transformer leakage inductance and some lead inductances are absorbed into the resonant inductance L. Therefore, the rectifier is especially suitable for high-frequency applications such as dc-to-dc converters. A dc-to-dc converter can be obtained by replacing the sinusoidal input voltage source v_R by a dc-to-ac inverter with a parallel-resonant circuit. Figure 4.13(b) and (c) depicts the equivalent circuits of the rectifier for the intervals when the diode is OFF and ON, respectively.

Steady-state current and voltage waveforms explaining the principle of operation of the rectifier are shown in Fig. 4.13(d). When the diode is OFF, the inductor L and the capacitor C form a series-resonant circuit. The voltage across the resonant circuit is the difference between the sinusoidal input voltage v and the dc output voltage V_O. Consequently, the current through inductor L and the capacitor C is a portion of a sinusoid. This current shapes the voltage across the capacitor and the diode $v_C = v_D$, in accordance with the equation $i_C = C dv_C/dt$. At $\omega_o t = 0$, the derivative of the capacitor voltage dv_C/dt is zero because the capacitor current is zero. The capacitor current then becomes negative, and therefore the capacitor and diode voltage v_C gradually decreases. When the capacitor current crosses zero, the capacitor voltage reaches its maximum value. Next, the capacitor current is positive, causing the capacitor and diode voltage to increase gradually. When the diode voltage reaches the diode threshold voltage, the diode turns on. As seen, the diode turns on and off at low dv_D/dt, reducing switching losses and noise.

When the diode is ON, the capacitor is shorted out and the inductor current i_L is driven by the voltage difference $v - V_O$. Once the diode current i_D reaches zero, the turn-off transition of the diode will commence. In the case of the *pn* junction diode, the reverse recovery should be considered. In the Class E rectifier, the reverse-recovery current is low because $| di_D/dt |$ before turn-off is low. During the storage time, this current is a portion of a sine wave and the voltage is about 0.7 V (0.4 V for Schottky diodes). After the storage time, the diode current decreases to zero and the diode voltage is held low by the resonant capacitor C, reducing the turn-off power loss due to the reverse recovery.

4.3.2 Assumptions

The analysis of the rectifier operation is based on the equivalent circuits of Fig. 4.13(b) and (c) and the following assumptions:

1) The diode is ideal, that is, it has zero threshold voltage and zero on-resistance, infinite off-resistance, and zero minority carrier charge lifetime in the case of the *pn* junction diode.
2) The filter capacitance C_f is large enough that the output voltage is ripple-free. Consequently, the C_f-R_L circuit can be replaced by a dc voltage sink.
3) The rectifier is driven by an ideal sinusoidal voltage source.
4) The operating frequency f is equal to the resonant frequency $f_o = 1/(2\pi\sqrt{LC})$.

4.3.3 Characteristics

The following definitions are used in the subsequent analysis:

- the angular resonant frequency

$$\omega_o = \frac{1}{\sqrt{LC}} \tag{4.49}$$

- the characteristic impedance of the resonant circuit

$$Z_o = \sqrt{\frac{L}{C}} = \omega_o L = \frac{1}{\omega_o C} \tag{4.50}$$

- the normalized load resistance or the loaded quality factor

$$Q = \frac{R_L}{\omega_o L} = \omega_o C R_L = \frac{R_L}{Z_o} \tag{4.51}$$

- the ac-to-dc voltage transfer function

$$M_{VR} = \frac{V_O}{V_{rms}}. \tag{4.52}$$

According to assumption 3), the input voltage is sinusoidal and given by

$$v_R = V_{Rm} sin(\omega_o t + \phi). \tag{4.53}$$

Hence, the voltage across the secondary side of the transformer becomes

$$v = \frac{v_R}{n} = V_m sin(\omega_o t + \phi), \tag{4.54}$$

where $V_m = \sqrt{2} V_{rms} = V_{Rm}/n$ and n is the transformer turns ratio.

Consider first the time interval $0 < \omega_o t \leq 2\pi(1 - D)$ during which the diode is OFF. The equivalent circuit of the rectifier for this interval is shown in Fig. 4.13(b). It follows from Fig. 4.13(d) that the boundary conditions at $\omega_o t = 0$ are $i_L(0) = 0$ and $v_C(0) = 0$. Applying the Laplace transform to (4.54) yields

$$V(s) = V_m \frac{\omega_o cos\phi + s sin\phi}{s^2 + \omega_o^2}. \tag{4.55}$$

Hence, one arrives at the current through the inductor L and the capacitor C in the s-domain

$$I_L(s) = \frac{V(s) - \frac{V_O}{s}}{sL + \frac{1}{sC}} = \frac{sV(s) - V_O}{L(s^2 + \omega_o^2)} = \frac{V_m(s\omega_o cos\phi + s^2 sin\phi)}{L(s^2 + \omega_o^2)^2} - \frac{V_O}{L(s^2 + \omega_o^2)}$$

$$= \frac{\sqrt{2} V_O(s\omega_o cos\phi + s^2 sin\phi)}{M_{VR} L(s^2 + \omega_o^2)^2} - \frac{V_O}{L(s^2 + \omega_o^2)}$$

$$= \frac{V_O}{\omega_o L M_{VR}} \left[\frac{\sqrt{2}\omega_o(s\omega_o cos\phi + s^2 sin\phi)}{(s^2 + \omega_o^2)^2} - \frac{\omega_o M_{VR}}{s^2 + \omega_o^2} \right] \tag{4.56}$$

and in the time domain

$$i_C = i_L = \frac{I_O Q}{\sqrt{2} M_{VR}} \left[\omega_o t sin(\omega_o t + \phi) + \left(sin\phi - \sqrt{2} M_{VR} \right) sin\omega_o t \right]. \tag{4.57}$$

This yields the voltage across the capacitor and the diode

$$v_D = v_C = \frac{1}{\omega_o C}\int_0^{\omega_o t} i_C d(\omega_o t)$$
$$= V_O\left\{\frac{1}{\sqrt{2}M_{VR}}\left[cos\phi sin\omega_o t - \omega_o t cos(\omega_o t + \phi)\right] + cos\omega_o t - 1\right\} \tag{4.58}$$

and the voltage across the inductor

$$v_L = v - v_C - V_O$$
$$= \frac{\sqrt{2}V_O}{M_{VR}}\left[\frac{1}{2}\omega_o t cos(\omega_o t + \phi) + \frac{1}{2}cos\phi sin\omega_o t + \left(sin\phi - \frac{M_{VR}}{\sqrt{2}}\right) cos\omega_o t\right]. \tag{4.59}$$

The voltage across the capacitor and the diode reaches zero at $\omega_o t = 2\pi(1 - D)$. Hence, from (4.58) the voltage transfer function is

$$M_{VR} = \sqrt{2}\frac{\pi(1 - D)sin2\pi D sin\phi + [\frac{1}{2}sin2\pi D + \pi(1 - D)cos2\pi D]cos\phi}{cos2\pi D - 1}. \tag{4.60}$$

This relation can also be derived by equating the average value of capacitor current i_C to zero. Since the average value of a voltage across an inductor is zero, the average value of the diode voltage V_D is equal to $-V_O$. Thus, using (4.58)

$$V_D = -V_O = \frac{1}{2\pi}\int_0^{2\pi(1-D)} v_D d(\omega_o t)$$
$$= \frac{V_O}{\sqrt{2}\pi M_{VR}}\{[1 - cos2\pi D + \pi(1 - D)sin2\pi D]cos\phi$$
$$- \left[\pi(1 - D)cos2\pi D + \frac{1}{2}sin2\pi D\right] sin\phi - \frac{M_{VR}}{\sqrt{2}}[sin2\pi D + 2\pi(1 - D)]\}. \tag{4.61}$$

Hence,

$$M_{VR} = \sqrt{2}\frac{[\frac{1}{2}sin2\pi D + \pi(1 - D)cos2\pi D]sin\phi - [1 - cos2\pi D + \pi(1 - D)sin2\pi D]cos\phi}{2\pi D - sin2\pi D}. \tag{4.62}$$

Figure 4.14 shows plots of M_{VR} versus D and Q. The transfer function M_{VR} decreases from ∞ to 0 as D is increased from 0 to 1, which happens when Q is decreased from ∞ to 0. Note that M_{VR} is almost a linear function of Q.

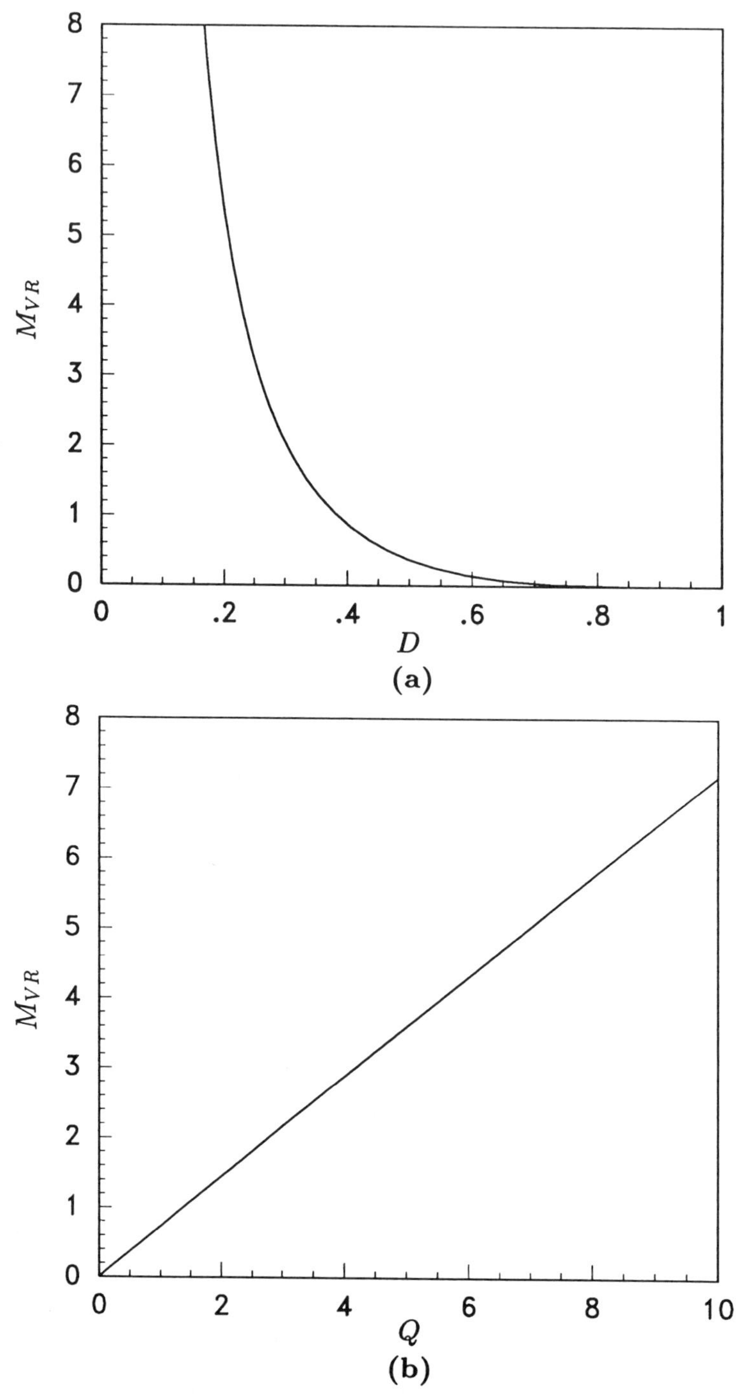

Figure 4.14: Voltage transfer function M_{VR} versus diode ON duty ratio D and normalized load resistance Q. (a) M_{VR} as a function of D. (b) M_{VR} as a function of Q.

Equating the right-hand sides of (4.60) and (4.62) produces a relationship between the initial phase ϕ and the duty cycle D.

$$tan\phi = \frac{[4+4\pi^2 D(1-D)]cos2\pi D + 2\pi(2D-1)sin2\pi D + sin^2 2\pi D - 4}{2\pi(1-D) - [1+4\pi^2 D(1-D)]sin2\pi D - [2\pi(1-D) - sin2\pi D]cos2\pi D}. \tag{4.63}$$

Figure 4.15 shows a plot of ϕ as a function of D.

Consider now the time interval $2\pi(1-D) < \omega_o t \leq 2\pi$ during which the diode is ON. The equivalent circuit of the rectifier for this interval is shown in Fig. 4.13(c). From (4.57), the inductor current at $\omega_o t = 2\pi(1-D)$ is

$$i_L[2\pi(1-D)] = 2QI_O\left[\pi D - \frac{\sqrt{2}sin\pi D sin(\phi - \pi D)}{M_{VR}}\right]. \tag{4.64}$$

Referring to Fig. 4.13(c) and using (4.52) and (4.54), one obtains the voltage across the inductor

$$v_L = v - V_O = V_m sin(\omega_o t + \phi) - V_O = V_O\left[\frac{\sqrt{2}sin(\omega_o t + \phi)}{M_{VR}} - 1\right], \tag{4.65}$$

and, using (4.64), the current through the inductor and the diode

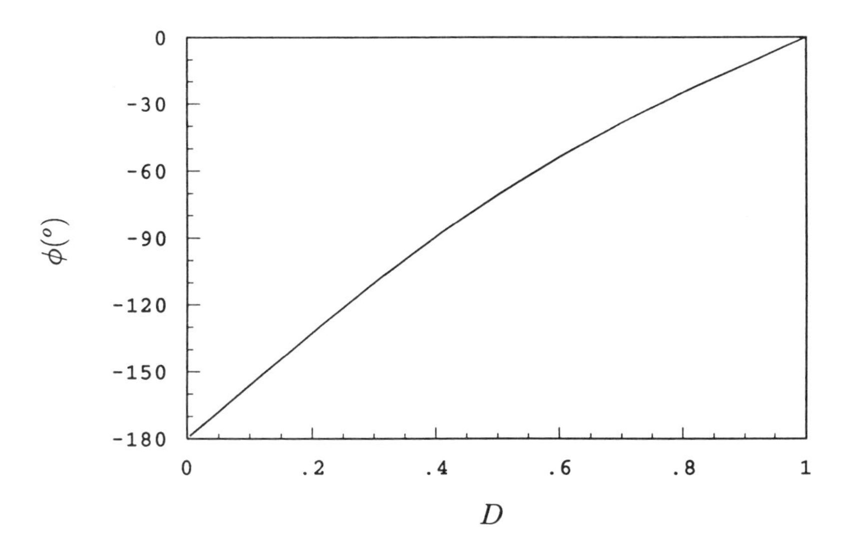

Figure 4.15: Initial phase of the input voltage ϕ as a function of the diode ON duty cycle D.

$$
\begin{aligned}
i_D = i_L &= \frac{1}{\omega_o L}\int_{2\pi(1-D)}^{\omega_o t} v_L d(\omega_o t) + i_L[2\pi(1-D)] \\
&= \frac{V_O}{\omega_o L}\left[2\pi(1-D) - \omega_o t - \sqrt{2}\frac{cos(\omega_o t+\phi) - cos(\phi - 2\pi D)}{M_{VR}}\right] \\
&\quad + i_L[2\pi(1-D)] \qquad (4.66) \\
&= I_O Q\left\{\left[2\pi(1-D) - \omega_o t - \sqrt{2}\frac{cos(\omega_o t+\phi) - cos(\phi - 2\pi D)}{M_{VR}}\right]\right. \\
&\quad \left. + \frac{i_L[2\pi(1-D)]}{QI_O}\right\} \\
&= \frac{\sqrt{2}I_O Q}{M_{VR}}\left[cos\phi - cos(\omega_o t+\phi) - \frac{M_{VR}}{\sqrt{2}}(\omega_o t - 2\pi)\right].
\end{aligned}
$$

Because the average current through the capacitor C for the steady-state operation is 0, the dc output current I_O is equal to the average value of the diode current. Thus,

$$
\begin{aligned}
I_O &= \frac{1}{2\pi}\int_{2\pi(1-D)}^{2\pi} i_D d(\omega_o t) \\
&= \frac{I_O Q}{\sqrt{2}\pi M_{VR}}[(cos2\pi D - 1)sin\phi + (2\pi D - sin2\pi D)cos\phi \\
&\quad + \sqrt{2}\pi^2 D^2 M_{VR}]. \qquad (4.67)
\end{aligned}
$$

Rearrangement of this expression yields the relationship among Q, D, and ϕ

$$
Q = \frac{M_{VR}}{\frac{(cos2\pi D - 1)sin\phi}{\sqrt{2}\pi} + \sqrt{2}\left(D - \frac{sin2\pi D}{2\pi}\right)cos\phi + \pi D^2 M_{VR}}, \qquad (4.68)
$$

where D and ϕ are related by (4.63). Variations of D with Q are illustrated in Fig. 4.16.

The waveforms of the diode current i_D and the diode voltage v_D normalized with respect to the dc output current I_O and the dc output voltage V_O are, respectively,

$$
\frac{i_D}{I_O} = \begin{cases} 0, & \text{for } 0 < \omega_o t \le 2\pi(1-D) \\ \frac{\sqrt{2}Q}{M_{VR}}[cos\phi - cos(\omega_o t+\phi) - \frac{M_{VR}}{\sqrt{2}}(\omega_o t - 2\pi)], & \text{for } 2\pi(1-D) < \omega_o t \le 2\pi. \end{cases}
$$

$$
\frac{v_D}{V_O} = \begin{cases} \frac{1}{\sqrt{2}M_{VR}}[cos\phi sin\omega_o t - \omega_o t cos(\omega_o t+\phi)] + cos\omega_o t - 1, & \text{for } 0 < \omega_o t \le 2\pi(1-D) \\ 0, & \text{for } 2\pi(1-D) < \omega_o t \le 2\pi. \end{cases} \qquad (4.69)
$$

These waveforms are shown in Fig. 4.13(d).

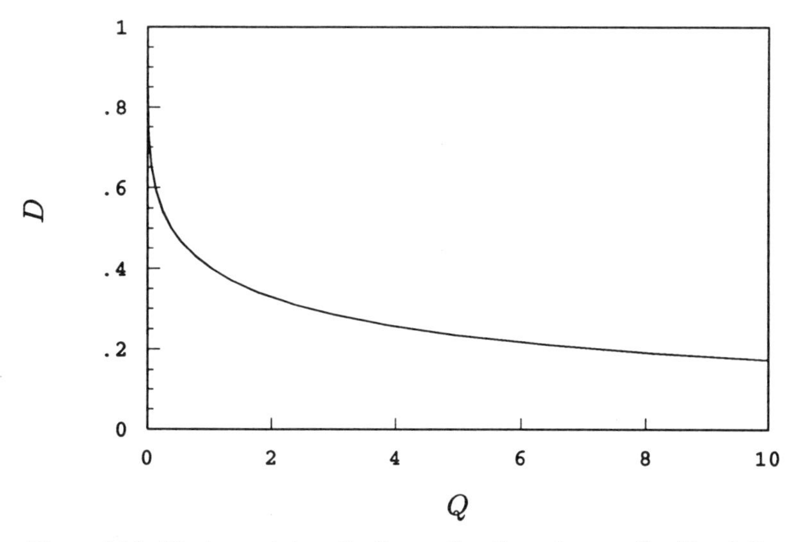

Figure 4.16: Diode ON duty ratio D as a function of normalized load Q.

4.3.4 Input Impedance

The input power of the rectifier contains only the power of the fundamental component because the input voltage is sinusoidal. Consequently, the input impedance of the rectifier Z_i can be represented as a parallel combination of the input resistance R_i and the input inductance L_i, both defined at the operating frequency f. An equivalent circuit of the rectifier is shown in Fig. 4.17.

Referring to Fig. 4.17, the fundamental component of the rectifier's input current can be written as

$$i_{R1} = i_{Ri} + i_{Li} = I_{Rim} sin(\omega_o t + \phi) - I_{Lim} cos(\omega_o t + \phi), \tag{4.70}$$

where I_{Rim} and I_{Lim} are the amplitudes of currents through the input resistance R_i and the input inductance L_i, respectively. The input resistance is

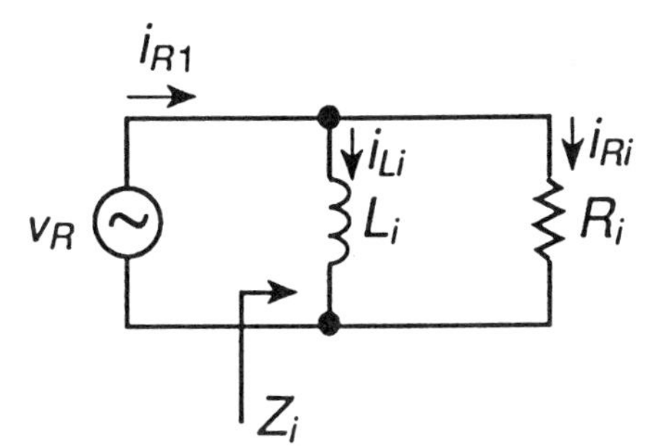

Figure 4.17: Equivalent circuit of the input impedance of the rectifier at the operating frequency f.

defined as $R_i = V_m/I_{Rim}$ and the input reactance is defined as $X_{Li} = \omega_o L_i = V_m/I_{Lim}$. If the power loss is neglected, the dc output power

$$P_O = \frac{V_O^2}{R_L} \tag{4.71}$$

equals the ac input power

$$P_i = \frac{V_m^2}{2R_i}. \tag{4.72}$$

Hence, using (4.52) the input resistance R_i at the fundamental frequency f normalized with respect to the load resistance R_L is

$$\frac{R_i}{R_L} = \frac{1}{M_{VR}^2}. \tag{4.73}$$

From (4.51) and (4.74), one obtains the input resistance R_i normalized with respect to $Z_o = \omega_o L$

$$\frac{R_i}{Z_o} = \frac{QR_i}{R_L} = \frac{Q}{M_{VR}^2}. \tag{4.74}$$

Figure 4.18 shows plots of R_i/R_L and R_i/Z_o as functions of D and Q.

Substitution of (4.57) and (4.66) into the Fourier series formula yields the amplitude of the current through the input inductance

$$\begin{aligned} I_{Lim} &= -\frac{1}{\pi}\int_0^{2\pi} i_L cos(\omega_o t + \phi) d(\omega_o t) \\ &= \frac{\sqrt{2} I_O Q}{\pi M_{VR}} \left\{ \frac{\pi(5D-1)}{4} - \left[\frac{3sin4\pi D}{16} - \frac{\pi(1-D)(1+sin^2 2\pi D)}{2}\right] sin^2\phi \right. \\ &\quad - \left[sin2\pi D - \frac{5sin4\pi D}{16} - \frac{\pi(1-D)cos^2 2\pi D}{2}\right] cos^2\phi \\ &\quad + \left[(1-cos2\pi D)cos2\pi D + \frac{\pi(1-D)sin4\pi D}{2}\right] sin\phi cos\phi \\ &\quad + \frac{M_{VR}}{\sqrt{2}}\left[2\pi D cos2\pi D - \pi(1-D) - \left(1 + \frac{cos2\pi D}{2}\right) sin2\pi D\right] sin\phi \\ &\quad \left. + \frac{M_{VR}}{\sqrt{2}}\left[1 - cos2\pi D + \left(\frac{sin2\pi D}{2} - 2\pi D\right) sin2\pi D\right] cos\phi \right\}. \end{aligned} \tag{4.75}$$

Because

$$I_{Lim} = \frac{V_m}{\omega_o L_i} = \frac{\sqrt{2} V_O}{\omega_o L_i M_{VR}} = \frac{\sqrt{2} I_O R_L}{\omega_o L_i M_{VR}} = \frac{\sqrt{2} I_O Q L}{M_{VR} L_i}, \tag{4.76}$$

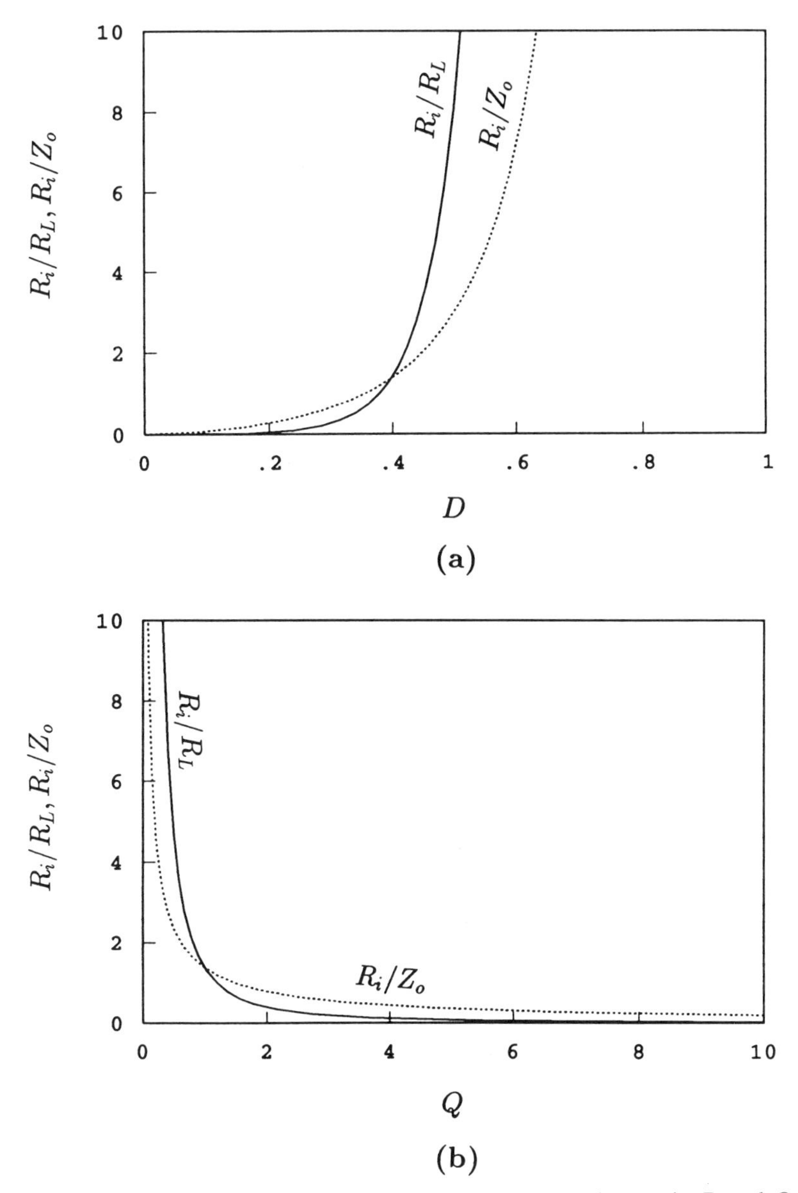

Figure 4.18: R_i/R_L and R_i/Z_o as functions of the diode ON duty ratio D and Q. (a) R_i/R_L and R_i/Z_o versus D. (b) R_i/R_L and R_i/Z_o versus Q.

(4.76) leads to the input inductance at the fundamental frequency

$$\frac{L_i}{L} = \frac{\pi}{\frac{\pi(5D-1)}{4} - asin^2\phi - bcos^2\phi + csin\phi cos\phi + \frac{M_{VR}}{\sqrt{2}}(dsin\phi + ecos\phi)}, \tag{4.77}$$

where

$$a = \left[\frac{3cos2\pi D}{8} - \frac{\pi(1-D)sin2\pi D}{2}\right] sin2\pi D - \frac{\pi(1-D)}{2}, \quad (4.78)$$

$$b = sin2\pi D - \frac{5sin4\pi D}{16} - \frac{\pi(1-D)cos^2 2\pi D}{2}, \quad (4.79)$$

$$c = (1 - cos2\pi D)cos2\pi D + \frac{\pi(1-D)sin4\pi D}{2}, \quad (4.80)$$

$$d = 2\pi Dcos2\pi D - \pi(1-D), - \left(1 + \frac{cos2\pi D}{2}\right) sin2\pi D \quad (4.81)$$

and

$$e = 1 - cos2\pi D + \left(\frac{sin2\pi D}{2} - 2\pi D\right) sin2\pi D. \quad (4.82)$$

Figure 4.19(a) and (b) illustrates L_i/L as functions of D and Q, respectively.

4.3.5 Diode Stresses

The peak value of the diode current I_{DM} occurs at the end of the ON interval $\omega_o t = 2\pi(1-D)$ for $D < 0.34$ (i.e., $M_{VR} > \sqrt{2}$) and at the maximum value of the waveform given by (4.70) for $D \geq 0.34$ ($M_{VR} \leq \sqrt{2}$). The peak diode current is

$$\frac{I_{DM}}{I_O} = \begin{cases} 2Q[\pi D + \frac{\sqrt{2}sin\pi Dsin(\pi D - \phi)}{M_{VR}}], & \text{for } 0 < D < 0.34 \\ \frac{\sqrt{2}Q}{M_{VR}}[cos\phi + cos(arcsin\frac{M_{VR}}{\sqrt{2}}) \\ + \frac{M_{VR}}{\sqrt{2}}(\pi + \phi + arcsin\frac{M_{VR}}{\sqrt{2}})], & \text{for } 0.34 \leq D \leq 1. \end{cases} \quad (4.83)$$

Figure 4.20(a) and (b) shows plots of I_{DM}/I_O versus D and versus Q, respectively. It can be seen that I_{DM}/I_O increases from 2 to ∞ as Q increases from 0 to ∞.

To determine the maximum of the diode current waveform, (4.70) is differentiated with respect to $\omega_o t$ and the result is set to be equal to zero. This maximum occurs when $sin(\omega_o t_{imax} + \phi) = M_{VR}/\sqrt{2}$. Hence,

$$\omega t_{imax} = \pi - \phi - arcsin\frac{M_{VR}}{\sqrt{2}}, \quad \text{for} \quad M_{VR} \leq \sqrt{2}. \quad (4.84)$$

For $M_{VR} > \sqrt{2}$, the peak value of the diode current occurs at the end of the ON interval

$$\omega_o t_{imax} = 2\pi(1-D). \quad (4.85)$$

For $D < 0.34$, $M_{VR} > \sqrt{2}$ and for $D \geq 0.34$, $M_{VR} \leq \sqrt{2}$. Substitution of (4.85) or (4.86) into (4.70) yields (4.84).

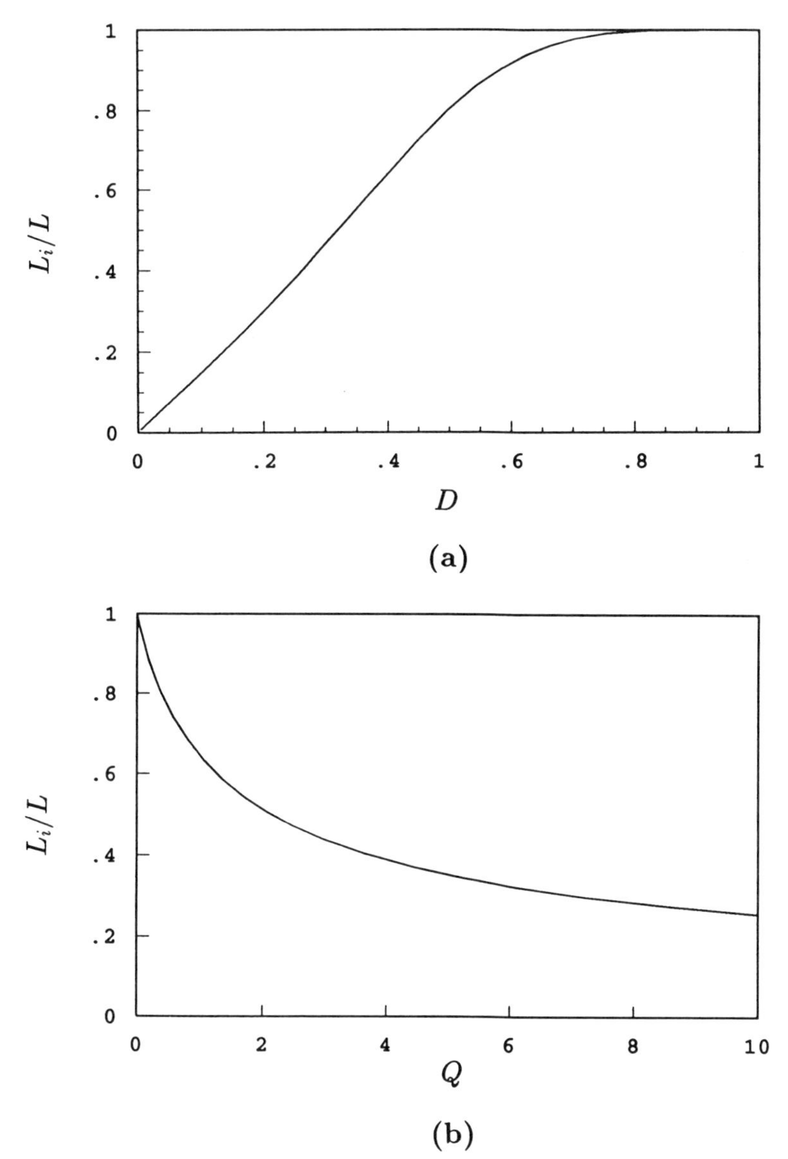

Figure 4.19: Ratio L_i/L versus the diode ON duty ratio D and the normalized load resistance Q. (a) L_i/L versus D. (b) L_i/L versus Q.

To calculate the maximum value of the diode reverse voltage V_{DM}, (4.70) is differentiated with respect to $\omega_o t$ and the result is equated to zero, resulting in a transcendental equation

$$tan(\omega_o t_{vmax}) = -\frac{\omega_o t_{vmax}}{\frac{\omega_o t_{vmax}}{tan\phi} - \frac{\sqrt{2}M_{VR}}{sin\phi} + 1}. \tag{4.86}$$

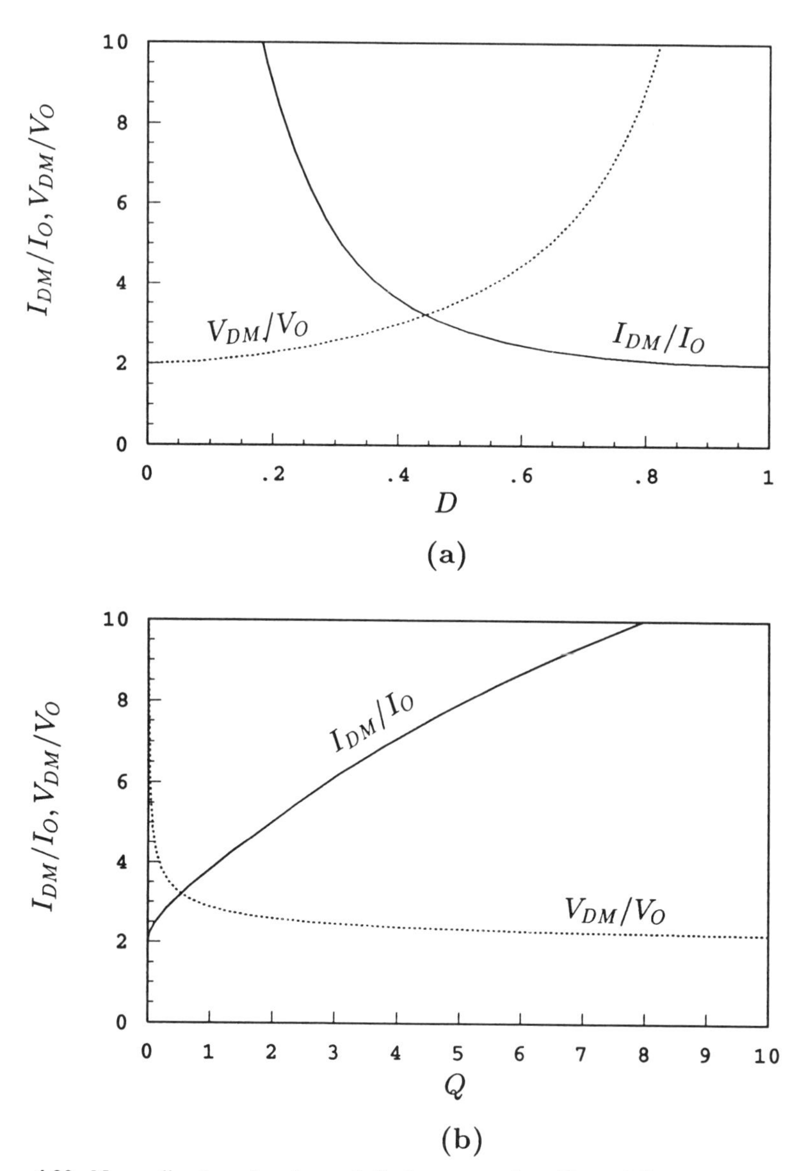

Figure 4.20: Normalized peak values of diode current I_{DM}/I_O and diode reverse voltage V_{DM}/V_O as functions of D and Q. (a) I_{DM}/I_O and V_{DM}/V_O versus D. (b) I_{DM}/I_O and V_{DM}/V_O versus Q.

This equation was solved using the Newton–Raphson method. Substitution of the solution into (4.70) gives V_{DM}.

The peak values of the diode reverse voltage V_{DM} can be found by differentiating the diode voltage waveform in (4.70) and setting the result equal to zero. The results can be obtained only in numerical form. Figure 4.20(a) and

(b) illustrates V_{DM}/V_O against D and Q, respectively. As D increases from 0 to 1, or Q decreases from ∞ to 0, V_{DM}/V_O increases from 2 to ∞.

The normalized output-power capability is

$$c_p = \frac{P_O}{I_{DM}V_{DM}} = \frac{I_O V_O}{I_{DM}V_{DM}}. \tag{4.87}$$

This factor has been computed and plotted versus D and Q in Fig. 4.6(a) and (b), respectively. The maximum value of c_p occurs at $D = 0.5$ and equals 0.0999. Factor c_p first increases with Q, reaches its maximum at $Q = 0.3884$, and then decreases with Q.

The numerical values of the rectifier parameters at selected values of D are given in Table 4.2.

4.3.6 Parameters for *D* = 0.5

The rectifier has the following parameters for $D = 0.5$:

$$tan\phi = -\frac{\pi^2 + 8}{2\pi} \tag{4.88}$$

$$sin\phi = -\frac{\pi^2 + 8}{\sqrt{(\pi^2 + 8)^2 + 4\pi^2}} \tag{4.89}$$

$$cos\phi = \frac{2\pi}{\sqrt{(\pi^2 + 8)^2 + 4\pi^2}} \tag{4.90}$$

$$\phi = -arctan\frac{\pi^2 + 8}{2\pi} = -1.2327 \text{ rad} = -70.63^\circ \tag{4.91}$$

$$Q = \frac{R_L}{\omega_o L} = \omega_o C R_L = \frac{4\pi^3}{\pi^4 + 16\pi^2 + 64} \approx 0.3884 \tag{4.92}$$

$$M_{VR} = \frac{V_O}{V_{rms}} = \frac{\pi^2}{\sqrt{2}\sqrt{(\pi^2 + 8)^2 + 4\pi^2}} \approx 0.3684 \tag{4.93}$$

$$\frac{R_i}{R_L} = \frac{1}{M_{VR}^2} = \frac{2[(\pi^2 + 8)^2 + 4\pi^2]}{\pi^4} \approx 7.3669 \tag{4.94}$$

$$\frac{R_i}{\omega_o L} = \frac{Q}{M_{VR}^2} = \frac{8(\pi^4 + 20\pi^2 + 64)}{\pi(\pi^4 + 16\pi^2 + 64)} \approx 2.8613 \tag{4.95}$$

$$\frac{L_i}{L} = \frac{8(\pi^4 + 20\pi^2 + 64)}{11\pi^4 + 196\pi^2 + 576} \approx 0.8014 \tag{4.96}$$

$$\frac{I_{DM}}{I_O} = \frac{143\pi^3}{5(\pi^4 + 16\pi^2 + 64)} \approx 2.777 \tag{4.97}$$

$$\frac{V_{DM}}{V_O} \approx 3.601. \tag{4.98}$$

Table 4.2. Parameters of Class E resonant low *dv/dt* rectifier

D	ϕ (°)	Q	I_{DM}/I_O	V_{DM}/V_O	R_i/R_L	$R_i/\omega_o L$	L_i/L	M_{VR}	c_p
0	−180.00	∞	∞	2	0	0	0	∞	0
0.1	−156.02	31.4243	19.8718	2.0911	0.0020	0.0623	0.1471	22.4662	0.0240
0.2	−132.42	7.2214	9.5261	2.3125	0.0351	0.2532	0.2980	5.3403	0.0454
0.25	−121.00	4.2424	7.3015	2.4550	0.0964	0.4091	0.3787	3.2205	0.0558
0.3	−109.97	2.6117	5.7288	2.6184	0.2383	0.6224	0.4633	2.0485	0.0666
0.4	−89.30	1.0283	3.6355	3.0266	1.2976	1.3343	0.6394	0.8779	0.0909
0.5	−70.63	0.3884	2.7777	3.6012	7.3669	2.8613	0.8014	0.3684	0.0999
0.6	−53.84	0.1301	2.4062	4.4726	52.219	6.7915	0.9174	0.1384	0.0922
0.7	−38.68	0.0349	2.2065	5.9399	580.22	20.2764	0.9769	0.0415	0.0763
0.75	−31.62	0.0157	2.1397	7.1198	2584.6	40.7502	0.9901	0.0197	0.0656
0.8	−24.88	0.0061	2.0881	8.8936	15813.5	96.5825	0.9966	0.0080	0.0538
0.9	−12.12	0.0004	2.0219	17.7783	4174485	1476.2380	0.9999	0.0005	0.0278
1	0	0	2	∞	∞	∞	1	0	0

4.3.7 Design Example

Example 4.2

Design a transformerless version of the Class E low dv/dt resonant rectifier shown in Fig. 4.13(a) to meet the following specifications: $V_O = 5$ V, $I_O = 0$ to 0.1 A, and $f = 1$ MHz. Estimate the minimum value of the filter capacitance C_f needed to keep the output voltage ripple below 5% of the output voltage if the ESR of the filter capacitor is $r_C = 0.025\ \Omega$.

Solution: The dc load resistance R_L varies from its minimum value

$$R_{Lmin} = \frac{V_O}{I_{Omax}} = \frac{5}{0.1} = 50\ \Omega \tag{4.99}$$

to ∞. The dc output power varies from its maximum value

$$P_{Omax} = V_O I_{Omax} = 5 \times 0.1 = 0.5\ \text{W} \tag{4.100}$$

to 0. Let us assume $D = 0.5$ at full load resistance $R_{Lmin} = 50\ \Omega$ because the power-output capability takes on the maximum value at $D = 0.5$. Hence, from (4.94)

$$L = \frac{R_{Lmin}}{\omega_o Q} = \frac{50}{0.3884 \times 2 \times \pi \times 10^6} = 20.5\ \mu\text{H} \tag{4.101}$$

and

$$C = \frac{Q}{\omega_o R_{Lmin}} = \frac{0.3884}{2 \times \pi \times 10^6 \times 50} = 1.236\ \text{nF}. \tag{4.102}$$

From (4.95), the amplitude of the input voltage can be found as

$$V_m = \frac{\sqrt{2} V_O}{M_{VR}} = \frac{\sqrt{2} \times 5}{0.3684} = 19.19\ \text{V}. \tag{4.103}$$

The maximum values of the diode current and voltage occur at full load resistance R_{Lmin}. From (4.99) and (4.99), one can calculate

$$I_{DM} = 2.777 I_{Omax} = 2.777 \times 0.1 = 277\ \text{mA} \tag{4.104}$$

and

$$V_{DM} = 3.601 V_O = 3.601 \times 5 = 18\ \text{V}. \tag{4.105}$$

The current through the filter capacitor is the ac component of the current through the inductor i_L. It can be seen in Fig. 4.13(d) that for $D = 0.5$ this ac component can be approximated by a sinusoidal current with an amplitude $I_{Cfm} = I_{DM} - I_O$. For the sinusoidal approximation, the peak-to-peak ripple voltage of the filter capacitor is

$$V_r = 2I_{Cfm}\sqrt{\frac{1}{\omega^2 C_f^2} + r_C^2}, \tag{4.106}$$

from which the minimum value of the filter capacitance is obtained as

$$C_f = \frac{1}{\omega}\sqrt{\frac{1}{\frac{V_r^2}{4(I_{DM}-I_O)^2} - r_C^2}} = \frac{1}{2\pi \times 10^6}\sqrt{\frac{1}{\frac{(0.05\times 5)^2}{4(0.277-0.1)^2} - 0.025^2}} = 225.5 \text{ nF}. \tag{4.107}$$

Let $C_f = 300$ nF and the voltage rating greater than 10 V.

4.4 SUMMARY

- The diode junction capacitance, the winding capacitance of the filter inductor, and the winding capacitance of the transformer (if any), which couples the rectifier circuit to the ac source, are absorbed into the capacitance connected in parallel with the diode.
- In Class E low dv/dt rectifiers, the diode turns on at low dv/dt and very high di/dt, and turns off at zero dv/dt and low $\mid di/dt \mid$, yielding high efficiency and low switching noise.
- The waveform of the diode current has a step change at turn-on.
- The absolute value of the derivative of the diode current at turn-off is limited by the external circuit, reducing the reverse-recovery current and noise. For these reasons, the rectifier is suitable for high-frequency operation.
- The diode ON duty cycle D decreases from 1 to 0 as $\omega C R_L$ increases from 0 to ∞.
- The value of D is independent of the output filter components.
- The normalized diode peak current I_{DM}/I_O increases with $\omega C R_L$, but the absolute value of I_{DM} decreases with R_L at a constant dc output voltage V_O.
- The normalized diode peak voltage V_{DM}/V_O decreases with $\omega C R_L$.
- The maximum value of the power-output capability c_p occurs at $D = 0.5$.
- The ac-to-dc current transfer function of the Class E low dv/dt rectifier with a parallel capacitor decreases with $\omega C R_L$.
- The diode current and voltage waveforms in the Class E rectifier are time-reversed images of the corresponding transistor waveforms in a Class E inverter for optimum operation.
- Parameters of the rectifier depend on the values of $\omega C R_L$.
- The diode ON duty ratio D decreases with $\omega C R_L$.

- At fixed values of the operating frequency $f = \omega/(2\pi)$ and the capacitance C, D decreases from its maximum value D_{max} to 0 as the load resistance R_L is increased from R_{Lmin} to ∞.
- The capacitance C should be chosen in such a way that the desired value of D_{max} occurs at R_{Lmin}.
- In the Class E resonant rectifier, the diode ON duty cycle D can be in the range from 0 to 1.
- Unlike in peak rectifiers, D is independent of the filter capacitance C_f.
- The capacitance C_f is determined by the specified level of the ripple voltage.
- The decoupling of the diode ON duty cycle D and the ripple voltage is an important advantage of the Class E resonant rectifier because the diode may operate with much lower peak currents than in the peak rectifier.
- The diode current contains a relatively low amount of harmonics, reducing EMI levels.
- The input impedance of the Class rectifiers has both resistive and reactive components, which are dependent on the load resistance.
- The Class E low dv/dt rectifier with a parallel capacitor acts as an impedance inverter.

4.5 REFERENCES

1. R. J. Gutmann, "Application of RF circuit design principles to distributed power converters," *IEEE Trans. Ind. Electron. Contr. Instrum.*, vol. IECI-27, pp. 156–164, August 1980.
2. W. C. Bowman, F. M. Magalhaes, W. B. Suiter, and N. G. Ziesse, "Resonant rectifier circuits," U.S. Patent no. 4,685,041, August 4, 1987.
3. W. A. Nitz, W. C. Bowman, F. T. Dickens, F. M. Magalhaes, W. Strauss, W. B. Suiter, and N. G. Ziesse, "A new family of resonant rectifier circuits for high frequency dc-dc converter applications," in *Proc. IEEE Applied Power Electronics Conf.*, New Orleans, LA, February 1–5, 1988, pp. 12–22.
4. M. K. Kazimierczuk and J. Jóźwik, "Class E resonant rectifiers," in *Proc. 31st Midwest Symp. Circuits and Systems*, St. Louis, MO, Aug. 10–12, 1988, pp. 138–141.
5. M. K. Kazimierczuk and J. Jóźwik, "Class E zero-voltage-switching rectifier with a series capacitor," *IEEE Trans. Circuits Syst.*, vol. CAS–36, pp. 926–928, June 1989.
6. M. K. Kazimierczuk, "Class E low dv_D/dt rectifier," *Proc. Inst. Elec. Eng., Pt. B, Electric Power Appl.*, vol. 136, pp. 257–262, Nov. 1989.

7. M. K. Kazimierczuk and J. Jóźwik, "Class E zero-voltage-switching and zero-current-switching rectifiers," *IEEE Trans. Circuits Syst.*, vol. CAS–37, pp. 436–444, Mar. 1990.
8. M. K. Kazimierczuk, "Analysis of Class E zero-voltage-switching rectifier," *IEEE Trans. Circuits Syst.*, vol. CAS–37, pp. 747–755, June 1990.
9. M. K. Kazimierczuk and K. Puczko, "Class E low dv/dt synchronous rectifier with controlled duty ratio and output voltage," *IEEE Trans. Circuits Syst.*, vol. CAS–38, pp. 1165–1172, Oct. 1991.
10. M. K. Kazimierczuk and W. Szaraniec, "Analysis of a Class E rectifier with a series capacitor," *Proc. Inst. Elec. Eng., Pt. G, Circuits, Systems and Devices,* vol. 139, no. 3, pp. 269–276, June 1992.
11. A. Ivascu, M. K. Kazimierczuk, and S. Birca-Galateanu, " Class E resonant low dv/dt rectifier," *IEEE Trans. Circuits Syst.*, vol. CAS–39, pp. 604–613, Aug. 1992.
12. A. Reatti, M. K. Kazimierczuk, and R. Redl, "Class E full-wave low dv/dt rectifier," *IEEE Trans. Circuits Syst.,* vol. CAS–40, pp. 73–85, Jan. 1993.
13. A. Reatti and M. K. Kazimierczuk, "Efficiency of the transformer version of Class E half-wave low dv_D/dt rectifier," *IEEE International Symposium on Circuits and Systems*, Chicago, IL, May 3–6, 1993, vol. 4 pp. 2331–2334.
14. A. Reatti and M. K. Kazimierczuk, "Comparison of efficiency of the Class E and Class D rectifiers," *36th IEEE Midwest Symposium on Circuits and Systems*, Detroit, MI, August 16–18, 1993, pp. 871–874.

4.6 REVIEW QUESTIONS

4.1 List the advantages of Class E low dv/dt rectifiers.

4.2 Sketch the waveforms of the diode current and voltage in the Class E low dv/dt rectifier.

4.3 Sketch the waveforms of the diode current and voltage in the Class E resonant low dv/dt rectifier.

4.4 What is the range of the diode on-duty cycle D in Class E low dv/dt rectifiers?

4.5 Does the diode on-duty cycle D depend on the filter components?

4.6 Does the diode on-duty cycle D depend on the load resistance?

4.7 Does the diode junction capacitance adversely affect the Class E low dv/dt operation?

4.8 Does the diode lead inductance adversely affect the Class E low dv/dt operation?

4.9 How do the input resistance and reactance of the Class E low dv/dt rectifiers depend on the load resistance?

4.7 PROBLEMS

4.1 Derive Equation (4.14).

4.2 Design a Class E low dv/dt rectifier of Fig. 4.1(a) with the following specifications: $V_O = 10$ V, $I_O = 0$ to 2 A, and $f = 500$ kHz.

4.3 Design a transformerless version of the Class E rectifier of Fig. 4.13(a) to meet the following specifications: $V_O = 10$ V, $I_O = 0$ to 2 A, and $f = 500$ kHz.

CHAPTER 5

CLASS E LOW *di/dt* RECTIFIERS

5.1 INTRODUCTION

In this chapter we consider Class E low di/dt rectifiers [1]–[7]. They are counterparts of Class E zero-current-switching (ZCS) inverters, discussed in Chapter 14. Specifically, the diode current and voltage waveforms in the Class E low di/dt rectifiers are time-reversed images of the corresponding transistor current and voltage waveforms in Class E ZCS inverters. The advantage of Class E low dv/dt rectifiers is that the diode turns on and off at low di/dt and turns off at low dv/dt. First, the circuits of the rectifiers will be presented and the principle of operation will be described. Then, the mathematical description will be carried out and design equations will be derived. Finally, design examples will be given.

5.2 LOW *di/dt* RECTIFIER WITH A PARALLEL INDUCTOR

5.2.1 Circuit Description

The basic circuit of a Class E low di/dt rectifier with a parallel inductor [4], [5] is shown in Fig. 5.1(a). It consists of a diode, an inductor L, and a filter capacitor C_f. The inductance L is relatively small so that the ac component of its current is large compared to the dc component of i_L (which is equal to $-I_O$). The capacitor C_f and the load resistance R_L form a first-order low-pass output filter, which reduces the ripple voltage below a specified level. Assuming that the ripple voltage is much lower than the dc component of the output voltage V_O, the R_L-C_f filter may be modeled by a dc voltage source

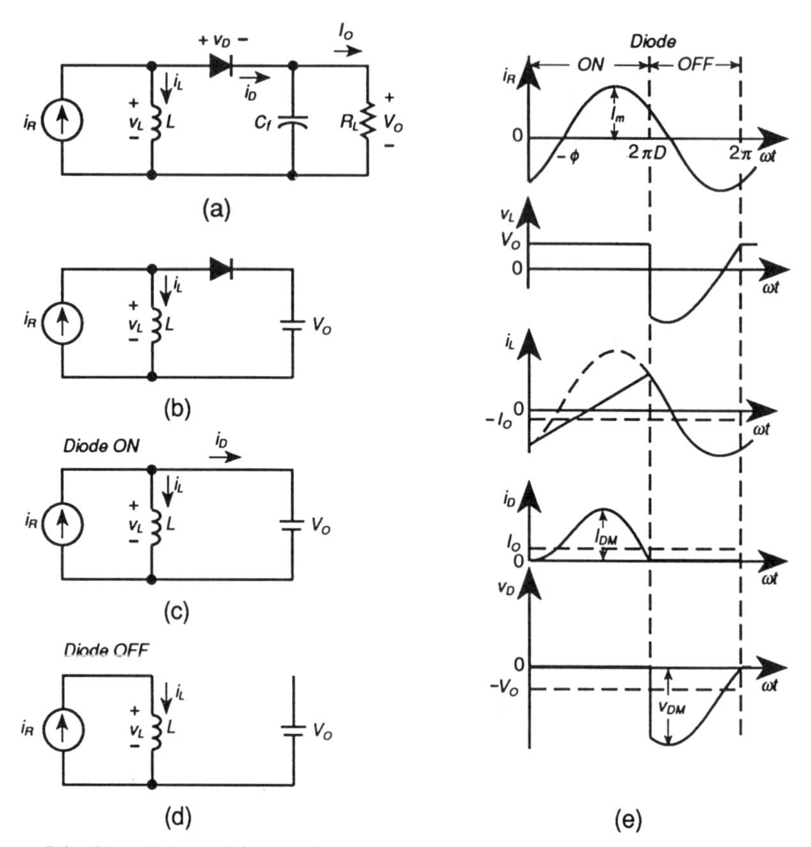

Figure 5.1: Class E low di/dt rectifier with a parallel inductor. (a) Circuit. (b) Model. (c) Model for the interval when the diode is on. (d) Model for the time interval when the diode is off. (e) Current and voltage waveforms.

V_O, as shown in Fig. 5.1(b). The rectifier is driven by a sinusoidal current source i_R. In practice, in resonant dc-to-dc converters the rectifier is driven by an inverter which contains a series-resonant circuit at its output (for example, a Class E or Class D inverter). The loaded quality factor Q_L of this resonant circuit at a full load resistance R_{Lmin} of the rectifier is usually low (e.g., $Q_L = 5$) in order to obtain high efficiency of the resonant circuit. Hence, the input current i_R of the rectifier may differ from an ideal sine wave. The analysis of such a case is very difficult to carry out and, therefore, an ideal sinusoidal input current will be assumed.

The current and voltage waveforms, explaining the circuit operation, are shown in Fig. 5.1(e). The input current i_R is a sine wave. The diode is ON during the interval $0 < \omega t < 2\pi D$, where D is the diode ON duty cycle. The model of the rectifier for this time interval is shown in Fig. 5.1(c). The voltage across the inductor L is equal to V_O. Therefore, the inductor current $i_L = (V_O/L)t + i_L(0)$ increases linearly. The diode current i_D is the difference between the input current i_R and the inductor current i_L. When the diode current reaches 0, the diode turns off. The derivative of the diode current di_D/dt is 0 at turn-on and its absolute value is relatively small at turn-off. For this reason, switching losses and switching noise are reduced at both transitions and the reverse-recovery effect at turn-off is minimized.

The diode is off during the time interval $2\pi D \leq \omega t \leq 2\pi$. The model of the rectifier for this time interval is shown in Fig. 5.1(d). The inductor current i_L is equal to the input current i_R. The voltage v_L across the inductor L is, therefore, part of a sine wave. The diode reverse voltage v_D is equal to the difference between V_O and v_L. When the diode reverse voltage v_D reaches zero (or more exactly -0.7 V for a silicon *pn* junction diode), the diode turns on. The diode reverse voltage v_D has a step change at turn-off and a low absolute value of the derivative dv_D/dt at turn-on. The switching loss and the switching noise at turn-on are reduced to a negligible level. The average value of the diode current i_D is equal to the dc output current I_O, and the average value of the diode reverse voltage v_D is equal to the dc output voltage V_O.

The diode ON duty cycle D depends on the load resistance R_L, the inductance L, and the operating frequency f. At fixed values of L and f, D decreases from its maximum value D_{max} to 0 as R_L increases from a full load R_{Lmin} to infinity. The value of D_{max} can be large, for example, 0.75. At given values of R_{Lmin} and f, the inductance L sets D_{max} and the capacitance C_f sets the ripple voltage. For instance, in voltage-driven peak rectifiers with a capacitive output filter, D_{max} is typically less than 0.05 in order to obtain a low ripple voltage. Therefore, the amount of harmonics generated by such narrow pulses of i_D is substantial, causing large conducted and radiated electromagnetic interference (EMI), which may adversely affect a supplied system or environment. Moreover, if the pulses of i_D are so narrow, the peak diode current I_{DM} is many times larger than I_O. Therefore, a diode with a large junction area is required.

The diode junction capacitance C_j is not included in the rectifier topology. This capacitance and the inductance L form a parasitic resonant circuit when the diode is off. If C_j were zero, the diode voltage would have a step change ΔV when the diode turns off, as shown in Fig. 5.1(e). Since C_j is not zero, the resonant circuit L–C_j will initiate parasitic oscillations with the voltage amplitude ΔV. If the damping factor of oscillations is low, the overshoot of the diode reverse voltage v_D may be close to 100%, that is, v_D may rise to almost $2\Delta V$. This may require a diode with a higher breakdown voltage. Moreover, if the oscillations will decay before the diode turns on, the turn-off power loss is $P_{turn-off} = (1/2)fC_j(\Delta V)^2$, assuming a linear capacitance C_j.

5.2.2 Assumptions

The analysis of the rectifier shown in Fig. 5.1(a) is carried out under the following assumptions:

1) The diode is an ideal device with zero threshold voltage, zero on resistance, zero junction capacitance, and infinite off resistance.
2) The output filter capacitance C_f is large enough so that the ripple of the output voltage can be neglected.
3) The rectifier is driven by an ideal sinusoidal current source.

5.2.3 Component Values

The basic equations for the rectifier model shown in Fig. 5.1(b) are

$$i_D = i_R - i_L \tag{5.1}$$

and

$$v_D = v_L - V_O. \tag{5.2}$$

The input current of the rectifier is a sine wave, according to assumption 3),

$$i_R = I_m sin(\omega t + \phi), \tag{5.3}$$

where I_m and ϕ are the amplitude and the initial phase, respectively.

Figure 5.1(c) shows the model of the rectifier when the diode is on, that is, for $0 < \omega t < 2\pi D$. The diode current at turn-on is zero. Therefore, from (5.1) and (5.3) the initial current of the inductor L is $i_L(0) = i_R(0) = I_m sin\phi$. Since the voltage across the diode is zero, $v_L = V_O = I_O R_L$. Thus, the current through the inductor L is

$$i_L = \frac{1}{\omega L}\int_0^{\omega t} v_L d(\omega t) + i_L(0) = \frac{V_O \omega t}{\omega L} + I_m sin\phi. \tag{5.4}$$

Hence, from (5.1) the current through the diode can be expressed as

$$i_D = i_R - i_L = I_m[sin(\omega t + \phi) - sin\phi] - \frac{V_O \omega t}{\omega L}. \tag{5.5}$$

According to Fig. 5.1(e),

$$\frac{di_D}{d(\omega t)}|_{\omega t=0} = 0. \tag{5.6}$$

Thus,

$$I_m = \frac{V_O}{\omega L cos\phi} = \frac{R_L I_O}{\omega L cos\phi}. \tag{5.7}$$

Substitution of this relationship into (5.5) yields the normalized waveform of the diode current

$$\frac{i_D}{I_O} = \begin{cases} \frac{R_L}{\omega L}[sin\omega t - \omega t - (1 - cos\omega t)tan\phi], & \text{for } 0 < \omega t < 2\pi D \\ 0, & \text{for } 2\pi D \leq \omega t \leq 2\pi, \end{cases} \tag{5.8}$$

where $tan\phi$ and $R_L/\omega L$ are given by (5.9) and (5.14), respectively. According to Fig. 5.1(e), $i_D(2\pi D) = 0$. Hence, (5.8) gives the initial phase as a function of the diode ON duty cycle D

$$tan\phi = \frac{sin2\pi D - 2\pi D}{1 - cos2\pi D}. \tag{5.9}$$

Figure 5.2 shows the plot of ϕ versus D. As D increases from 0 to 1, ϕ decreases from 0 to $-90°$. For $D = 0.5$, (5.9) becomes

$$tan\phi = -\frac{\pi}{2}, \tag{5.10}$$

which gives $\phi = -arctan(\pi/2) = -1.004$ rad $= -57.52°$. From (5.10) and trigonometric relationships,

$$sin\phi = -\frac{\pi}{\sqrt{\pi^2 + 4}} \tag{5.11}$$

$$cos\phi = \frac{2}{\sqrt{\pi^2 + 4}}. \tag{5.12}$$

The average value of the diode current equals the dc output current. Thus, using (5.8)

$$I_O = \frac{1}{2\pi}\int_0^{2\pi D} i_D d(\omega t) = \frac{I_O R_L}{2\pi\omega L}[1 - 2\pi^2 D^2 - cos2\pi D + (sin2\pi D - 2\pi D)tan\phi]. \tag{5.13}$$

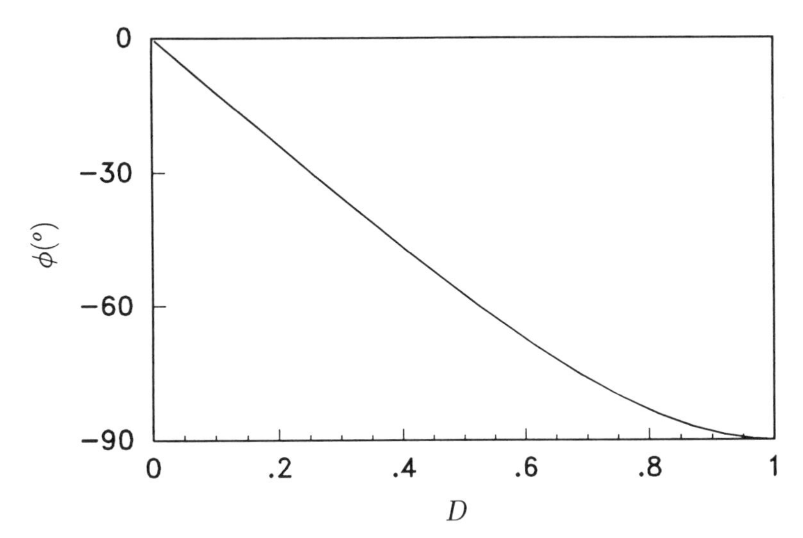

Figure 5.2: Initial phase ϕ of the input current i_R as a function of the diode ON duty cycle D.

Substituting (5.9) into this expression, one arrives at the relationship between the diode ON duty cycle D and the circuit components

$$\frac{R_L}{\omega L} = \frac{2\pi}{\left[1 - 2\pi^2 D^2 - cos2\pi D + \frac{(sin2\pi D - 2\pi D)^2}{1 - cos2\pi D}\right]}. \tag{5.14}$$

A plot of D versus $R_L/\omega L$ is shown in Fig. 5.3. It can be seen that D decreases from 1 to 0 as R_L increases from 0 to infinity at a constant value of ωL. For $D = 0.5$, $R_L/\omega L = \pi$.

Figure 5.1(d) shows the model of the rectifier when the diode is off, that is, for $2\pi D \leq \omega t \leq 2\pi$. The current through the inductor L is

$$i_L = i_R = I_m sin(\omega t + \phi). \tag{5.15}$$

Hence, using (5.7) one obtains the voltage across the inductor L

$$\begin{aligned} v_L &= \omega L \frac{di_L}{d(\omega t)} = \omega L I_m cos(\omega t + \phi) \\ &= V_O \frac{cos(\omega t + \phi)}{cos\phi} = V_O(cos\omega t - tan\phi sin\omega t). \end{aligned} \tag{5.16}$$

Substitution of this into (5.2) yields the diode voltage waveform

$$\frac{v_D}{V_O} = \begin{cases} 0, & \text{for } 0 \leq \omega t < 2\pi D, \\ cos\omega t - tan\phi sin\omega t - 1, & \text{for } 2\pi D \leq \omega t < 2\pi. \end{cases} \tag{5.17}$$

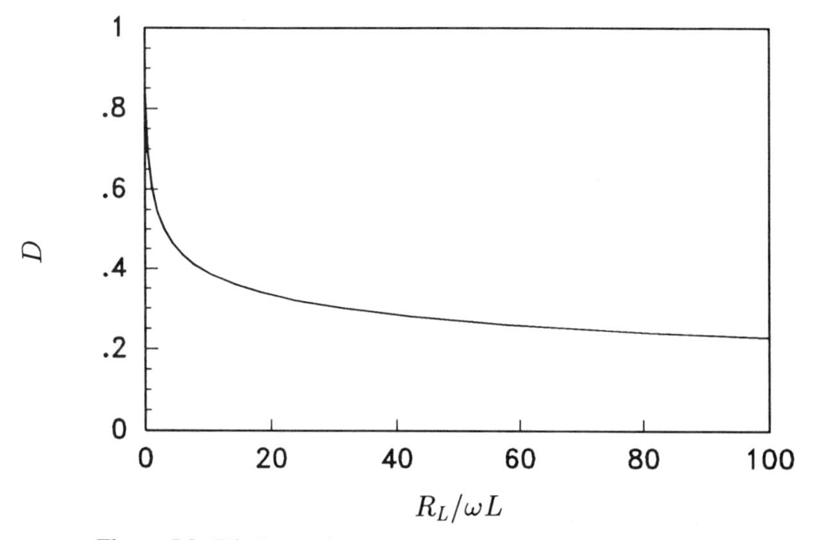

Figure 5.3: Diode ON duty cycle D as a function of $R_L/\omega L$.

5.2.4 Device Stresses

Differentiating (5.8) and setting the result equal to zero, one obtains the peak value of the diode current I_{DM}. It occurs at $\omega t_{im} = -2\phi$ and is given by

$$\frac{I_{DM}}{I_O} = \frac{2R_L}{\omega L}(\phi - tan\phi), \tag{5.18}$$

where $R_L/\omega L$ is given by (5.14) and ϕ is given by (5.9) and is expressed in radians. For $D = 0.5$, $I_{DM}/I_O = \pi^2 + 2\pi\phi = 3.562$. Figure 5.4(a) shows the plot of I_{DM}/I_O as a function of D, and Fig. 5.4(b) shows the plot of I_{DM}/I_O as a function of $R_L/\omega L$. As D decreases from 1 to 0, which happens when $R_L/\omega L$ increases from 0 to ∞, I_{DM}/I_O increases from 2 to ∞. Note, however, that I_O decreases with R_L at a fixed value of V_O, and therefore I_{DM} decreases with R_L.

The value of the peak diode current I_{DM} is also a peak-to-peak value of an ac current that flows through the filter capacitor C_f. Hence, the peak-to-peak ripple voltage on the output voltage can be conservatively estimated as

$$V_r = I_{DM}\left(\frac{1}{\omega C_f} + r_C\right), \tag{5.19}$$

where r_C is the ESR of the filter capacitor.

To determine the peak value of the diode reverse voltage V_{DM}, two cases should be considered. For $D \leq 0.72$, V_{DM} occurs at the maximum value of v_D. Differentiating (5.17) and setting the result equal to zero gives the time at which the maximum occurs; this time is $\omega t_{vm} = \pi - \phi$. For $D \geq 0.72$, V_{DM} occurs when the diode turns off, that is, $\omega t_{vm} = 2\pi D$. Hence, from (5.17)

$$\frac{V_{DM}}{V_O} = \begin{cases} \frac{1}{cos\phi} + 1, & \text{for } 0 \leq D \leq 0.72 \\ 1 - cos2\pi D + tan\phi\, sin2\pi D, & \text{for } 0.72 < D \leq 1. \end{cases} \tag{5.20}$$

For $D = 0.5$, $V_{DM}/V_O = \frac{1}{2}\sqrt{\pi^2 + 4} + 1 = 2.862$. Figure 5.4(a) depicts V_{DM}/V_O as a function of D, and Fig. 5.4(b) depicts V_{DM}/V_O as a function of $R_L/\omega L$. It can be seen that V_{DM}/V_O decreases from ∞ to 2 as D decreases from 1 to 0, or as $R_L/\omega L$ increases from 0 to ∞.

Using (5.18) and (5.20), the power-output capability can easily be calculated from the expression

$$c_p = \frac{P_O}{I_{DM}V_{DM}} = \frac{I_O V_O}{I_{DM}V_{DM}}. \tag{5.21}$$

Figure 5.5(a) illustrates c_p as a function of D and Fig. 5.5(b) as a function of $R_L/\omega L$. The maximum value of c_p occurs at $D = 0.5$ and equals 0.0981.

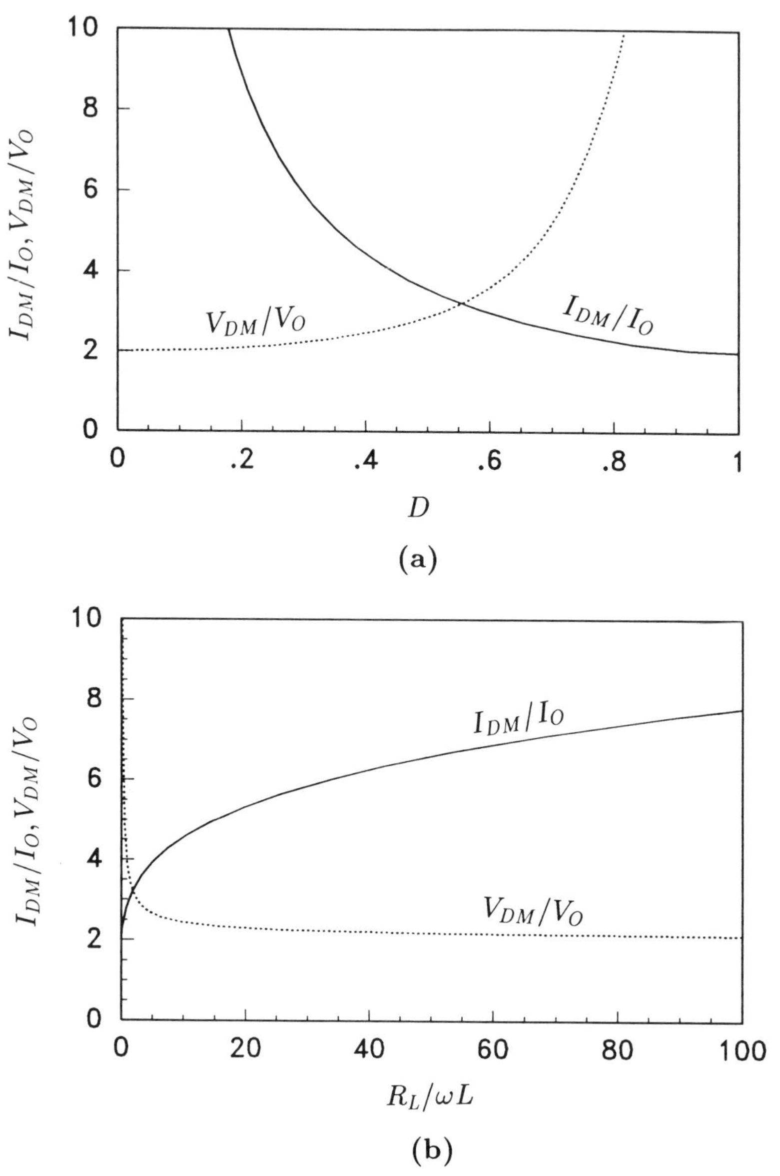

Figure 5.4: Peak values of the diode current and voltage. (a) I_{DM}/I_O and V_{DM}/V_O versus D. (b) I_{DM}/I_O and V_{DM}/V_O versus $R_L/\omega L$.

5.2.5 Input Impedance

The input power of the rectifier contains only the fundamental component because the input current i_R is sinusoidal. For this reason, it is sufficient to determine the input impedance Z_i of the rectifier at the fundamental frequency f. This impedance consists of an input resistance R_i and an input

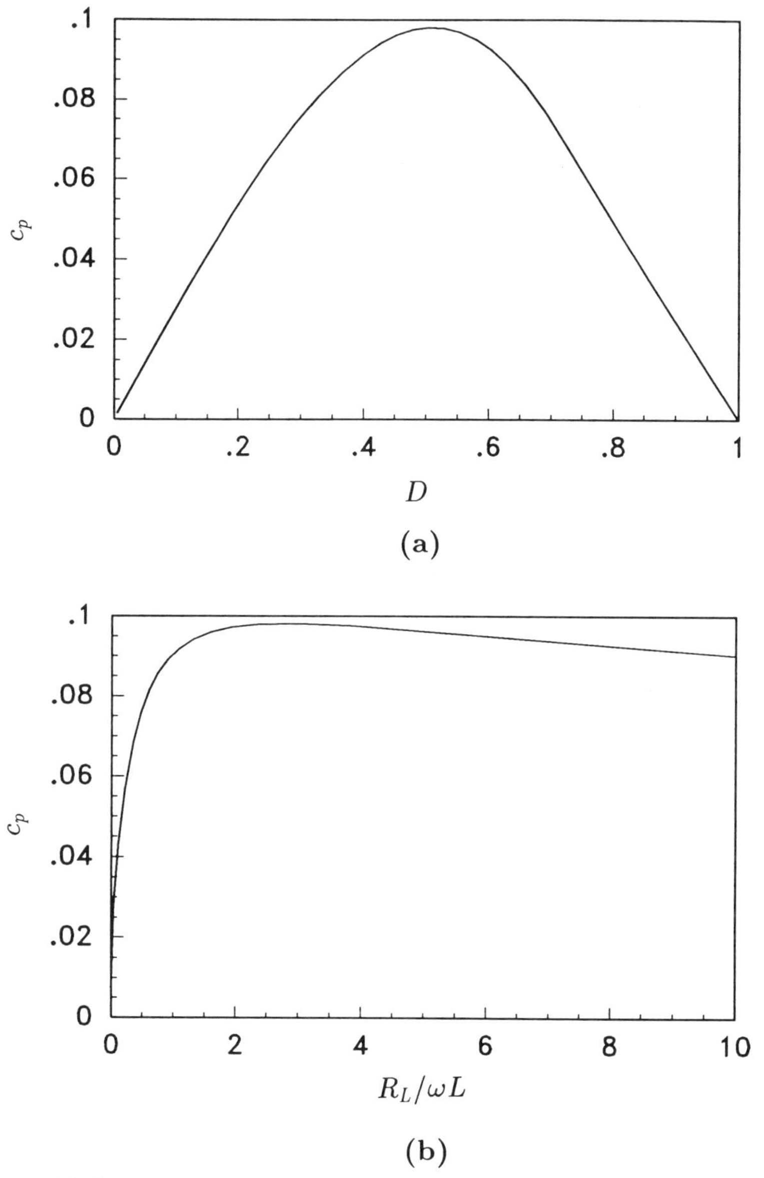

Figure 5.5: Power output capability c_p. (a) c_p versus D. (b) c_p versus $R_L/\omega L$.

inductance L_i, as shown in Fig. 5.6. From (5.16), the input voltage of the rectifier of Fig. 5.1(a) is

$$v_R = v_L = \begin{cases} V_O, & \text{for } 0 < \omega t \le 2\pi D \\ V_O \frac{cos(\omega t + \phi)}{cos\phi}, & \text{for } 2\pi D < \omega t \le 2\pi. \end{cases} \tag{5.22}$$

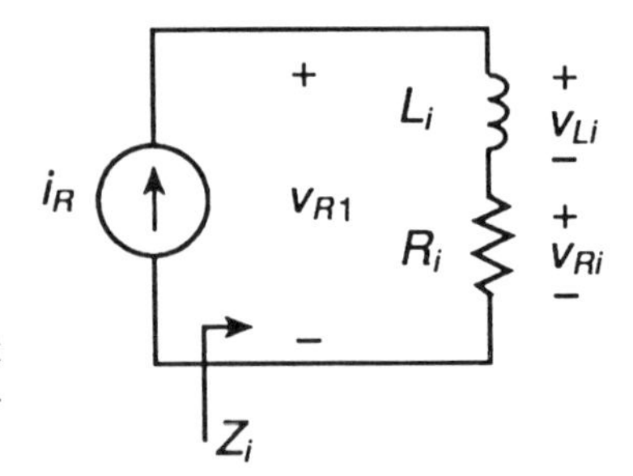

Figure 5.6: Equivalent circuit of the input impedance of the rectifier at the fundamental frequency f.

The fundamental component v_{R1} of this voltage is

$$v_{R1} = v_{Ri} + v_{Li} = V_{Rim} sin(\omega t + \phi) + V_{Lim} cos(\omega t + \phi), \tag{5.23}$$

where V_{Rim} is the amplitude of the voltage v_{Ri} across the input resistance R_i and V_{Lim} is the amplitude of the voltage v_{Li} across the input inductance L_i at the fundamental frequency f. Using (5.22) and the Fourier series,

$$\begin{aligned} V_{Rim} &= \frac{1}{\pi}\int_0^{2\pi} v_L sin(\omega t + \phi) d(\omega t) \\ &= \frac{V_O}{\pi}\left[cos\phi - cos(2\pi D + \phi) + \frac{sin^2\phi - sin^2(2\pi D + \phi)}{2cos\phi}\right]. \end{aligned} \tag{5.24}$$

For $D = 0.5$, $V_{Rim}/V_O = 4/(\pi\sqrt{\pi^2 + 4}) = 0.3419$. From (5.7) and (5.24), the input resistance of the rectifier at the fundamental frequency f is

$$R_i = \frac{V_{Rim}}{I_m} = \frac{\omega L}{\pi}\left\{[cos\phi - cos(2\pi D + \phi)]cos\phi + \frac{1}{2}[sin^2\phi - sin^2(2\pi D + \phi)]\right\}. \tag{5.25}$$

The normalized input resistance of the rectifier at the fundamental frequency f is

$$\frac{R_i}{R_L} = \frac{cos^2\phi}{2\pi^2}[1 - 2\pi^2 D^2 - cos 2\pi D + (sin 2\pi D - 2\pi D) tan\phi]^2. \tag{5.26}$$

From (5.3) and (5.8), the input resistance R_i normalized with respect to ωL at the fundamental frequency f is

$$\frac{R_i}{\omega L} = \frac{cos^2\phi}{\pi}[1 - 2\pi^2 D^2 - cos 2\pi D + (sin 2\pi D - 2\pi D) tan\phi]. \tag{5.27}$$

For $D = 0.5$, $R_i/(\omega L) = 8/[\pi(\pi^2 + 4)] = 0.1836$. The plots of $R_i/\omega L$ versus D and $R_L/\omega L$ are shown in Fig. 5.7. The maximum value of $R_i/\omega L$ is 0.2525

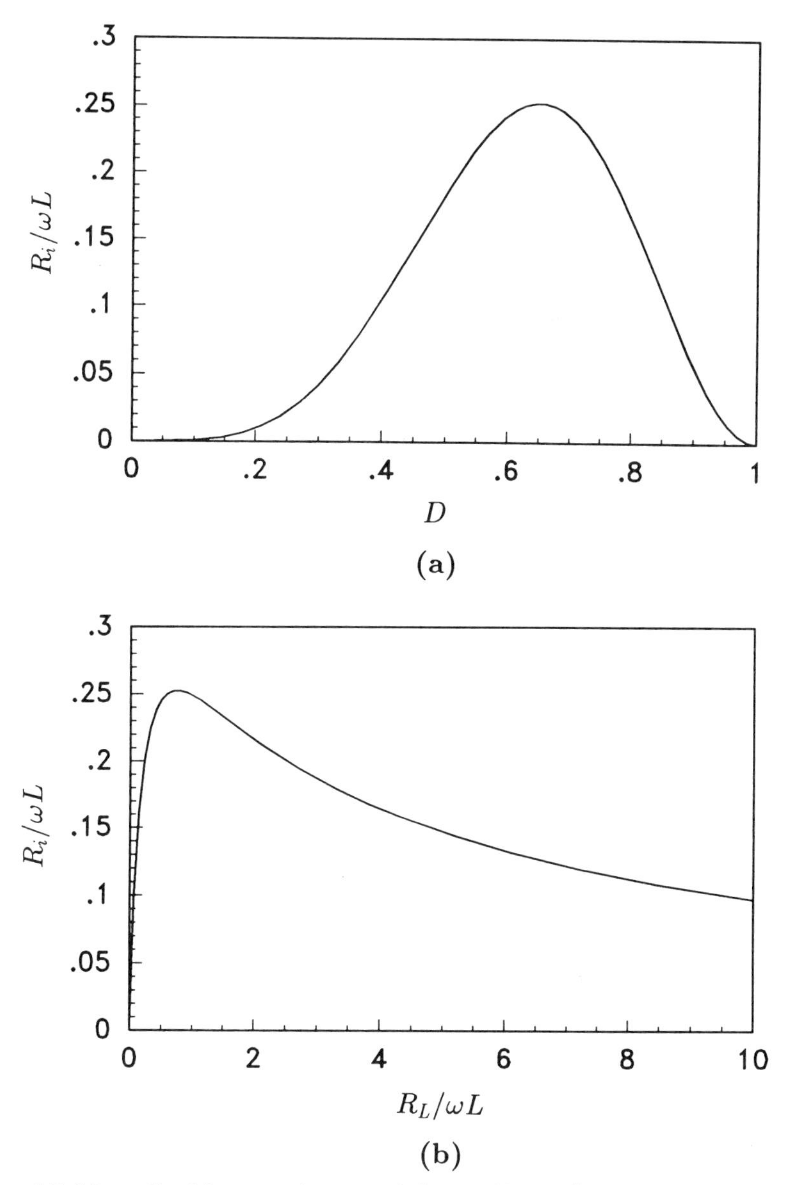

Figure 5.7: Normalized input resistance of the rectifier $R_i/\omega L$ at the fundamental frequency f. (a) $R_i/\omega L$ as a function of D. (b) $R_i/\omega L$ as a function of $R_L/\omega L$.

and occurs at $D = 0.65$ or $R_L/\omega L = 0.7546$. For $D \leq 0.65$ or $R_L/\omega \geq 0.7546$, $R_i/\omega L$ decreases as $R_L/\omega L$ increases. Thus, the rectifier acts as an impedance inverter. It is therefore compatible with a Class E inverter in which the switch turns on at zero voltage only for load resistances ranging from zero to a maximum value.

For $D = 0.5$, $R_i/R_L = 8/[\pi^2(\pi^2+4)] = 0.05844$. Figure 5.8(a) shows a plot of R_i/R_L versus D, and Fig. 5.8(b) shows a plot of R_i/R_L versus $R_L/\omega L$. As seen, R_i/R_L decreases from 2 to 0 as D decreases from 1 to 0, or as $R_L/\omega L$ increases from 0 to ∞.

Combining (5.14) and (5.25), one arrives at (5.26). Using (5.14),

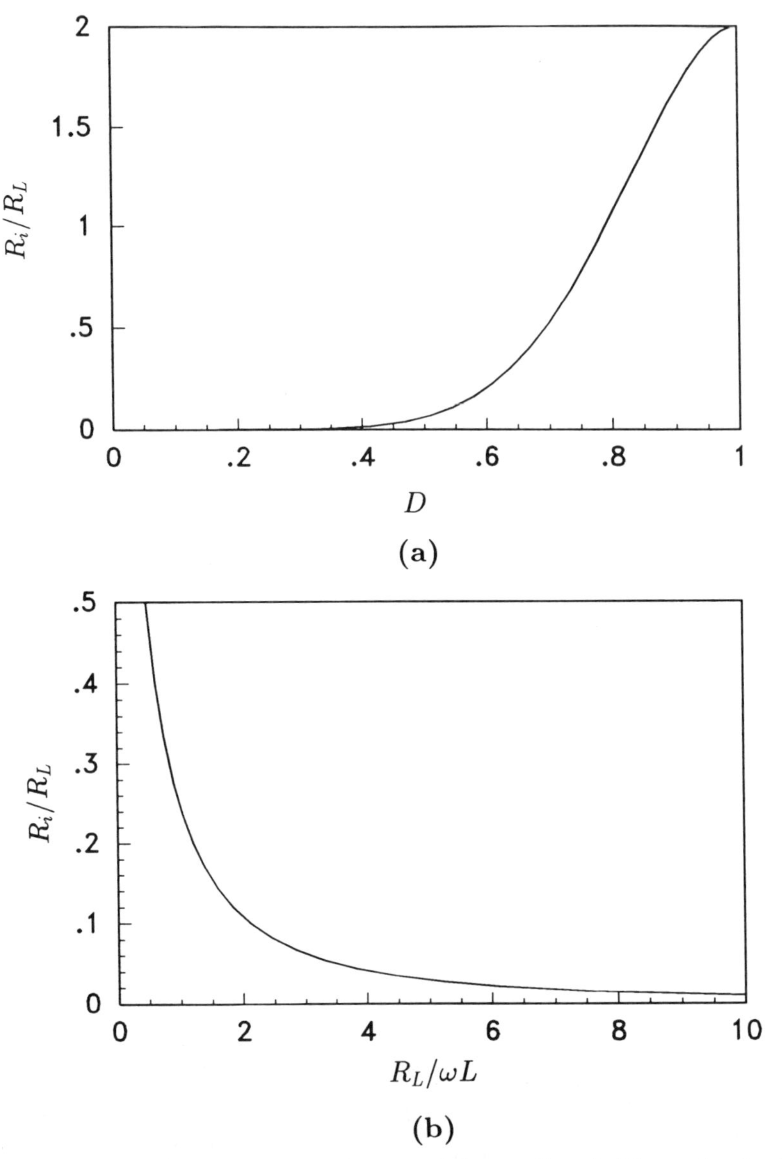

Figure 5.8: Normalized input resistance of the rectifier R_i/R_L at the fundamental frequency f. (a) R_i/R_L as a function of D. (b) R_i/R_L as a function of $R_L/\omega L$.

$$V_{Lim} = \frac{1}{\pi}\int_0^{2\pi} v_R cos(\omega t + \phi) d(\omega t)$$
$$= \frac{V_O}{\pi}\left[sin(2\pi D + \phi) - \frac{1}{2}sin\phi + \frac{4\pi(1-D) - sin(4\pi D + 2\phi)}{4cos\phi}\right]. \tag{5.28}$$

For $D = 0.5$, $V_{Lim} = (\pi^2 + 12)/(4\sqrt{\pi^2+4})V_O = 1.4681V_O$. From (5.7) and (5.28), the input reactance at the fundamental frequency f is

$$X_i = \omega L_i = \frac{V_{Lim}}{I_m}$$
$$= \frac{\omega L}{\pi}\left\{\pi(1-D) + cos\phi sin(2\pi D + \phi) - \frac{1}{4}[sin2\phi + sin(4\pi D + 2\phi)]\right\}. \tag{5.29}$$

Rearrangement of this expression produces (5.27).

The normalized input inductance at the fundamental frequency f is given by

$$\frac{L_i}{L} = \frac{1}{\pi}\left\{\pi(1-D) + cos\phi sin(2\pi D + \phi) - \frac{1}{4}[sin2\phi + sin(4\pi D + 2\phi)]\right\}. \tag{5.30}$$

For $D = 0.5$, $L_i/L = (\pi^2 + 12)/[2(\pi^2 + 4)] = 0.7884$. The plots of L_i/L as a function of D and L_i/L as a function of $R_L/\omega L$ are depicted in Fig. 5.9. It can be seen that L_i/L increases from 0 to 1 as D decreases from 1 to 0, or as $R_L/\omega L$ increases from 0 to ∞.

5.2.6 Current and Voltage Transfer Functions

The ac-to-dc current transfer function is

$$M_{IR} = \frac{I_O}{I_{rms}} = \frac{\sqrt{2}cos\phi}{R_L/\omega L}, \tag{5.31}$$

where I_{rms} is the rms value of the input current, ϕ is given by (5.9), and $R_L/\omega L$ is given by (5.14). Figure 5.10(a) depicts M_{IR} as a function of D, and Fig. 5.10(b) depicts it as a function of $R_L/\omega L$. As seen, M_{IR} decreases from $\sqrt{2}$ to 0, as D decreases from 1 to 0, or as $R_L/\omega L$ increases from 0 to ∞. The ac input power of the rectifier is $P_i = V_{Ri\,rms}^2/R_i$, where $V_{Ri\,rms}$ is the rms value of the v_{R1}. The dc output power of the rectifier is $P_O = V_O^2/R_L$. Assuming that the efficiency of the rectifier is 100%, $P_O = P_i$, or $V_{Ri\,rms}^2/R_i = V_O^2/R_L$. Hence, the voltage transfer function of the rectifier is

$$M_{VR} = \frac{V_O}{V_{Ri\,rms}} = \sqrt{\frac{R_L}{R_i}}, \tag{5.32}$$

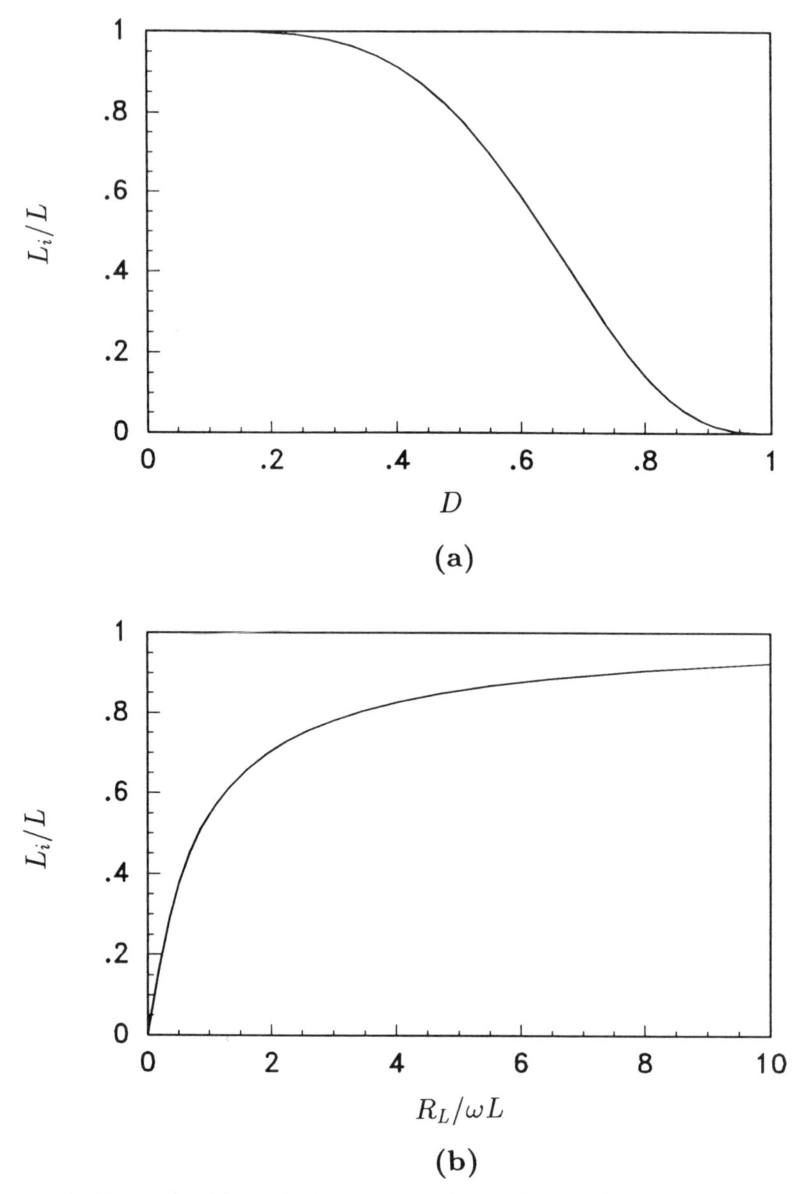

Figure 5.9: Normalized input inductance of the rectifier L_i/L at the fundamental frequency f. (a) L_i/L as a function of D. (b) L_i/L as a function of $R_L/\omega L$.

where R_L/R_i may be computed from (5.26). For $D = 0.5$, $M_{VR} = \sqrt{\pi^2(\pi^2+4)/8} = 4.1365$. Figure 5.11 illustrates M_{VR} as a function of D and as a function of $R_L/\omega L$. It is seen that M_{VR} increases from 0 to ∞ as D decreases from 1 to 0, or as $R_L/\omega L$ increases from 0 to ∞. The parameters of the rectifier are given in Table 5.1 for various values of D.

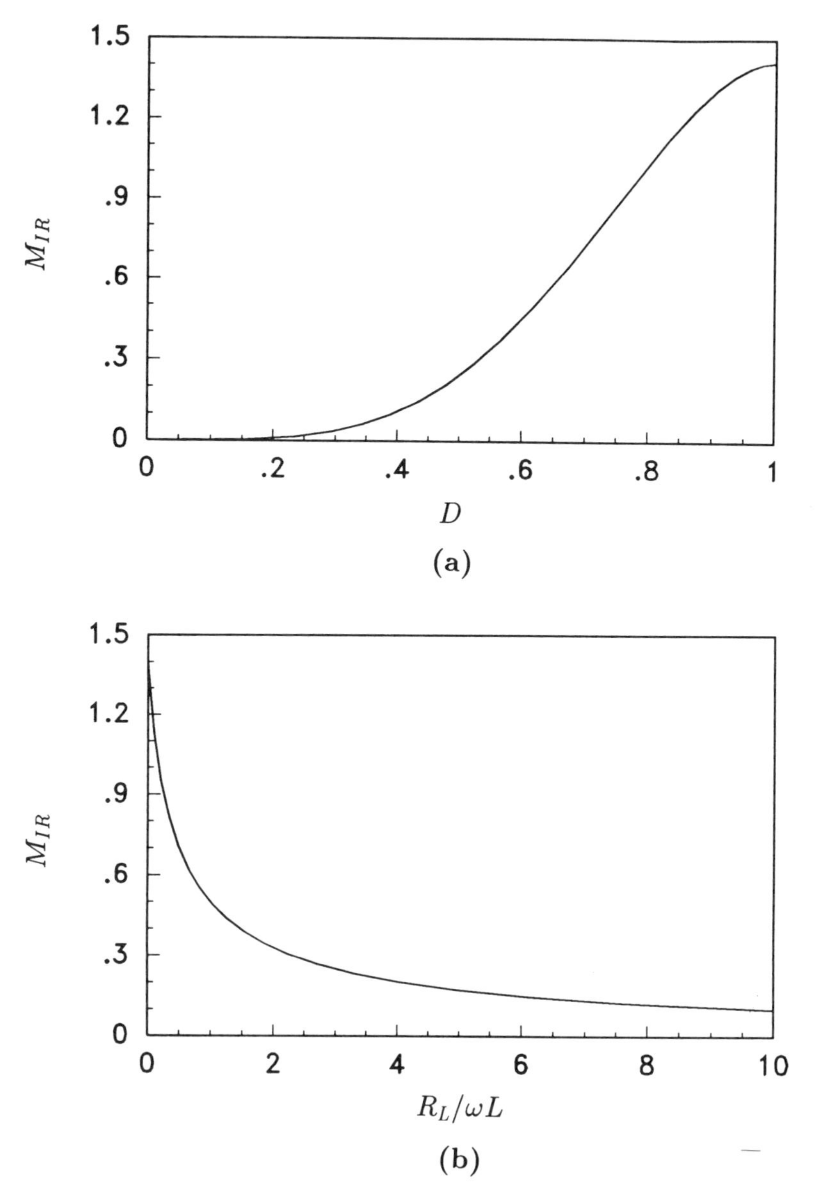

Figure 5.10: Current transfer function M_{IR}. (a) M_{IR} versus D. (b) M_{IR} versus $R_L/\omega L$.

The actual conducting diode can be modeled by a dc voltage source V_F representing the forward voltage drop across the diode. The voltage source V_F can be added in series with V_O in Fig. 5.1(c). Hence, the effective dc output voltage is $V_{Oeff} = V_O + V_F$, and the effective load resistance is $R_{Leff} = V_{Oeff}/I_O = R_L + V_F/I_O$. The conduction power loss is $P_D = V_F I_O$, and the rectifier efficiency is $\eta = P_O/(P_O + P_D) = 1/(1 + V_F/V_O)$.

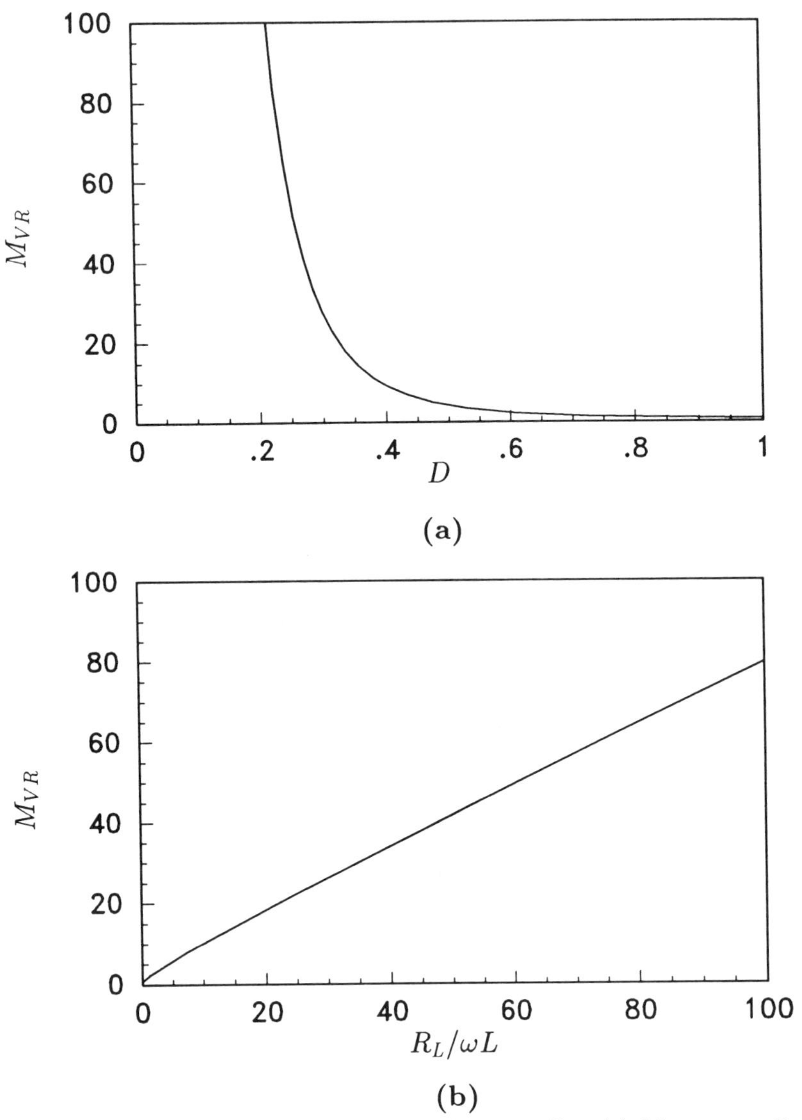

Figure 5.11: Voltage transfer function M_{VR} of the rectifier. (a) M_{VR} versus D. (b) M_{VR} versus $R_L/\omega L$.

5.2.7 Design Example

Example 5.1

Design a Class E low di/dt rectifier with a parallel inductor shown in Fig. 5.1(a) to meet the following specifications: $V_O = 5$ V, $R_L = 10\ \Omega$ to infinity, $f = 0.5$ MHz, and $V_r/V_O \leq 5\%$. Assume $D = 0.75$ at $R_{Lmin} = 10\ \Omega$.

Table 5.1. Parameters of Class E Low *di/dt* Rectifier with a Parallel Inductor

D	ϕ	$R_L/\omega L$	I_{DM}/I_O	V_{DM}/V_O	R_i/R_L	L_i/L	M_{VR}	M_{IR}
0	0	∞	∞	2	0	1	∞	0
0.1	−11.98	2865.33	17.777	2.022	$2.333 \cdot 10^{-7}$	0.9999	2070.5	$4.828 \cdot 10^{-4}$
0.2	−23.86	171.88	8.889	2.093	$5.662 \cdot 10^{-5}$	0.9966	132.90	$7.525 \cdot 10^{-3}$
0.25	−29.72	68.217	7.111	2.151	$3.242 \cdot 10^{-4}$	0.9901	55.540	$1.800 \cdot 10^{-2}$
0.3	−35.51	31.612	5.927	2.228	$1.326 \cdot 10^{-3}$	0.9767	27.458	$3.642 \cdot 10^{-2}$
0.4	−46.79	8.9733	4.447	2.460	$1.165 \cdot 10^{-2}$	0.9155	9.2668	0.1079
0.5	−57.52	3.1416	3.562	2.862	$5.844 \cdot 10^{-2}$	0.7884	4.1365	0.2909
0.6	−67.46	1.2083	2.976	3.608	0.2014	0.5909	2.2284	0.4487
0.7	−76.25	0.4655	2.566	5.207	0.5214	0.3541	1.3849	0.7270
0.75	−80.07	0.2789	2.407	6.712	0.7645	0.2405	1.1437	0.8744
0.8	−83.41	0.1580	2.274	8.918	1.0565	0.1427	0.9729	1.0273
0.9	−88.25	0.03338	2.079	19.404	1.6788	0.0228	0.7718	1.2957
1	−90	0	2	∞	2	0	0.7071	1.4142

Solution: Consider the case for the full power, that is, $R_L = 10\ \Omega$. From Table 5.1,

$$L = \frac{R_L}{0.2789\omega} = \frac{10}{0.2789 \times 2 \times \pi \times 0.5 \times 10^6} = 11.41\ \mu\text{H}. \tag{5.33}$$

The maximum value of the dc load current is

$$I_{Omax} = \frac{V_O}{R_{Lmin}} = \frac{5}{10} = 0.5\ \text{A}. \tag{5.34}$$

From Table 5.1, $I_{DM}/I_O = 2.407$ at $D = 0.75$. Thus, the maximum value of the diode peak current is

$$I_{DMmax} = 2.407 I_{Omax} = 2.407 \times 0.5 = 1.204\ \text{A}. \tag{5.35}$$

From Table 5.1, $V_{DM}/V_O = 6.712$ at $D = 0.75$; hence, the maximum value of the diode peak reverse voltage is

$$V_{DMmax} = 6.712 V_O = 6.712 \times 5 = 33.56\ \text{V}. \tag{5.36}$$

Using (5.31) and Table 5.1, the amplitude of the input current is

$$I_{m(max)} = \frac{\sqrt{2} I_{Omax}}{M_{IR}} = \frac{\sqrt{2} \times 0.5}{0.8744} = 0.81\ \text{A}. \tag{5.37}$$

Assuming that the ESR of the filter capacitor is $r_C = 20\ \text{m}\Omega$, and using (5.19) the minimum filter capacitance can be calculated as

$$C_f = \frac{I_{DM}}{\omega(V_r - I_{DM} r_C)} = \frac{1.204}{2 \times \pi \times 0.5 \times 10^6 (0.05 \times 5 - 1.204 \times 0.02)} = 1.7\ \mu\text{F}. \quad (5.38)$$

5.3 LOW *di/dt* RECTIFIER WITH A SERIES INDUCTOR

5.3.1 Principle of Operation

A circuit of a Class E low di/dt rectifier with a series inductor [6] is shown in Fig. 5.12(a). It consists of a diode, a series inductor L, and a large filter capacitor C_f. The inductance L shapes the diode current so that the diode turns on at zero di_D/dt and turns off at low $| di_D/dt |$, reducing switching losses, switching noise, and reverse-recovery current. Furthermore, a large value of the diode ON switch duty cycle D (e.g., $D = 0.5$) can be obtained at a full load resistance $R_{Lmin} = V_O/I_{Omax}$, reducing the amount of harmonics generated by the diode current pulses, while still maintaining a low level of the ripple voltage at the output. This is because D is independent of C_f. The filter capacitor C_f and the load resistance R_L form a first-order low-pass filter to smooth out the output voltage. This filter can be modeled by a dc voltage source V_O, as shown in Fig. 5.12(b). The rectifier is driven by a sinusoidal voltage source v_R and can be coupled to the ac source with an isolation transformer. An important advantage of the rectifier topology is that the transformer leakage inductance (if any) and the diode lead inductance are absorbed into L. The circuit may be designed in such a way that L is formed by parasitic inductances. The diode junction capacitance C_j is not included in the analysis of the rectifier circuit. It may cause ringing in the diode voltage when the diode is OFF, as shown in Fig. 5.12(e) by a broken line. A negative dc output voltage V_O may be obtained by reversing the diode polarity.

Idealized current and voltage waveforms in the rectifier are shown in Fig. 5.12(e). The input voltage v_R is a sine wave. The voltage across the diode/inductor combination is equal to $v_R - V_O$. This voltage occurs across L when the diode is ON, and across the diode when the diode is OFF. The diode is ON for $0 < \omega t \leq 2\pi D$, and the model of the rectifier is shown in Fig. 5.12(c). The current i_D through the diode and the inductance L is determined by the voltage $v_R - V_O$. Therefore, i_D contains two components: one sinusoidal and the other linearly decreasing. According to equation $v_L = L di_D/dt$, the diode turns on at zero di_D/dt because v_L is zero at turn-on. The diode current i_D slowly increases when v_L is positive, reaches its peak value I_{DM} when v_L returns to zero, and decreases to zero when v_L is negative. The diode turns off when i_D reaches zero. Because v_L has a negative, finite value just before turn-off, the diode turns off at low $| di_D/dt |$, reducing the peak value of the diode reverse-recovery current.

The diode is OFF for $2\pi D < \omega t \leq 2\pi$, and the model of the rectifier is shown in Fig. 5.12(d). The voltage across the diode is $v_R - V_O$. When the

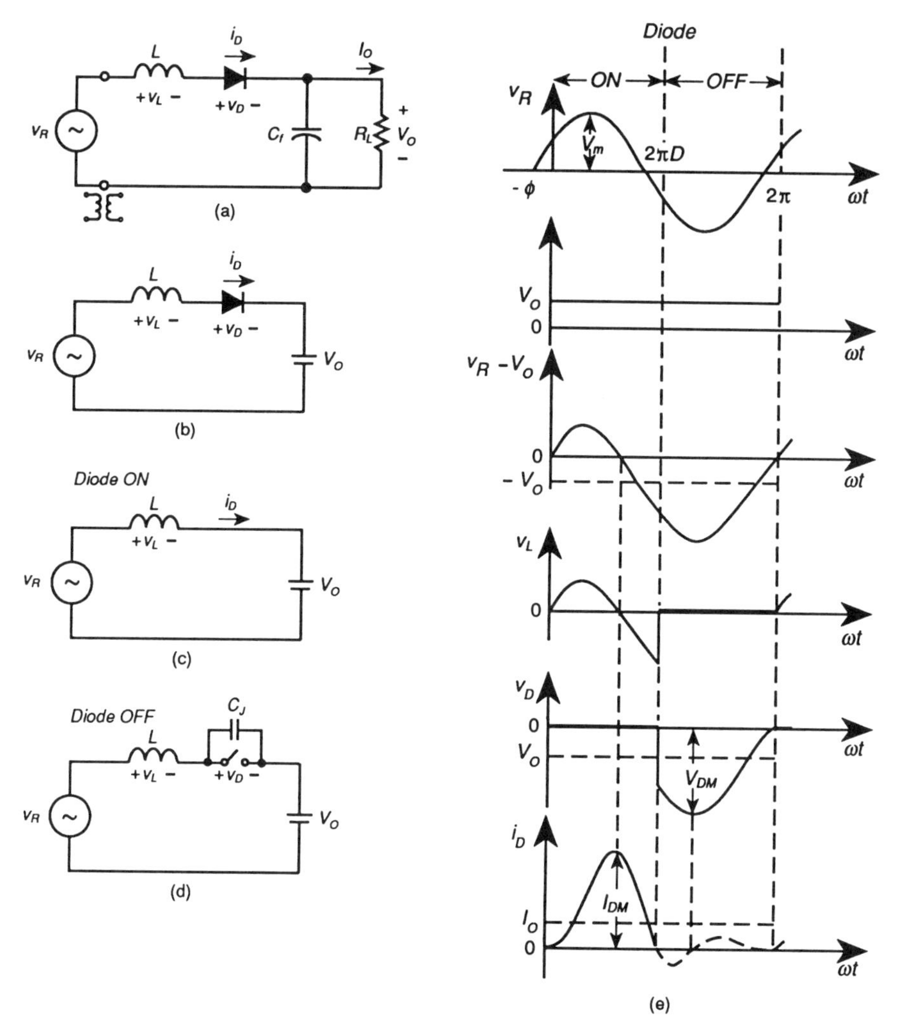

Figure 5.12: Class E low di/dt rectifier with a series inductor. (a) Circuit. (b) Model. (c) Model when the diode is ON. (d) Model when the diode is OFF. (e) Current and voltage waveforms.

diode voltage reaches its threshold voltage, the diode turns on. The turn-on transition occurs at low dv_D/dt, reducing switching noise. The diode ON switch duty cycle D decreases with $R_L/\omega L$ and is independent of the filter capacitance C_f. For given values of the full load resistance R_{Lmin} and the frequency f, the maximum value of D is determined by the inductance L. The voltage ripple at the output V_r is independently determined by the filter capacitance C_f. A rough estimate of this voltage is given by (5.19).

When the diode is OFF, its junction capacitance C_j and the inductance form a parasitic series-resonant circuit. The resonant frequency is $f_o = 1/(2\pi\sqrt{LC_j})$, and the characteristic impedance is $Z_o = \sqrt{L/C_j}$. A step change of the diode voltage ΔV results in a current in the resonant circuit $i = (\Delta V/Z_o) sin\omega_o t$. This current causes a voltage across the junction capacitance $v_D = -\Delta V(1 - cos\omega_o t)$. Hence, the peak value of the diode voltage is $V_{DM} = 2\Delta V$. If the oscillations decay before the diode turns on, the turn-off switching loss is $P_{turn-off} = (1/2) f C_j \Delta V^2$.

Figure 5.13 shows a family of Class E low di/dt rectifiers. The inductance L is low so that the ac component of the inductor current is high compared with I_O. In all the rectifiers, the transformer leakage inductance L_ℓ is absorbed into L. The rectifiers of Fig. 5.13(b) and (c) with a large inductance L are Class D voltage-driven rectifiers, discussed in Chapter 3. At full load resistance R_L, the following condition is usually satisfied: $R_L/(\omega L) << 1$. Hence, the ac ripple of the inductor current is small compared to I_O, and the input current of the rectifier is a square wave $\pm I_O$, as in the Class D voltage-driven rectifier. However, as $R_L/(\omega L)$ is increased the ac ripple of the inductor current is large and the circuit becomes a Class E rectifier. Inductances on

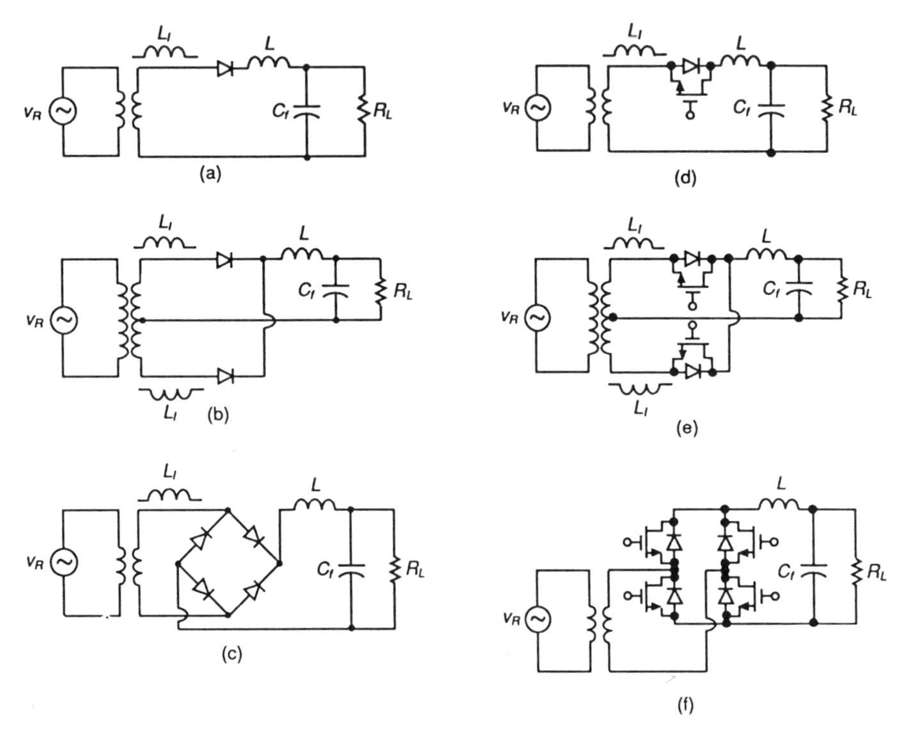

Figure 5.13: Class E conventional and regulated synchronous low di/dt rectifiers. (a) Half-wave rectifier. (b) Transformer center-tapped rectifier. (c) Full-bridge rectifier. (d) Regulated synchronous half-wave rectifier. (e) Regulated synchronous transformer center-tapped rectifier. (f) Regulated synchronous full-bridge rectifier.

the ac-input side and on the dc-output side of each diode have the same effect for the discontinuous conduction mode, but the inductance on the dc-output side and the inductance on the primary side of the transformer, if any, is present for both diodes. The equivalent inductance for each diode is $L + L_\ell$. If the diodes are replaced by controllable switches such as power MOSFETs, Class E regulated synchronous rectifiers are obtained, as shown in Fig. 5.13(d), (e), and (f). In these rectifiers, the conduction angle of the switches can be controlled to regulate the dc output voltage V_O.

5.3.2 Assumptions

The analysis of the rectifier of Fig. 5.12(a) is performed under the following assumptions:

1) The diode is ideal, that is, its threshold voltage, ON resistance, junction capacitance, and minority carrier lifetime are zero and its OFF resistance is infinity.
2) The output filter capacitance C_f is large enough so that the ripple of the output voltage is much lower than the dc output voltage V_O.
3) The transformer leakage inductance, if any, and the diode lead inductance are absorbed into L.
4) The rectifier is driven by an ideal voltage source.

5.3.3 Component Values

The input voltage of the rectifier is sinusoidal and given by

$$v_R = V_m sin(\omega t + \phi), \tag{5.39}$$

where V_m and ϕ are the amplitude and the initial phase, respectively. Figure 5.12(c) shows the model of the rectifier for $0 < \omega t \le 2\pi D$ when the diode is ON. The voltage across the inductor L is

$$v_L = v_R - V_O = V_m sin(\omega t + \phi) - V_O. \tag{5.40}$$

Because $v_L(0) = 0$,

$$V_O = I_O R_L = V_m sin\phi. \tag{5.41}$$

From (5.41), (5.40) becomes

$$v_L = V_O \left(\frac{sin\omega t}{tan\phi} + cos\omega t - 1 \right). \tag{5.42}$$

Thus,

$$i_D = i_L = \frac{1}{\omega L} \int_0^{\omega t} v_L d(\omega t) = \frac{I_O R_L}{\omega L} \left(sin\omega t - \omega t - \frac{cos\omega t - 1}{tan\phi} \right). \tag{5.43}$$

Because $i_D(2\pi D) = 0$, (5.43) leads to the relationship between the phase ϕ and the diode ON–switch duty cycle D

$$tan\phi = \frac{1 - cos2\pi D}{2\pi D - sin2\pi D} \tag{5.44}$$

illustrated in Fig. 5.14. For $D = 0.5$, $tan\phi = 2/\pi$, $\phi = 32.48° = 0.5669$ rad, $sin\phi = 2/\sqrt{\pi^2 + 4}$, and $cos\phi = \pi/\sqrt{\pi^2 + 4}$.

The average diode current equals the dc output current

$$I_O = \frac{1}{2\pi}\int_0^{2\pi D} i_D d(\omega t) = \frac{I_O R_L}{2\pi\omega L}\left(1 - 2\pi^2 D^2 - cos2\pi D + \frac{2\pi D - sin2\pi D}{tan\phi}\right). \tag{5.45}$$

Simplifying this expression, one arrives at the relationship between the diode ON-switch duty cycle D and the normalized load resistance

$$\frac{R_L}{\omega L} = \frac{2\pi}{1 - 2\pi^2 D^2 - cos2\pi D + \frac{2\pi D - sin2\pi D}{tan\phi}}. \tag{5.46}$$

The duty cycle D is plotted in Fig. 5.15 versus $R_L/(\omega L)$. As $R_L/\omega L$ increases from 0 to ∞, D decreases from 1 to 0. At $D = 0.5$, $R_L/\omega L = \pi$.

Figure 5.12(d) shows the model of the rectifier for $2\pi D < \omega t \le 2\pi$ when the diode is OFF. The voltage across the diode is

$$\begin{aligned} v_D &= v_R - V_O = V_m sin(\omega t + \phi) - V_O \\ &= V_O\left[\frac{sin(\omega t + \phi)}{sin\phi} - 1\right] = V_O\left(cos\omega t - \frac{sin\omega t}{tan\phi} - 1\right). \end{aligned} \tag{5.47}$$

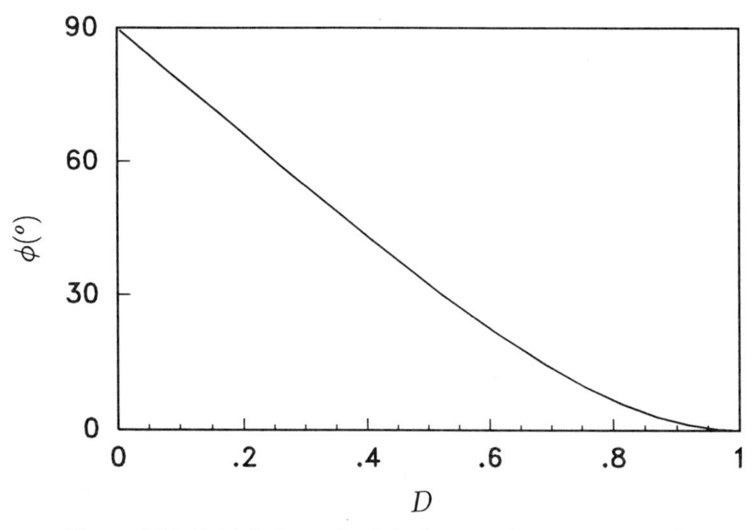

Figure 5.14: Initial phase ϕ of the input voltage v_R versus D.

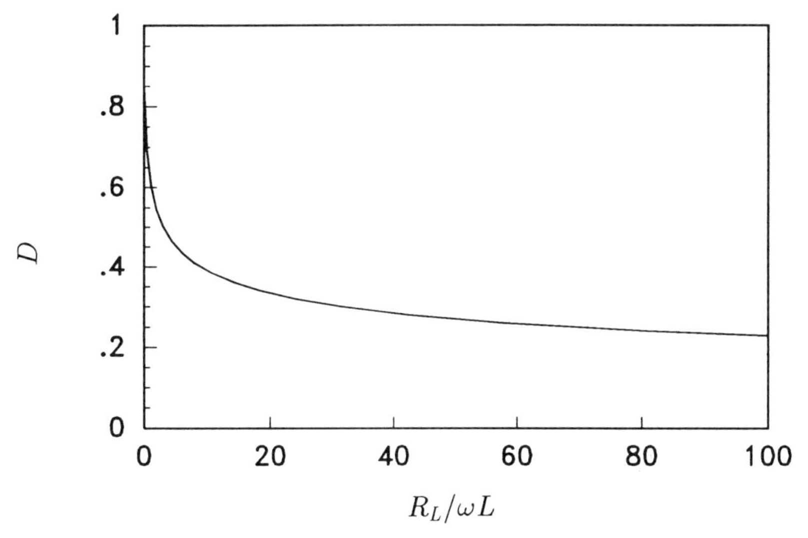

Figure 5.15: Diode ON–switch duty cycle D versus $R_L/\omega L$.

5.3.4 Diode Waveforms

The normalized waveforms of the diode current and voltage are

$$\frac{i_D}{I_O} = \begin{cases} \frac{R_L}{\omega L}\left(sin\omega t - \omega t - \frac{cos\omega t - 1}{tan\phi}\right), & \text{for} \quad 0 < \omega t \leq 2\pi D \\ 0, & \text{for} \quad 2\pi D < \omega t \leq 2\pi \end{cases} \tag{5.48}$$

$$\frac{v_D}{V_O} = \begin{cases} 0, & \text{for} \quad 0 < \omega t \leq 2\pi D \\ \frac{sin\omega t}{tan\phi} + cos\omega t - 1, & \text{for} \quad 2\pi D < \omega t \leq 2\pi. \end{cases} \tag{5.49}$$

These waveforms are shown in Fig. 5.12(e) for $D = 0.5$.

5.3.5 Peak Diode Current and Voltage

The diode peak current I_{DM} may be determined by setting $di_D/d(\omega t) = 0$. It occurs at $\omega t_{im} = \pi - 2\phi$ and is given by

$$\frac{I_{DM}}{I_O} = \frac{R_L}{\omega L}\left(\frac{2}{tan\phi} + 2\phi - \pi\right), \tag{5.50}$$

where ϕ is expressed in radians. At $D = 0.5$, $I_{DM}/I_O = 2\pi\phi = 3.562$. Figure 5.16 depicts the plots of I_{DM}/I_O versus D and versus $R_L/\omega L$. I_{DM}/I_O increases from 2 to ∞ as D decreases from 1 to 0, or as $R_L/\omega L$ increases from 0 to ∞. Since $I_O = V_O/R_L$ decreases with R_L at a constant value of V_O, the absolute value of I_{DM} decreases with $R_L/\omega L$.

The peak value of the diode reverse voltage occurs at $\omega t_{vm} = 3\pi/2 - \phi$ for $D \leq 0.72$ and at $\omega t_{vm} = 2\pi D$ for $D > 0.72$. It is expressed by

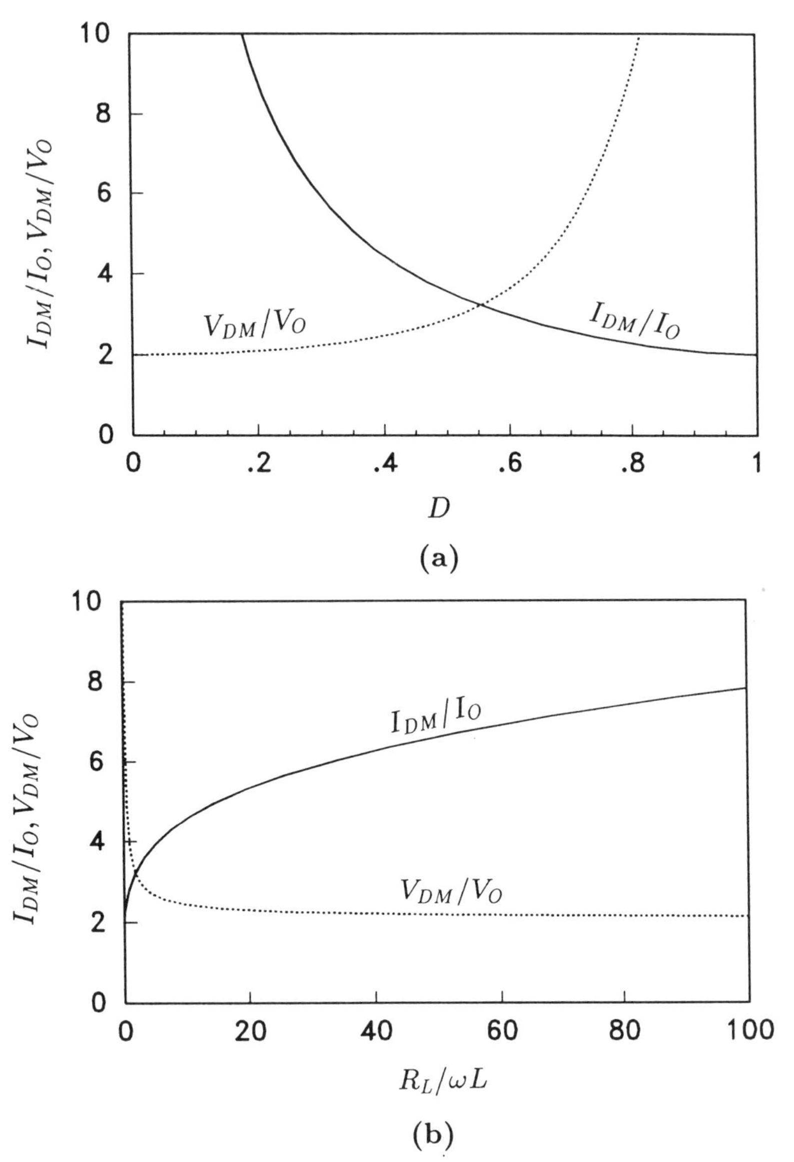

Figure 5.16: Normalized peak diode current I_{DM}/I_O and reversed voltage V_{DM}/V_O. (a) I_{DM}/I_O and V_{DM}/V_O versus D. (b) I_{DM}/I_O and V_{DM}/V_O versus $R_L/\omega L$.

$$\frac{V_{DM}}{V_O} = \begin{cases} \frac{1}{sin\phi} + 1, & \text{for} \quad D \leq 0.72 \\ 1 - cos2\pi D - \frac{sin2\pi D}{tan\phi}, & \text{for} \quad D > 0.72. \end{cases} \tag{5.51}$$

For $D = 0.5$, $V_{DM}/V_O = \sqrt{\pi^2 + 4}/2 + 1 = 2.862$. Figure 5.16 shows plots of V_{DM}/V_O versus D and versus $R_L/\omega L$. V_{DM}/V_O decreases from ∞ to 2 as D decreases from 1 to 0, or as $R_L/\omega L$ increases from 0 to ∞.

The power-output capability is given by

$$c_p = \frac{P_O}{I_{DM}V_{DM}} = \frac{I_O V_O}{I_{DM}V_{DM}} \tag{5.52}$$

and is illustrated in Fig. 5.5. The maximum value of c_p occurs at $D = 0.5$ or $R_L/\omega L = \pi$ and equals 0.0981.

5.3.6 Voltage Transfer Function

The ac-to-dc voltage transfer function is given by

$$M_{VR} = \frac{V_O}{V_{rms}} = \sqrt{2} sin\phi, \tag{5.53}$$

where V_{rms} is the rms value of the input voltage v_R. For $D = 0.5$, $M_{VR} = 2\sqrt{2}/\sqrt{\pi^2 + 4} = 0.7595$. Figure 5.17 shows plots of M_{VR} versus D and $R_L/\omega L$. M_{VR} increases from 0 to $\sqrt{2}$ as D decreases from 1 to 0, or as $R_L/\omega L$ increases from 0 to ∞.

5.3.7 Input Impedance

Since the input voltage v_R is a sine wave, the input power contains only the fundamental component. In addition, the fundamental-frequency component of the input current is the largest; the harmonic components decrease with increasing harmonic order. Therefore, it suffices to find the input impedance Z_i at the operating frequency f. This impedance consists of an input resistance R_i and an input inductance L_i connected in parallel, as shown in Fig. 5.18. The fundamental component of the input current is

$$i_{R1} = i_{Ri} + i_{Li} = I_{Rim} sin(\omega t + \phi) - I_{Lim} cos(\omega t + \phi), \tag{5.54}$$

where I_{Rim} and I_{Lim} are the amplitudes of the currents i_{Ri} and i_{Li} through R_i and L_i, respectively. Neglecting power losses in the rectifier, the output power $P_O = V_O^2/R_L$ equals the input power $P_i = V_m^2/(2R_i)$, that is, $V_O^2/R_L = V_m^2/(2R_i)$. Hence, from (5.41) one obtains the input resistance R_i at the operating frequency f normalized with respect to the load resistance R_L

$$\frac{R_i}{R_L} = \frac{1}{2}\left(\frac{V_m}{V_O}\right)^2 = \frac{1}{2sin^2\phi}. \tag{5.55}$$

At $D = 0.5$, $R_i/R_L = (\pi^2 + 4)/8 = 1.7337$. Figure 5.19 shows plots of R_i/R_L versus D and versus $R_L/\omega L$. R_i/R_L decreases from infinity to 0.5 as D decreases from ∞ to 0, or as $R_L/\omega L$ increases from 0 to ∞. Product of (5.46) and (5.55) results in the normalized input resistance of the rectifier

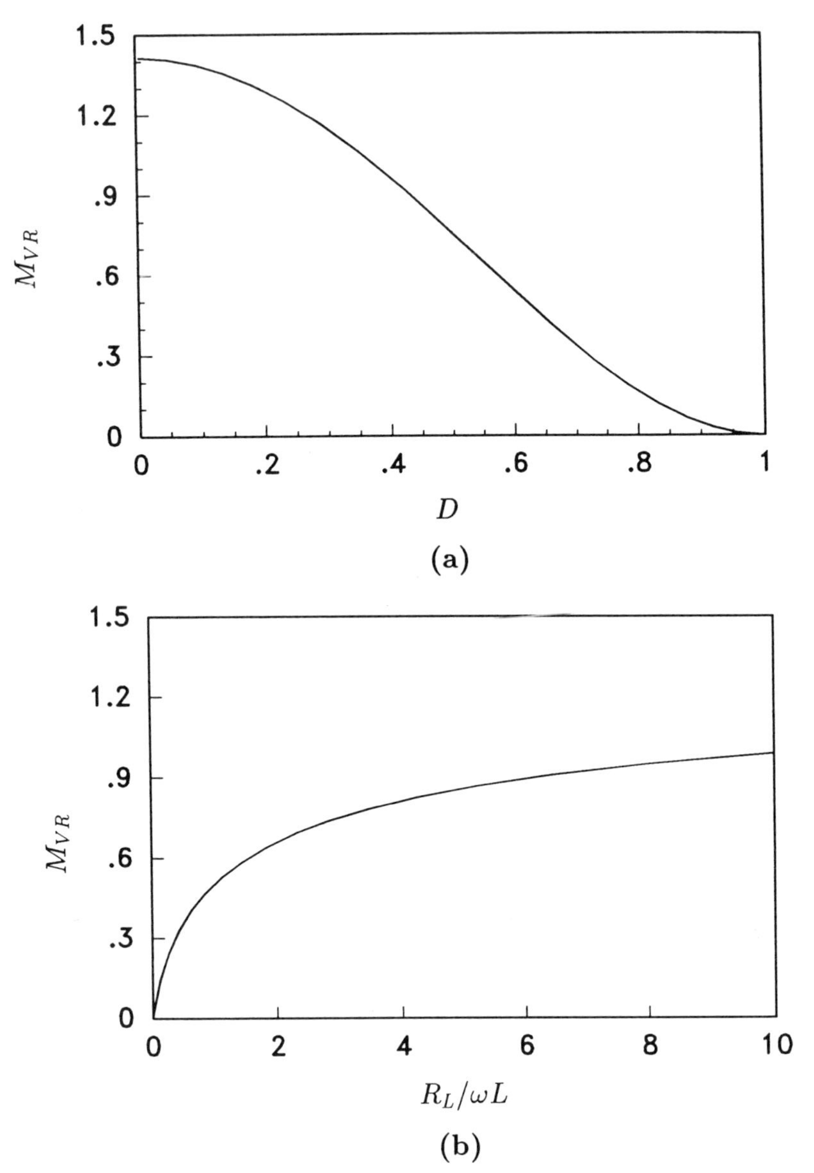

Figure 5.17: Voltage transfer function M_{VR}. (a) M_{VR} versus D. (b) M_{VR} versus $R_L/\omega L$.

$$\frac{R_i}{\omega L} = \frac{\pi}{\left(1 - 2\pi^2 D^2 - cos2\pi D + \frac{2\pi D - sin2\pi D}{tan\phi}\right) sin^2\phi}. \tag{5.56}$$

Plots of $R_i/\omega L$ versus D and versus $R_L/\omega L$ are depicted in Fig. 5.20.

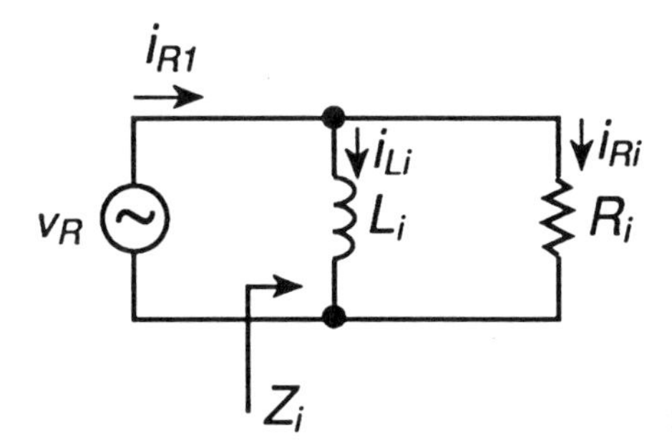

Figure 5.18: Model of the input impedance of the rectifier Z_i at the operating frequency f.

Substitution of (5.48) into the Fourier formula yields

$$\begin{aligned} I_{Lim} &= \frac{1}{\pi}\int_0^{2\pi D} i_D cos(\omega t + \phi) d(\omega t) \\ &= \frac{I_O R_L}{4\pi\omega L sin\phi}\{4\pi D + 4\pi D cos 2\pi D - 4\pi D cos[2(\pi D + \phi)] \\ &- sin 2\phi - 4 sin 2\pi D + sin(4\pi D + 2\phi)\}. \end{aligned} \tag{5.57}$$

Hence, using (5.41) the input reactance of the rectifier at the operating frequency f is

$$\begin{aligned} X_i &= \omega L_i = \frac{V_m}{I_{Lim}} \\ &= \frac{4\pi\omega L}{4\pi D + 4\pi D cos 2\pi D - 4\pi D cos[2(\pi D + \phi)] - sin 2\phi - 4 sin 2\pi D + sin(4\pi D + 2\phi)}, \end{aligned} \tag{5.58}$$

which gives the input inductance L_i at the operating frequency normalized with respect to L

$$\frac{L_i}{L} = \frac{4\pi}{4\pi D + 4\pi D cos 2\pi D - 4\pi D cos[2(\pi D + \phi)] - sin 2\phi - 4 sin 2\pi D + sin(4\pi D + 2\phi)}. \tag{5.59}$$

For $D = 0.5$, $L_i/L = 2(\pi^2 + 4)/(\pi^2 - 4) = 4.726$. Figure 5.21 shows plots of L_i/L versus D and versus $R_L/\omega L$. L_i/L increases from 1 to ∞ as D decreases from 1 to 0, or as $R_L/\omega L$ increases from 0 to ∞.

Table 5.2 gives the numerical values of the rectifier parameters for various values of D.

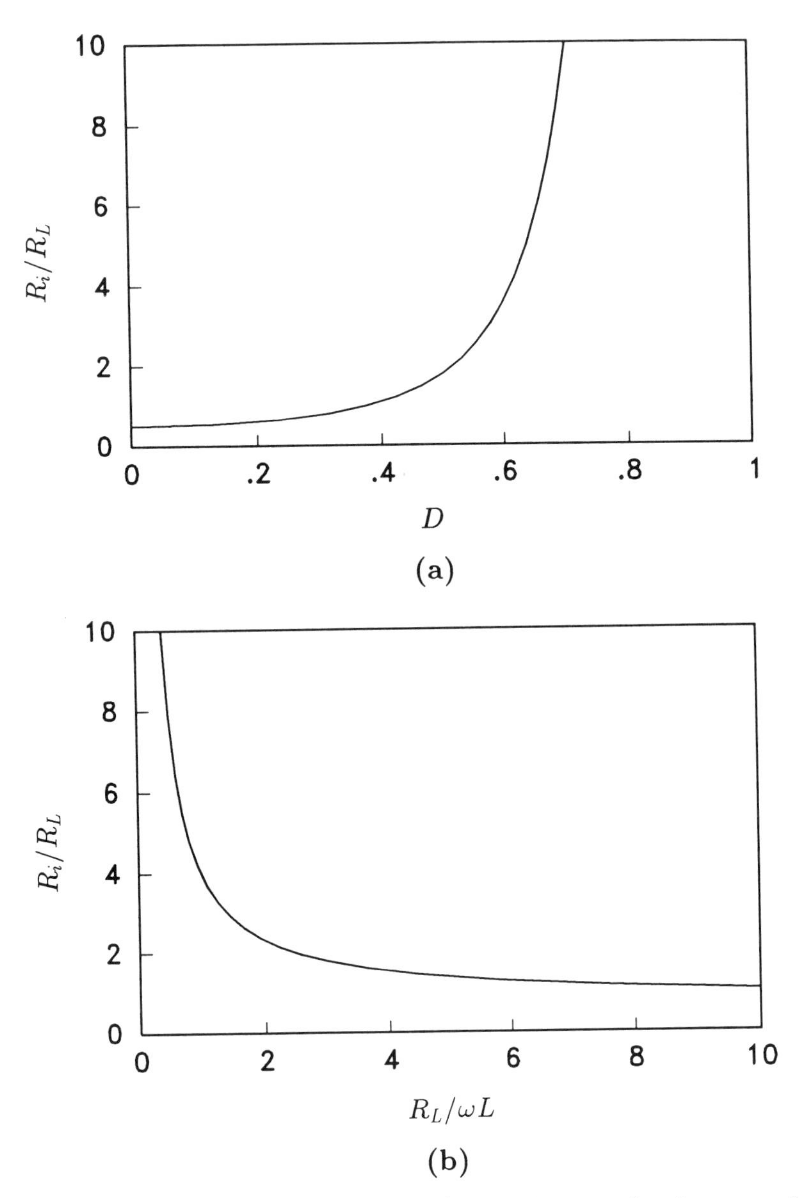

Figure 5.19: Normalized input resistance R_i/R_L at the operating frequency f. (a) R_i/R_L versus D. (b) R_i/R_L versus $R_L/\omega L$.

5.3.8 Design Example

Example 5.2

Design a Class E low di/dt rectifier with a series inductance to meet the following specifications: $V_O = 5$ V, $R_L = 25\ \Omega$ to infinity, $f = 150$ kHz, and

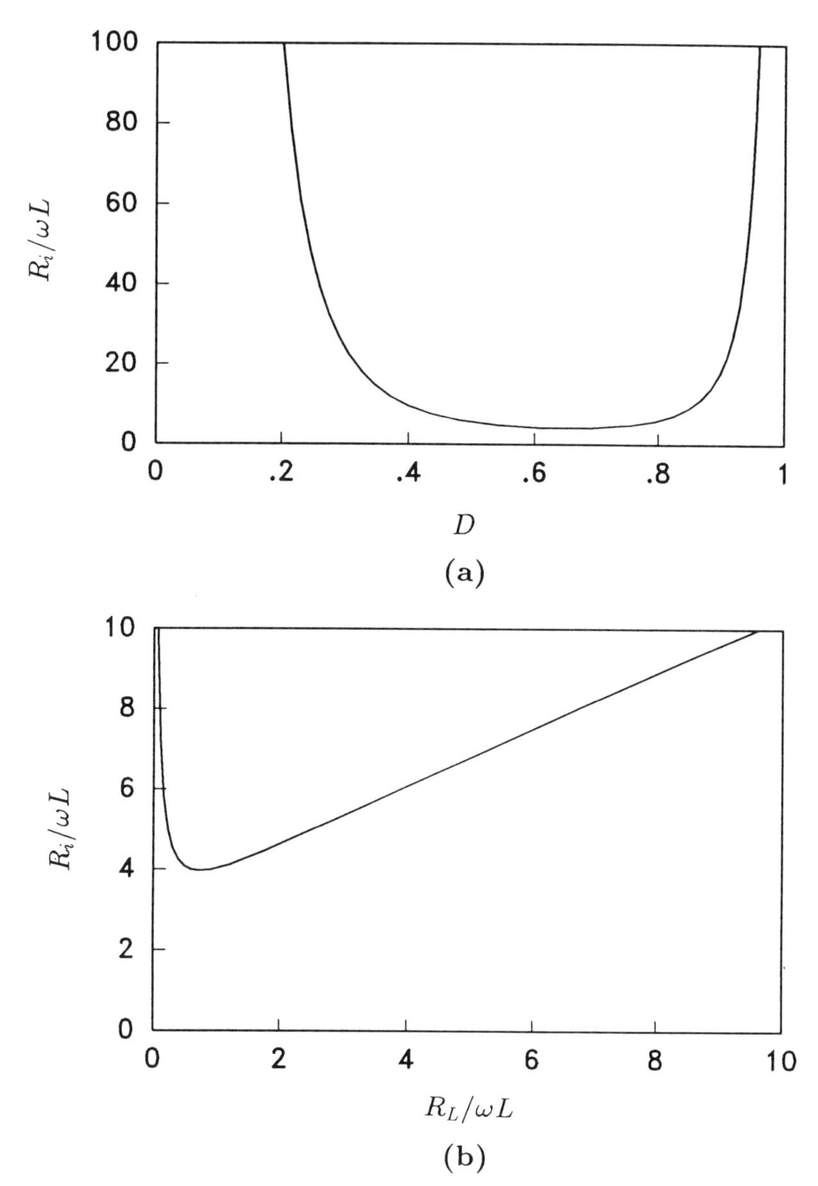

Figure 5.20: Normalized input resistance $R_i/\omega L$ at the operating frequency f. (a) $R_i/\omega L$ versus D. (b) $R_i/\omega L$ versus $R_L/\omega L$.

$V_r/V_O \leq 5\%$. Assume the maximum value of the diode ON duty cycle D_{max} that occurs at R_{Lmin} to be 0.5.

Solution: It is sufficient to design the circuit for the minimum value of the load resistance $R_{Lmin} = 25\ \Omega$ because the peak values of all currents and voltages decrease with increasing R_L. The maximum value of the load current is $I_{Omax} = V_O/R_{Lmin} = 0.2$ A. From (5.46),

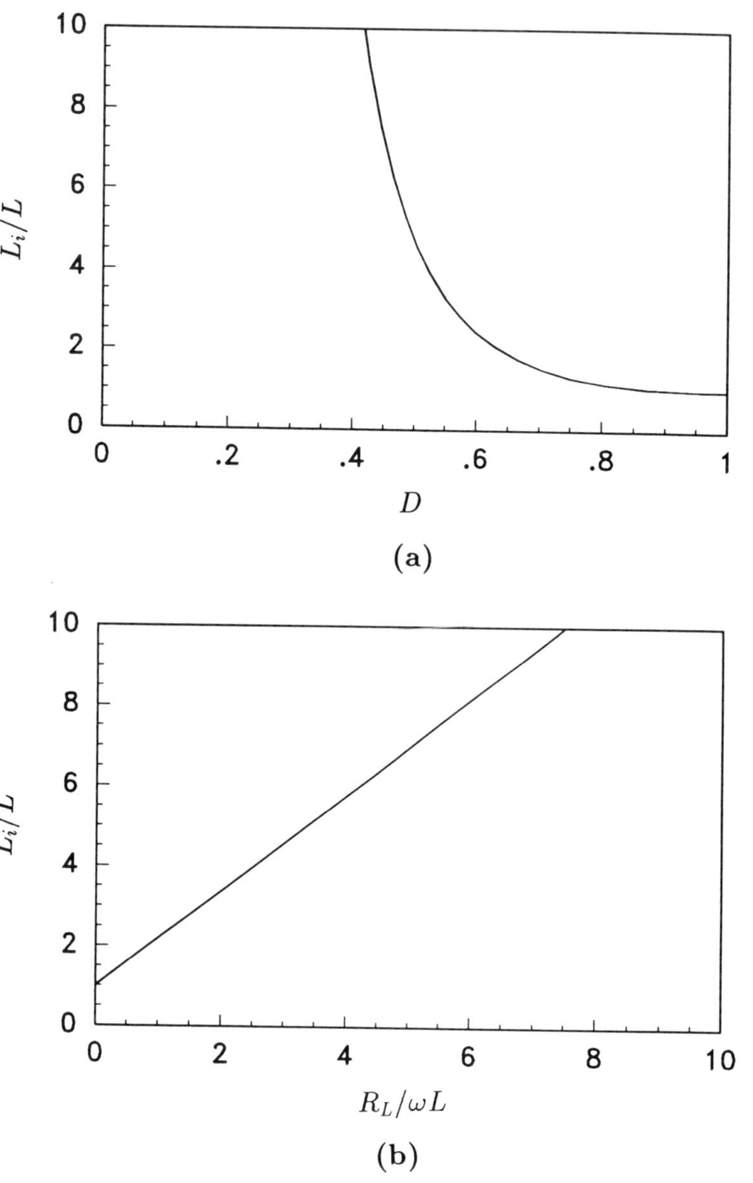

Figure 5.21: Normalized input inductance L_i/L at the operating frequency f. (a) L_i/L versus D. (b) L_i/L versus $R_L/\omega L$.

$$L = \frac{R_{Lmin}}{2\pi^2 f} = \frac{25}{2\pi^2 \times 150 \times 10^3} = 8.443\ \mu\text{H}. \tag{5.60}$$

From Table 5.2, the peak values of the diode and inductor current and the diode reverse voltage are

Table 5.2. Parameters of Class E Low *di/dt* Rectifier with a Series Inductor

D	ϕ	$R_L/\omega L$	I_{DM}/I_O	V_{DM}/V_O	R_i/R_L	$R_i/\omega L$	L_i/L	M_{VR}
0	90	∞	∞	2	0.5	∞	∞	1.4142
0.05	84.00	46288.7695	35.556	2.006	0.5055	23399.871	278873.97	1.4065
0.1	78.02	2864.4343	17.777	2.022	0.5225	1496.7294	8842.0029	1.3834
0.15	72.06	556.4025	11.582	2.051	0.5524	307.3634	1194.4158	1.3455
0.2	66.14	171.8842	8.889	2.093	0.5978	102.7504	293.8265	1.2814
0.25	60.28	68.2173	7.111	2.151	0.6629	45.2203	100.8816	1.2137
0.3	54.50	31.6111	5.927	2.228	0.7545	23.8504	42.9442	1.1346
0.35	48.80	16.2459	5.081	2.329	0.8833	14.3491	21.2838	1.0455
0.4	43.21	8.9733	4.447	2.460	1.0646	9.5697	11.8312	0.9484
0.45	37.77	4.7016	3.955	2.633	1.4003	6.5836	6.5837	0.8451
0.5	32.48	3.1416	3.562	2.862	1.7337	5.4466	4.7259	0.7595
0.55	26.40	1.7611	3.242	3.173	2.5282	4.4526	3.0983	0.6289
0.6	22.55	1.2083	2.976	3.608	3.4013	4.1098	2.4443	0.5208
0.65	17.98	0.7546	2.754	4.240	5.2484	3.9609	1.9009	0.4365
0.7	13.75	0.4655	2.566	5.207	8.8497	4.1199	1.5483	0.3170
0.75	9.93	0.2789	2.407	6.712	16.8156	4.6900	1.3167	0.2439
0.8	6.59	0.1580	2.274	8.918	37.9185	5.9908	1.1665	0.1477
0.85	3.36	0.06916	2.164	12.482	145.4796	10.0562	1.0607	0.0829
0.9	1.75	0.03338	2.079	19.404	534.7070	17.8465	1.0233	0.03525
1	0	0	2	∞	∞	∞	1	0

$$I_{DM} = 3.5621 I_O = 3.5621 \times 0.2 = 0.712 \text{ A} \quad (5.61)$$

and

$$V_{DM} = 2.862 V_O = 2.862 \times 5 = 14.3 \text{ V}, \quad (5.62)$$

respectively. From (5.53) and Table 5.2, the rms value of the input voltage is

$$V_{rms} = \frac{V_O}{M_{VR}} = \frac{5}{0.7395} = 6.761 \text{ V}. \quad (5.63)$$

Using Table 5.2, the input resistance and the input inductance are

$$R_i = 1.7337 R_{Lmin} = 1.7337 \times 25 = 43.3 \ \Omega \quad (5.64)$$

and

$$L_i = 4.726 L = 4.726 \times 8.44 \times 10^{-6} = 39.89 \ \mu\text{H}. \quad (5.65)$$

Assume that the ESR of the filter capacitor is $r_C = 20$ mΩ. From (5.19), the minimum filter capacitance is

$$C_f = \frac{I_{DM}}{\omega(V_r - I_{DM} r_C)} = \frac{0.712}{2 \times \pi \times 150 \times 10^3 (0.05 \times 5 - 0.712 \times 0.02)} = 3.2 \ \mu\text{F}. \quad (5.66)$$

5.4 SUMMARY

- The Class E low di/dt rectifiers consist of only three components: an inductor, a diode, and a filter capacitor.
- The rectifiers are driven by an ac current source.
- The diode turns on at zero di_D/dt and low $|dv_D/dt|$, and turns off at low $|di_D/dt|$ and very high $|dv_D/dt|$. Therefore, switching losses, switching noise, and reverse-recovery effects are reduced.
- The diode ON duty cycle may be large (e.g., $D = 0.75$), resulting in a low harmonic content and a low peak value of the diode current.
- In the Class E low di/dt rectifier with a series inductor, the diode lead inductance and the transformer leakage inductance are absorbed into the series inductance.
- A disadvantage of the rectifiers is that the diode junction capacitance is not included in the rectifier topology, causing parasitic oscillations when the diode is off. These oscillations can be damped with a snubber but the power dissipation increases.
- The rectifier has been successfully applied in a high-performance resonant dc-dc power converter.

5.5 REFERENCES

1. M. K. Kazimierczuk and J. Jóźwik, "Class E resonant rectifiers," Proc. 31st Midwest Symp. Circuits and System, St. Louis, MO, August 10–12, 1988, pp. 138–141.
2. N. Mohan, T. M. Undeland, and W. P. Robbins, *Power Electronics: Converters, Applications and Design*, New York: John Wiley & Sons, 1989, pp. 26–32.
3. M. K. Kazimierczuk and J. Jóźwik, "Class E zero-voltage-switching and zero-current-switching rectifiers," *IEEE Trans. Circuits Syst.*, vol. CAS-37, pp. 436–444, March 1990.
4. J. Jóźwik and M. K. Kazimierczuk, "Class E zero-current-switching rectifier with a parallel inductor," Proc. Natl. Aerospace and Electronics Conf. (NAECON'89), Dayton, OH, May 22–26, 1989, vol. 1, pp. 233–239.
5. M. K. Kazimierczuk and J. Jóźwik, "Analysis and design of Class E zero-current-switching rectifier," *IEEE Trans. Circuits Syst.*, vol. CAS–37, pp. 1000–1009, August 1990.
6. M. K. Kazimierczuk and W. Szaraniec, "Analysis of Class E low di/dt rectifier with a series inductor," *IEEE Trans. Aerospace and Electronic Systems,* vol. AES-29, pp. 228–287, January 1993.
7. A. Ivascu, M. K. Kazimierczuk, and S. Birca-Galateanu, "Class E low di/dt rectifier," *Proc. Inst. Elec. Eng., Pt. G, Circuits, Systems and Devices,* vol. 141, pp. 417–423, December 1993.

5.6 REVIEW QUESTIONS

5.1 What is the value of di/dt at the diode turn-on transition?

5.2 What is the value of di/dt at the diode turn-off transition?

5.3 What is the value of dv/dt at the diode turn-on transition?

5.4 What is the value of dv/dt at the diode turn-off transition?

5.5 Is the leakage inductance of the transformer included in the topology of the rectifier with a parallel inductor?

5.6 Is the leakage inductance of the transformer included in the topology of the rectifier with a series inductor?

5.7 Is the diode capacitance included in the topologies of Class E low di/dt rectifiers?

5.8 Is the diode on-duty cycle dependent on the load resistance in Class E low di/dt rectifiers?

5.9 Is the diode on-duty cycle dependent on the output filter components in Class E low di/dt rectifiers?

5.7 PROBLEMS

5.1 Perform step-by-step integration in equation (5.24).

5.2 Design a Class E low di/dt rectifier with a parallel inductor shown in Fig. 5.1(a). The following specifications should be met: $V_O = 20$ V, $R_L = 10\ \Omega$ to infinity, $f = 0.5$ MHz, and $D = 0.5$ at the full load.

5.3 Design a Class E low di/dt rectifier with a series inductance that satisfies the following specifications: $V_O = 15$ V, $R_L = 10\ \Omega$ to infinity, and $f =$ 200 kHz. Assume that the maximum value of the diode ON duty cycle D_{max} cannot exceed 0.5.

PART II

INVERTERS

CHAPTER 6

CLASS D SERIES RESONANT INVERTER

6.1 INTRODUCTION

Class D dc-ac resonant inverters, also called Class D resonant amplifiers, were invented in 1959 by Baxandall [1] and have been widely used in various applications [2]–[12] to convert dc energy into ac energy. Examples of applications of resonant inverters are dc-dc resonant converters, radio transmitters, solid-state electronic ballasts for fluorescent lamps, high-frequency electric process heating applied in induction welding, surface hardening, soldering and annealing, induction sealing for tamper-proof packaging, fiber-optics production, and dielectric heating for plastic welding. Class D inverters can be classified into two groups:

- Class D voltage-source (or voltage-switching) inverters
- Class D current-source (or current-switching) inverters

Class D voltage-switching inverters are fed by a dc voltage source. They employ a series-resonant circuit or a resonant circuit that is derived from the series-resonant circuit. If the loaded quality factor is sufficiently high, the current through the resonant circuit is sinusoidal and the currents through the switches are half-wave sinusoids. The voltages across the switches are square waves.

In contrast, the Class D current-switching inverters are fed by a dc current source. They include a parallel-resonant circuit or a resonant circuit that is derived from the parallel resonant circuit. The voltage across the resonant circuit is sinusoidal for high values of the loaded quality factor. The voltages across the switches are half-wave sinusoids, and the currents through the switches are square waves.

One of the main advantages of Class D voltage-switching inverters is the low voltage across the transistors, which is equal to the supply voltage. This makes them suitable for high-voltage applications, where, for example, a 220 V or 277 V rectified line voltage is used to supply the inverters. In addition, low-voltage MOSFETs can be used. Such MOSFETs have low on-resistances, reducing conduction losses and operating junction temperatures; this yields high efficiencies. The MOSFET's on-resistance r_{DS} increases considerably with increasing junction temperature. This causes the conduction loss $r_{DS}I_{rms}^2$ to increase, where I_{rms} is the rms value of the drain current. Typically, r_{DS} doubles as the temperature rises by 100°C (for example, from 25°C to 125°C), doubling the conduction loss. The MOSFET's on-resistance r_{DS} increases with temperature T because both the mobility of electrons $\mu_n \approx K_1/T^{2.5}$ and the mobility of holes $\mu_p \approx K_2/T^{2.7}$ decrease with T over a temperature range of 100 to 400 K, where K_1 and K_2 are constants. In many applications, the output power or the output voltage should be controlled by varying the operating frequency f (FM control).

In this chapter we study Class D half-bridge and full-bridge series resonant inverters. The design procedure of inverters is illustrated by detailed examples. Also presented are relationships among inverters and rectifiers.

6.2 CIRCUIT DESCRIPTION

A circuit of the Class D voltage-switching half-bridge series-resonant inverter (SRI) [1]–[12] is shown in Fig. 6.1. It consists of two bidirectional switches S_1 and S_2 and a series-resonant circuit L-C-R_i. Each switch is composed of a transistor and an antiparallel diode. The MOSFET's intrinsic body-drain *pn* junction diode may be used as an antiparallel diode in the case of inductive load, as will be discussed shortly. The switch can conduct either positive or negative current. However, it can only take voltages higher than about −1 V. A positive or negative switch current can flow through the transistor if the transistor is ON. If the transistor is OFF, the switch can conduct only a negative current that flows through the diode. The transistors are driven by nonoverlapping rectangular-wave voltages v_{GS1} and v_{GS2} with a dead time at the operating frequency $f = 1/T$. Switches S_1 and S_2 are alternately ON and OFF, with a duty ratio of 50% or slightly less. The dead time is the time interval when both switching devices are off. Input resistance R_i is an ac load to which the ac power is to be delivered. If the inverter is a part of a dc-dc resonant converter, R_i represents an input resistance of a rectifier.

Equivalent circuits of the Class D inverter are shown in Fig. 6.1(b)–(d). In Fig. 6.1(b), the MOSFETs are modeled by switches whose on-resistances are r_{DS1} and r_{DS2}. Resistance r_L is the equivalent series resistance (ESR) of the physical inductor L and resistance r_C is the equivalent series resistance

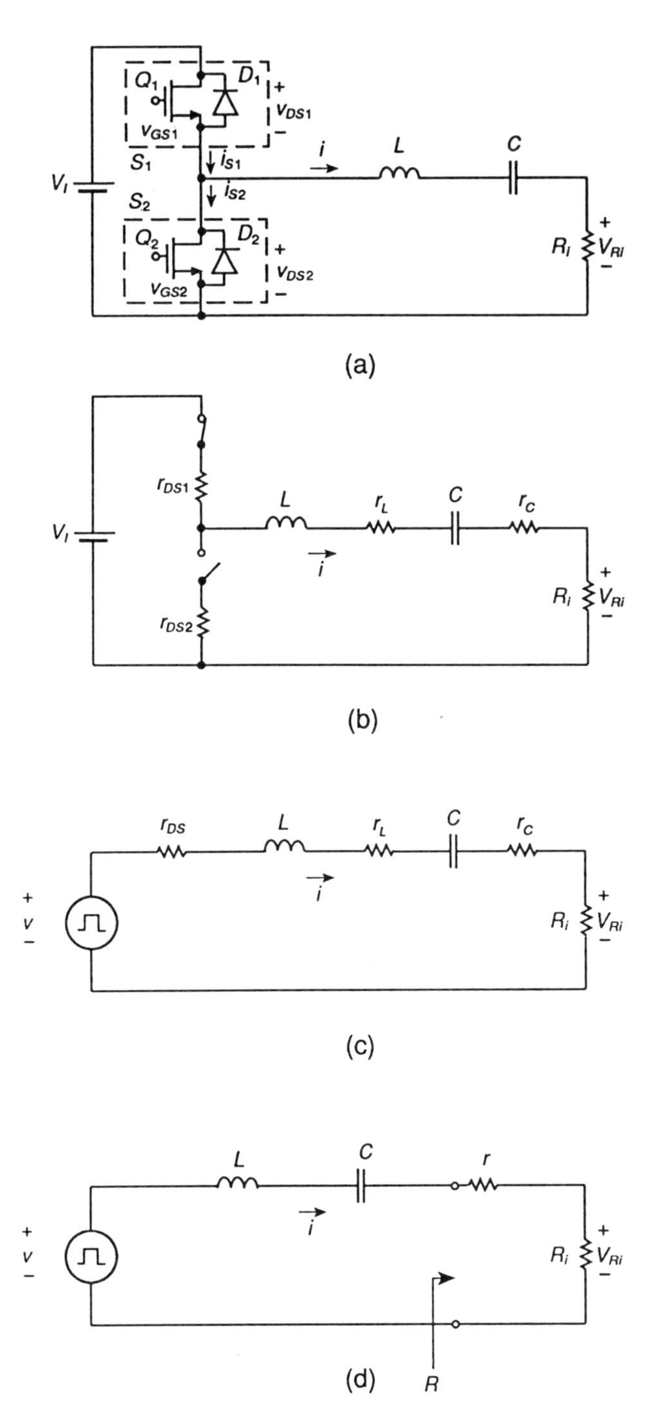

Figure 6.1: Class D voltage-source half-bridge inverter with a series-resonant circuit. (a) Circuit. (b)–(d) Equivalent circuits.

of the physical capacitor C. In Fig. 6.1(c), $r_{DS} \approx (r_{DS1} + r_{DS2})/2$ represents the average equivalent on-resistance of the MOSFETs. In Fig. 6.1(d), the total parasitic resistance is represented by $r \approx r_{DS} + r_L + r_C$, which yields the overall resistance $R = R_i + r \approx R_i + r_{DS} + r_L + r_C$.

6.3 PRINCIPLE OF OPERATION

The principle of operation of the Class D inverter is explained by the waveforms sketched in Fig. 6.2. The voltage at the input of the series-resonant circuit is a square wave of magnitude V_I. If the loaded quality factor $Q_L = \sqrt{L/C}/R$ of the resonant circuit is high enough (e.g., $Q_L \geq 2.5$), the current i through this circuit is nearly a sine wave. Only at $f = f_o$, the MOSFETs turn on and off at zero current, resulting in zero switching losses and high

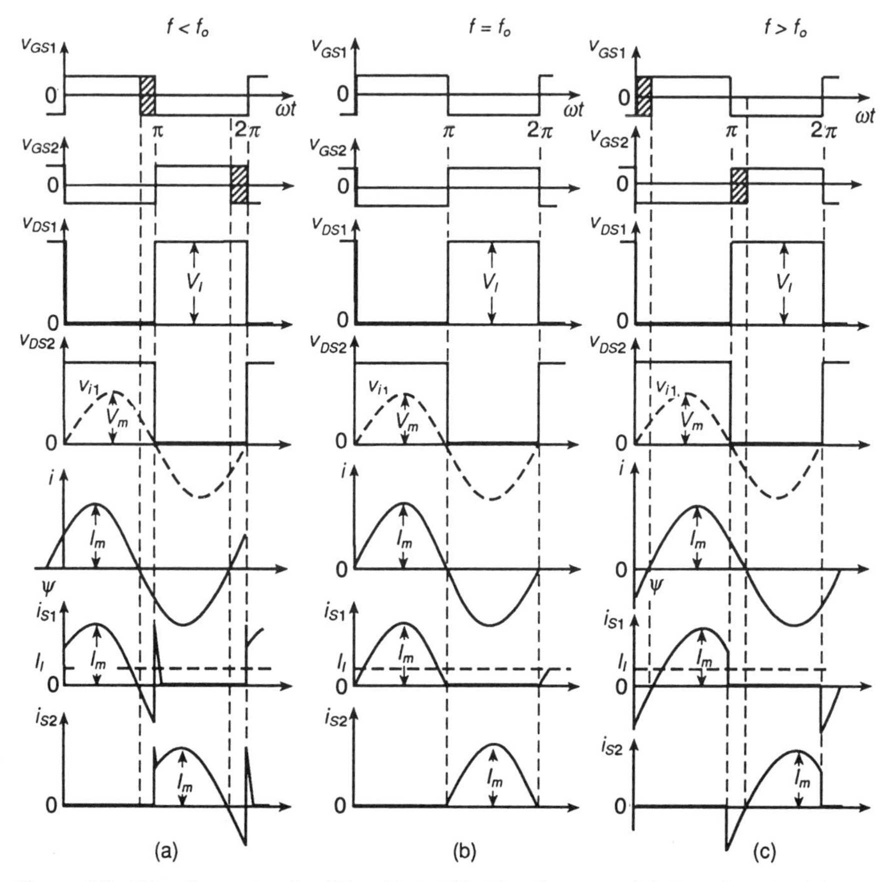

Figure 6.2: Waveforms in the Class D half-bridge inverter. (a) For $f < f_o$. (b) For $f = f_o$. (c) For $f > f_o$.

efficiency. In this case, the antiparallel diode never conducts. In many applications, the operating frequency f is not equal to the resonant frequency $f_o = 1/(2\pi\sqrt{LC})$ because the output power or the output voltage is often controlled by varying the operating frequency f (FM control). Figure 6.2(a), (b), and (c) shows the waveforms for $f < f_o$, $f = f_o$, and $f > f_o$, respectively. The tolerance of the gate drive voltage is indicated by the shaded areas. Each transistor should be turned off for $f < f_o$ or turned on for $f > f_o$ during the time interval when the switch current is negative. During this time interval, the switch current can flow through the antiparallel diode. To prevent cross conduction (also called shoot-through current), the waveforms of the drive voltages v_{GS1} and v_{GS2} should be nonoverlapping and have a sufficient dead time (not shown in Fig. 6.2). At turn-off, MOSFETs have a delay time and BJTs have a storage time. If the dead time is too short, one transistor still remains on while the other turns on. Consequently, both transistors are ON at the same time and the power supply V_I is short-circuited by the small transistor on-resistances r_{DS1} and r_{DS2}. For this reason, cross-conduction current pulses of magnitude $I_{pk} = V_I/(r_{DS1} + r_{DS2})$ flow through the transistors. For example, if $V_I = 200$ V and $r_{DS1} = r_{DS2} = 0.5\ \Omega$, $I_{pk} = 200$ A! The excessive current stresses may cause immediate failure of the devices. The dead time should not be too long, as will be discussed in Sections 6.3.1 and 6.3.2. The maximum dead time increases as f/f_o increases for $f > f_o$ or decreases for $f < f_o$ because the time interval during which the switch current is negative becomes longer. The shortest dead time must be at $f = f_o$. There are commercial IC drivers available that have an adjustable dead time, for example, UC 2525 (Unitrode).

6.3.1 Operation Below Resonance

For $f < f_o$, the series-resonant circuit represents a capacitive load. It means that the current through the resonant circuit i leads the fundamental component v_{i1} of the voltage v_{DS2} by the phase angle $|\ \psi\ |$, where $\psi < 0$. Therefore, the switch current is positive after switch turn-on and is negative before switch turn-off. The conduction sequence of the semiconductor devices is Q_1-D_1-Q_2-D_2. Notice that the current of the resonant circuit is diverted from the diode of one switch to the transistor of the other switch (Fig. 6.1). This causes a lot of problems as will be explained shortly. Consider the turn-on of switch S_2 in Fig. 6.1. Prior to this transition, the current i flows through antiparallel diode D_1 of switch S_1. When transistor Q_2 is turned on by the drive voltage v_{GS2}, v_{DS2} is decreased, causing v_{DS1} to increase. Therefore, diode D_1 turns off and the current i is diverted from D_1 to Q_2. There are three detrimental effects at turn-on of the MOSFET:

1) Reverse recovery of the antiparallel diode of the opposite switch.
2) Discharging the transistor output capacitance.
3) Miller's effect.

The most severe drawback of operation below resonance is the diode reverse-recovery stress when the diode turns off. The MOSFET's intrinsic body-drain diode is a minority carrier device. Each diode turns off at a very large dv/dt and therefore at a very large di/dt, generating a high reverse-recovery current spike (turned upside down). This spike flows through the other transistor because it cannot flow through the resonant circuit. The resonant inductor L does not allow for abrupt current changes. Consequently, the spikes occur in the switch current waveform at both the turn-on and turn-off transitions of the switch. The magnitude of these spikes can be much (e.g., 10 times) higher than the magnitude of the steady-state switch current. High current spikes may destroy the transistors and always cause a considerable increase in switching losses and noise. During a part of the reverse-recovery interval, the diode voltage increases from −1 V to V_I, and both the diode current and voltage are simultaneously high, causing a high reverse-recovery power loss.

The turn-off failure of the power MOSFETs may be caused by the *second breakdown* of the parasitic bipolar transistor. This parasitic bipolar transistor is an integral part of a power MOSFET structure. The body region serves as the base of the parasitic BJT, the source as the BJT emitter, and drain as the BJT collector. If the body-drain diode is forward biased just prior to the sudden application of drain voltage, the turn-on process of the parasitic bipolar transistor may be initiated by the reverse-recovery current of the antiparallel diode. The second breakdown of the parasitic BJT may destroy the power MOSFET structure. The second breakdown voltage is usually one-half of the voltage at which the device fails if the diode has not been forward biased, that is, the manufacturer-rated voltage capability V_{DSS}. For the reasons given above, operation at $f < f_o$ should be avoided if power MOSFETs are used as switches.

The current spikes can be reduced, for example, by adding Schottky antiparallel diodes if V_I is low. Schottky diodes have low breakdown voltages, typically below 100 V. Since the forward voltage of the Schottky diode is lower than that of the *pn* junction body diode, most if not all of the negative switch current flows through the Schottky diode, reducing the reverse-recovery current of the *pn* junction body diode. Another solution is to add a diode in series with the MOSFET and an ultrafast diode in parallel with the series combination of the MOSFET and diode. This arrangement does not allow the intrinsic diode to conduct and thereby to store the excess minority charge. However, the higher parts count, additional cost, and the voltage drop across the series diode that reduces the efficiency are undesirable. Also, the peak voltages of the transistor and the series diode may become much higher than V_I. Consequently, transistors with a higher permissible voltage and, therefore, a higher on-resistance should be used, increasing conduction loss. This method may reduce but cannot eliminate the spikes. Snubbers can be used to slow down the switching process, and reverse-recovery spikes can be reduced by connecting small inductances in series with the power MOSFETs.

For $f < f_o$, the turn-off switching loss is zero, but the turn-on switching loss is not zero. The transistors are turned on at a high voltage, equal to V_I. When the transistor is turned on, its output capacitance is discharged, causing a switching loss. Suppose that the upper MOSFET is initially ON and the output capacitance C_{out} of the upper transistor is initially discharged. When the upper transistor is turned off, the energy drawn from the dc input voltage source V_I to the inverter to charge the output capacitance C_{out} from 0 to V_I is

$$W_I = \int_0^T V_I i_{Cout} dt = V_I \int_0^T i_{Cout} dt = V_I Q, \tag{6.1}$$

where i_{Cout} is the charging current of the output capacitance and Q is the charge transferred from the source V_I to the capacitor. This charge equals the integral of the current i_{Cout} over the charging time interval which is usually much shorter than period T of the operating frequency f. Equation (6.1) holds true for both linear and nonlinear capacitances. If the transistor output capacitance is assumed to be linear, $Q = C_{out} V_I$ and (6.1) becomes

$$W_I = C_{out} V_I^2. \tag{6.2}$$

The energy stored in the linear output capacitance at the voltage V_I is

$$W_C = \frac{1}{2} C_{out} V_I^2. \tag{6.3}$$

The charging current flows through a resistance that consists of the on-resistance of the bottom MOSFET and lead resistances. The energy dissipated in this resistance is

$$W_R = W_I - W_C = \frac{1}{2} C_{out} V_I^2, \tag{6.4}$$

which is the same amount of energy as that stored in the capacitance. Note that charging a linear capacitance from a dc voltage source through a resistance requires twice the energy that is stored in the capacitance. When the upper MOSFET is turned on, its output capacitance is discharged through the on-resistance of the upper MOSFET, dissipating energy in that transistor. Thus, the energy dissipated during charging and discharging the transistor output capacitance is

$$W_{sw} = C_{out} V_I^2. \tag{6.5}$$

Accordingly, the turn-on switching loss per transistor is

$$P_{ton} = \frac{W_C}{T} = \frac{1}{2} f C_{out} V_I^2. \tag{6.6}$$

The total power loss associated with charging and discharging the transistor output capacitance of each MOSFET is

$$P_{sw} = \frac{W_{sw}}{T} = fC_{out}V_I^2. \tag{6.7}$$

Charging and discharging process of the output capacitance of the bottom transistor is similar. In reality, the drain-source *pn* step junction capacitance is nonlinear. An analysis of turn-on switching loss with a nonlinear transistor output capacitance is given in Section 6.7.2.

Another effect that should be considered at turn-on of the MOSFET is Miller's effect. Since the gate-source voltage increases and the drain-source voltage decreases during the turn-on transition, Miller's effect is significant, increasing the transistor input capacitance and the gate drive charge and power requirements, and reducing the turn-on switching speed.

An advantage of operation below resonance is that the transistors are turned off at nearly zero voltage, resulting in *zero turn-off switching loss*. For example, the drain-source voltage v_{DS1} is held at about -1 V by the antiparallel diode D_1 when i_{S1} is negative. During this time interval, transistor Q_1 is turned off by drive voltage v_{GS1}. The drain-to-source voltage v_{DS1} is almost zero and the drain current is low during the MOSFET turn-off, yielding the zero turn-off switching loss in the MOSFET. Since v_{DS1} is constant, Miller's effect is absent during turn-off, the transistor input capacitance is not increased by Miller's effect, the gate drive requirement is reduced, and the turn-off switching speed is enhanced. In summary, for $f < f_o$, there is a turn-on switching loss in the transistor and a turn-off (reverse-recovery) switching loss in the diode. The transistor turn-off and the diode turn-on are lossless.

As already mentioned, the drive voltages v_{GS1} and v_{GS2} are nonoverlapping and have a dead time. However, this dead time should not be made too long. If transistor Q_1 is turned off too early when the switch current i_{S1} is still positive, diode D_1 cannot conduct and diode D_2 turns on, decreasing v_{DS2} to -0.7 V and increasing v_{DS1} to V_I. When the current through D_2 reaches zero, diode D_1 turns on, v_{DS1} decreases to -0.7 V, and v_{DS2} increases to V_I. These two additional transitions of each of the switch voltages would result in switching losses. Note that only the turn-on transition of each switch is *forced* and directly controllable by the driver, while the turn-off transition is caused by the turn-on of the opposite transistor (i.e., it is automatic).

In very high power applications, thyristors with antiparallel diodes can be used as switches in Class D SRI topologies. An advantage of such a solution is that, for operation below resonance, thyristors are turned off naturally when the switch current crosses zero. Thyristors, however, require more complicated and powerful drive circuitry, and their operating frequency range is limited to about 20 kHz. Such relatively low frequencies make the size and weight of resonant components large, which increases conduction losses.

6.3.2 Operation Above Resonance

For $f > f_o$, the series-resonant circuit represents an inductive load and the current i lags behind the voltage v_{i1} by the phase angle ψ, where $\psi > 0$. Hence, the switch current is negative after turn-on (for part of the switch "on" interval) and positive before turn-off. The conduction sequence of the semiconductor devices is D_1-Q_1-D_2-Q_2. Consider the turn-off of switch S_1. When transistor Q_1 is turned off by the drive voltage v_{GS1}, v_{DS1} increases, causing v_{DS2} to decrease. As v_{DS2} reaches -0.7 V, D_2 turns on and the current i is diverted from transistor Q_1 to diode D_2. Thus, the turn-off switch transition is *forced* by the driver, while the turn-on transition is caused by the turn-off transition of the opposite transistor, not by the driver. Only the turn-off transition is directly controllable.

The transistors are turned on at *zero voltage*. In fact, there is a small negative voltage of the antiparallel diode, but this voltage is negligible in comparison to the input voltage V_I. For example, transistor Q_2 is turned on by v_2 when i_{S2} is negative. Voltage v_{DS2} is maintained at about -1 V by the antiparallel diode D_2 during the transistor turn-on transition. Therefore, the turn-on switching loss is eliminated, Miller's effect is absent, transistor input capacitance is not increased by Miller's effect, the gate drive power is low, and the turn-on switching speed is high. The diodes turn on at a very low di/dt. The diode reverse-recovery current is a fraction of a sine wave and becomes a part of the switch current when the switch current is positive. Therefore, the antiparallel diodes can be slow, and MOSFET's body-drain diodes are sufficiently fast as long as the reverse-recovery time is less than one-half of the cycle. The diode voltage is kept at a low voltage of the order of 1 V by the transistor in the on-state during the reverse-recovery interval, reducing the diode reverse-recovery power loss. The transistor can be turned on not only when the switch current is negative but also when the switch current is positive and the diode is still conducting because of the reverse recovery. Therefore, the range of the on-duty cycle of the gate-source voltages and the dead time can be larger. If, however, the dead time is too long, the current will be diverted from the recovered diode D_2 to diode D_1 of the opposite transistor until transistor Q_2 is turned on, causing extra transitions of both switch voltages, current spikes, and switching losses.

For $f > f_o$, the turn-on switching loss is zero, but there is a turn-off loss in the transistor. Both the switch voltage and current waveforms overlap during turn-off, causing a turn-off switching loss. Also, Miller's effect is considerable, increasing the transistor input capacitance, the gate drive requirements, and reducing the turn-off speed. An approximated analysis of the turn-off switching loss is given in Section 6.7.3. In summary, for $f > f_o$, there is a turn-off switching loss in the transistor, while turn-on of the transistor and the diode are lossless. The turn-off switching loss can be eliminated by adding a shunt capacitor to one of the transistors and using a dead time in the drive voltages, as discussed in Chapter 10.

6.4 TOPOLOGIES OF CLASS D VOLTAGE-SOURCE INVERTERS

Figure 6.3 shows a Class D voltage-switching inverter with various resonant circuits. These resonant circuits are derived from a series-resonant circuit. In Fig. 6.3(b), C_c is a large coupling capacitor, which can also be connected in series with the load resistance. The resonant frequency (i.e., the boundary between the capacitive and inductive load) for the circuits of Fig. 6.3(b)–(g) depends on the load. The resonant circuit shown in Fig. 6.3(b) is employed in a parallel resonant converter. The circuit of Fig. 6.3(d) is used in a series-parallel resonant converter. The circuit of Fig. 6.3(e) is used in a CLL resonant converter. Resonant circuits of Figs. 6.3(a), (f), and (g) supply a sinusoidal output current. Therefore, they are compatible with current-driven rectifiers. The inverters of Figs. 6.3(b)–(e) produce a sinusoidal voltage output and are compatible with voltage-driven rectifiers. A high-frequency transformer can be inserted in places indicated in Fig. 6.3.

Half-bridge topologies of the Class D voltage-switching inverter are depicted in Fig. 6.4. They are equivalent for ac components to the basic topology of Fig. 6.1. Figure 6.4(a) shows a half-bridge inverter with two dc voltage sources. The bottom voltage source $V_I/2$ acts as a short circuit for the current through the resonant circuit, resulting in the circuit of Fig. 6.4(b). A drawback of this circuit is that the load current flows through internal resistances of the dc voltage sources, reducing efficiency. In Fig. 6.4(c), filter capacitors $C_f/2$ act as short circuits for the ac component. The dc voltage across each of them is $V_I/2$, but the ac power is dissipated in the ESRs of the capacitors. An equivalent circuit for the inverter of Fig. 6.4(c) is shown in Fig. 6.4(d). This is a useful circuit if the dc power supply contains a voltage doubler. An advantage of this circuit in high-voltage applications is that the voltage stress across the filter capacitors is lower than in the basic circuit of Fig. 6.1. In Fig. 6.4(e), the resonant capacitor is split into two halfs, which are connected in parallel for the ac component. This is possible because the dc input voltage source V_I acts as a short circuit for the ac component for the upper capacitor. The disadvantage of all transformerless versions of half-bridge inverter of Fig. 6.4 is that the load resistance R_i is not grounded.

6.5 ANALYSIS

6.5.1 Assumptions

The analysis of the Class D inverter of Fig. 6.1 is based on the equivalent circuit of Fig. 6.1(d) and the following assumptions:

1) The transistor and diode form a resistive switch whose on-resistance is linear, the parasitic capacitances of the switch are neglected, and the switching times are zero.

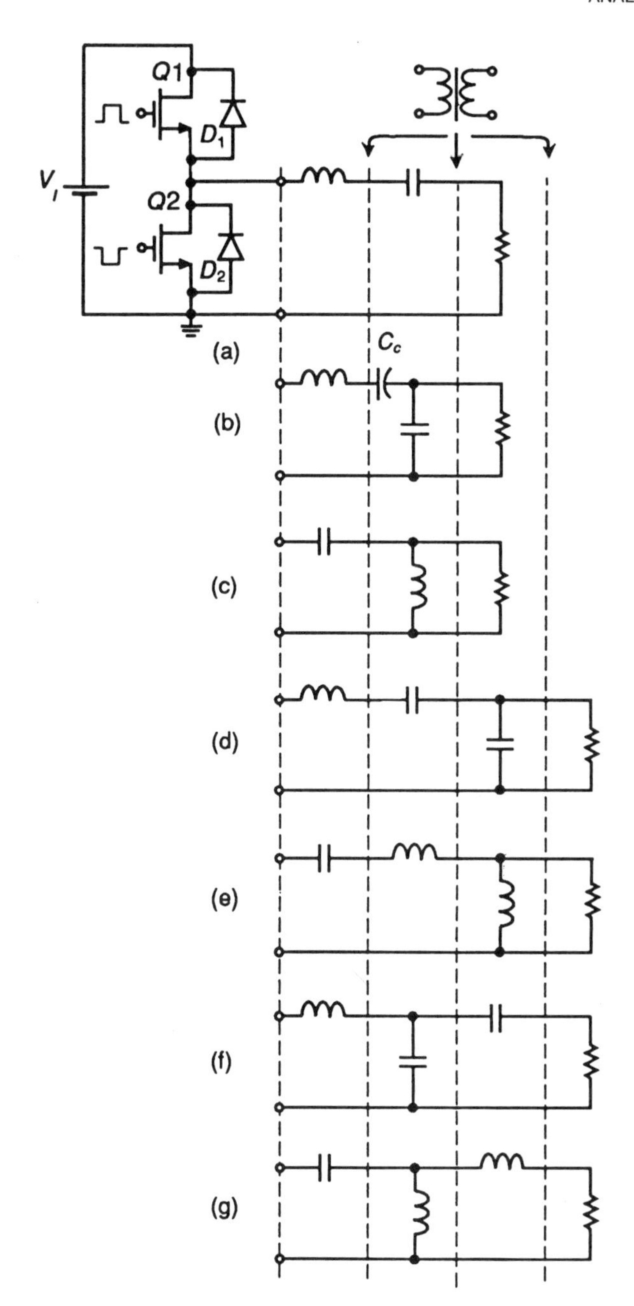

Figure 6.3: Class D voltage-switching inverter with various resonant circuits.

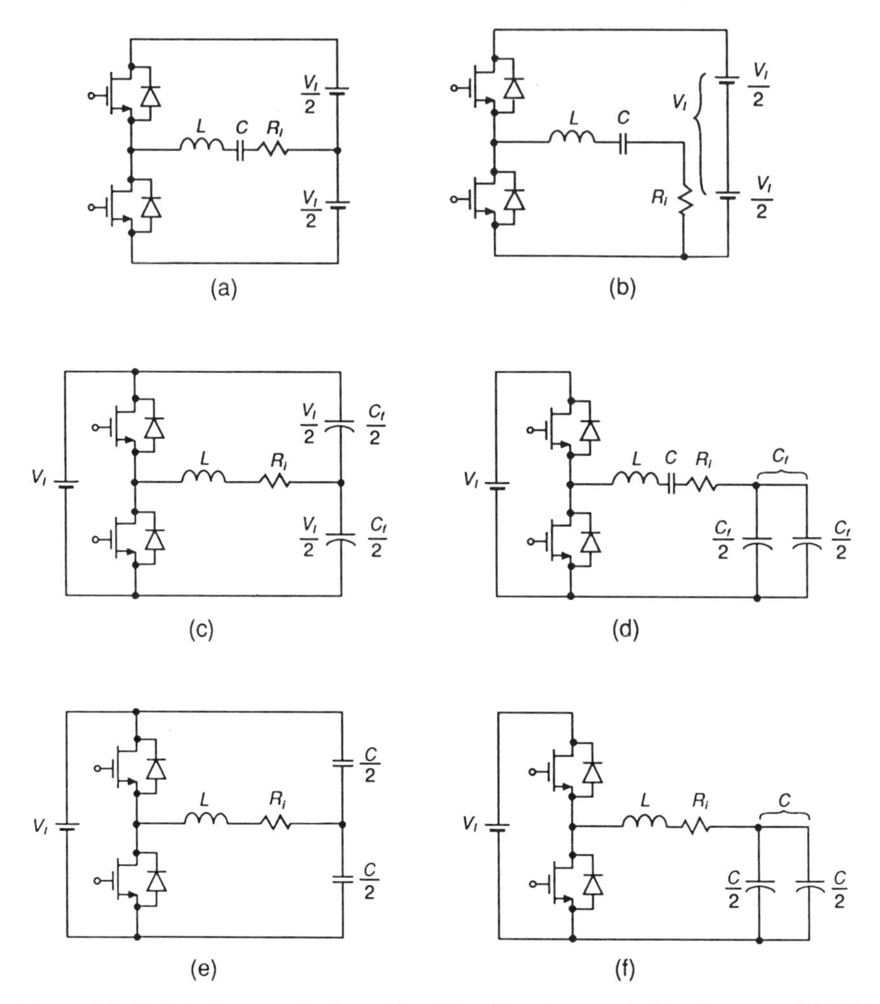

Figure 6.4: Half-bridge topologies of the Class D voltage-switching inverter. (a) With two dc voltage sources. (b) Equivalent circuit of inverter of Fig. 6.4(a). (c) With two filter capacitors. (d) Equivalent circuit of inverter of Fig. 6.4(c). (e) With a resonant capacitor split into two halfs. (f) Equivalent circuit of inverter of Fig. 6.4(e).

2) The elements of the series-resonant circuit are passive, linear, time invariant, and do not have parasitic reactive components.
3) The loaded quality factor Q_L of the series-resonant circuit is high enough so that the current i through the resonant circuit is sinusoidal.

6.5.2 Series-Resonant Circuit

The parameters of the series-resonant circuit are defined as follows:

- the resonant frequency

$$\omega_o = \frac{1}{\sqrt{LC}} \tag{6.8}$$

- the characteristic impedance

$$Z_o = \sqrt{\frac{L}{C}} = \omega_o L = \frac{1}{\omega_o C} \tag{6.9}$$

- the loaded quality factor

$$Q_L = \frac{\omega_o L}{R} = \frac{1}{\omega_o CR} = \frac{Z_o}{R} = \frac{\sqrt{\frac{L}{C}}}{R} \tag{6.10}$$

- the unloaded quality factor

$$Q_o = \frac{\omega_o L}{r} = \frac{1}{\omega_o Cr} = \frac{Z_o}{r} \tag{6.11}$$

where

$$r = r_{DS} + r_L + r_C \tag{6.12}$$

and

$$R = R_i + r. \tag{6.13}$$

The loaded quality factor is defined as

$$\begin{aligned} Q_L &\equiv 2\pi \frac{\text{Total energy stored at resonant frequency}}{\text{Energy dissipated per cycle at resonant frequency}} \\ &= 2\pi \frac{W_s}{T_o P_R} = 2\pi \frac{f_o W_s}{P_R} = \frac{\omega_o W_s}{P_{Ri} + P_r} = \frac{Q}{P_R}, \end{aligned} \tag{6.14}$$

where W_s is the total energy stored in the resonant circuit at the resonant frequency $f_o = 1/T_o$, $Q = \omega_o W_s$ is the reactive power of inductor L or capacitor C at the resonant frequency f_o. The total energy stored in the resonant circuit at any frequency is given by

$$\begin{aligned} w_s(\omega, t) &= w_L(\omega, t) + w_C(\omega, t) = \frac{1}{2} L I_m^2 sin^2(\omega t - \psi) + \frac{1}{2} C V_{Cm}^2 sin^2(\omega t - \psi - 90^\circ) \\ &= \frac{1}{2}[L I_m^2 sin^2(\omega t - \psi) + C V_{Cm}^2 cos^2(\omega t - \psi)]. \end{aligned} \tag{6.15}$$

For steady-state operation at the resonant frequency f_o, the total instantaneous energy stored in the resonant circuit is constant and equal to the maximum energy stored in the inductor

$$W_s = W_{Lmax} = \frac{1}{2}LI_m^2 \tag{6.16}$$

or, using (6.8), in the capacitor

$$W_s = W_{Cmax} = \frac{1}{2}CV_{Cm}^2 = \frac{1}{2}C\frac{I_m^2}{(\omega_o C)^2} = \frac{1}{2}\frac{I_m^2}{(C\omega_o^2)} = \frac{1}{2}LI_m^2. \tag{6.17}$$

Substitution of (6.16) and (6.17) into (6.14) produces

$$Q_L = \frac{\pi f_o LI_m^2}{P_R} = \frac{\pi f_o CV_{Cm}^2}{P_R}. \tag{6.18}$$

The reactive power of the inductor at f_o is $Q = (1/2)V_{Lm}I_m = (1/2)\omega_o LI_m^2$ and of the capacitor is $Q = (1/2)I_m V_{Cm} = (1/2)\omega_o CV_{Cm}^2$. Thus, the quality factor can be defined as a ratio of the reactive power of the inductor or a capacitor to the true power dissipated in the form of heat in R. The total power dissipated in $R = R_i + r$ is

$$P_R = \frac{1}{2}RI_m^2 = \frac{1}{2}(R_i + r)I_m^2. \tag{6.19}$$

Substitution of (6.16) or (6.17), and (6.19) into (6.14) gives (6.10).

For $R_i = 0$,

$$P_R = P_r = \frac{1}{2}rI_m^2, \tag{6.20}$$

and the unloaded quality factor is defined as

$$Q_o \equiv \frac{\omega_o W_s}{P_r} \tag{6.21}$$

resulting in (6.11). Similarly, the quality factor of the inductor is

$$Q_{Lo} \equiv \frac{\omega_o W_s}{P_{rL}} = \frac{\omega_o L}{r_L}, \tag{6.22}$$

and the capacitor is

$$Q_{Co} \equiv \frac{\omega_o W_s}{P_{rC}} = \frac{1}{\omega_o C r_C}. \tag{6.23}$$

6.5.3 Input Impedance of the Series-Resonant Circuit

The input impedance of the series-resonant circuit is

$$\mathbf{Z} = R + j\left(\omega L - \frac{1}{\omega C}\right) = R\left[1 + jQ_L\left(\frac{\omega}{\omega_o} - \frac{\omega_o}{\omega}\right)\right]$$
$$= Z_o\left[\frac{R}{Z_o} + j\left(\frac{\omega}{\omega_o} - \frac{\omega_o}{\omega}\right)\right] = Ze^{j\psi} = R + jX \tag{6.24}$$

where

$$Z = R\sqrt{1 + Q_L^2\left(\frac{\omega}{\omega_o} - \frac{\omega_o}{\omega}\right)^2} = Z_o\sqrt{\left(\frac{R}{Z_o}\right)^2 + \left(\frac{\omega}{\omega_o} - \frac{\omega_o}{\omega}\right)^2}$$
$$= Z_o\sqrt{\frac{1}{Q_L^2} + \left(\frac{\omega}{\omega_o} - \frac{\omega_o}{\omega}\right)^2} \tag{6.25}$$

$$\psi = arctan\left[Q_L\left(\frac{\omega}{\omega_o} - \frac{\omega_o}{\omega}\right)\right] \tag{6.26}$$

$$R = Zcos\psi \tag{6.27}$$

$$X = Zsin\psi. \tag{6.28}$$

Notice that the reactance of the resonant circuit becomes zero at the resonant frequency f_o. From (6.26),

$$cos\psi = \frac{1}{\sqrt{1 + Q_L^2(\frac{\omega}{\omega_o} - \frac{\omega_o}{\omega})^2}}. \tag{6.29}$$

Figure 6.5 shows a three-dimensional representation of Z/Z_o as a function of the normalized frequency f/f_o and the normalized load resistance $R/Z_o = 1/Q_L$. Plots of Z/Z_o and ψ against f/f_o at fixed values of R/Z_o are graphed in Fig. 6.6. For $f < f_o$, ψ is less than zero, which means that the resonant circuit

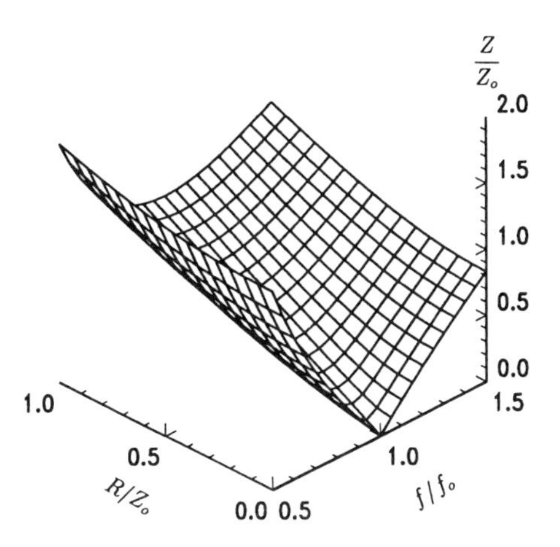

Figure 6.5: Z/Z_o as a function of f/f_o and $R/Z_o = 1/Q_L$.

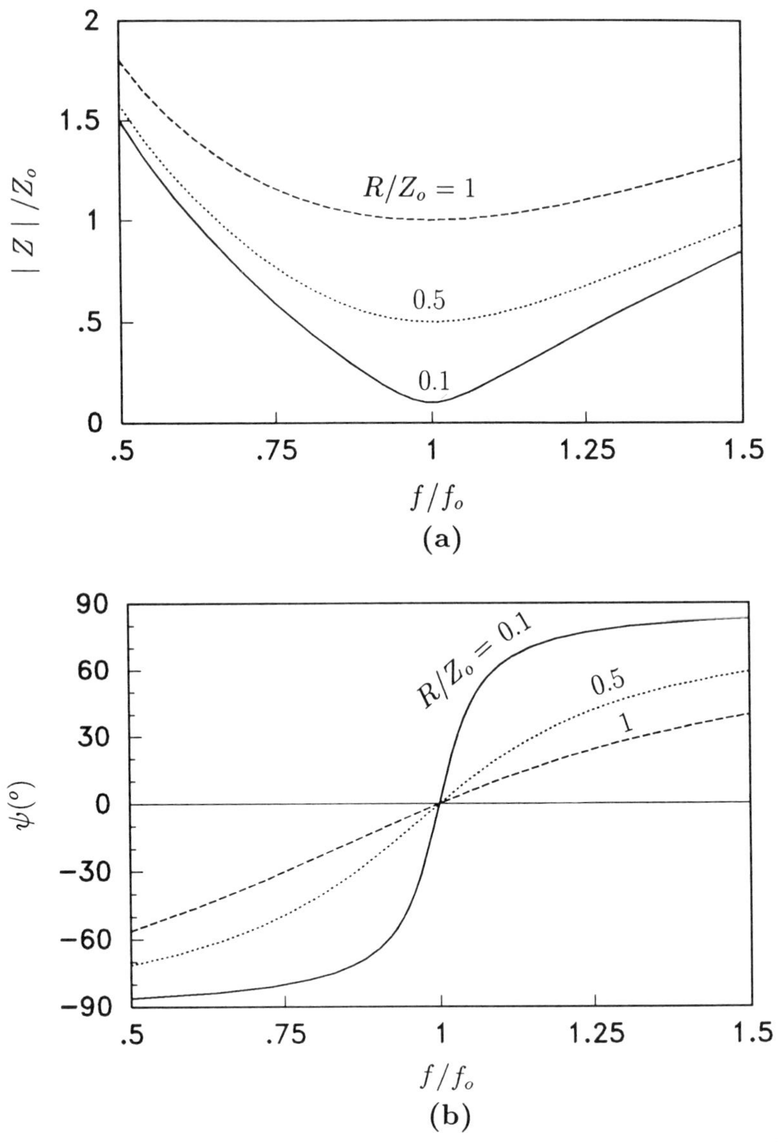

Figure 6.6: Input impedance of the series-resonant circuit against frequency f/f_o at fixed values of the normalized load resistance $R/Z_o = 1/Q_L$. (a) Modulus of the normalized input impedance Z/Z_o against f/f_o at constant values of R/Z_o. (b) Phase of the input impedance ψ against f/f_o at constant values of $R/Z_o = 1/Q_L$.

represents a capacitive load to the switching part of the inverter. For $f > f_o$, ψ is greater than zero, which indicates that the resonant circuit represents an inductive load. The magnitude of impedance Z is usually normalized with respect to R, but it is not a good normalization if R is variable and resonant components L and C are fixed.

6.5.4 Currents, Voltages, and Powers

Referring to Fig. 6.1(d), the input voltage of the series-resonant circuit is a square-wave

$$v = \begin{cases} V_I, & \text{for} \quad 0 < \omega t \le \pi \\ 0, & \text{for} \quad \pi < \omega t \le 2\pi. \end{cases} \tag{6.30}$$

This voltage can be expanded into Fourier series

$$\begin{aligned} v &= \frac{V_I}{2} + \frac{2V_I}{\pi} \sum_{n=1}^{\infty} \frac{1-(-1)^n}{2n} sin(n\omega t) \\ &= V_I \left(\frac{1}{2} + \frac{2}{\pi} sin\omega t + \frac{2}{3\pi} sin3\omega t + \frac{2}{5\pi} sin5\omega t + \cdots \right). \end{aligned} \tag{6.31}$$

The fundamental component of voltage v is

$$v_{i1} = V_m sin\omega t, \tag{6.32}$$

where its amplitude is given by

$$V_m = \frac{2V_I}{\pi} \approx 0.637 V_I. \tag{6.33}$$

This leads to the rms value of voltage v_{i1}

$$V_{rms} = \frac{V_m}{\sqrt{2}} = \frac{\sqrt{2}V_I}{\pi} \approx 0.45 V_I. \tag{6.34}$$

If the operating frequency f is close to the resonant frequency f_o, the impedance of the resonant circuit is very high for higher harmonics and, therefore, the current through the resonant circuit is approximately sinusoidal equal to the fundamental component

$$i = I_m sin(\omega t - \psi), \tag{6.35}$$

where from (6.28), (6.29), and (6.33)

$$\begin{aligned} I_m &= \frac{V_m}{Z} = \frac{2V_I}{\pi Z} = \frac{2V_I cos\psi}{\pi R} = \frac{2V_I}{\pi R \sqrt{1 + Q_L^2 (\frac{\omega}{\omega_o} - \frac{\omega_o}{\omega})^2}} \\ &= \frac{2V_I}{\pi Z_o \sqrt{(\frac{R}{Z_o})^2 + (\frac{\omega}{\omega_o} - \frac{\omega_o}{\omega})^2}}. \end{aligned} \tag{6.36}$$

Figure 6.7 shows a three-dimensional representation of $I_m Z_o / V_I$ as a function of f/f_o and R/Z_o. Plots of $I_m Z_o / V_I$ against f/f_o at fixed values of R/Z_o are

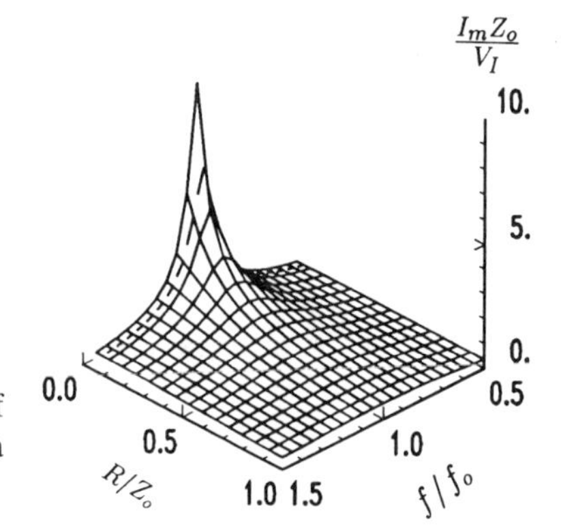

Figure 6.7: Normalized amplitude $I_m Z_o / V_I$ of the current through the resonant circuit as a function of f/f_o and $R/Z_o = 1/Q_L$.

depicted in Fig. 6.8. It can be seen that high values of $I_m Z_o / V_I$ occur at the resonant frequency f_o and low total resistance R.

The output voltage is also sinusoidal

$$v_{Ri} = iR_i = V_{Rim} sin(\omega t - \psi). \tag{6.37}$$

The input current of the inverter i_I equals the current through the switch S_1 and is given by

$$i_I = i_{S1} = \begin{cases} I_m sin(\omega t - \psi), & \text{for} \quad 0 < \omega t \leq \pi \\ 0, & \text{for} \quad \pi < \omega t \leq 2\pi. \end{cases} \tag{6.38}$$

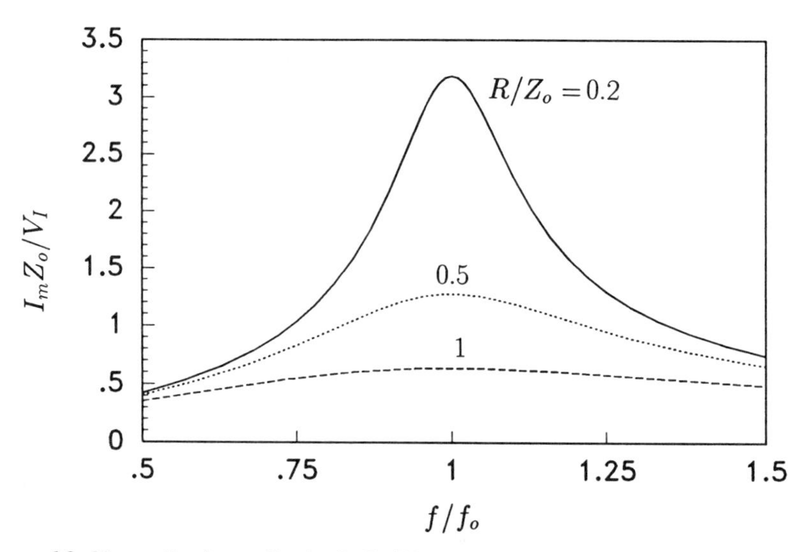

Figure 6.8: Normalized amplitude $I_m Z_o / V_I$ of the current in the resonant circuit against f/f_o at fixed values of $R/Z_o = 1/Q_L$.

Hence, from (6.25), (6.28), and (6.33), one obtains the dc component of the input current

$$I_I = \frac{1}{2\pi}\int_0^{2\pi} i_{S1} d(\omega t) = \frac{I_m}{2\pi}\int_0^{\pi} sin(\omega t - \psi) d(\omega t) = \frac{I_m cos\psi}{\pi} = \frac{V_m cos\psi}{\pi Z}$$
$$= \frac{2V_I cos\psi}{\pi^2 Z} = \frac{2V_I cos^2\psi}{\pi^2 R} = \frac{2V_I R}{\pi^2 Z^2} = \frac{I_m}{\pi\sqrt{1 + Q_L^2(\frac{\omega}{\omega_o} - \frac{\omega_o}{\omega})^2}}$$
$$= \frac{2V_I}{\pi^2 R[1 + Q_L^2(\frac{\omega}{\omega_o} - \frac{\omega_o}{\omega})^2]}. \tag{6.39}$$

At $f = f_o$,

$$I_I = \frac{I_m}{\pi} = \frac{2V_I}{\pi^2 R} \approx \frac{V_I}{5R}. \tag{6.40}$$

The dc input power can be expressed as

$$P_I = I_I V_I = \frac{2V_I^2 cos^2\psi}{\pi^2 R} = \frac{2V_I^2}{\pi^2 R[1 + Q_L^2(\frac{\omega}{\omega_o} - \frac{\omega_o}{\omega})^2]}$$
$$= \frac{2V_I^2 R}{\pi^2 Z_o^2[(\frac{R}{Z_o})^2 + (\frac{\omega}{\omega_o} - \frac{\omega_o}{\omega})^2]}. \tag{6.41}$$

At $f = f_o$,

$$P_I = \frac{2V_I^2}{\pi^2 R} \approx \frac{V_I^2}{5R}. \tag{6.42}$$

Using (6.36), one arrives at the output power

$$P_{Ri} = \frac{I_m^2 R_i}{2} = \frac{2V_I^2 R_i cos^2\psi}{\pi^2 R^2} = \frac{2V_I^2 R_i}{\pi^2 R^2[1 + Q_L^2(\frac{\omega}{\omega_o} - \frac{\omega_o}{\omega})^2]}$$
$$= \frac{2V_I^2 R_i}{\pi^2 Z_o^2[(\frac{R}{Z_o})^2 + (\frac{\omega}{\omega_o} - \frac{\omega_o}{\omega})^2]}. \tag{6.43}$$

At $f = f_o$,

$$P_{Ri} = \frac{2V_I^2 R_i}{\pi^2 R^2} \approx \frac{V_I^2 R_i}{5R^2}. \tag{6.44}$$

Figure 6.9 depicts $P_{Ri}Z_o^2/(V_I^2 R_i)$ as a function of f/f_o and R/Z_o. The normalized output power $P_{Ri}Z_o^2/(V_I^2 R_i)$ is plotted as a function of f/f_o at different values of R/Z_o in Fig. 6.10. The maximum output power occurs at the resonant frequency f_o and low total resistance R.

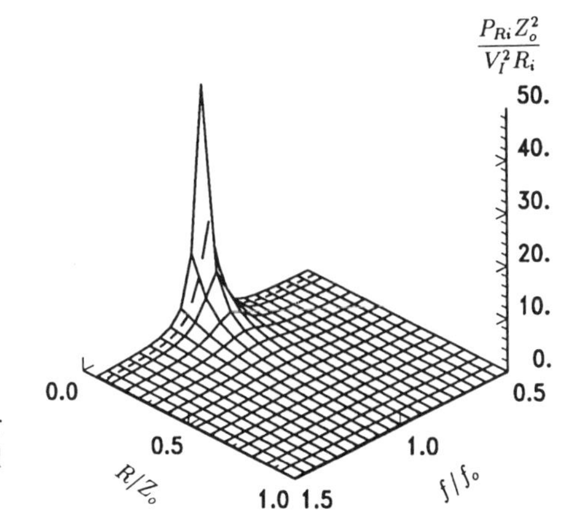

Figure 6.9: Normalized output power $P_{Ri}Z_o^2/V_I^2R_i$ as a function of f/f_o and $R/Z_o = 1/Q_L$.

6.5.5 Current and Voltage Stresses

The peak voltage across each switch is equal to the dc input voltage

$$V_{SM} = V_I. \tag{6.45}$$

The maximum value of the switch peak currents and the maximum amplitude of the current through the resonant circuit occurs at $f = f_o$. Hence, from (6.36)

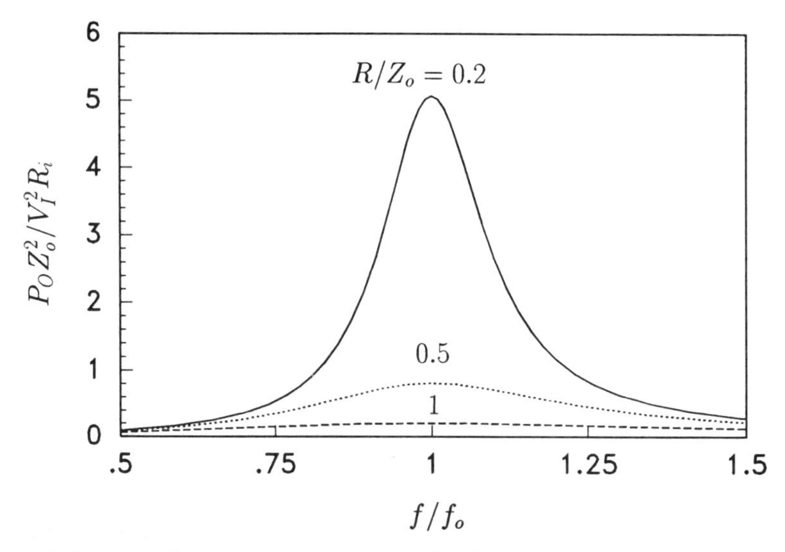

Figure 6.10: Normalized output power $P_{Ri}Z_o^2/V_I^2R_i$ as a function of f/f_o at fixed values of $R/Z_o = 1/Q_L$.

$$I_{SM} = I_{mr} = \frac{2V_I}{\pi R}. \tag{6.46}$$

The amplitude of the voltage across the capacitor C is obtained from (6.36)

$$V_{Cm} = \frac{I_m}{\omega C} = \frac{2V_I}{\pi(\frac{\omega}{\omega_o})\sqrt{(\frac{R}{Z_o})^2 + (\frac{\omega}{\omega_o} - \frac{\omega_o}{\omega})^2}}. \tag{6.47}$$

A three-dimensional representation of V_{Cm}/V_I is shown in Fig. 6.11. Figure 6.12 depicts plots of V_{Cm}/V_I as a function of f/f_o at fixed values of R/Z_o.

Likewise, the amplitude of the voltage across the inductor L is expressed as

$$V_{Lm} = \omega L I_m = \frac{2V_I(\frac{\omega}{\omega_o})}{\pi\sqrt{(\frac{R}{Z_o})^2 + (\frac{\omega}{\omega_o} - \frac{\omega_o}{\omega})^2}}. \tag{6.48}$$

Figure 6.13 shows V_{Lm}/V_I as a function of f/f_o and R/Z_o. Plots of V_{Lm}/V_I against f/f_o at constant values of R/Z_o are displayed in Fig. 6.14. At $f = f_o$,

$$V_{Cm(max)} = V_{Lm(max)} = Z_o I_{mr} = Q_L V_m = \frac{2V_I Q_L}{\pi}. \tag{6.49}$$

The maximum voltage stresses of the resonant components occur at the resonant frequency $f \approx f_o$, a maximum dc input voltage $V_I = V_{Imax}$, and a maximum loaded quality factor Q_L. Actually, the maximum value of V_{Lm} occurs slightly above f_o and of V_{Cm} slightly below f_o. However, this effect is negligible for practical purposes (see Problem 6.7). At the resonant frequency $f = f_o$, the amplitudes of the voltages across the resonant inductor

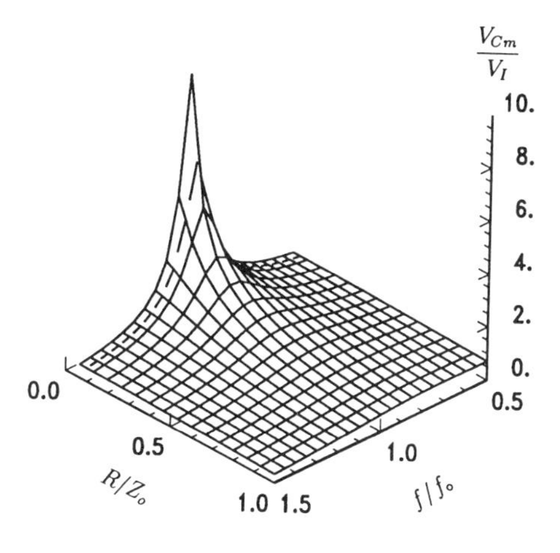

Figure 6.11: Normalized amplitude V_{Cm}/V_I of the voltage across the resonance capacitor C as a function of f/f_o and $R/Z_o = 1/Q_L$.

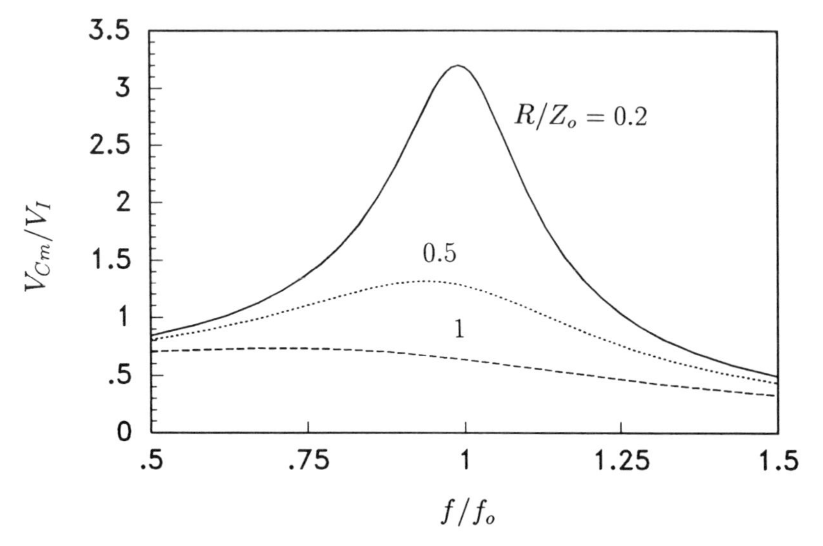

Figure 6.12: Normalized amplitude V_{Cm}/V_I of the voltage across the resonance capacitor C against f/f_o at fixed values of $R/Z_o = 1/Q_L$.

and resonant capacitor are Q_L times higher than the amplitude V_m of the fundamental component of the voltage at the input of the resonant circuit, which is equal to the amplitude of the output voltage V_{Rim}.

6.5.6 Operation Under Short-Circuit and Open-Circuit Conditions

The Class D inverter with a series-resonant circuit can operate safely with an open circuit at the output. However, it is prone to catastrophic failure if

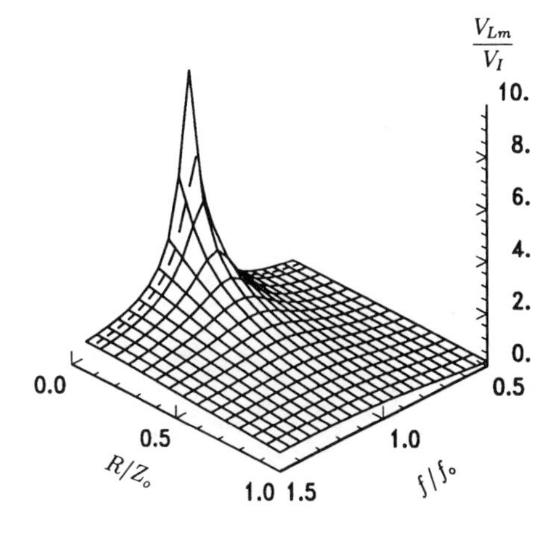

Figure 6.13: Normalized amplitude V_{Lm}/V_I of the voltage across the resonance inductor L as a function of f/f_o and $R/Z_o = 1/Q_L$.

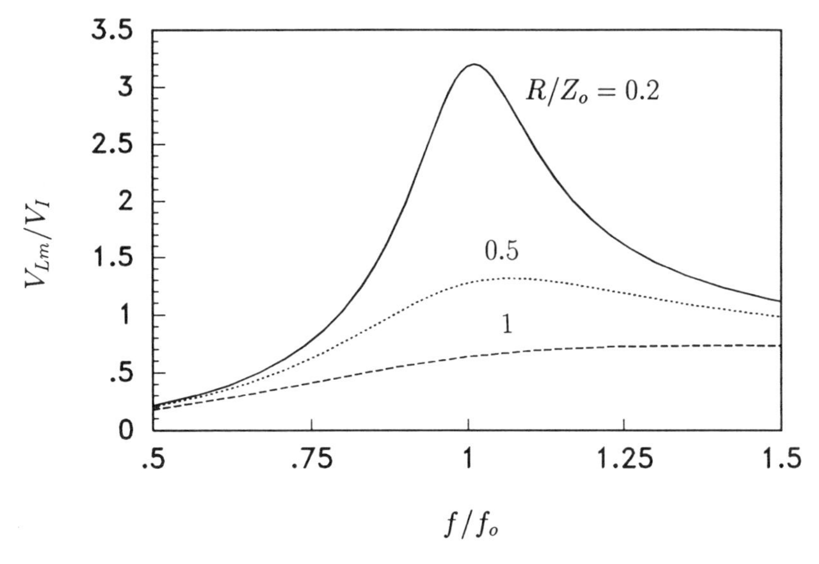

Figure 6.14: Normalized amplitude V_{Lm}/V_I of the voltage across the resonance inductor L against f/f_o at fixed values of $R/Z_o = 1/Q_L$.

the output is short-circuited at f close to f_o. If $R_i = 0$, the amplitude of the current through the resonant circuit and the switches is

$$I_m = \frac{2V_I}{\pi r\sqrt{1 + (\frac{Z_o}{r})^2(\frac{\omega}{\omega_o} - \frac{\omega_o}{\omega})^2}}. \tag{6.50}$$

The maximum value of I_m occurs at $f = f_o$ and is given by

$$I_{SM} = I_{mr} = \frac{2V_I}{\pi r}, \tag{6.51}$$

and the amplitudes of the voltages across the resonant components L and C are

$$V_{Cm} = V_{Lm} = \frac{I_{mr}}{\omega_o C} = \omega_o L I_{mr} = Z_o I_{mr} = \frac{2V_I Z_o}{\pi r} = \frac{2V_I Q_o}{\pi}. \tag{6.52}$$

For instance, if $V_I = 320$ V and $r = 2\ \Omega$, $I_{SM} = I_{mr} = 102$ A and $V_{Cm} = V_{Lm}$ $= 80$ kV! Thus, excessive current in the switches and the resonant circuit as well as the excessive voltages across L and C can lead to catastrophic failure of the inverter.

6.6 VOLTAGE TRANSFER FUNCTION

Class D inverters can be functionally divided into two parts: the switching part and the resonant part. A block diagram of an inverter is shown in Fig. 6.15. The switching part is comprised of a dc input voltage source V_I and a set of switches. The switches are controlled to produce a square-wave voltage v. Since a resonant circuit forces a sinusoidal current, only the power of the fundamental component is transferred from the switching part to the resonant part. Therefore, it is sufficient to consider only the fundamental component of the voltage v given by (6.32). A voltage transfer function of the switching part can be defined as

$$M_{Vs} \equiv \frac{V_{rms}}{V_I}, \tag{6.53}$$

where V_{rms} is the rms value of the fundamental component v_{i1} of the voltage v. The resonant part of an inverter converts square-wave voltage v into sinusoidal current or voltage signal. Because the dc input source V_I and switches S_1 and S_2 form a nearly ideal ac voltage source, many resonant circuits can be connected in parallel. If the resonant circuit is loaded by a resistance R_i, a voltage transfer function of the resonant part is

$$\mathbf{M_{Vr}} \equiv \frac{\mathbf{V_{Ri}}}{\mathbf{V_1}} = M_{Vr}e^{j\varphi}, \tag{6.54}$$

where $\mathbf{V_1}$ is the phasor of voltage v_{i1} and $\mathbf{V_{Ri}}$ is the phasor of the sinusoidal output voltage v_{Ri} across R_i. The modulus of $\mathbf{M_{Vr}}$ is

$$M_{Vr} = \frac{V_{Ri}}{V_{rms}}, \tag{6.55}$$

where V_{Ri} is the rms value of v_{Ri}. A voltage transfer function of the entire inverter is defined as a product of (6.53) and (6.55)

$$M_{VI} = M_{Vs}M_{Vr} = \frac{V_{Ri}}{V_I}. \tag{6.56}$$

Let us consider the half-bridge circuit. Using (6.34), one arrives at the

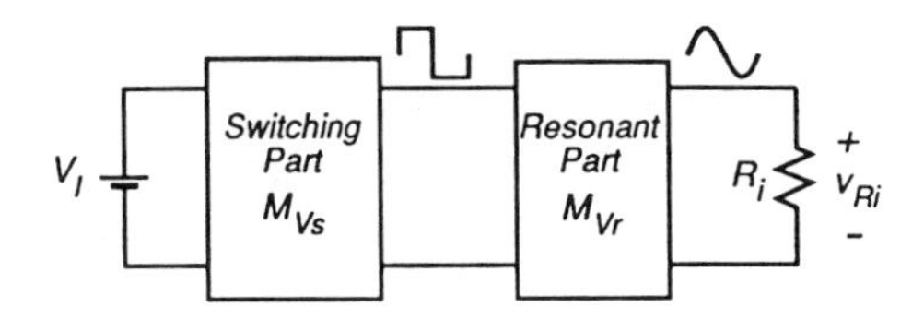

Figure 6.15: Block diagram of a Class D inverter.

voltage transfer function from the input of the inverter to the input of the series-resonant circuit

$$M_{Vs} = \frac{\sqrt{2}}{\pi} = 0.45. \tag{6.57}$$

The ac-to-ac voltage transfer function of the series-resonant circuit is

$$\begin{aligned} \mathbf{M_{Vr}} &= \frac{\mathbf{V_{Ri}}}{\mathbf{V_1}} = \frac{R_i}{R_i + r + j(\omega L - \frac{1}{\omega C})} = \frac{\eta_{Ir}}{1 + jQ_L(\frac{\omega}{\omega_o} - \frac{\omega_o}{\omega})} \\ &= M_{Vr}e^{j\varphi} \end{aligned} \tag{6.58}$$

from which

$$M_{Vr} = \frac{V_{Ri}}{V_{rms}} = \frac{\eta_{Ir}}{\sqrt{1 + Q_L^2(\frac{\omega}{\omega_o} - \frac{\omega_o}{\omega})^2}} \tag{6.59}$$

$$\varphi = -arctan\left[Q_L\left(\frac{\omega}{\omega_o} - \frac{\omega_o}{\omega}\right)\right] \tag{6.60}$$

where $\eta_{Ir} = R_i/(R_i+r) = R_i/R$ is the efficiency of the inverter taking into account conduction losses only (see Section 6.7.1). Figure 6.16 illustrates (6.59) in a three-dimensional space. Rearrangement of (6.59) yields

$$\frac{f}{f_o} = \frac{\sqrt{1 - (M_{Vr}/\eta_{Ir})^2 + 4Q_L^2(M_{Vr}/\eta_{Ir})^2} - \sqrt{1 - (M_{Vr}/\eta_{Ir})^2}}{2Q_L(M_{Vr}/\eta_{Ir})}, \quad \text{for} \quad \frac{f}{f_o} \leq 1 \tag{6.61}$$

$$\frac{f}{f_o} = \frac{\sqrt{1 - (M_{Vr}/\eta_{Ir})^2 + 4Q_L^2(M_{Vr}/\eta_{Ir})^2} + \sqrt{1 - (M_{Vr}/\eta_{Ir})^2}}{2Q_L(M_{Vr}/\eta_{Ir})}, \quad \text{for} \quad \frac{f}{f_o} \geq 1. \tag{6.62}$$

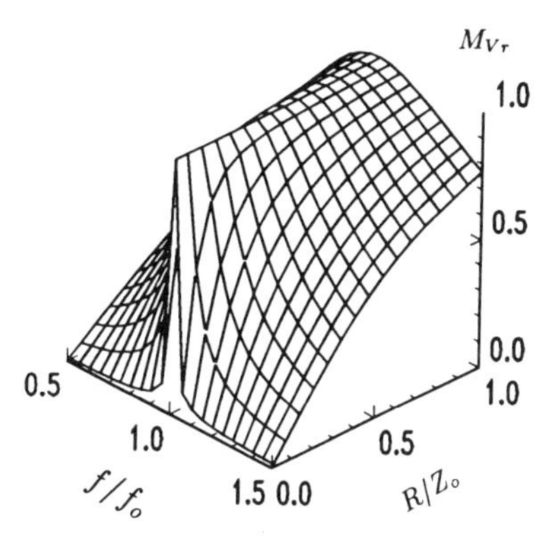

Figure 6.16: Three-dimensional representation of the M_{Vr} as a function of f/f_o and R/Z_o.

Expressions (6.59), (6.61), and (6.62) are functions of three normalized variables: Q_L, f/f_o, and M_{Vr}. These functions are illustrated in Fig. 6.17. Figure 6.17(a) shows M_{Vr} versus f/f_o at $\eta_{Ir} = 95\%$ and various values of $R/Z_o = 1/Q_L$; it is a family of resonance curves. It can be seen that M_{Vr} increases with f/f_o for $f \leq f_o$ and M_{Vr} decreases with f/f_o for $f \geq f_o$. Two families of characteristics for $f \leq f_o$ are displayed in Fig. 6.17(b) and (c). Figure 6.17(b) shows M_{Vr} as a function of R/Z_o at fixed values of f/f_o. As R/Z_o is increased and f/f_o is held constant, M_{Vr} increases. Figure 6.17(c) plots f/f_o against R/Z_o at constant values of M_{Vr}. To maintain a constant value of M_{Vr} as R/Z_o is increased, f/f_o should be decreased. The characteristics for $f \geq f_o$ are plotted in Fig. 6.17(d) and (e). Figure 6.17(d) depicts M_{Vr} versus R/Z_o at fixed values of f/f_o. As R/Z_o is increased at constant values of f/f_o, M_{Vr} increases. Figure 6.17(e) plots f/f_o against R/Z_o at various values of M_{Vr}. To maintain a fixed value of M_{Vr} as R/Z_o is increased, f/f_o should be increased. It is interesting to note that the frequency range required to maintain a constant value of M_{Vr} over a wide range of the normalized load resistance R/Z_o is narrower at higher values of M_{Vr} for both $f \leq f_o$ and $f \geq f_o$.

Comparison of (6.36) and (6.59) yields

$$I_m = \frac{2V_I M_{Vr}}{\pi R \eta_{Ir}}. \tag{6.63}$$

It can be seen that at constant M_{Vr} the amplitude of the output current of the series resonant inverter I_m is inversely proportional to the resistance R.

Combining (6.57) and (6.59) gives the magnitude of the dc-to-ac voltage transfer function for the Class D series resonant inverter

$$M_{VI} = \frac{V_{Ri}}{V_I} = \frac{V_{Ri}}{V_{rms}}\frac{V_{rms}}{V_I} = M_{Vs}M_{Vr} = \frac{\sqrt{2}\eta_{Ir}}{\pi\sqrt{1 + Q_L^2(\frac{\omega}{\omega_o} - \frac{\omega_o}{\omega})^2}}. \tag{6.64}$$

The maximum value of M_{VI} occurs at $f/f_o = 1$ and equals $M_{VImax} = \sqrt{2}\eta_{Ir}/\pi = 0.45\eta_{Ir}$. Thus, the values of M_{VI} range from zero to $0.45\eta_{Ir}$.

6.7 EFFICIENCY

6.7.1 Conduction Losses

The conduction loss for the power MOSFET is

$$P_{rDS} = \frac{I_m^2 r_{DS}}{4}, \tag{6.65}$$

for the resonant inductor is

$$P_{rL} = \frac{I_m^2 r_L}{2}, \tag{6.66}$$

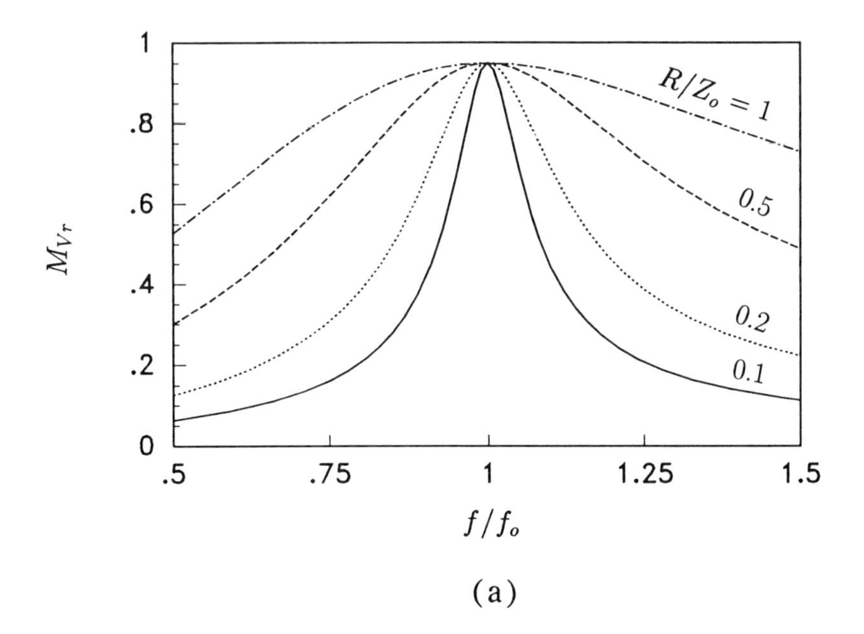

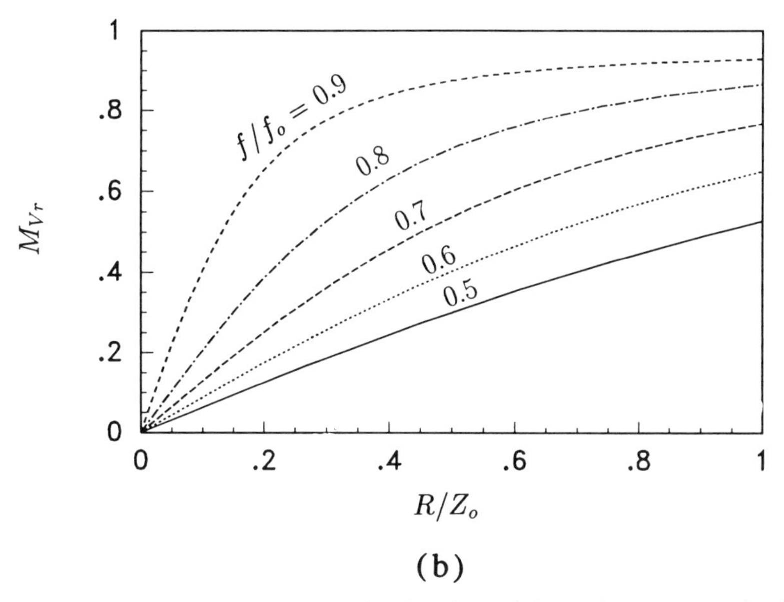

Figure 6.17: Steady-state voltage transfer function of the series-resonant circuit at $\eta_{Ir} = 95\%$. (a) Magnitude of voltage transfer function M_{Vr} versus normalized operating frequency f/f_o at fixed values of normalized load resistance R/Z_o. (b) M_{Vr} versus R/Z_o at fixed values of f/f_o for $f \leq f_o$. (c) M_{Vr} versus R/Z_o at fixed values of f/f_o for $f \geq f_o$. (d) f/f_o versus R/Z_o at fixed values of M_{Vr} for $f \leq f_o$. (e) f/f_o versus R/Z_o at fixed values of M_{Vr} for $f \geq f_o$.

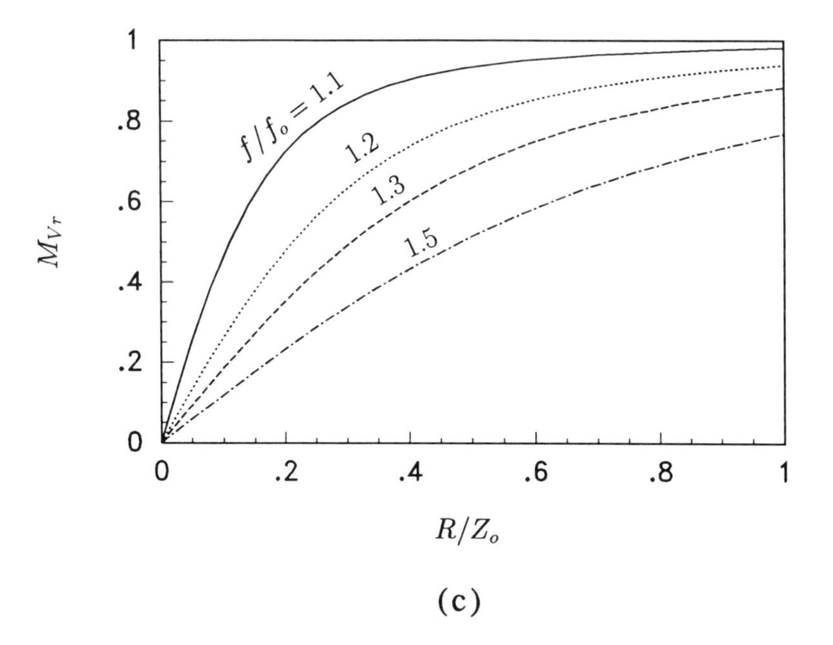

(c)

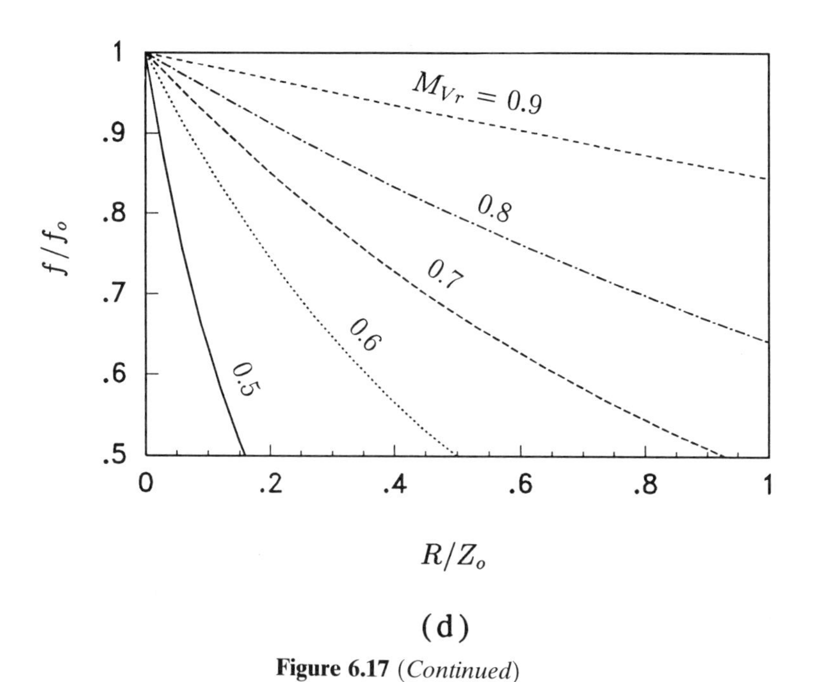

(d)

Figure 6.17 (*Continued*)

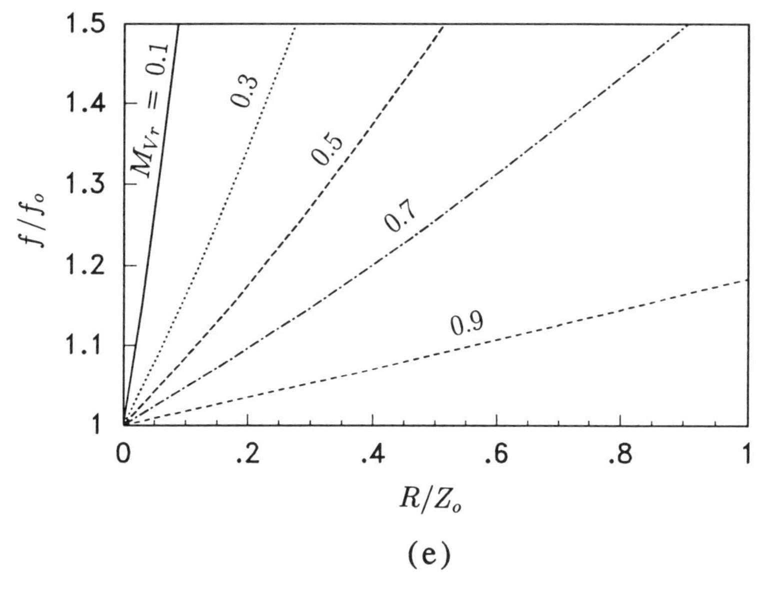

Figure 6.17 (*Continued*)

and for the resonant capacitor is

$$P_{rC} = \frac{I_m^2 r_C}{2}. \tag{6.67}$$

Hence, the conduction power loss in both transistors and the resonant circuit is

$$P_r = 2P_{rDS} + P_{rL} + P_{rC} = \frac{I_m^2(r_{DS} + r_L + r_C)}{2} = \frac{I_m^2 r}{2}. \tag{6.68}$$

Neglecting switching and the gate-drive losses and using (6.41) and (6.43), one obtains the efficiency of the inverter determined by the conduction losses only

$$\eta_{Ir} = \frac{P_{Ri}}{P_I} = \frac{P_{Ri}}{P_{Ri} + P_r} = \frac{R_i}{R_i + r} = \frac{1}{1 + \frac{r}{R_i}} = 1 - \frac{r}{R_i + r}$$
$$= 1 - \frac{1}{1 + \frac{R_i}{r}} = 1 - \frac{Q_L}{Q_o}. \tag{6.69}$$

Note that in order to achieve high efficiency, the ratio of the load resistance R_i to the parasitic resistance r must be high.

The turn-on loss for operation below resonance is given in the next section, and the turn-off loss for operation above resonance is given in Section

6.7.3. Expressions for the efficiency for the two cases are also given in those sections.

6.7.2 Turn-On Switching Loss

For the operation below resonance, the turn-off switching loss is zero; however, there is a turn-on switching loss. This loss is associated with charging and discharging the output capacitances of the MOSFETs. The diode junction capacitance is

$$C_j(v_D) = \frac{C_{j0}}{(1 - \frac{v_D}{V_B})^m} = \frac{C_{j0} V_B^m}{(V_B - v_D)^m}, \quad \text{for} \quad v_D \leq V_B, \tag{6.70}$$

where C_{j0} is the junction capacitance at $v_D = 0$ and m is the grading coefficient; $m = 1/2$ for step junctions and $m = 1/3$ for graded junctions. The barrier potential is

$$V_B = V_T ln\left(\frac{N_A N_D}{n_i^2}\right), \tag{6.71}$$

where n_i is the intrinsic carrier density (1.5×10^{10} cm^{-3} for silicon at 25°C), N_A is the acceptor concentration, and N_D is the donor concentration. The thermal voltage is

$$V_T = \frac{kT}{q} = \frac{T}{11,609} \text{ (V)}, \tag{6.72}$$

where $k = 1.38 \times 10^{-23}$ J/K is the Boltzmann's constant, $q = 1.602 \times 10^{-19}$ C is the charge per electron, and T is the absolute temperature in K. For p^+n diodes, a typical value of the acceptor concentration is $N_A = 10^{16}$ cm^{-3}, and a typical value of the donor concentration is $N_D = 10^{14}$ cm^{-3}, which gives $V_B = 0.57$ V. The zero-voltage junction capacitance is given by

$$C_{j0} = A\sqrt{\frac{\varepsilon_r \varepsilon_o q}{2V_B(\frac{1}{N_D} + \frac{1}{N_A})}} \approx A\sqrt{\frac{\varepsilon_r \varepsilon_o q N_D}{2V_B}}, \quad \text{for} \quad N_D << N_A, \tag{6.73}$$

where A is the junction area in cm^2, $\varepsilon_r = 11.7$ for silicon, and $\varepsilon_o = 8.85 \times 10^{-14}$ (F/cm). Hence, $C_{j0}/A = 3.1234 \times 10^{-16}\sqrt{N_D}$ (F/cm^2). For instance, if $N_D = 10^{14}$, $C_{j0}/A \approx 3$ nF/cm^2. Typical values of C_{j0} are of the order of 1 nF for power diodes.

The MOSFET's drain-source capacitance C_{ds} is the capacitance of the body-drain pn step junction diode. Setting $v_D = -v_{DS}$ and $m = 1/2$, one obtains from (6.70)

$$C_{ds}(v_{DS}) = \frac{C_{j0}}{\sqrt{1+\frac{v_{DS}}{V_B}}} = C_{j0}\sqrt{\frac{V_B}{v_{DS}+V_B}}, \quad \text{for} \quad v_{DS} \geq -V_B. \tag{6.74}$$

Hence,

$$\frac{C_{ds1}}{C_{ds2}} = \sqrt{\frac{v_{DS2}+V_B}{v_{DS1}+V_B}} \approx \sqrt{\frac{v_{DS2}}{v_{DS1}}}, \tag{6.75}$$

where C_{ds1} is the drain-source capacitance at v_{DS1} and C_{ds2} is the drain-source capacitance at v_{DS2}. Manufacturers of power MOSFETs usually specify the capacitances $C_{oss} = C_{gd} + C_{ds}$ and $C_{rss} = C_{gd}$ at $V_{DS} = 25$ V, $V_{GS} = 0$ V, and $f = 1$ MHz. Thus, the drain-source capacitance at $V_{DS} = 25$ V can be found as $C_{ds(25V)} = C_{oss} - C_{rss}$. The interterminal capacitances of MOSFETs are essentially independent of frequency. From (6.75), the drain-source capacitance at the dc voltage V_I is

$$C_{ds(V_I)} = C_{ds(25V)}\sqrt{\frac{25+V_B}{V_I+V_B}} \approx \frac{5C_{ds(25V)}}{\sqrt{V_I}}(\text{F}). \tag{6.76}$$

The drain-source capacitance at $v_{DS} = 0$ is

$$C_{j0} = C_{ds(25V)}\sqrt{\frac{25}{V_B}+1} \approx 6.7C_{ds(25V)} \tag{6.77}$$

for $V_B = 0.57$ V. Also,

$$C_{ds}(v_{DS}) = C_{ds(V_I)}\sqrt{\frac{V_I+V_B}{v_{DS}+V_B}} \approx C_{ds(V_I)}\sqrt{\frac{V_I}{v_{DS}}}. \tag{6.78}$$

Using (6.74) and $dQ_j = C_{ds}dv_{DS}$, the charge stored in the drain-source junction capacitance at v_{DS} can be found as

$$\begin{aligned} Q_j(v_{DS}) &= \int_{-V_B}^{v_{DS}} dQ_j = \int_{-V_B}^{v_{DS}} C_{ds}(v_{DS})dv_{DS} \\ &= C_{j0}\sqrt{V_B}\int_{-V_B}^{v_{DS}} \frac{dv_{DS}}{\sqrt{v_{DS}+V_B}} = 2C_{j0}\sqrt{V_B(v_{DS}+V_B)} \\ &= 2C_{j0}V_B\sqrt{1+\frac{v_{DS}}{V_B}} = 2(v_{DS}+V_B)C_{ds}(v_{DS}) \approx 2v_{DS}C_{ds}(v_{DS}), \end{aligned} \tag{6.79}$$

which, by substituting (6.76) at $v_{DS} = V_I$, simplifies to

$$Q_j(V_I) = 2V_I C_{ds(V_I)} = 10C_{ds(25V)}\sqrt{V_I}(\text{V}). \tag{6.80}$$

Hence, the energy transferred from the dc input source V_I to the output capacitance of the upper MOSFET after the upper transistor is turned off is

$$W_I = \int_{-V_B}^{V_I} vidt = V_I \int_{-V_B}^{V_I} idt = V_I Q_j(V_I)$$
$$= 2V_I^2 C_{ds(V_I)} = 10\sqrt{V_I^3} C_{ds(25V)} (\text{W}). \quad (6.81)$$

Using $dW_j = (1/2)Q_j dv_{DS}$ and (6.79), the energy stored in the drain–source junction capacitance C_{ds} at v_{DS} is

$$W_j(v_{DS}) = \frac{1}{2}\int_{-V_0}^{v_{DS}} Q_j dv_{DS} = C_{j0}\sqrt{V_B}\int_{-V_0}^{v_{DS}} \sqrt{v_{DS}+V_B} dv_{DS}$$
$$= \frac{2}{3}C_{j0}\sqrt{V_B}(v_{DS}+V_B)^{\frac{3}{2}} = \frac{2}{3}C_{ds}(v_{DS})(v_{DS}+V_B)^2 \approx \frac{2}{3}C_{ds}(v_{DS})v_{DS}^2. \quad (6.82)$$

Hence, from (6.76) the energy stored in the drain-source junction capacitance at $v_{DS} = V_I$ is

$$W_j(V_I) = \frac{2}{3}C_{ds(V_I)}V_I^2 = \frac{10}{3}C_{ds(25V)}\sqrt{V_I^3}(\text{J}). \quad (6.83)$$

This energy is lost as heat when the transistor turns on and the capacitor is discharged through r_{DS}, resulting in the turn-on switching power loss per transistor

$$P_{tron} = \frac{W_j(V_I)}{T} = fW_j(V_I) = \frac{2}{3}fC_{j0}\sqrt{V_B}(V_I+V_B)^{\frac{3}{2}}$$
$$= \frac{2}{3}fC_{ds(V_I)}V_I^2 = \frac{10}{3}fC_{ds(25V)}\sqrt{V_I^3}(\text{W}). \quad (6.84)$$

Figure 6.18 illustrates C_{ds}, Q_j, and W_j as functions of v_{DS} given by (6.74), (6.79), and (6.82).

Using (6.81) and (6.83), one arrives at the energy lost in the resistances of the charging path during the charging process of the capacitance C_{ds}

$$W_{char}(V_I) = W_I(V_I) - W_j(V_I) = \frac{4}{3}C_{ds(V_I)}V_I^2 = \frac{20}{3}C_{ds(25V)}\sqrt{V_I^3}(\text{J}), \quad (6.85)$$

and the corresponding power associated with charging the capacitance C_{ds} is

$$P_{char} = \frac{W_{char}(V_I)}{T} = fW_{char}(V_I) = \frac{4}{3}fC_{ds(V_I)}V_I^2 = \frac{20}{3}fC_{ds(25V)}\sqrt{V_I^3}(\text{J}). \quad (6.86)$$

From (6.81), one arrives at the total switching power loss per transistor

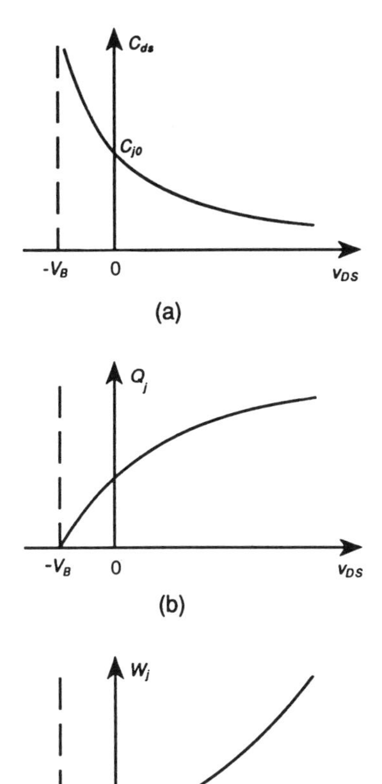

Figure 6.18: Plots of C_{ds}, Q_j, and W_j versus v_{DS}. (a) C_{ds} versus v_{DS}. (b) Q_j versus v_{DS}. (c) W_j versus v_{DS}.

$$\begin{aligned} P_{sw} &= \frac{W(V_I)}{T} = fW_I(V_I) = 2fC_{j0}V_I\sqrt{V_B(V_I + V_B)} \\ &= 2fC_{ds(V_I)}V_I^2 = 10fC_{ds(25V)}\sqrt{V_I^3}(\text{W}). \end{aligned} \tag{6.87}$$

The switching loss associated with charging and discharging an equivalent linear capacitance C_{eq} is $P_{sw} = fC_{eq}V_I^2$. Hence, from (6.87)

$$C_{eq} = 2C_{ds(V_I)} = \frac{10C_{ds(V_I)}}{\sqrt{V_I}}. \tag{6.88}$$

Example 6.1

For MTP5N40 MOSFETs, it is given in data sheets that $C_{oss} = 300$ pF and $C_{rss} = 80$ pF at $V_{DS} = 25$ V and $V_{GS} = 0$ V. These MOSFETs are to be used in a Class D half-bridge series resonant inverter that is operated at frequency

$f = 100$ kHz and fed by dc voltage source $V_I = 350$ V. Calculate the drain-source capacitance at the dc supply voltage V_I, the drain-source capacitance at $v_{DS} = 0$, the charge stored in the drain-source junction capacitance at V_I, the energy transferred from the dc input source V_I to the output capacitance of a MOSFET during turn-on transition, the energy stored in the drain-source junction capacitance C_{ds} at V_I, the turn-on switching power loss, and the total switching power loss per transistor in the inverter operating below resonance. Assume $V_B = 0.57$ V.

Solution: Using data sheets,

$$C_{ds(25V)} = C_{oss} - C_{rss} = 300 - 80 = 220 \text{ pF}. \tag{6.89}$$

From (6.76), one obtains the drain-source capacitance at the dc supply voltage $V_I = 350$ V

$$C_{ds(V_I)} = \frac{5C_{ds(25V)}}{\sqrt{V_I}} = \frac{5 \times 220}{\sqrt{350}} = 59 \text{ pF}. \tag{6.90}$$

Equation (6.77) gives the drain-source capacitance at $v_{DS} = 0$

$$C_{j0} = 6.7C_{ds(25V)} = 6.7 \times 220 \times 10^{-12} = 1474 \text{ pF}. \tag{6.91}$$

The charge stored in the drain-source junction capacitance at $V_I = 350$ V is obtained from (6.80)

$$Q_j(V_I) = 2V_I C_{ds(V_I)} = 2 \times 350 \times 59 \times 10^{-12} = 41.3 \text{ nC}. \tag{6.92}$$

The energy transferred from the input voltage source V_I to the inverter is calcultated from (6.81) as

$$W_I(V_I) = V_I Q_j(V_I) = 350 \times 41.3 \times 10^{-9} = 14.455\ \mu\text{J}, \tag{6.93}$$

and the energy stored in the drain-source junction capacitance C_{ds} at V_I is calculated from (6.83) as

$$W_j(V_I) = \frac{10}{3} C_{ds(25V)} \sqrt{V_I^3} = \frac{10}{3} \times 220 \times 10^{-12} \sqrt{350^3} = 4.8\ \mu\text{J}. \tag{6.94}$$

Using (6.86), the power associated with charging the capacitance C_{ds} is calculated as

$$P_{char} = \frac{20}{3} f C_{ds(25V)} \sqrt{V_I^3} = \frac{20}{3} \times 10^5 \times 220 \times 10^{-12} \sqrt{350^3} = 0.96 \text{ W}. \tag{6.95}$$

From (6.84), the turn-on switching power loss per transistor for operation below resonance is

$$P_{tron} = \frac{10}{3} f C_{ds(25V)} \sqrt{V_I^3} = \frac{10}{3} \times 10^5 \times 220 \times 10^{-12} \sqrt{350^3} = 0.48 \text{ W}. \quad (6.96)$$

Usinig (6.87), one arrives at the total switching power loss per transistor for operation below resonance

$$P_{sw} = 10 f C_{ds(25V)} \sqrt{V_I^3} (\text{W}) = 10 \times 10^5 \times 220 \times 10^{-12} \sqrt{350^3} = 1.44 \text{ W}. \quad (6.97)$$

Note that $P_{tron} = \frac{1}{3} P_{sw}$ and $P_{char} = \frac{2}{3} P_{sw}$. The equivalent linear capacitance is $C_{eq} = 2C_{ds(V_I)} = 2 \times 59 = 118$ pF.

The overall power dissipation in the Class D inverter is

$$P_T = P_r + 2P_{sw} + 2P_G = \frac{r I_m^2}{2} + 20 C_{ds(25V)} \sqrt{V_I^3} + 2 f Q_g V_{GSpp}, \quad (6.98)$$

where P_G is the gate-drive power and is given by (2.117). Hence, the efficiency of the half-bridge inverter for operation below resonance is

$$\eta_I = \frac{P_{Ri}}{P_{Ri} + P_T} = \frac{P_{Ri}}{P_{Ri} + P_r + 2P_{sw} + 2P_G}. \quad (6.99)$$

6.7.3 Turn-Off Switching Loss

For the operation above resonance, the turn-on switching loss is zero, but there is a turn-off switching loss. The switch current and voltage waveforms during turn-off for $f > f_o$ are sketched in Fig. 6.19. These waveforms were observed in various Class D experimental circuits. Notice that the voltage v_{DS2} increases slowly at its low values and much faster at its high values. This is because the MOSFET output capacitance is highly nonlinear, and it is much higher at low voltage v_{DS2} than at high voltage v_{DS2}. The current that charges this capacitance is approximately constant. The drain-to-source voltage v_{DS2} during voltage rise time t_r can be approximated by a parabolic function

$$v_{DS2} = a(\omega t)^2. \quad (6.100)$$

Because $v_{DS2}(\omega t_r) = V_I$, one obtains

$$a = \frac{V_I}{(\omega t_r)^2}. \quad (6.101)$$

Hence, (6.100) becomes

$$v_{DS2} = \frac{V_I (\omega t)^2}{(\omega t_r)^2}. \quad (6.102)$$

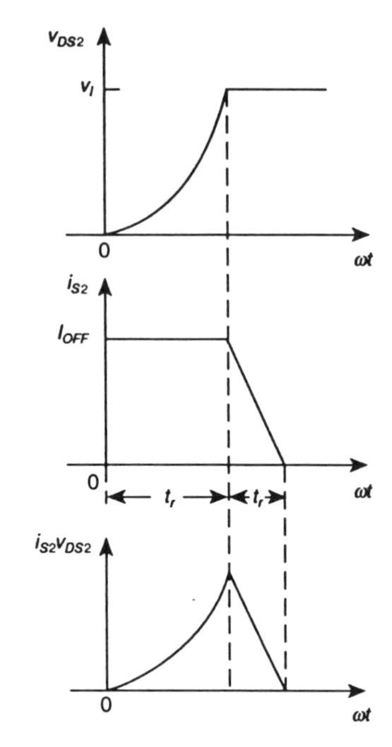

Figure 6.19: Waveforms of v_{DS2}, i_{DS2}, and $i_{S2}v_{DS2}$ during turn-off for $f > f_o$.

The switch current during rise time t_r is a small portion of a sinusoid and can be approximated by a constant

$$i_{S2} = I_{OFF}. \tag{6.103}$$

The average value of the power loss associated with voltage rise time t_r is

$$\begin{aligned} P_{tr} &= \frac{1}{2\pi}\int_0^{2\pi} i_{S2}v_{DS2}d(\omega t) = \frac{V_I I_{OFF}}{2\pi(\omega t_r)^2}\int_0^{\omega t_r}(\omega t)^2 d(\omega t) \\ &= \frac{\omega t_r V_I I_{OFF}}{6\pi} = \frac{f t_r V_I I_{OFF}}{3} = \frac{t_r V_I I_{OFF}}{3T}. \end{aligned} \tag{6.104}$$

The switch current during fall time t_f can be approximated by a ramp function

$$i_{S2} = I_{OFF}\left(1 - \frac{\omega t}{\omega t_f}\right), \tag{6.105}$$

and the drain-to-source voltage is

$$v_{DS2} = V_I, \tag{6.106}$$

which yields the average value of the power loss associated with current fall time t_f

$$P_{tf} = \frac{1}{2\pi}\int_0^{2\pi} i_{S2}v_{DS2}d(\omega t) = \frac{V_I I_{OFF}}{2\pi}\int_0^{\omega t_f}\left(1 - \frac{\omega t}{\omega t_f}\right)d(\omega t)$$
$$= \frac{\omega t_f V_I I_{OFF}}{4\pi} = \frac{f t_f V_I I_{OFF}}{2} = \frac{t_f V_I I_{OFF}}{2T}. \quad (6.107)$$

Hence, the turn-off switching loss is

$$P_{toff} = P_{tr} + P_{tf} = f V_I I_{OFF}\left(\frac{t_r}{3} + \frac{t_f}{2}\right). \quad (6.108)$$

Usually t_r is much longer than t_f. The overall power dissipation in the Class D half-bridge inverter is

$$P_T = P_r + 2P_{toff} + 2P_G = \frac{rI_m^2}{2} + f V_I I_{OFF}\left(\frac{2t_r}{3} + t_f\right) + 2fQ_g V_{GSpp}. \quad (6.109)$$

Hence the efficiency of the inverter for operation above resonance is

$$\eta_I = \frac{P_{Ri}}{P_{Ri} + P_T} = \frac{P_{Ri}}{P_{Ri} + P_r + 2P_{toff} + 2P_G}. \quad (6.110)$$

6.8 DESIGN EXAMPLE

A design procedure of a series resonant inverter is illustrated by means of an example for a half-bridge SRI.

Example 6.2

Design a Class D half-bridge inverter of Fig. 6.1 that meets the following specifications: $V_I = 100$ V, $P_{Ri} = 50$ W, and $f = 110$ kHz. Assume $Q_L = 5.5$, $\psi = 30°$ (i.e., $cos^2\psi = 0.75$), and the efficiency $\eta_{Ir} = 90\%$. The converter employs IRF621 MOSFETs (International Rectifier) with $r_{DS} = 0.5\ \Omega$, $C_{ds(25V)} = 110$ pF, and $Q_g = 11$ nC. Check the initial assumption about η_{Ir} using $Q_{Lo} = 300$ and $Q_{Co} = 1200$. Estimate switching losses and gate-drive power loss assuming $V_{GSpp} = 15$ V.

Solution: From (6.69), the dc input power of the inverter is

$$P_I = \frac{P_{Ri}}{\eta_{Ir}} = \frac{50}{0.9} = 55.56 \text{ W}. \quad (6.111)$$

Using (6.41), the overall resistance of the inverter can be calculated as

$$R = \frac{2V_I^2}{\pi^2 P_I}cos^2\psi = \frac{2 \times 100^2}{\pi^2 \times 55.56} \times 0.75 = 27.35\ \Omega. \quad (6.112)$$

Relationships (6.69) and (6.13) give the load resistance

$$R_i = \eta_{Ir} R = 0.9 \times 27.35 = 24.62\ \Omega \tag{6.113}$$

and the maximum total parasitic resistance of the inverter

$$r = R - R_i = 27.35 - 24.62 = 2.73\ \Omega. \tag{6.114}$$

The dc supply current is obtained from (6.41)

$$I_I = \frac{P_I}{V_I} = \frac{55.56}{100} = 0.556\ \text{A}. \tag{6.115}$$

The peak value of the switch current is

$$I_m = \sqrt{\frac{2P_{Ri}}{R_i}} = \sqrt{\frac{2 \times 50}{24.62}} = 2.02\ \text{A}, \tag{6.116}$$

and from (6.45) the peak value of the switch voltage is equal to the input voltage

$$V_{SM} = V_I = 100\ \text{V}. \tag{6.117}$$

Using (6.26), one arrives at the ratio f/f_o at full load

$$\frac{f}{f_o} = \frac{1}{2}\left(\frac{tan\psi}{Q_L} + \sqrt{\frac{tan^2\psi}{Q_L^2} + 4}\right) = \frac{1}{2}\left(\frac{0.5774}{5.5} + \sqrt{\frac{0.5774^2}{5.5^2} + 4}\right) = 1.054, \tag{6.118}$$

from which

$$f_o = \frac{f}{(f/f_o)} = \frac{110 \times 10^3}{1.054} = 104.4\ \text{kHz}. \tag{6.119}$$

The values of the reactive components of the resonant circuit are calculated from (6.10) as

$$L = \frac{Q_L R}{\omega_o} = \frac{5.5 \times 27.35}{2\pi \times 104.4 \times 10^3} = 229.3\ \mu\text{H} \tag{6.120}$$

$$C = \frac{1}{\omega_o Q_L R} = \frac{1}{2\pi \times 104.4 \times 10^3 \times 5.5 \times 27.35} = 10\ \text{nF}. \tag{6.121}$$

From (6.9),

$$Z_o = \sqrt{\frac{L}{C}} = \sqrt{\frac{229.3 \times 10^{-6}}{10 \times 10^{-9}}} = 151.4\ \Omega. \tag{6.122}$$

The maximum voltage stresses for the resonant components can be approximated using (6.49)

$$V_{Cm(max)} = V_{Lm(max)} = \frac{2V_I Q_L}{\pi} = \frac{2 \times 100 \times 5.5}{\pi} = 350 \text{ V}. \quad (6.123)$$

Once the values of the resonant components are known, the parasitic resistance of the inverter can be recalculated. From (6.22) and (6.23),

$$r_L = \frac{\omega L}{Q_{Lo}} = \frac{2\pi \times 110 \times 10^3 \times 229.3 \times 10^{-6}}{300} = 0.53 \ \Omega \quad (6.124)$$

and

$$r_C = \frac{1}{\omega C Q_{Co}} = \frac{1}{2\pi \times 110 \times 10^3 \times 10 \times 10^{-9} \times 1200} = 0.1 \ \Omega. \quad (6.125)$$

Thus, the parasitic resistance is

$$r = r_{DS} + r_L + r_C = 0.5 + 0.53 + 0.1 = 1.13 \ \Omega. \quad (6.126)$$

From (6.65), the conduction loss in each MOSFET is

$$P_{rDS} = \frac{I_m^2 r_{DS}}{4} = \frac{2.02^2 \times 0.5}{4} = 0.51 \text{ W}. \quad (6.127)$$

Using (6.66), the conduction loss in the resonant inductor L is

$$P_{rL} = \frac{I_m^2 r_L}{2} = \frac{2.02^2 \times 0.53}{2} = 1.08 \text{ W}. \quad (6.128)$$

From (6.67), the conduction loss in the resonant capacitor C is

$$P_{rC} = \frac{I_m^2 r_C}{2} = \frac{2.02^2 \times 0.1}{2} = 0.204 \text{ W}. \quad (6.129)$$

Hence, one obtains the overall conduction loss

$$P_r = 2P_{rDS} + P_{rL} + P_{rC} = 2 \times 0.51 + 1.08 + 0.204 = 2.304 \text{ W}. \quad (6.130)$$

The efficiency η_{Ir} associated with the conduction losses only at full power is

$$\eta_{Ir} = \frac{P_{Ri}}{P_{Ri} + P_r} = \frac{50}{50 + 2.304} = 95.6\%. \quad (6.131)$$

Using (2.117) and assuming the peak-to-peak gate-source voltage $V_{GSpp} = 15$ V, the gate-drive power loss in both MOSFETs is

$$2P_G = 2fQ_g V_{GSpp} = 2 \times 110 \times 10^3 \times 11 \times 10^{-9} \times 15 = 0.036 \text{ W}. \quad (6.132)$$

The sum of the conduction losses and the gate-drive power loss is

$$P_{LS} = P_r + 2P_G = 2.304 + 0.036 = 2.34 \text{ W}. \tag{6.133}$$

The turn-on conduction loss is zero because the inverter is operated above resonance. The efficiency of the inverter associated with the conduction loss and the gate-drive power at full power is

$$\eta_I = \frac{P_{Ri}}{P_{Ri} + P_{LS}} = \frac{50}{50 + 2.34} = 95.53\%. \tag{6.134}$$

6.9 CLASS D FULL-BRIDGE SERIES RESONANT INVERTER

6.9.1 Currents, Voltages, and Powers

A circuit of a Class D full-bridge voltage-switching series resonant inverter is depicted in Fig. 6.20. It consists of four controllable switches and a series resonant circuit. Current and voltage waveforms for the inverter are shown in Fig. 6.21. Notice that the voltage at the input of the resonant circuit is two times higher than in the case of the half-bridge inverter. The averaged resistance of the on-resistances of power MOSFETs is $r_S = (r_{DS1} + r_{DS2} + r_{DS3} + r_{DS4})/2 \approx 2r_{DS}$. The total parasitic resistance is represented by

$$r \approx 2r_{DS} + r_L + r_C, \tag{6.135}$$

which yields the overall resistance

$$R = R_i + r \approx R_i + 2r_{DS} + r_L + r_C. \tag{6.136}$$

Referring to Fig. 6.21, the input voltage of the series-resonant circuit is a square-wave described by

$$v = \begin{cases} V_I, & \text{for} \quad 0 < \omega t \le \pi \\ -V_I, & \text{for} \quad \pi < \omega t \le 2\pi. \end{cases} \tag{6.137}$$

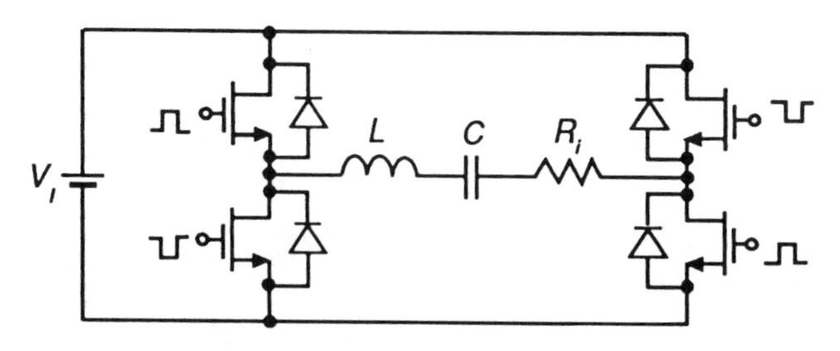

Figure 6.20: Class D voltage-switching full-bridge inverter with a series-resonant circuit.

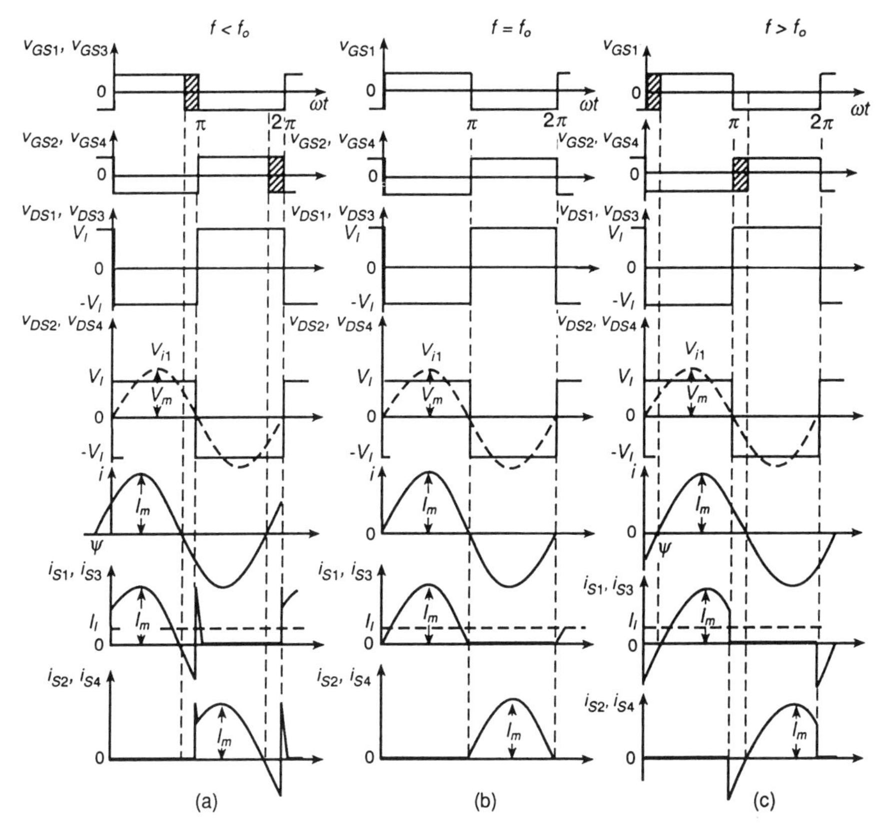

Figure 6.21: Waveforms in the Class D full-bridge inverter. (a) For $f < f_o$. (b) For $f = f_o$. (c) For $f > f_o$.

The Fourier expansion of this voltage is

$$v = \frac{4V_I}{\pi}\sum_{n=1}^{\infty}\frac{1-(-1)^n}{2n} sinn\omega t$$

$$= V_I\left(\frac{4}{\pi}sin\omega t + \frac{4}{3\pi}sin3\omega t + \frac{4}{5\pi}sin5\omega t + ...\right). \tag{6.138}$$

The fundamental component of voltage v is

$$v_{i1} = V_m sin\omega t, \tag{6.139}$$

where its amplitude is given by

$$V_m = \frac{4V_I}{\pi} \approx 1.273V_I. \tag{6.140}$$

Hence, one obtains the rms value of v_{i1}

$$V_{rms} = \frac{V_m}{\sqrt{2}} = \frac{2\sqrt{2}V_I}{\pi} \approx 0.9V_I. \tag{6.141}$$

The current through the switches S_1 and S_4 is

$$i_{S1} = i_{S4} = \begin{cases} I_m sin(\omega t - \psi), & \text{for} \quad 0 < \omega t \leq \pi \\ 0, & \text{for} \quad \pi < \omega t \leq 2\pi, \end{cases} \tag{6.142}$$

and the current through the switches S_2 and S_3 is

$$i_{S2} = i_{S3} = \begin{cases} 0, & \text{for} \quad 0 < \omega t \leq \pi \\ -I_m sin(\omega t - \psi), & \text{for} \quad \pi < \omega t \leq 2\pi. \end{cases} \tag{6.143}$$

The input current of the inverter is

$$i_I = i_{S1} + i_{S3}. \tag{6.144}$$

The cycle of the input current is two times higher than the operating frequency. Hence, from (6.25), (6.28), and (6.140), one obtains the dc component of the input current

$$\begin{aligned} I_I &= \frac{1}{\pi}\int_0^{\pi} i_{S1} d(\omega t) = \frac{I_m}{\pi}\int_0^{\pi} sin(\omega t - \psi) d(\omega t) = \frac{2I_m cos\psi}{\pi} = \frac{2V_m cos\psi}{\pi Z} \\ &= \frac{8V_I cos\psi}{\pi^2 Z} = \frac{8V_I cos^2\psi}{\pi^2 R} = \frac{8V_I R}{\pi^2 Z^2} = \frac{2I_m}{\pi\sqrt{1 + Q_L^2(\frac{\omega}{\omega_o} - \frac{\omega_o}{\omega})^2}} \\ &= \frac{8V_I}{\pi^2 R[1 + Q_L^2(\frac{\omega}{\omega_o} - \frac{\omega_o}{\omega})^2]}. \end{aligned} \tag{6.145}$$

At $f = f_o$,

$$I_I = \frac{2I_m}{\pi} = \frac{8V_I}{\pi^2 R}. \tag{6.146}$$

The dc input power is

$$\begin{aligned} P_I = I_I V_I &= \frac{8V_I^2 cos^2\psi}{\pi^2 R} = \frac{8V_I^2}{\pi^2 R[1 + Q_L^2(\frac{\omega}{\omega_o} - \frac{\omega_o}{\omega})^2]} \\ &= \frac{8V_I^2 R}{\pi^2 Z_o^2[(\frac{R}{Z_o})^2 + (\frac{\omega}{\omega_o} - \frac{\omega_o}{\omega})^2]}. \end{aligned} \tag{6.147}$$

At $f = f_o$,

$$P_I = \frac{8V_I^2}{\pi^2 R}. \tag{6.148}$$

The current through the series-resonant circuit is given by (6.35). From (6.28), (6.29), and (6.140), its amplitude can be found as

$$I_m = \frac{V_m}{Z} = \frac{4V_I}{\pi Z} = \frac{4V_I cos\psi}{\pi R} = \frac{4V_I}{\pi R\sqrt{1 + Q_L^2(\frac{\omega}{\omega_o} - \frac{\omega_o}{\omega})^2}}$$

$$= \frac{4V_I}{\pi Z_o\sqrt{(\frac{R}{Z_o})^2 + (\frac{\omega}{\omega_o} - \frac{\omega_o}{\omega})^2}}. \quad (6.149)$$

At $f = f_o$,

$$I_{SM} = I_{mr} = \frac{4V_I}{\pi R}. \quad (6.150)$$

The voltage stress of every switch is

$$V_{SM} = V_I. \quad (6.151)$$

From (6.149), one obtains the amplitude of the voltage across the capacitor C

$$V_{Cm} = \frac{I_m}{\omega C} = \frac{4V_I}{\pi(\frac{\omega}{\omega_o})\sqrt{(\frac{R}{Z_o})^2 + (\frac{\omega}{\omega_o} - \frac{\omega_o}{\omega})^2}}. \quad (6.152)$$

Similarly, the amplitude of the voltage across the inductor L is

$$V_{Lm} = \omega L I_m = \frac{4V_I(\frac{\omega}{\omega_o})}{\pi\sqrt{(\frac{R}{Z_o})^2 + (\frac{\omega}{\omega_o} - \frac{\omega_o}{\omega})^2}}. \quad (6.153)$$

At $f = f_o$,

$$V_{Cm} = V_{Lm} = Z_o I_{mr} = Q_L V_m = \frac{4V_I Q_L}{\pi}. \quad (6.154)$$

The output power is obtained from (6.149)

$$P_{Ri} = \frac{I_m^2 R_i}{2} = \frac{8V_I^2 R_i cos^2\psi}{\pi^2 R^2} = \frac{8V_I^2 R_i}{\pi^2 R^2[1 + Q_L^2(\frac{\omega}{\omega_o} - \frac{\omega_o}{\omega})^2]}$$

$$= \frac{8V_I^2 R_i}{\pi^2 Z_o^2[(\frac{R}{Z_o})^2 + (\frac{\omega}{\omega_o} - \frac{\omega_o}{\omega})^2]}. \quad (6.155)$$

At $f = f_o$,

$$P_{Ri} = \frac{8V_I^2 R_i}{\pi^2 R^2}. \quad (6.156)$$

6.9.2 Efficiency

The conduction losses in every transistor, the resonant inductor, and the resonant capacitor are given by (6.65), (6.66), and (6.67), respectively. The conduction power loss in four transistors and the resonant circuit is

$$P_r = \frac{I_m^2 r}{2} = \frac{I_m^2(2r_{DS} + r_L + r_C)}{2}. \quad (6.157)$$

The gate drive power P_G is given by (2.117). The turn-on switching loss per transistor for operation below resonance P_{sw} is given by (6.87). The overall power dissipation in the Class D inverter for operation below resonance is

$$P_T = P_r + 4P_{sw} + 4P_G = \frac{rI_m^2}{2} + 40C_{ds(25V)}\sqrt{V_I^3} + 4fQ_g V_{GSpp}. \quad (6.158)$$

Hence, the efficiency of the full-bridge inverter for operation below resonance is

$$\eta_I = \frac{P_{Ri}}{P_{Ri} + P_T} = \frac{P_{Ri}}{P_{Ri} + P_r + 4P_{sw} + 4P_G}. \quad (6.159)$$

The turn-off switching loss per transistor for operation above resonance P_{toff} is given by (6.108). The overall power dissipation in the inverter operating above resonance is

$$P_T = P_r + 4P_{toff} + 4P_G = \frac{rI_m^2}{2} + fV_I I_{OFF}\left(\frac{4t_r}{3} + 2t_f\right) + 4fQ_g V_{GSpp}, \quad (6.160)$$

resulting in the efficiency of the Class D full-bridge series-resonant inverter operating above resonance

$$\eta_I = \frac{P_{Ri}}{P_{Ri} + P_T} = \frac{P_{Ri}}{P_{Ri} + P_r + 4P_{toff} + 4P_G}. \quad (6.161)$$

6.9.3 Operation Under Short-Circuit and Open-Circuit Conditions

The Class D inverter with a series-resonant circuit can operate safely with an open circuit at the output. However, it is prone to catastrophic failure if the output is short-circuited at f close to f_o. If $R_i = 0$, the amplitude of the current through the resonant circuit and the switches is

$$I_m = \frac{4V_I}{\pi r\sqrt{1 + (\frac{Z_o}{r})^2(\frac{\omega}{\omega_o} - \frac{\omega_o}{\omega})^2}}. \quad (6.162)$$

The maximum value of I_m occurs at $f = f_o$ and is given by

$$I_{mr} = \frac{4V_I}{\pi r}. \tag{6.163}$$

The amplitudes of the voltages across the resonant components L and C are

$$V_{Cm} = V_{Lm} = \frac{I_{mr}}{\omega_o C} = \omega_o L I_{mr} = Z_o I_{mr} = \frac{4V_I Z_o}{\pi r} = \frac{4V_I Q_o}{\pi}. \tag{6.164}$$

6.9.4 Voltage Transfer Function

Using (6.141), one obtains the voltage transfer function from the input of the inverter to the input of the series-resonant circuit

$$M_{Vs} = \frac{V_{rms}}{V_I} = \frac{2\sqrt{2}}{\pi} = 0.9. \tag{6.165}$$

The product of (6.59) and (6.165) yields the magnitude of the dc-to-ac voltage transfer function for the Class D full-bridge series-resonant inverter

$$M_{VI} = \frac{V_{Ri}}{V_I} = \frac{V_{Ri}}{V_{rms}}\frac{V_{rms}}{V_I} = M_{Vs}M_{Vr} = \frac{2\sqrt{2}\eta_{Ir}}{\pi\sqrt{1 + Q_L^2(\frac{\omega}{\omega_o} - \frac{\omega_o}{\omega})^2}}. \tag{6.166}$$

The maximum value of M_{VI} occurs at $f/f_o = 1$ and equals $M_{VImax} = 2\sqrt{2}\eta_{Ir}/\pi = 0.9\eta_{Ir}$. Thus, the values of M_{VI} range from zero to $0.9\eta_{Ir}$.

Example 6.3

Design a Class D full-bridge inverter of Fig. 6.20 to meet the following specifications: $V_I = 270$ V, $P_{Ri} = 500$ W, and $f = 110$ kHz. Assume $Q_L = 5.3$, $\psi = 30°$ (i.e., $cos^2\psi = 0.75$) and the efficiency $\eta_{Ir} = 94$ %. Neglect switching losses.

Solution: The input power of the inverter is

$$P_I = \frac{P_{Ri}}{\eta_{Ir}} = \frac{500}{0.9} = 555.6 \text{ W}. \tag{6.167}$$

The overall resistance of the inverter can be obtained from (6.155)

$$R = \frac{8V_I^2}{\pi^2 P_I}cos^2\psi = \frac{8 \times 270^2}{\pi^2 \times 555.6} \times 0.75 = 79.8\ \Omega. \tag{6.168}$$

Following the design procedure of Example 6.2, one obtains:

$$R_i = \eta_{Ir} R = 0.94 \times 79.8 = 75\ \Omega \quad (6.169)$$

$$r = R - R_i = 79.8 - 75 = 4.8\ \Omega \quad (6.170)$$

$$I_I = \frac{P_I}{V_I} = \frac{555.6}{270} = 2.06\ \text{A} \quad (6.171)$$

$$I_m = \sqrt{\frac{2P_{Ri}}{R_i}} = \sqrt{\frac{2 \times 500}{75}} = 3.65\ \text{A} \quad (6.172)$$

$$\frac{f}{f_o} = \frac{1}{2}\left(\frac{tan\psi}{Q_L} + \sqrt{\frac{tan^2\psi}{Q_L^2} + 4}\right)$$
$$= \frac{1}{2}\left(\frac{0.5774}{5.3} + \sqrt{\frac{0.5774^2}{5.3^2} + 4}\right) = 1.056 \quad (6.173)$$

$$f_o = \frac{f}{(f/f_o)} = \frac{110 \times 10^3}{1.056} = 104.2\ \text{kHz} \quad (6.174)$$

$$L = \frac{Q_L R}{\omega_o} = \frac{5.3 \times 79.8}{2\pi \times 104.2 \times 10^3} = 646\ \mu\text{H} \quad (6.175)$$

$$C = \frac{1}{\omega_o Q_L R} = \frac{1}{2\pi \times 104.2 \times 10^3 \times 5.3 \times 79.8} = 3.61\ \text{nF} \quad (6.176)$$

$$Z_o = \sqrt{\frac{L}{C}} = \sqrt{\frac{646 \times 10^{-6}}{3.61 \times 10^{-9}}} = 423\ \Omega. \quad (6.177)$$

From (6.164), the maximum voltage stresses for the resonant components are

$$V_{Cm} = V_{Lm} = \frac{4V_I Q_L}{\pi} = \frac{4 \times 270 \times 5.3}{\pi} = 1822\ \text{V}. \quad (6.178)$$

Referring to (6.151), the peak value of the switch voltage is equal to the input voltage

$$V_{SM} = V_I = 270\ \text{V}. \quad (6.179)$$

6.10 RELATIONSHIPS AMONG INVERTERS AND RECTIFIERS

Figures 6.22 to 6.24 show the relationships among various inverters and rectifiers, namely, Class D, push-pull, and bridge topologies. The symbols used in these figures are as follows:

1) i_o is an ac current load formed by a series-resonant circuit.
2) v_o is an ac voltage load formed by a parallel-resonant circuit.
3) I_O is a dc current load comprised of a large filter inductance L_f con-

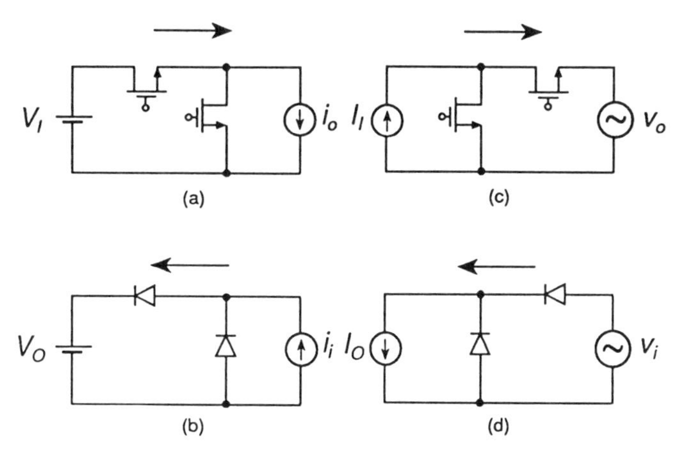

Figure 6.22: Relationships among Class D resonant inverters and rectifiers. (a) Class D voltage-fed current-loaded resonant inverter. (b) Class D current-driven voltage-loaded half-wave rectifier. (c) Class D current-fed voltage-loaded resonant inverter. (d) Class D voltage-driven current-loaded rectifier.

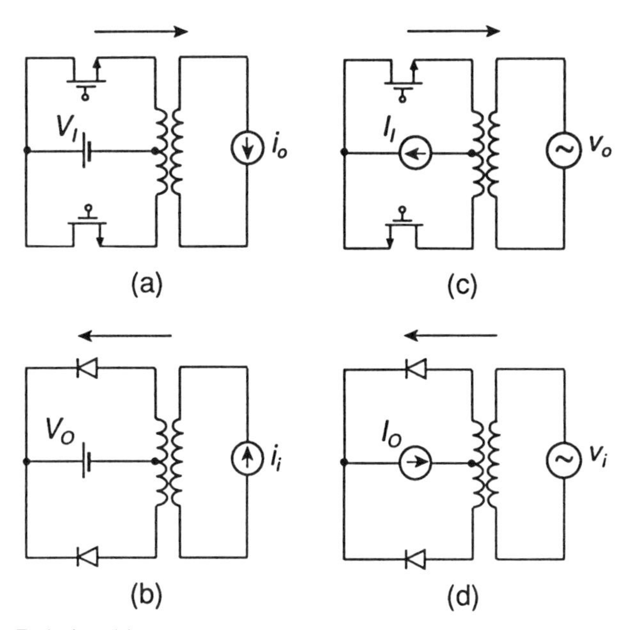

Figure 6.23: Relationships among push-pull resonant inverters and transformer center-tapped rectifiers. (a) Push-pull voltage-fed current-loaded resonant inverter. (b) Transformer center-tapped current-driven voltage-loaded rectifier. (c) Push-pull current-fed voltage-loaded resonant inverter. (d) Transformer center-tapped voltage-driven current-loaded rectifier.

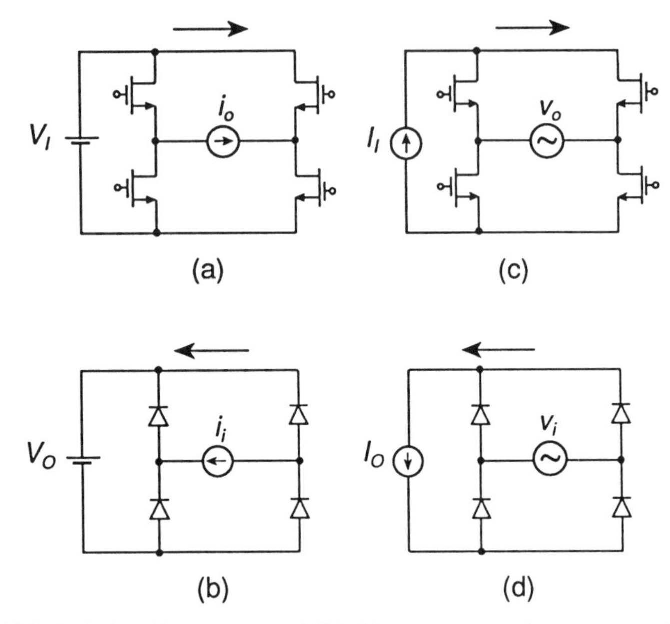

Figure 6.24: Relationships among full-bridge resonant inverters and rectifiers. (a) Full-bridge voltage-fed current-loaded resonant inverter. (b) Full-bridge current-driven voltage-loaded rectifier. (c) Full-bridge current-fed voltage-loaded resonant inverter. (d) Full-bridge voltage-driven current-loaded rectifier.

nected in series with either a load resistance R_i or a parallel combination of a large filter capacitance C_f and a load resistance R_i.

4) V_O is a dc voltage load composed of a large filter capacitance C_f and a load resistance R_i.
5) I_I is a dc current source comprised of a dc voltage source V_I connected in series with a large inductance L_f.

The *bilateral inversion* can be used to derive a rectifier from an inverter. The principles of this inversion are as follows:

1) The ac current load i_o or the ac voltage load v_o is replaced by an ac current source i_i or an ac voltage source v_i, respectively.
2) The dc current source I_I or the dc voltage source V_I is replaced by a dc current load I_O or a dc voltage V_O, respectively.
3) Transistors are replaced by diodes connected such that the current flows in the opposite direction.

A reversed procedure can be used to derive an inverter from a rectifier. In each of the figures, inverters are dual and rectifiers are dual. In all the circuits in the left column, the semiconductor device voltage waveform is a square wave and the semiconductor device current waveform is a half sine wave. In all the circuits in the right column, the waveforms are dual, that is, the semiconductor device voltage waveform is a half sine wave and the semiconductor device current is a square wave. Each circuit contains either one voltage source and one current load or one current source and one voltage load.

6.11 SUMMARY

- The maximum voltage across the switches in both Class D half-bridge and full-bridge inverters is low and equal to the dc input voltage V_I. Therefore, they are suitable for applications in off-line inverters.
- Operation with a capacitive load (i.e., below resonance) is not recommended. The antiparallel diodes turn off at a high di/dt. If the MOSFET's body-drain *pn* junction diode (or any *pn* junction diode) is used as an antiparallel diode, it generates high reverse-recovery current spikes. These spikes occur in the switch current waveforms at both the switch turn-on and turn-off, and may destroy the transistor. The reverse-recovery spikes may initiate the turn-on of the parasitic BJT in the MOSFET structure and may cause the MOSFET to fail due to the second breakdown of parasitic BJT. The current spikes can be reduced by adding a Schottky antiparallel diode (if V_I is below 100 V), or a series diode and an antiparallel diode.
- For operation below resonance, the transistors are turned on at a high voltage equal to V_I and the transistor output capacitance is short-circuited by a low transistor on-resistance, dissipating the energy stored in that capacitance. Therefore, the turn-on switching loss is high, Miller's effect is significant, the transistor input capacitance is high, the gate drive power is high, and the turn-on transition speed is reduced.
- The operation with an inductive load (i.e., above resonance) is preferred. The antiparallel diodes turn off at a low di/dt. Therefore, MOSFET's body-drain *pn* junction diodes can be used as antiparallel diodes because they do not generate reverse-recovery current spikes and are sufficiently fast.
- For operation above resonance, the transistors turn on at zero voltage. For this reason, the turn-on switching loss is reduced, Miller's effect is absent, the transistor input capacitance is low, the gate drive power is low, and turn-on speed is high. However, the turn-off is lossy.
- The efficiency is high at light loads because R_i/r increases with increasing R_i [see Equation (6.69)].
- The inverter can operate safely with an open circuit at the output.

- There is a risk of catastrophic failure if the the output is short-circuited at the operating frequency f close to the resonant frequency f_o.
- The input voltage of the resonant circuit in the Class D full-bridge inverter is a square wave whose low level is $-V_I$ and whose high level is V_I. The peak-to-peak voltage across the resonant circuit in the full-bridge inverter is two times higher than in the half-bridge inverter. Therefore, the output power of the full-bridge inverter is four times higher than in the half-bridge inverter at the same load, the dc input voltage, and f/f_o.
- The dc voltage source V_I and the switches form an ideal square-wave voltage source; therefore, many loads can be connected between the two switches and ground and operated without mutual interactions.
- Although this chapter has focused on MOSFETs, other power switches can be used, such as BJTs, thyristors, MCTs, GTOs, and IGBTs.

6.12 REFERENCES

1. P. J. Baxandall, "Transistor sine-wave *LC* oscillators, some general considerations and new developments," *Proc. IEE*, vol. 106, Pt. B, suppl. 16, pp. 748–758, May 1959.
2. M. R. Osborne, "Design of tuned transistor power inverters," *Electron. Eng.*, vol. 40, no. 486, pp. 436–443, 1968.
3. W. J. Chudobiak and D. F. Page, "Frequency and power limitations of Class-D transistor inverter," *IEEE J. Solid-State Circuits*, vol. SC-4, pp. 25–37, Feb. 1969.
4. S. B. Dewan, A. Straughen, *Power Semiconductor Circuits*, New York: John Wiley & Sons, 1975, ch. 7.2 and 7.3, pp. 358–400.
5. M. Kazimierczuk and J. S. Modzelewski, "Drive-transformerless Class-D voltage-switching tuned power inverter," *Proc. IEEE*, vol. 68, pp. 740–741, June 1980.
6. H. L. Krauss, C. W. Bostian, and F. H. Raab, *Solid State Radio Engineering*, New York: John Wiley & Sons, ch. 14.1–2, pp. 432–448, 1980.
7. F. H. Raab, "Class-D power inverter load impedance for maximum efficiency," *RF Technology Expo'85 Conf.*, Anaheim, CA, Jan. 23–25, 1985, pp. 287–295.
8. M. H. Rashid, *Power Electronics: Circuits, Devices, and Applications*, Englewood Cliffs, NJ: Prentice-Hall, 1988, ch. 8.9, pp. 251–264.
9. N. Mohan, T. M. Undeland, and W. P. Robbins, *Power Electronics, Converters, Applications and Design*, New York: John Wiley & Sons, 1989, ch. 7. 4. 1, pp. 164–170.
10. M. K. Kazimierczuk, "Class D voltage-switching MOSFET power inverter," *IEE Proc., Pt. B, Electric Power Appl.*, vol. 138, pp. 286–296, Nov. 1991.
11. J. G. Kassakian, M. F. Schlecht, and G. C. Verghese, *Principles of Power Electronics*, Reading, MA: Addison-Wesley, 1991, ch. 9.2, pp. 202–212.
12. M. K. Kazimierczuk and W. Szaraniec, "Class D voltage-switching inverter with only one shunt capacitor," *IEE Proc., Pt. B, Electric Power Appl.*, vol. 139, pp. 449–456, Sept. 1992.

6.13 REVIEW QUESTIONS

6.1 Draw the inductive reactance X_L, capacitive reactance X_C, and total reactance $X_L - X_C$ versus frequency for the series-resonant circuit. What occurs at the resonance frequency?

6.2 What is the voltage across the switches in Class D half-bridge and full-bridge inverters?

6.3 What is the frequency range in which a series-resonant circuit represents a capacitive load to the switching part of the Class D series resonant inverter?

6.4 What are the disadvantages of operation of the Class D series resonant inverter with a capacitive load?

6.5 Is the turn-on switching loss of the power MOSFETs zero below resonance?

6.6 Is the turn-off switching loss of the power MOSFETs zero below resonance?

6.7 Is Miller's effect present at turn-on or turn-off below resonance?

6.8 What is the influence of zero-voltage switching on Miller's effect?

6.9 What is the frequency range in which a series-resonant circuit represents an inductive load to the switching part of the inverter?

6.10 What are the merits of operation of the Class D inverter with an inductive load?

6.11 Is the turn-on switching loss of the power MOSFETs zero above resonance?

6.12 Is the turn-off switching loss of the power MOSFETs zero above resonance?

6.13 What is the voltage stress of the resonant capacitor and inductor in half-bridge and full-bridge inverters?

6.14 What are the worst conditions for the voltage stresses of resonant components?

6.15 What happens when the output of the inverter is short-circuited?

6.16 Is the part-load efficiency of the SRI high?

6.14 PROBLEMS

6.1 A series-resonant circuit consists of an inductor $L = 84\ \mu H$ and a capacitor $C = 300$ pF. The ESRs of these components at the resonant frequency are $r_L = 1.4\ \Omega$ and $r_C = 50\ m\Omega$, respectively. The load resistance is $R_i = 200\ \Omega$. The resonant circuit is driven by a sinusoidal voltage source whose amplitude is $V_m = 100$ V. Find the resonant frequency f_o, characteristic impedance Z_o, loaded quality factor Q_L, unloaded quality factor Q_o, quality factor of the inductor Q_{Lo}, and quality factor of the capacitor Q_{Co}.

6.2 For the resonant circuit given in Problem 6.1, find the reactive power of the inductor Q and the total true power P_R.

6.3 For the resonant circuit given in Problem 6.1, find the voltage and current stresses for the resonant inductor and the resonant capacitor. Calculate also the reactive power of the resonant components.

6.4 Find the efficiency for the resonant circuit given in Problem 6.1. Is the efficiency dependent on the operating frequency?

6.5 Write general expressions for the instantaneous energy stored in the resonant inductor $w_L(t)$ and in the resonant capacitor $w_C(t)$, as well as the total instantaneous energy stored in the resonant circuit $w_t(t)$. Sketch these waveforms for $f = f_o$. Explain briefly how the energy is transferred between the resonant components.

6.6 A Class D half-bridge inverter is supplied by a dc voltage source of 350 to 400 V. Find the voltage stresses of the switches. Repeat the same problem for a Class D full-bridge inverter.

6.7 A series-resonant circuit which consists of a resistance $R = 25$ Ω, inductance $L = 100$ μH, and capacitance $C = 4.7$ nF is driven by a sinusoidal voltage source $v = 100 sin\omega t$ (V). The operating frequency can be changed over a wide range. Calculate exactly the maximum voltage stresses for the resonant components. Compare the results with voltages across the inductance and the capacitance at the resonant frequency.

6.8 Design a Class D half-bridge series resonant inverter that delivers to the load resistance power $P_{Ri} = 30$ W. The inverter is supplied from input voltage source $V_I = 180$ V. It is required that the operating frequency is $f = 210$ kHz. Neglect switching and drive-power losses.

CHAPTER 7

CLASS D PARALLEL RESONANT INVERTER

7.1 INTRODUCTION

In the Class D series resonant inverter discussed in the preceding chapter, the load resistance is connected in series with the LC components. Consequently, as the load resistance is increased, the current through the resonant circuit and the switches decreases, as does the output power. In this chapter, a Class D voltage-source parallel resonant inverter [1]–[5] is studied. Its basic characteristics are derived and illustrated. The load resistance in this inverter is connected in parallel with the resonant capacitor. As a result, if the load resistance is much higher than the reactance of the resonant capacitor, the current through the resonant inductor and the switches is almost independent of the load. As the load resistance is increased, the voltage across the resonant capacitor and the load increases, causing the output power to increase.

7.2 PRINCIPLE OF OPERATION

A circuit of a Class D voltage-source parallel resonant inverter (PRI) is shown in Fig. 7.1(a). It consists of two switches S_1 and S_2, a resonant inductor L, a resonant capacitor C, and a dc-blocking capacitor C_c. Resistance R_i represents a load to which the ac power is to be delivered and is connected in parallel with the resonant capacitor C. A dc-blocking capacitor C_c prevents dc current flow through the load resistance R_i. It can also be connected in series with resonant inductor L, but in this case a higher current flows through the equivalent series resistance (ESR) of capacitor C_c, reducing the

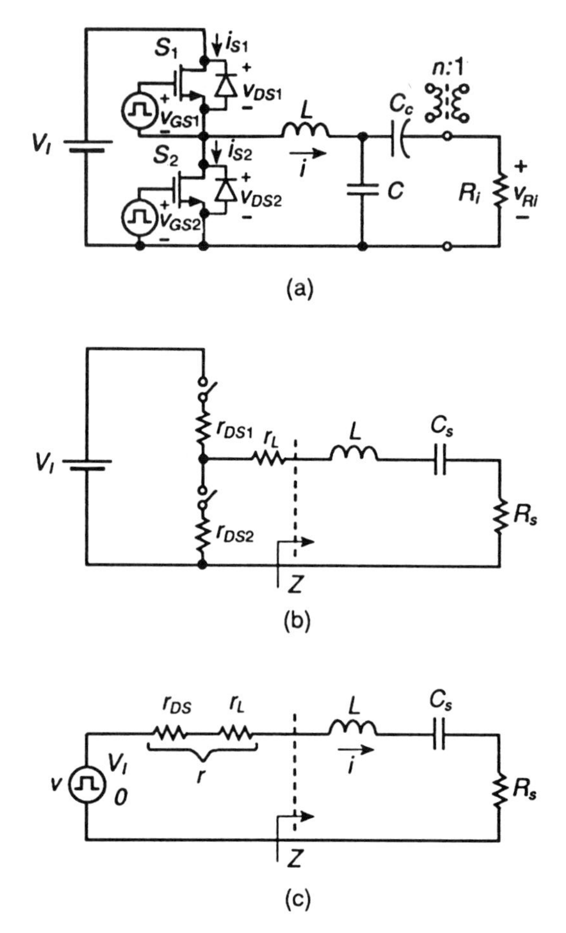

Figure 7.1: Circuit of the Class D parallel resonant inverter. (a) Circuit. (b) Transformation of the R_i-C circuit into the R_s-C_s circuit. (c) Equivalent circuit.

efficiency. The average voltage across capacitor C_c is equal to $V_I/2$. A transformer can be connected in parallel with resonant capacitor C. In this case, the capacitor C may be placed on either the primary or the secondary side of the transformer. If it is on the secondary side, the transformer leakage inductance is absorbed into the resonant inductance L. In the transformer version of the inverter, the blocking capacitor C_c must be placed in series with resonant inductance L to prevent a short circuit of the dc source V_I through the primary of the transformer if the the upper transistor is damaged in such a way that there is a short circuit between the drain and the source. The two bidirectional two-quadrant switches S_1 and S_2 and the dc input voltage source V_I form a square-wave voltage source that drives the resonant circuit L-C-R_i. Each switch consists of a transistor and an antiparallel diode. The MOSFET's body-drain *pn* junction diode can be used as an antiparallel diode for operation above resonance. The transistors are driven by rectangular-wave voltages v_{GS1} and v_{GS2}. Each switch is controllable only when its current is positive. Switches S_1 and S_2 are alternately turned ON and OFF at the switching frequency $f = \omega/2\pi$. Because of the turn-off delay time of power MOSFETs, the duty cycle of drive voltages v_{GS1} and v_{GS2} should be slightly less than 50% to avoid cross conduction.

Figure 7.1(b) shows an equivalent circuit of the Class D inverter. The MOSFETs are modeled by switches whose on-resistances are r_{DS1} and r_{DS2}. Resistance r_L represents the equivalent series resistance of inductor L. In practice, the series equivalent resistance r_C of capacitor C is usually very low and, therefore, is neglected in this analysis. The parallel combination of R_i and C of Fig. 7.1(a) is converted into a series combination of R_s and C_s, as shown in Fig. 7.1(b). In Fig. 7.1(c), the dc voltage source V_I and the switches S_1 and S_2 are replaced by a square-wave voltage source with a low level value of zero and a high level value of V_I. Resistance $r_{DS} = (r_{DS1} + r_{DS2})/2$ is the equivalent average on-resistance of the MOSFETs. Neglecting the parasitic resistances of the capacitors, the overall parasitic resistance is given by

$$r = r_{DS} + r_L. \tag{7.1}$$

Hence, the total resistance of the inverter is

$$R = R_s + r = R_s + r_{DS} + r_L. \tag{7.2}$$

Figure 7.2 shows the current and voltage waveforms for $f > f_r$, where $f_r = 1/(2\pi\sqrt{LC_s})$ is the resonant frequency of the L-C_s-R_s circuit. The input voltage v_{DS2} of the resonant circuit is a square wave with a low level value equal to zero and a high level value equal to V_I. The analysis is simplified by assuming sinusoidal currents in L, C, and R_i. This approximation is valid if the loaded quality factor Q_L of the resonant circuit is high (e.g., $Q_L \geq 2.5$). In this case, a nonlinear load such as a fluorescent lamp or a rectifier can be modeled by a linear impedance. If $Q_L < 2.5$, the inductor current waveform differs from a sine wave and an accurate analytical solution is more difficult to obtain. However, the predicted results are still qualitatively correct. A sinusoidal inductor current is assumed in the subsequent analysis. The inductor current i is conducted alternately by switches S_1 and S_2. Each transistor should be turned on when the switch current is negative and flows through the diode. The range of the dead time is indicated in Fig. 7.2 by the dashed areas.

For $f < f_r$, the phase shift $\psi < 0$, the resonant circuit represents a capacitive load for the voltage source v_i and the inductor current i leads the fundamental component of the voltage v. Therefore, the antiparallel diodes turn off at high di/dt, causing high current spikes in the switches and reducing efficiency and reliability. This problem can be alleviated by adding

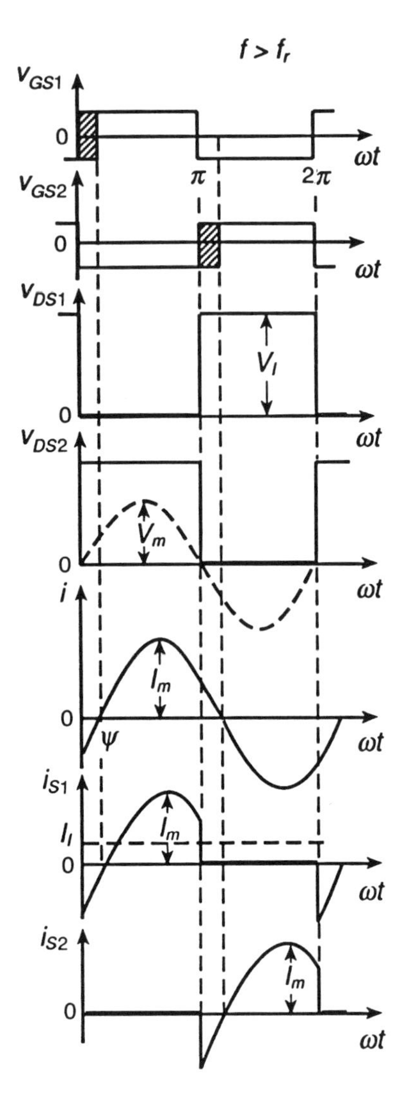

Figure 7.2: Waveforms in the Class D parallel resonant inverter for $f > f_r$.

external diodes; however, the efficiency will be reduced. For these reasons, only the operation above the resonant frequency is explored in the subsequent analysis.

For $f > f_r$, the phase shift $\psi > 0$, the resonant circuit represents an inductive load and the current i lags behind the fundamental component of voltage v. Hence, the switch current is negative after turn-on and positive before turn-off. Consider the turn-off of switch S_1. When transistor Q_1 is turned off by the drive voltage v_{GS1}, v_{DS1} increases, causing the decrease of v_{DS2}. As v_{DS2} reaches -0.7 V, D_2 turns on and therefore the current i is diverted from transistor Q_1 to diode D_2. The turn-off switch transition is forced by the driver, while the turn-on transition of the switch is caused by the turn-off transition of the opposite transistor, not by the driver. Only the turn-off transition is directly controllable by the driver. The transistor should be turned on by the driver when the switch current is negative and flows through the antiparallel diode. Therefore, the transistor is turned on at nearly zero voltage, reducing the turn-on switching loss to a negligible level.

7.3 ANALYSIS

7.3.1 Assumptions

The analysis of the inverter of Fig. 7.1(a) assumes

1) Each switch is a resistance r_{DS} when "on" and an open circuit when "off".
2) Switching losses are neglected.
3) The loaded quality factor Q_L of the resonant circuit is high enough so that the currents through inductance L, capacitance C, and resistance R_i are sinusoidal.
4) The coupling capacitance C_c is high enough so that its ac voltage ripple is negligible.
5) The output capacitances of MOSFETs are neglected.

7.3.2 Resonant Circuit

The resonant circuit in the inverter of Fig. 7.1(a) is a second-order low-pass filter and can be described by the following normalized parameters:

- the corner frequency (or the undamped natural frequency)

$$\omega_o = \frac{1}{\sqrt{LC}} \tag{7.3}$$

- the characteristic impedance

$$Z_o = \omega_o L = \frac{1}{\omega_o C} = \sqrt{\frac{L}{C}} \tag{7.4}$$

- the loaded quality factor at the corner frequency f_o

$$Q_L = \omega_o C R_i = \frac{R_i}{\omega_o L} = \frac{R_i}{Z_o} \tag{7.5}$$

- the resonant frequency that forms the boundary between capacitive and inductive loads

$$\omega_r = \frac{1}{\sqrt{LC_s}} \tag{7.6}$$

- the loaded quality factor at the resonant frequency f_r

$$Q_r = \frac{\omega_r L}{R_s} = \frac{1}{\omega_r C_s R_s}. \tag{7.7}$$

The damped natural frequency is $\omega_d = \omega_o\sqrt{1 - 1/(4Q_L^2)}$ for $Q_L \geq 1/2$.

Assuming a zero-reactance dc-blocking capacitor C_c and using (7.3) and (7.5), the input impedance of the resonant circuit is

$$\mathbf{Z} = j\omega L + \frac{R_i \frac{1}{j\omega C}}{R_i + \frac{1}{j\omega C}} = \frac{R_i[1 - (\frac{\omega}{\omega_o})^2 + j\frac{1}{Q_L}(\frac{\omega}{\omega_o})]}{1 + jQ_L(\frac{\omega}{\omega_o})} = Ze^{j\psi} = R_s + jX_s, \tag{7.8}$$

where

$$\frac{Z}{Z_o} = \sqrt{\frac{Q_L^2[1 - (\frac{\omega}{\omega_o})^2]^2 + (\frac{\omega}{\omega_o})^2}{1 + (Q_L\frac{\omega}{\omega_o})^2}} \tag{7.9}$$

$$\psi = arctan\left\{Q_L\left(\frac{\omega}{\omega_o}\right)\left[\left(\frac{\omega}{\omega_o}\right)^2 + \frac{1}{Q_L^2} - 1\right]\right\} \tag{7.10}$$

$$R_s = Zcos\psi \tag{7.11}$$

$$X_s = Zsin\psi. \tag{7.12}$$

From trigonometric relationships and (7.11),

$$cos\psi = \frac{1}{\sqrt{1 + \{Q_L(\frac{\omega}{\omega_o})[(\frac{\omega}{\omega_o})^2 + \frac{1}{Q_L^2} - 1]\}^2}}. \tag{7.13}$$

At $f = f_o$,

$$Z(f_o) = \frac{Z_o}{\sqrt{Q_L^2 + 1}} \approx \frac{Z_o^2}{R_i}, \quad \text{for } Q_L^2 \gg 1. \tag{7.14}$$

As R_i is increased, $Z(f_o)$ decreases. A three-dimensional representation of Z/Z_o is depicted in Fig. 7.3. Figure 7.4 shows plots of Z/Z_o and ψ versus f/f_o at various values of Q_L.

The resonant frequency f_r is defined as a frequency at which the phase shift ψ is zero. Hence, from (7.10), the ratio of the resonant frequency f_r to the the corner frequency f_o is

$$\frac{f_r}{f_o} = \sqrt{1 - \frac{1}{Q_L^2}}, \quad \text{for} \quad Q_L \geq 1. \tag{7.15}$$

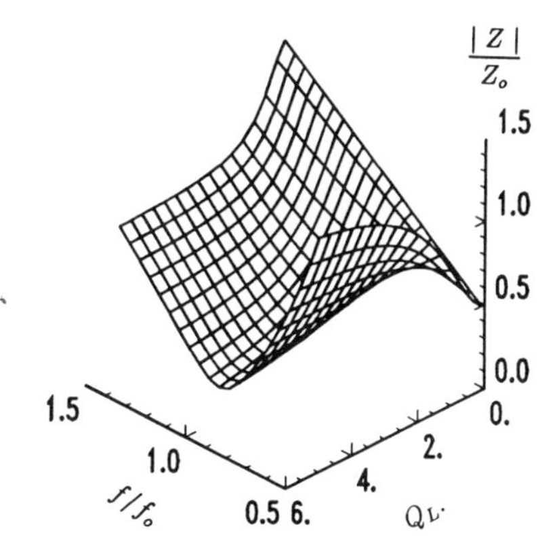

Figure 7.3: Three-dimensional representation of Z/Z_o versus f/f_o and Q_L.

Figure 7.5 illustrates f_r/f_o as a function of Q_L. Frequency f_r forms the boundary between inductive and capacitive loads. To achieve high efficiency and reliability, f should be higher than f_r under all operating conditions. The following conclusions can be drawn from (7.15):

1) For $Q_L \leq 1$, the resonant frequency f_r does not exist and the resonant circuit represents an inductive load at any operating frequency.
2) For $Q_L > 1$, f_r/f_o increases with Q_L. For $f > f_r$, the transistors are loaded by an inductive load and current i lags the voltage v_{DS2}, resulting in desirable operation. For $f < f_r$, the transistors are loaded by a capacitive load and the current i leads the voltage v_{DS2}, causing spikes in the switch currents due to the reverse recovery of the antiparallel diodes at turn-off.

The C-R_i parallel two-terminal network shown in Fig. 7.1(a) can be converted into an equivalent C_s-R_s series two-terminal network shown in Fig. 7.1(b), resulting in the basic topology of the Class D inverter with a series-resonant circuit. The input impedance of the C-R_i circuit at a given frequency f is

$$Z_i = \frac{\frac{R_i}{j\omega C}}{R_i + \frac{1}{j\omega C R_i}} = \frac{R_i}{1 + j\omega C R_i} = \frac{R_i}{1 + (\omega C R_i)^2} - j\frac{\frac{1}{\omega C}}{1 + \frac{1}{(\omega C R_i)^2}}$$
$$= \frac{R_i}{1 + q^2} + j\frac{X_C}{1 + \frac{1}{q^2}} = R_s + jX_{Cs} \tag{7.16}$$

where

$$q = \frac{X_{Cs}}{R_s} = \frac{1}{\omega C_s R_s} = \frac{R_i}{X_C} = \omega C R_i = \left(\frac{\omega}{\omega_o}\right) Q_L \tag{7.17}$$

is the reactance factor, $X_C = -1/(\omega C)$, and $X_{Cs} = -1/(\omega C_s)$. Hence, from (7.17) the relationships among R_i, R_s, C, and C_s at the frequency f are

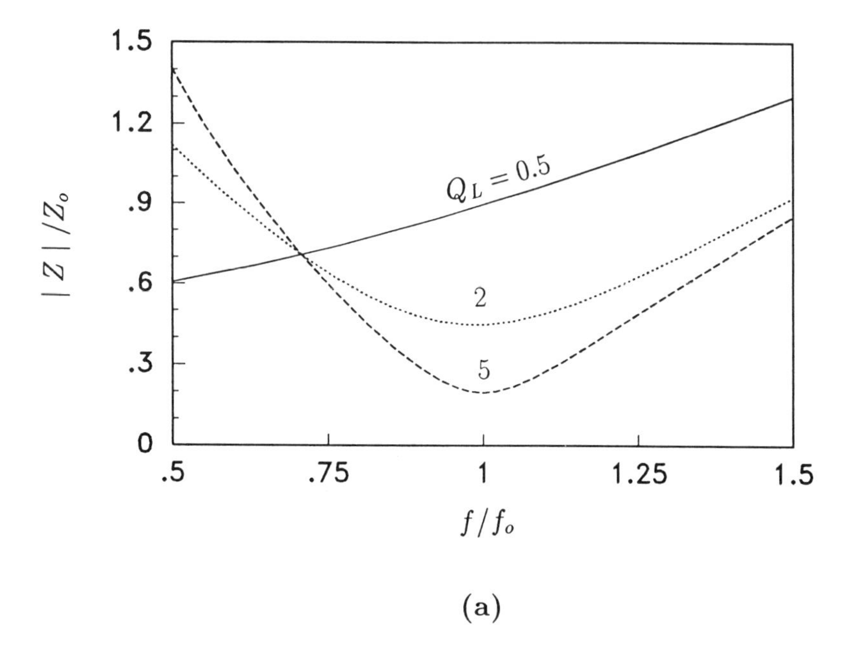

(a)

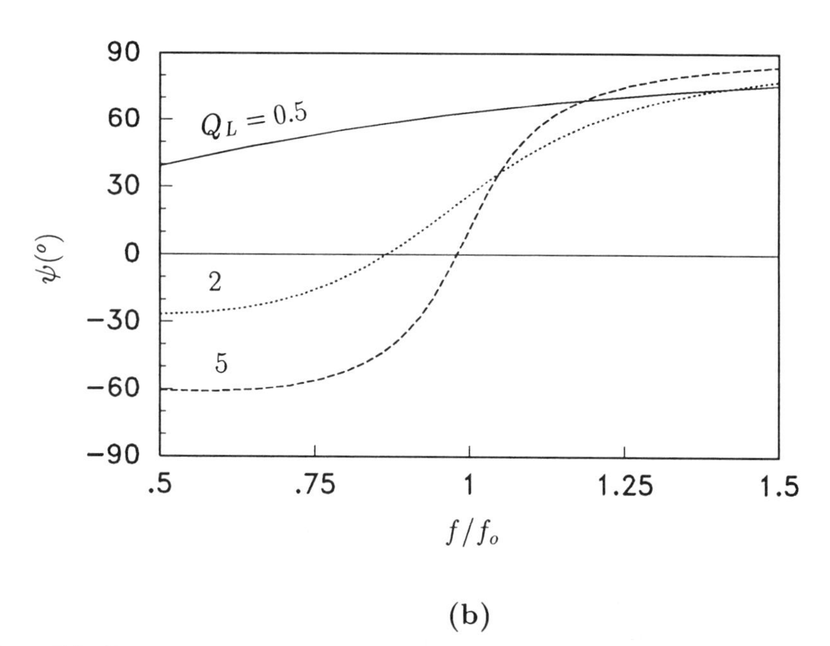

(b)

Figure 7.4: Characteristics of the input impedance of the resonant circuit. (a) Z/Z_o as a function of f/f_o at various values of Q_L. (b) ψ as a function of f/f_o at various values of Q_L.

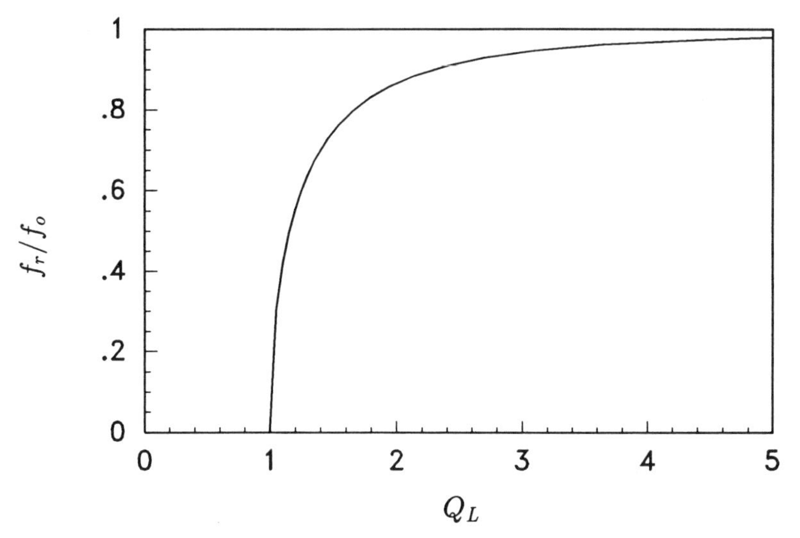

Figure 7.5: A plot of f_r/f_o versus Q_L.

obtained as

$$R_s = \text{Re}\{R_i||C\} = \frac{R_i}{1+q^2} = \frac{R_i}{1+\left(\frac{R_i}{X_C}\right)^2} = \frac{R_i}{1+\left(\frac{\omega}{\omega_o}\right)^2 Q_L^2}, \tag{7.18}$$

$$X_{Cs} = \text{Im}\{R_i||C\} = \frac{X_C}{1+\frac{1}{q^2}} = \frac{X_C}{1+\left(\frac{X_C}{R_i}\right)^2} = \frac{X_C}{1+\left(\frac{\omega_o}{\omega}\right)^2 \frac{1}{Q_L^2}}, \tag{7.19}$$

and

$$C_s = C\left(1+\frac{1}{q^2}\right) = C\left[1+\left(\frac{X_C}{R_i}\right)^2\right] = C\left[1+\left(\frac{\omega_o}{\omega}\right)^2 \frac{1}{Q_L^2}\right]. \tag{7.20}$$

If $q^2 << 1$, that is, $R_i^2 << X_C^2$,

$$R_s \approx R_i \tag{7.21}$$

and

$$X_{Cs} \approx \frac{R_i^2}{X_C}. \tag{7.22}$$

On the other hand, if $q^2 >> 1$, that is, $R_i^2 >> X_C^2$,

$$R_s \approx \frac{X_C^2}{R_i} \tag{7.23}$$

and

$$X_{Cs} \approx X_C. \tag{7.24}$$

Figure 7.6 shows plots of R_s and $| X_{Cs} |$ versus R_i at $| X_C | = 100\ \Omega$. As the parallel resistance R_i is increased from 0 to ∞, the series equivalent resistance R_s first increases starting from 0, reaches the maximum value equal to $| X_C | /2$ at $R_i = | X_C |$, and then decreases to 0. Since the series equivalent resistance R_s decreases with increasing parallel resistance R_i for $R_i > | X_C |$, the C-R_i circuit is called an *impedance inverter* [6]. The range of matching

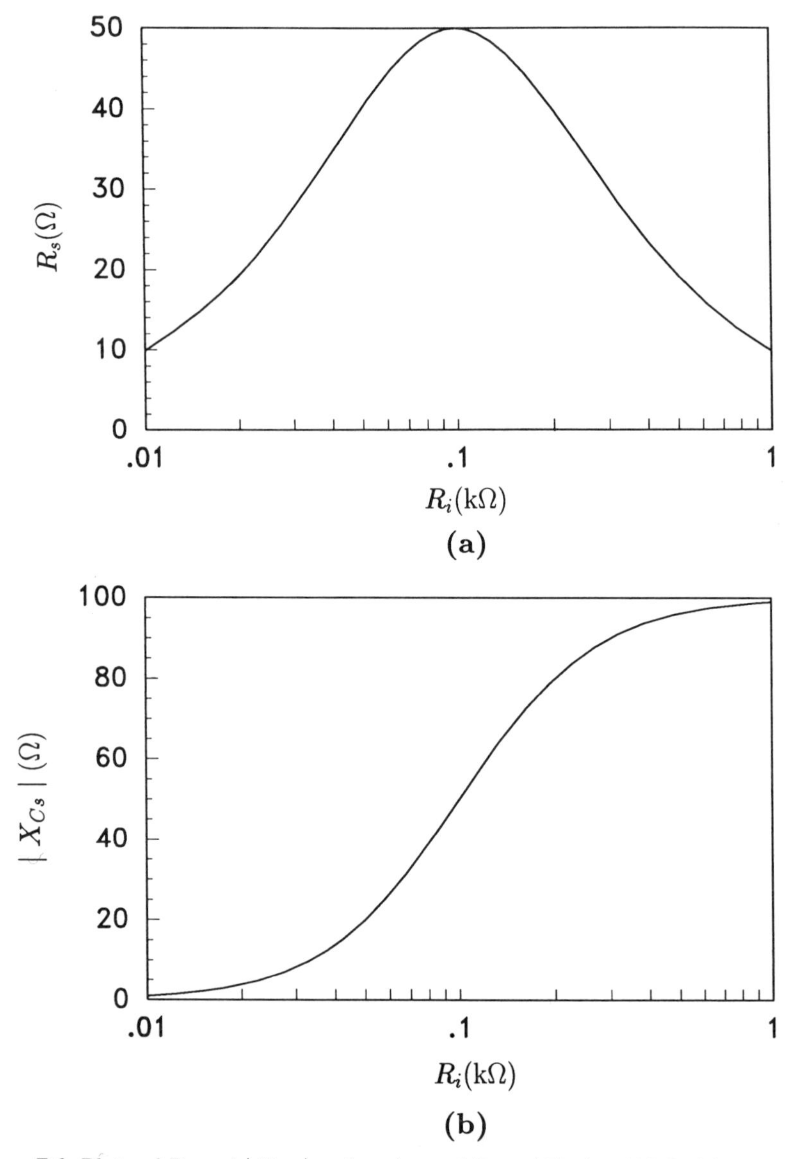

Figure 7.6: Plots of R_s and $| X_{Cs} |$ as functions of R_i at $| X_C | = 100\ \Omega$. (a) R_s versus R_i. (b) $| X_{Cs} |$ versus R_i.

R_i to R_s is determined by Q_L for a given normalized frequency f/f_o. The equivalent series reactance $| X_{Cs} |$ increases from 0 to $| X_C |$ when R_i is increased from 0 to ∞.

From (7.3), (7.5), (7.16), and (7.17), the loaded quality factors Q_L and Q_r are related by

$$Q_r = \frac{\omega_r L}{R_s} = \frac{1}{\omega_r C_s R_s} = \omega_r C R_i = Q_L \left(\frac{\omega_r}{\omega_o}\right) = \sqrt{Q_L^2 - 1}, \quad \text{for } Q_L \geq 1. \tag{7.25}$$

For $Q_L^2 >> 1$, $Q_r \approx Q_L$. Figure 7.7 shows a plot of Q_r versus Q_L. Substitution of (7.25) into (7.18), (7.19), and (7.20) yields the relationships among R_i, R_s, C, and C_s at $f = f_r$

$$R_{sr} = \frac{R_i}{1 + Q_r^2} = \frac{R_i}{Q_L^2} = \frac{Z_o^2}{R_i}, \tag{7.26}$$

$$X_{Csr} = \frac{X_C}{1 + \frac{1}{Q_r^2}}, \tag{7.27}$$

and

$$C_{sr} = \frac{C}{1 - \frac{1}{Q_L^2}}. \tag{7.28}$$

As Q_L is increased from 1 to ∞, R_{sr}/R_i decreases from 1 to 0 and C_{sr}/C decreases from ∞ to 1. As R_i is increased from $R_{i(min)} = Z_o$ to ∞, R_{sr} decreases from $R_{sr(max)} = Z_o$ to 0. The output power P_{Ri} at $f = f_r$ increases

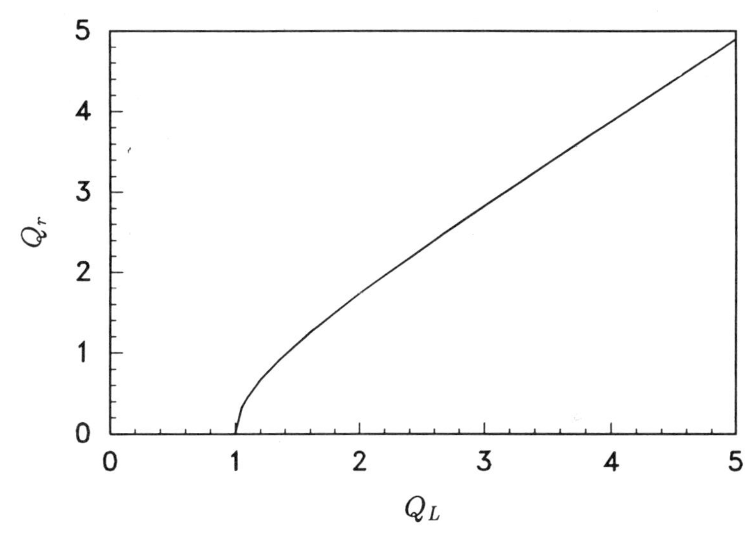

Figure 7.7: A plot of Q_r as a function of Q_L.

as R_i is increased because P_{Ri} increases as R_{sr} is decreased. Because C_{sr} decreases with R_i, the resonant frequency f_r increases as R_i is increased. For $Q_L = 1$, $C_{sr} = \infty$ and, therefore, $f_r = 0$.

7.3.3 Voltage Transfer Function

Referring to Figs. 7.1 and 7.2, the input voltage of the resonant circuit v is a square wave of magnitude V_I, given by

$$v = \begin{cases} V_I, & \text{for } 0 < \omega t \leq \pi \\ 0, & \text{for } \pi < \omega t \leq 2\pi. \end{cases} \tag{7.29}$$

The fundamental component of this voltage is

$$v_{i1} = V_m sin\omega t, \tag{7.30}$$

in which the amplitude of v_{i1} can be found from Fourier analysis as

$$V_m = \frac{2}{\pi} V_I = 0.6366 V_I. \tag{7.31}$$

Hence, one obtains the rms value of v_{i1}

$$V_{rms} = \frac{V_m}{\sqrt{2}} = \frac{\sqrt{2}}{\pi} V_I = 0.4502 V_I, \tag{7.32}$$

which leads to the voltage transfer function from V_I to the fundamental component at the input of the resonant circuit

$$M_{Vs} \equiv \frac{V_{rms}}{V_I} = \frac{\sqrt{2}}{\pi} = 0.4502. \tag{7.33}$$

According to Fig. 7.1(a), the voltage transfer function of the resonant circuit is

$$\mathbf{M_{Vr}} \equiv \frac{\mathbf{V_{Ri}}}{\sqrt{2} V_{rms}} = \frac{\frac{\frac{R_i}{j\omega C}}{R_i + \frac{1}{j\omega C}}}{j\omega L + \frac{\frac{R_i}{j\omega C}}{R_i + \frac{1}{j\omega C}}}$$

$$= \frac{1}{1 - (\frac{\omega}{\omega_o})^2 + j\frac{1}{Q_L}(\frac{\omega}{\omega_o})} = M_{Vr} e^{j\varphi}, \tag{7.34}$$

where

$$M_{Vr} = \frac{V_{Ri}}{V_{rms}} = \frac{1}{\sqrt{[1 - (\frac{\omega}{\omega_o})^2]^2 + \frac{1}{Q_L^2}(\frac{\omega}{\omega_o})^2}}, \tag{7.35}$$

$$\varphi = -arctan\left[\frac{\frac{1}{Q_L}(\frac{\omega}{\omega_o})}{1-(\frac{\omega}{\omega_o})^2}\right]. \qquad (7.36)$$

$\mathbf{V_{Ri}}$ is the phasor of the voltage across R_i, and V_{Ri} is the rms value of $\mathbf{V_{Ri}}$. Figure 7.8 shows a three-dimensional representation of the voltage transfer function of the resonant circuit given by (7.35). Cross-sectional views of this function are plotted in Fig. 7.9. From (7.35), $M_{Vr} = Q_L$ at $f/f_o = 1$ and

$$M_{Vr} \rightarrow \frac{1}{1-(\frac{\omega}{\omega_o})^2}, \qquad \text{as} \qquad Q_L \rightarrow \infty. \qquad (7.37)$$

With an open circuit at the output, M_{Vr} increases from 1 to ∞ as ω/ω_o is increased from 0 to 1, and M_{Vr} decreases from ∞ to 0 as ω/ω_o increases from 1 to ∞.

The maximum value of M_{Vr} is obtained by differentiating the quantity under the square-root sign with respect to f/f_o and setting the result equal to zero. Notice, however, that the maximum value of M_{Vr} occurs at the frequency equal to zero for $Q_L < 1/\sqrt{2}$. Hence, the normalized peak frequency is

$$\frac{f_{pk}}{f_o} = \begin{cases} 0, & \text{for} \quad 0 \le Q_L \le \frac{1}{\sqrt{2}} \\ \sqrt{1-\frac{1}{2Q_L^2}}, & \text{for} \quad Q_L > \frac{1}{\sqrt{2}}, \end{cases} \qquad (7.38)$$

resulting in the maximum magnitude of the voltage transfer function of the resonant circuit

$$M_{Vr(max)} = \begin{cases} 1, & \text{for} \quad 0 \le Q_L \le \frac{1}{\sqrt{2}} \\ \frac{Q_L}{\sqrt{1-\frac{1}{4Q_L^2}}}, & \text{for} \quad Q_L \ge \frac{1}{\sqrt{2}}. \end{cases} \qquad (7.39)$$

For $Q_L^2 >>1$, $f_{pk} \approx f_o$ and $M_{Vr(max)} \approx Q_L$.

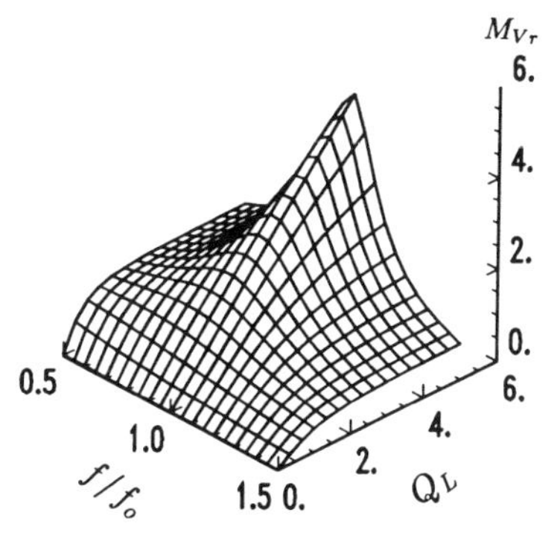

Figure 7.8: Three-dimensional representation of the voltage transfer function of the resonant circuit.

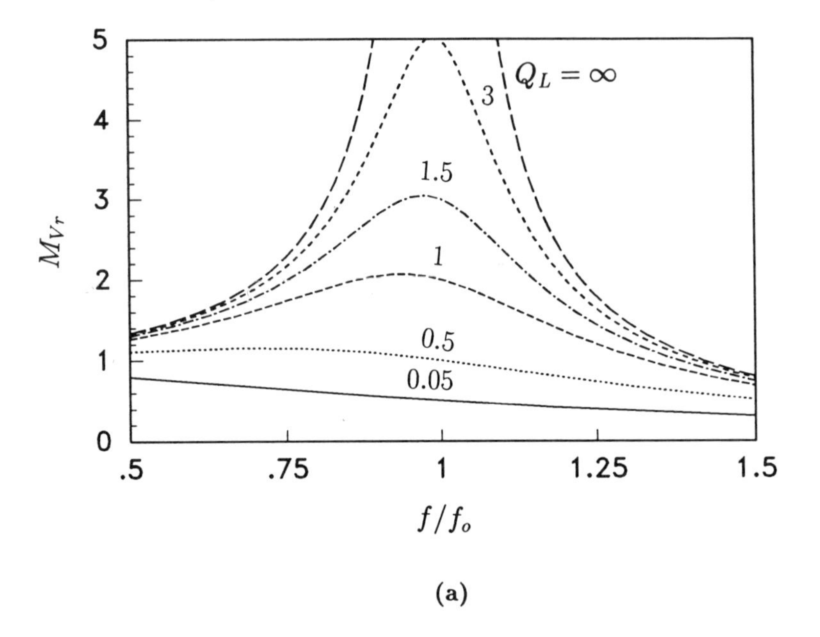

(a)

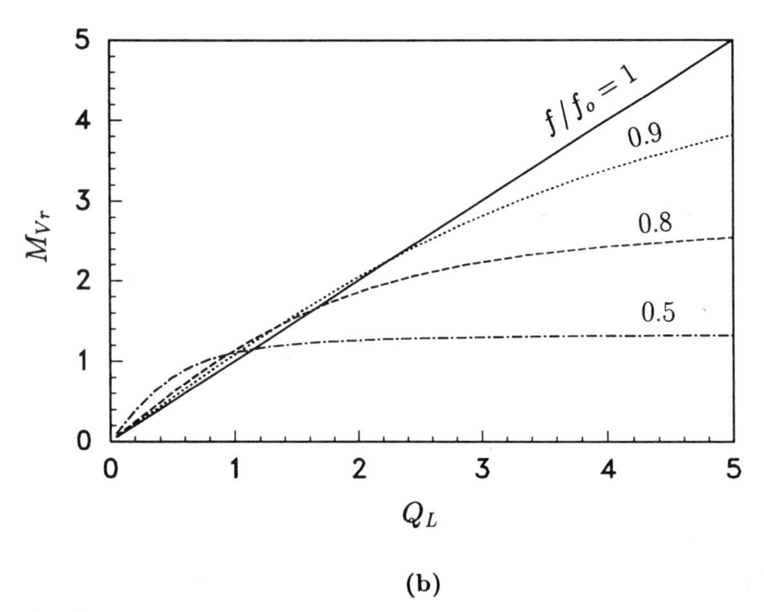

(b)

Figure 7.9: Frequency response characteristics of the resonant circuit (cross-sectional views of Fig. 7.8). (a) M_{Vr} against f/f_o at various values of Q_L. (b) M_{Vr} against Q_L at various values of f/f_o for $f/f_o \leq f_{pk}/f_o$. (c) M_{Vr} against Q_L at various values of f/f_o for $f/f_o > f_{pk}/f_o$. (d) f/f_o against Q_L at various values of M_{Vr}.

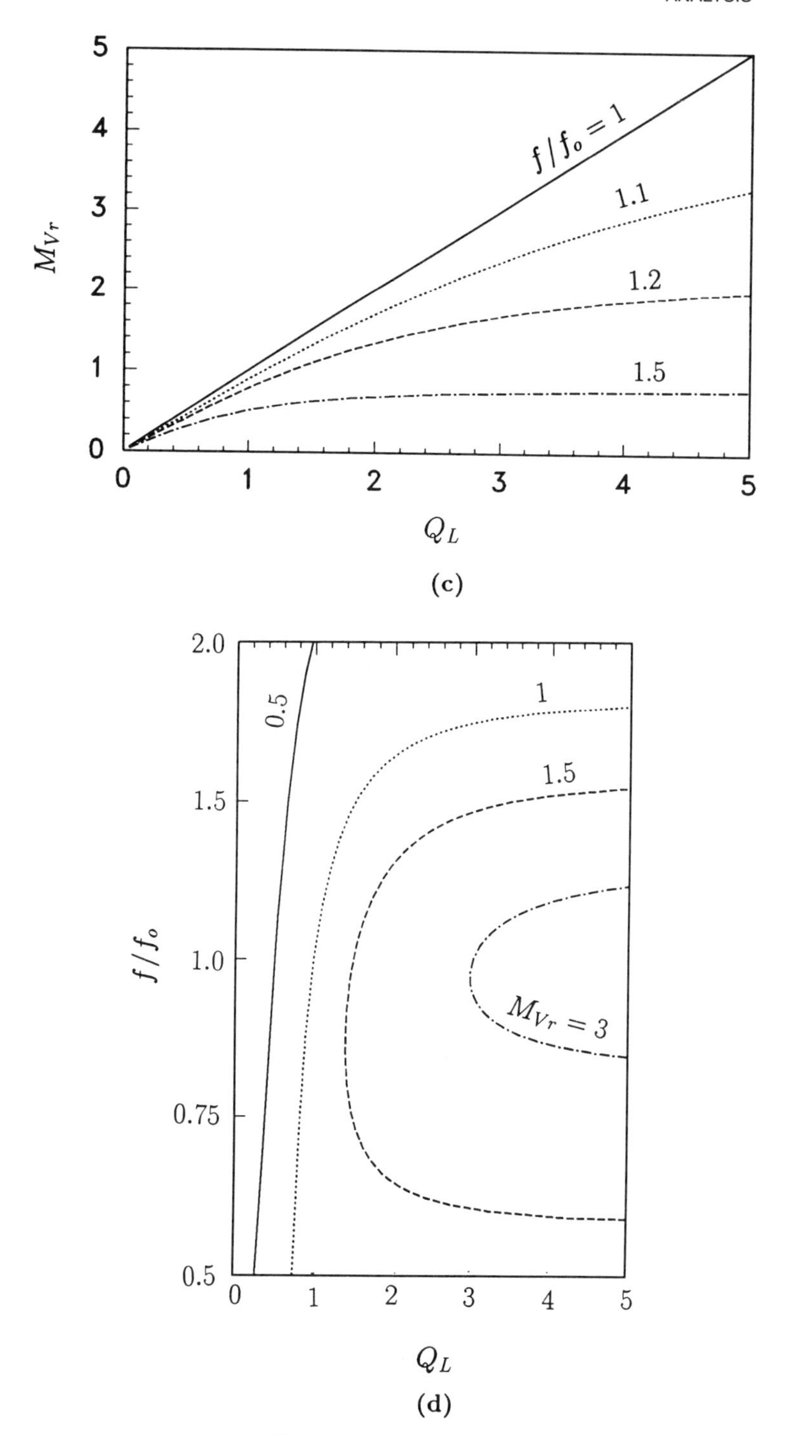

Figure 7.9 (*Continued*)

Equations (7.38) and (7.39) are illustrated in Fig. 7.10. As Q_L increases from $1/\sqrt{2}$ to ∞, f_{pk}/f increases from 0 to 1 and M_{Vr} increases from 1 to ∞.

Expression (7.35) can be rearranged to

$$\frac{f}{f_o} = \sqrt{1 - \frac{1}{2Q_L^2} + \sqrt{\left(1 - \frac{1}{2Q_L^2}\right)^2 + \frac{1}{M_{Vr}^2} - 1}}, \quad \text{for} \quad \frac{f}{f_o} \geq \frac{f_{pk}}{f_o} \tag{7.40}$$

$$\frac{f}{f_o} = \sqrt{1 - \frac{1}{2Q_L^2} - \sqrt{\left(1 - \frac{1}{2Q_L^2}\right)^2 + \frac{1}{M_{Vr}^2} - 1}}, \quad \text{for} \quad \frac{f}{f_o} < \frac{f_{pk}}{f_o}. \tag{7.41}$$

Note that $M_{Vr} \geq 1$ for $f/f_o \leq f_{pk}/f_o$ and $0 \leq M_{Vr} < \infty$ for $f/f_o > f_{pk}/f_o$. To obtain $M_{Vr} \geq 1$, the following condition must be satisfied:

$$Q_L \geq \frac{1}{\sqrt{2\left(1 - \sqrt{1 - \frac{1}{M_{Vr}^2}}\right)}}. \tag{7.42}$$

For example, $Q_{Lmin} \geq 1.93$ for $M_{Vr} = 2$ and $Q_{Lmin} \geq 3.97$ for $M_{Vr} = 4$. Thus, the minimum value of Q_L can be approximated by $Q_{Lmin} \approx M_{Vr}$. If $M_{Vr} \to \infty$, $Q_L \to \infty$. Theoretically, M_{Vr} ranges from 0 to infinity. However, high values of M_{Vr} require high values of Q_L, resulting in low efficiency of the resonant circuit. Therefore, in practice the value of Q_L at full load should be less than 10, which limits the range of M_{Vr} from 0 to 10.

For $2Q_L^2 >> 1$, (7.40) and (7.41) can be approximated by

$$\frac{f}{f_o} \approx \sqrt{1 + \frac{1}{M_{Vr}}}, \quad \text{for any } M_{Vr} \quad \text{and} \quad 1 \leq \frac{f}{f_o} < \infty \tag{7.43}$$

or

$$\frac{f}{f_o} \approx \sqrt{1 - \frac{1}{M_{Vr}}}, \quad \text{for} \quad M_{Vr} > 1 \quad \text{and} \quad 0 < \frac{f}{f_o} < 1. \tag{7.44}$$

It follows from the above relations and Fig. 7.9(d) that f/f_o should be held almost constant to maintain a fixed value of M_{Vr} for $2Q_L^2 >>1$, for example, for $Q_L > 2.5$. The frequency range required to regulate V_O against variations in R_i and V_I can be calculated from (7.41)–(7.44). For example, if $M_{Vr} = 0.5$ and a full power occurs at f close to f_o, f/f_o should be increased from 1 to $\sqrt{2}$ as R_i is increased from its minimum value to ∞.

The magnitude of the dc-to-ac voltage transfer function of the Class D inverter without losses is obtained from (7.33) and (7.35)

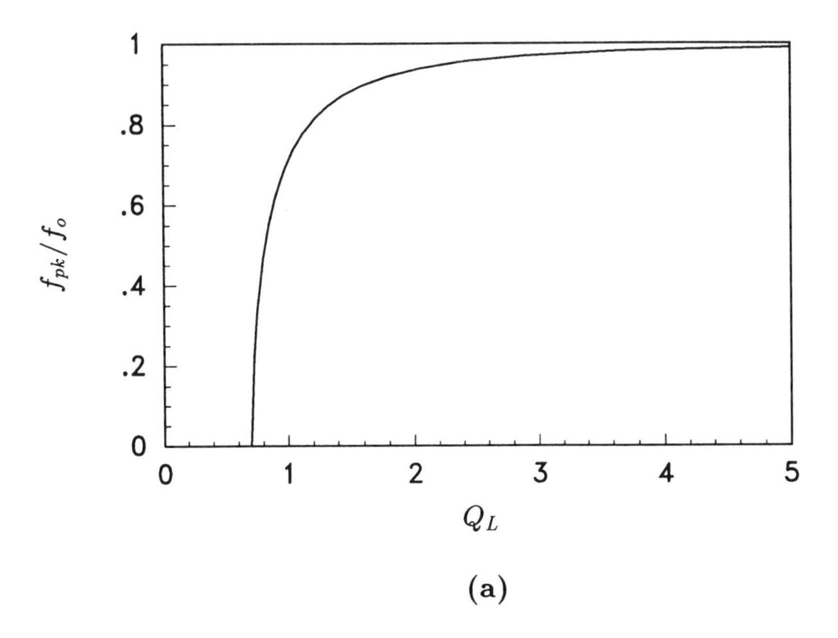

(a)

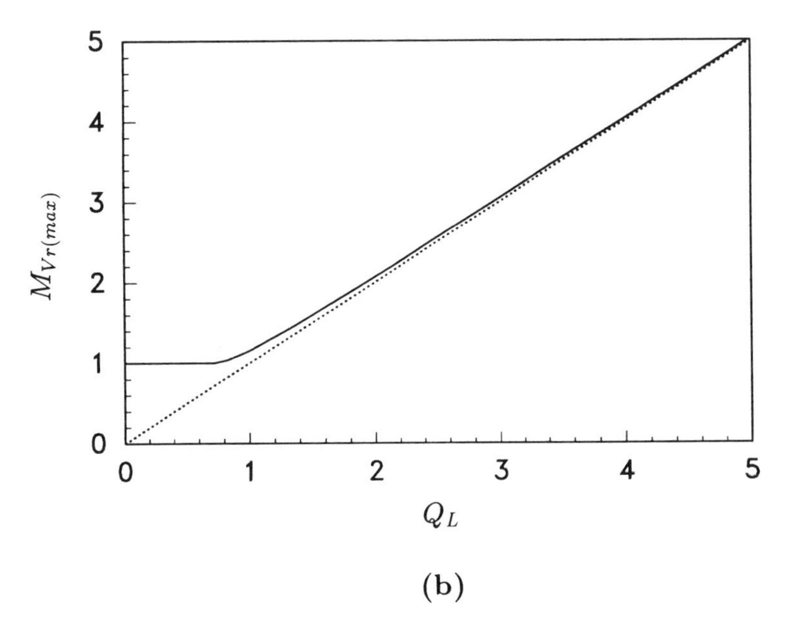

(b)

Figure 7.10: Ratio f_{pk}/f_o and maximum values of the voltage transfer function M_{Vr} versus Q_L. (a) f_{pk}/f_o versus Q_L. (b) $M_{Vr(max)}$ versus Q_L.

$$M_{VI} \equiv \frac{V_{Ri}}{V_I} = M_{Vs} M_{Vr} = \frac{\sqrt{2}}{\pi\sqrt{[1-(\frac{\omega}{\omega_o})^2]^2 + \frac{1}{Q_L^2}(\frac{\omega}{\omega_o})^2}}. \quad (7.45)$$

The range of M_{VI} is from 0 to ∞. The dc-to-ac voltage transfer function of the inverter with losses is

$$\mathbf{M_{VIa}} = \eta_I \mathbf{M_{VI}}, \quad (7.46)$$

where η_I is the efficiency of the inverter, determined in Section 7.3.5.

7.3.4 Currents, Voltages, and Powers

The current through the resonant inductor L is given by

$$i = I_m sin(\omega t - \psi), \quad (7.47)$$

where I_m is the amplitude of the inductor current equal to the peak value of the switch current I_{SM}. Assuming that $r << R_s$ and using (7.9) and (7.31),

$$I_m = I_{SM} = \frac{V_m}{Z} = \frac{2V_I}{\pi Z} = \frac{2V_I}{\pi Z_o}\sqrt{\frac{1+Q_L^2(\frac{\omega}{\omega_o})^2}{Q_L^2[1-(\frac{\omega}{\omega_o})^2]^2 + (\frac{\omega}{\omega_o})^2}}$$
$$= \frac{2V_I M_{Vr}\sqrt{1+Q_L^2(\frac{\omega}{\omega_o})^2}}{\pi Z_o Q_L} \approx \frac{2V_I M_{Vr}\left(\frac{\omega}{\omega_o}\right)}{\pi Z_o}, \quad \text{for} \quad Q_L^2 >> 1. \quad (7.48)$$

Using (7.43) and (7.44), one can approximate (7.48) by

$$I_m \approx \frac{2V_I M_{Vr}}{\pi Z_o}\sqrt{1 \pm \frac{1}{M_{Vr}}}, \quad \text{for} \quad Q_L^2 >> 1. \quad (7.49)$$

Thus, I_m is almost independent of the load for $Q_L >> 1$ for fixed values of M_{Vr}. The normalized amplitude $I_m Z_o/V_I$ is illustrated in three-dimensional space in Fig. 7.11. Figure 7.12 shows $I_m Z_o/V_I$ versus f/f_o and Q_L. The amplitude I_m increases with Q_L. Figure 7.13 shows plots of $I_m Z_o/V_I$ versus Q_L at fixed values of M_{Vr}. It can be seen that I_m is almost independent of the load resistance at a constant value of M_{Vr}. This is because the load resistance is much higher than the reactance of the capacitor C at a high value of Q_L and, therefore, almost all of the inductor current flows through the capacitor.

The maximum value of I_m occurs at $f = f_o$

$$I_{m(max)} = I_{SM(max)} = \frac{2V_I\sqrt{Q_L^2+1}}{\pi Z_o} \approx \frac{2V_I R_i}{\pi Z_o^2}, \quad \text{for} \quad Q_L^2 >> 1. \quad (7.50)$$

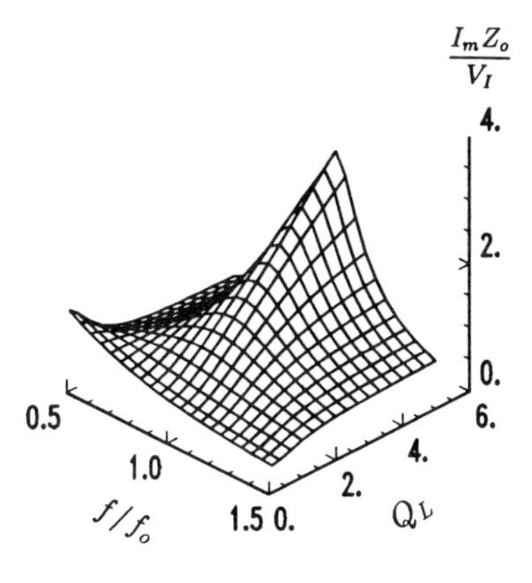

Figure 7.11: Three-dimensional representation of $I_m Z_o/V_I$ against f/f_o and Q_L.

Thus, I_m is directly proportional to R_i for $f = f_o$ and $Q_L^2 >> 1$. If f is increased from f_o, Z also increases [Fig. 7.4(a)], reducing I_m.

From (7.45), the amplitude of the voltage across the resonant capacitor C is

$$V_{Cm} = \sqrt{2}V_{Ri} = \sqrt{2}M_{VI}V_I = \frac{2V_I}{\pi\sqrt{[1-(\frac{\omega}{\omega_o})^2]^2 + [\frac{1}{Q_L}(\frac{\omega}{\omega_o})]^2}}. \quad (7.51)$$

The maximum value of V_{Cm} occurs at $f = f_o$,

$$V_{Cm(max)} = \frac{2V_I Q_L}{\pi}. \quad (7.52)$$

Figures 7.14 and 7.15 illustrate the dependence of V_{Cm}/V_I on f/f_o and Q_L in three and two-dimensional space. Using (7.48), the amplitude of the voltage across the resonant inductor L is found as

$$V_{Lm} = \omega L I_m = \left(\frac{\omega}{\omega_o}\right) Z_o I_m = \frac{2V_I}{\pi}\left(\frac{\omega}{\omega_o}\right)\sqrt{\frac{1+Q_L^2(\frac{\omega}{\omega_o})^2}{Q_L^2[1-(\frac{\omega}{\omega_o})^2]^2 + (\frac{\omega}{\omega_o})^2}}. \quad (7.53)$$

The maximum value of I_m occurs at $f = f_o$,

$$V_{Lm(max)} = \frac{2V_I\sqrt{Q_L^2+1}}{\pi}. \quad (7.54)$$

Figures 7.16 and 7.17 depict V_{Lm}/V_I versus f/f_o and Q_L.

The amplitude of the output current is obtained from (7.51)

$$I_{om} = \frac{V_{Cm}}{R_i} = \frac{2V_I}{\pi Z_o\sqrt{Q_L^2[1-(\frac{\omega}{\omega_o})^2]^2 + (\frac{\omega}{\omega_o})^2}}. \quad (7.55)$$

At $f = f_o$,

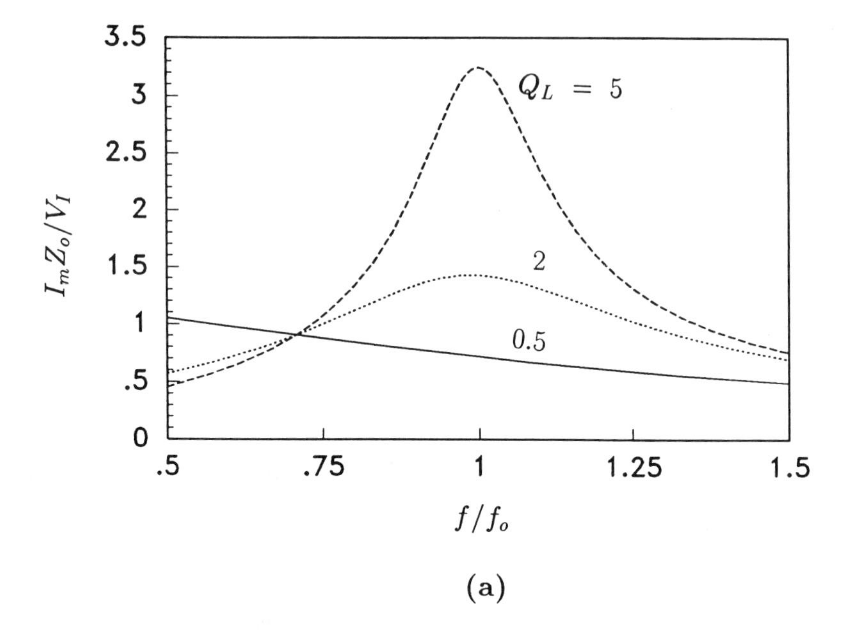

(a)

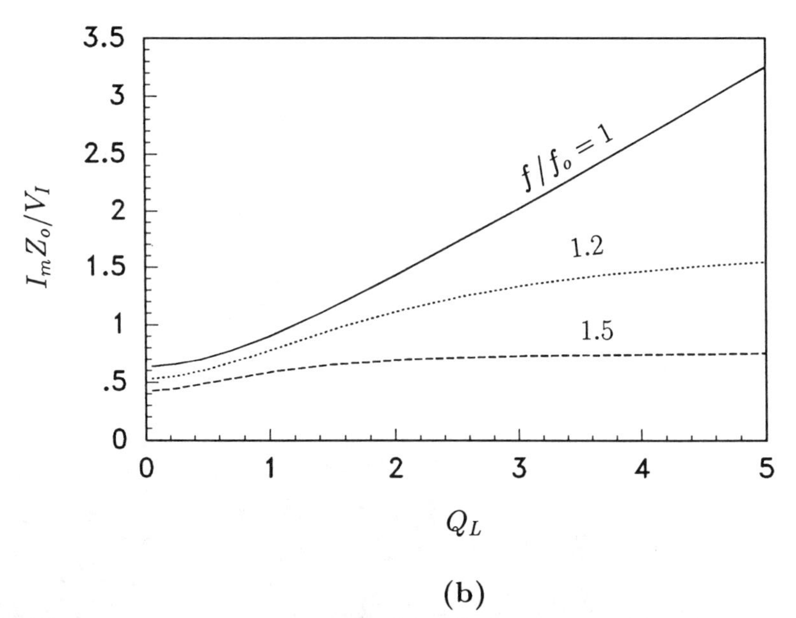

(b)

Figure 7.12: Normalized current amplitude $I_m \dot{Z}_o/V_I$ as functions of f/f_o and Q_L. (a) Normalized current versus f/f_o at various values of Q_L. (b) Normalized current versus Q_L at various values of f/f_o.

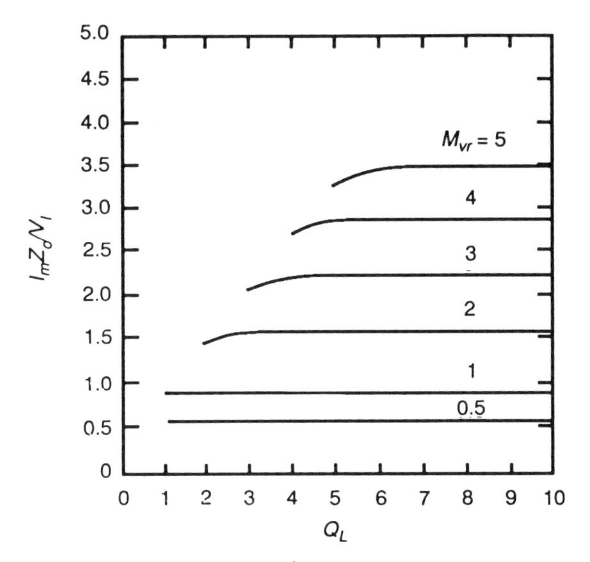

Figure 7.13: Normalized current $I_m Z_o/V_I$ versus Q_L at various values of M_{Vr}.

$$I_{om} = \frac{2V_I}{\pi Z_o} = \frac{2V_I}{\pi \omega_o L} = \frac{2V_I \omega_o C}{\pi}. \tag{7.56}$$

Note that I_{om} is independent of R_i at f_o. Therefore, the inverter is suitable for driving a negative load resistance such as a fluorescent lamp. Figures 7.18 and 7.19 show $I_{om}Z_o/V_I$ as functions of f/f_o and Q_L.

The output power is obtained from (7.55)

$$\begin{aligned} P_{Ri} &= \frac{R_i I_{om}^2}{2} = \frac{2V_I^2 R_i}{\pi^2 Z_o^2 \{Q_L^2[1-(\frac{\omega}{\omega_o})^2]^2 + (\frac{\omega}{\omega_o})^2\}} \\ &= \frac{2V_I^2 Q_L}{\pi^2 Z_o \{Q_L^2[1-(\frac{\omega}{\omega_o})^2]^2 + (\frac{\omega}{\omega_o})^2\}}. \end{aligned} \tag{7.57}$$

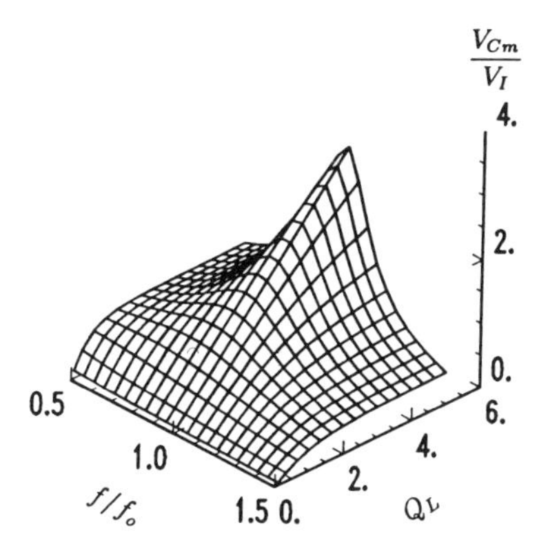

Figure 7.14: Three-dimensional representation of the normalized peak capacitor voltage V_{Cm}/V_I versus f/f_o and Q_L.

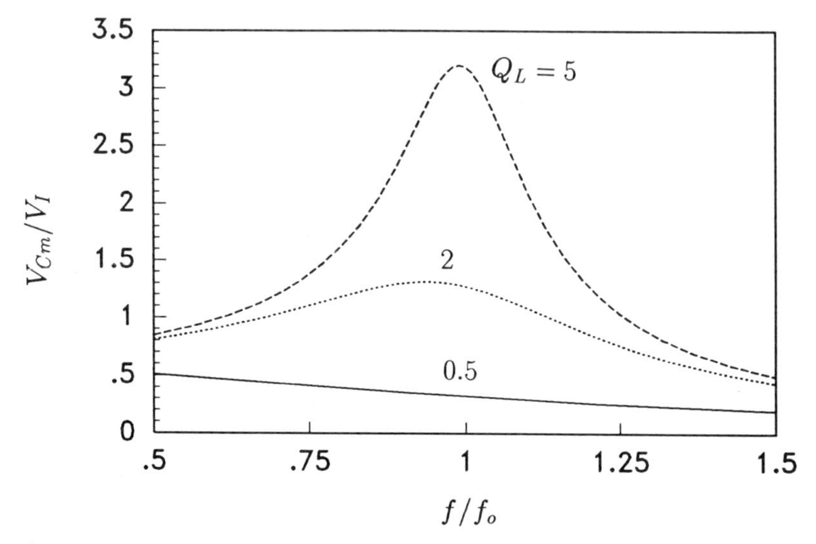

Figure 7.15: Normalized peak capacitor voltage V_{Cm}/V_I as a function of f/f_o at various values of Q_L.

Figure 7.20 shows a three-dimensional representation of $P_{Ri}Z_o/V_I^2$. The normalized output power $P_{Ri}Z_o/V_I^2$ is plotted versus f/f_o and Q_L in Fig. 7.21. At $f = f_o$,

$$P_{Ri} = \frac{2V_I^2 Q_L}{\pi^2 Z_o} = \frac{2V_I^2 R_i}{\pi^2 Z_o^2}. \tag{7.58}$$

Thus, P_{Ri} increases linearly with R_i at $f = f_o$. Therefore, the inverter can operate safely for load resistances R_i ranging from a short circuit to a maximum value of R_i limited by a maximum value of $I_{SM} = I_m$. The switching loss is estimated in Chapter 6.

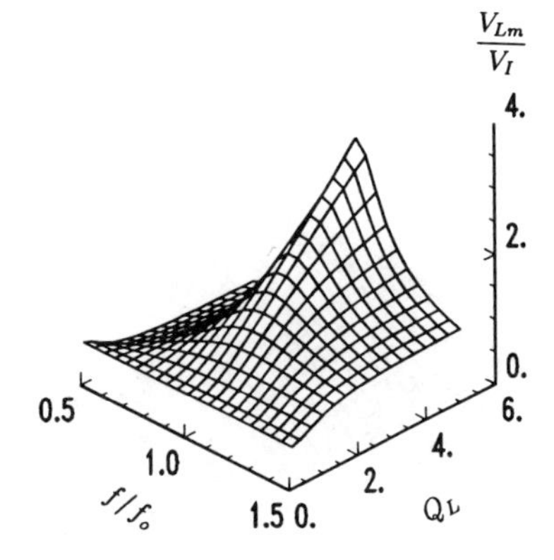

Figure 7.16: Three-dimensional representation of the normalized peak inductor voltage V_{Lm}/V_I versus f/f_o and Q_L.

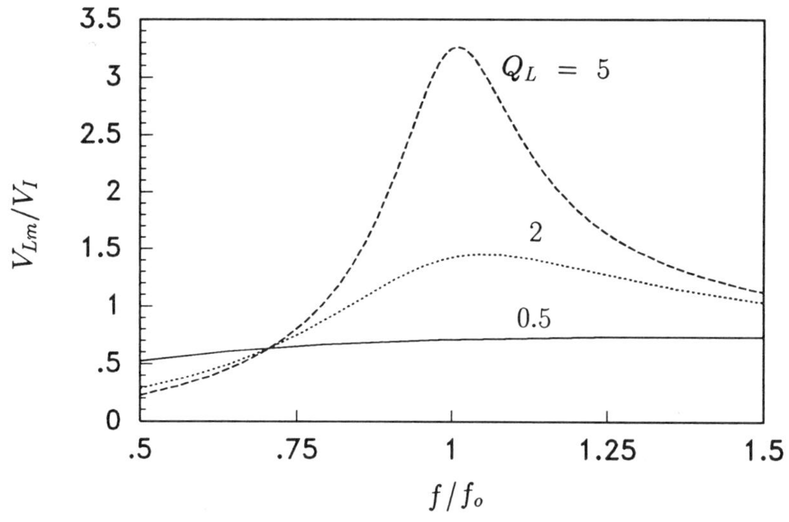

Figure 7.17: Normalized peak inductor voltage V_{Lm}/V_I as a function of f/f_o at various values of Q_L.

7.3.5 Efficiency

The conduction power loss in the MOSFETs and the inductor is

$$P_r = \frac{rI_m^2}{2} = \frac{2rV_I^2[1 + (Q_L\frac{\omega}{\omega_o})^2]}{\pi^2 Z_o^2\{Q_L^2[1 - (\frac{\omega}{\omega_o})^2]^2 + (\frac{\omega}{\omega_o})^2\}}, \quad (7.59)$$

where $r = r_{DS} + r_L$.

Neglecting switching losses, the conduction loss in the resonant capacitor and the coupling capacitor, and the gate-drive loss, $P_I = P_{Ri} + P_r$. Hence, from (7.59) one arrives at the efficiency of the inverter

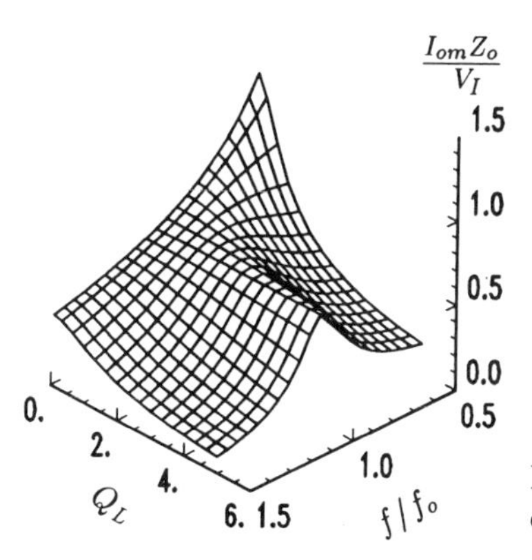

Figure 7.18: Three-dimensional representation of $I_{om}Z_o/V_I$ against f/f_o and Q_L.

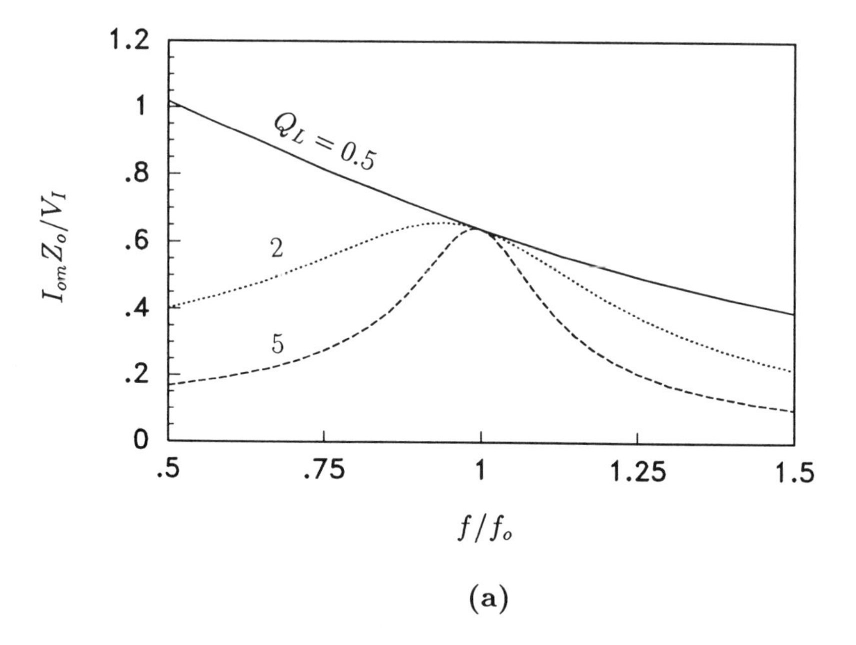

(a)

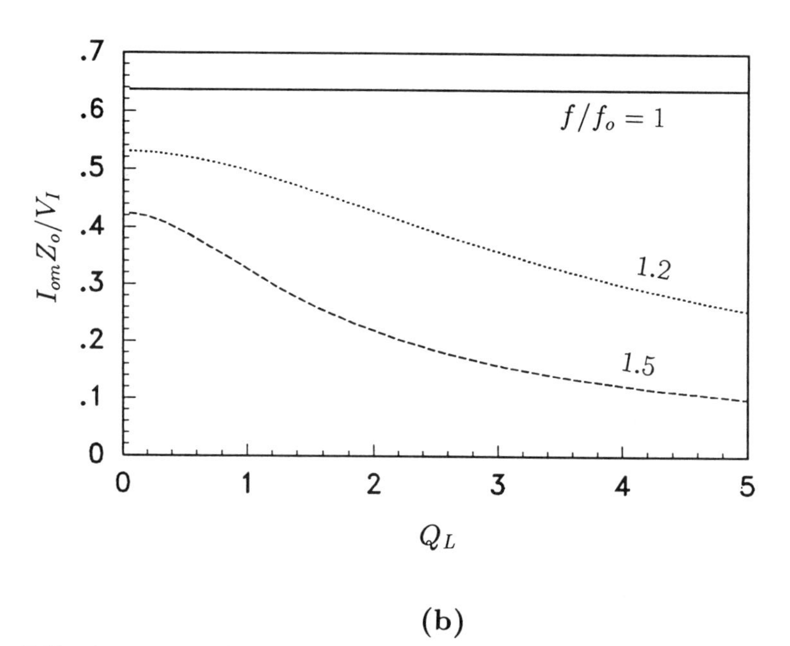

(b)

Figure 7.19: Normalized output current amplitude $I_{om}Z_o/V_I$ as functions of f/f_o and Q_L. (a) Normalized current versus f/f_o at various values of Q_L. (b) Normalized current versus Q_L at various values of f/f_o.

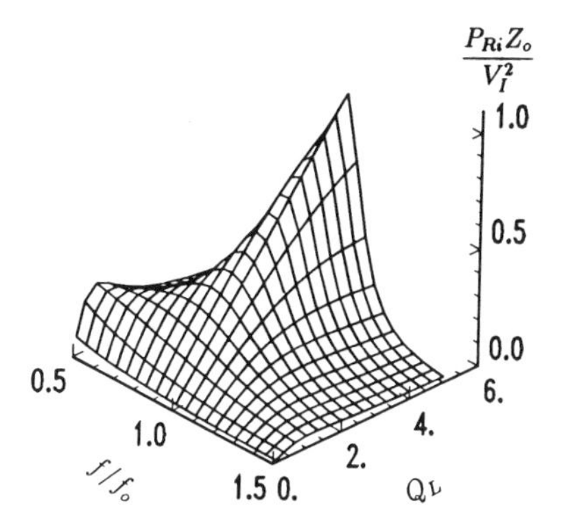

$$\eta_I = \frac{P_{Ri}}{P_I} = \frac{P_{Ri}}{P_{Ri} + P_r} = \frac{1}{1 + \frac{r}{R_i}[1 + (\frac{\omega}{\omega_o})^2 Q_L^2]} = \frac{1}{1 + \frac{r}{Z_o Q_L}[1 + (\frac{\omega}{\omega_o})^2 Q_L^2]}. \tag{7.60}$$

This expression can also be obtained from (7.57) and (7.59). For a given r, the maximum value of the efficiency occurs at a critical value of the loaded factor $Q_{Lcr} = 1/(\omega/\omega_o)$, which corresponds to $R_{icr} = 1/(\omega C) = Z_o/(\omega/\omega_o)$. Thus, the maximum efficiency is given by

$$\eta_{Imax} = \frac{1}{1 + 2r\omega C} = \frac{1}{1 + \frac{2r}{R_{icr}}} = \frac{1}{1 + 2(\frac{\omega}{\omega_o})(\frac{r}{Z_o})}. \tag{7.61}$$

Figure 7.22 shows a three-dimensional representation of the inverter efficiency η_I versus Q_L and f/f_o. The efficiency η_I is plotted in Fig. 7.23 as a function of f/f_o at fixed values of Q_L and as a function of Q_L at fixed values of f/f_o.

7.4 SHORT-CIRCUIT AND OPEN-CIRCUIT OPERATION

With a *short circuit* at the output ($R_i = 0$), capacitors C and C_c are connected in parallel, the voltage across these capacitors is $V_I/2$, the current in C is zero, and the transistors are loaded by inductor L, as shown in Fig. 7.24. The voltage across L is a square wave

$$v_L = \begin{cases} \frac{V_I}{2}, & \text{for } 0 < t \le \frac{T}{2} \\ -\frac{V_I}{2}, & \text{for } \frac{T}{2} < t \le T. \end{cases} \tag{7.62}$$

Hence, the inductor current is

$$i = \begin{cases} V_I t/(2L) + i(0), & \text{for } 0 < t \le \frac{T}{2} \\ -V_I t/(2L) + i(\frac{T}{2}), & \text{for } \frac{T}{2} < t \le T \end{cases} \tag{7.63}$$

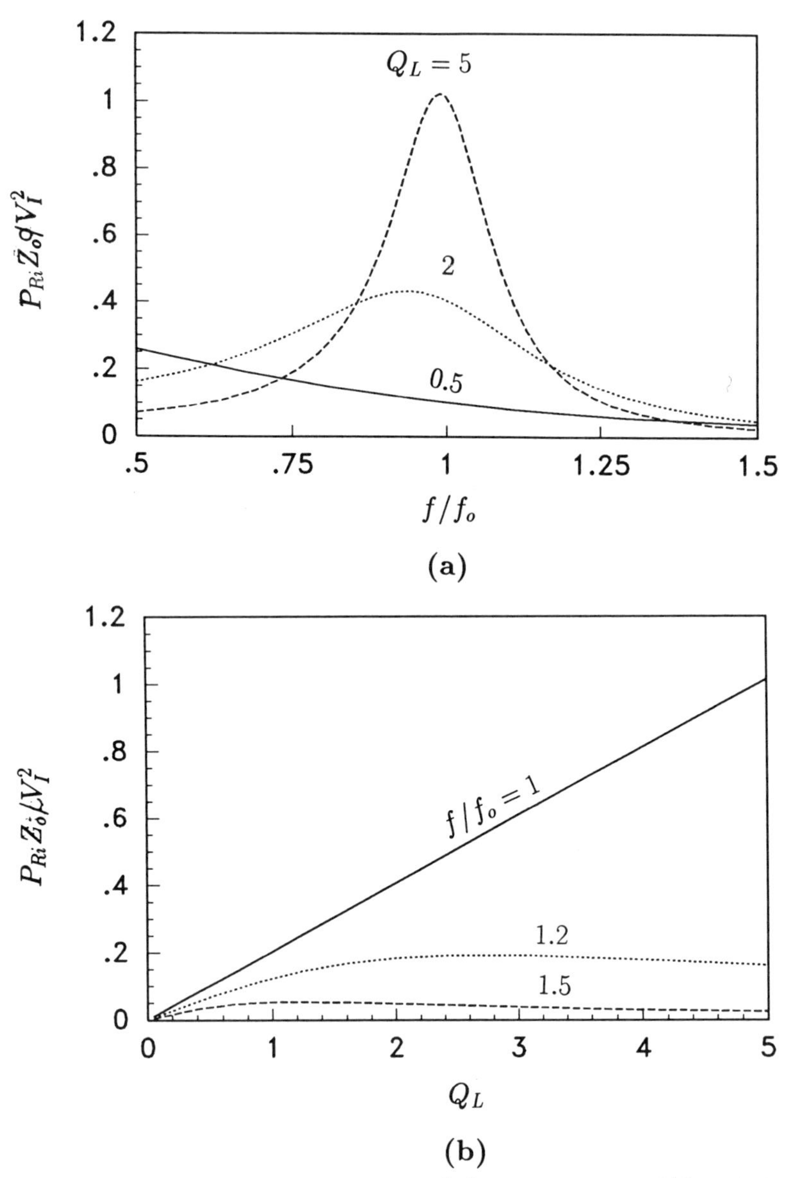

Figure 7.21: Normalized output power $P_{Ri}R_i/V_I^2$ as functions of f/f_o and Q_L. (a) Normalized output power versus f/f_o at various values of Q_L. (b) Normalized output power versus Q_L at various values of f/f_o.

where $i(0)$ and $i(T/2)$ are the boundary conditions and $T = 1/f$. The peak-to-peak inductor current is

$$\Delta i = i(T/2) - i(0) = \frac{V_I}{4Lf}, \tag{7.64}$$

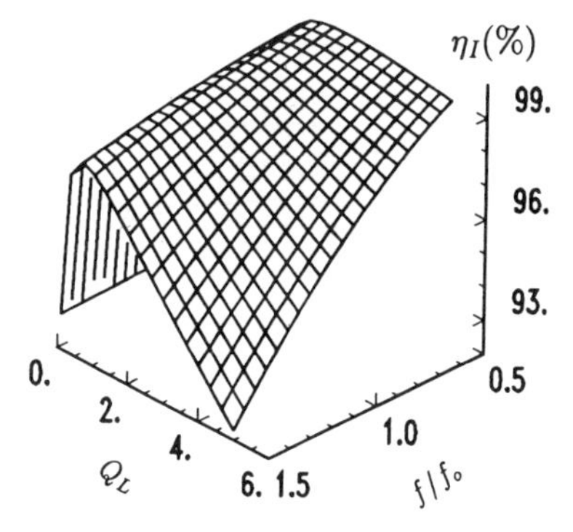

Figure 7.22: Three-dimensional representation of the efficiency η_I versus f/f_o and Q_L.

and the peak value of the inductor current I_{pk}, which is equal to the peak values of the switch currents I_{SM}, is

$$I_{pk} = I_{SM} = \frac{\Delta i}{2} = \frac{V_I}{8Lf}. \tag{7.65}$$

The inductor limits the current of the switches, providing inherent short-circuit protection.

An *open circuit* at the output ($R_i = \infty$) can lead to both 1) an excessive current through the resonant circuit and the transistors and 2) a very high voltage across the resonant capacitor C and the resonant inductor L. In this case, the transistors are loaded by a series-resonant circuit L-C-r. Hence, the resonant frequency f_r becomes equal to the corner frequency f_o. The input impedance of the L-C-r resonant circuit is

$$Z = r + j\left(\omega L - \frac{1}{\omega C}\right) = r + jZ_o\left(\frac{\omega}{\omega_o} - \frac{\omega_o}{\omega}\right) = Ze^{j\psi}, \tag{7.66}$$

where

$$Z = \sqrt{r^2 + Z_o^2\left(\frac{\omega}{\omega_o} - \frac{\omega_o}{\omega}\right)^2} \tag{7.67}$$

$$\psi = arctan\left[\frac{Z_o}{r}\left(\frac{\omega}{\omega_o} - \frac{\omega_o}{\omega}\right)\right]. \tag{7.68}$$

The amplitude I_m of current i is

$$I_m = \frac{V_m}{Z} = \frac{2V_I}{\pi Z} = \frac{2V_I}{\pi\sqrt{r^2 + Z_o^2(\frac{\omega}{\omega_o} - \frac{\omega_o}{\omega})^2}}. \tag{7.69}$$

The amplitude I_m depends on f/f_o. At $f = f_o$, $Z = r$ and I_m is

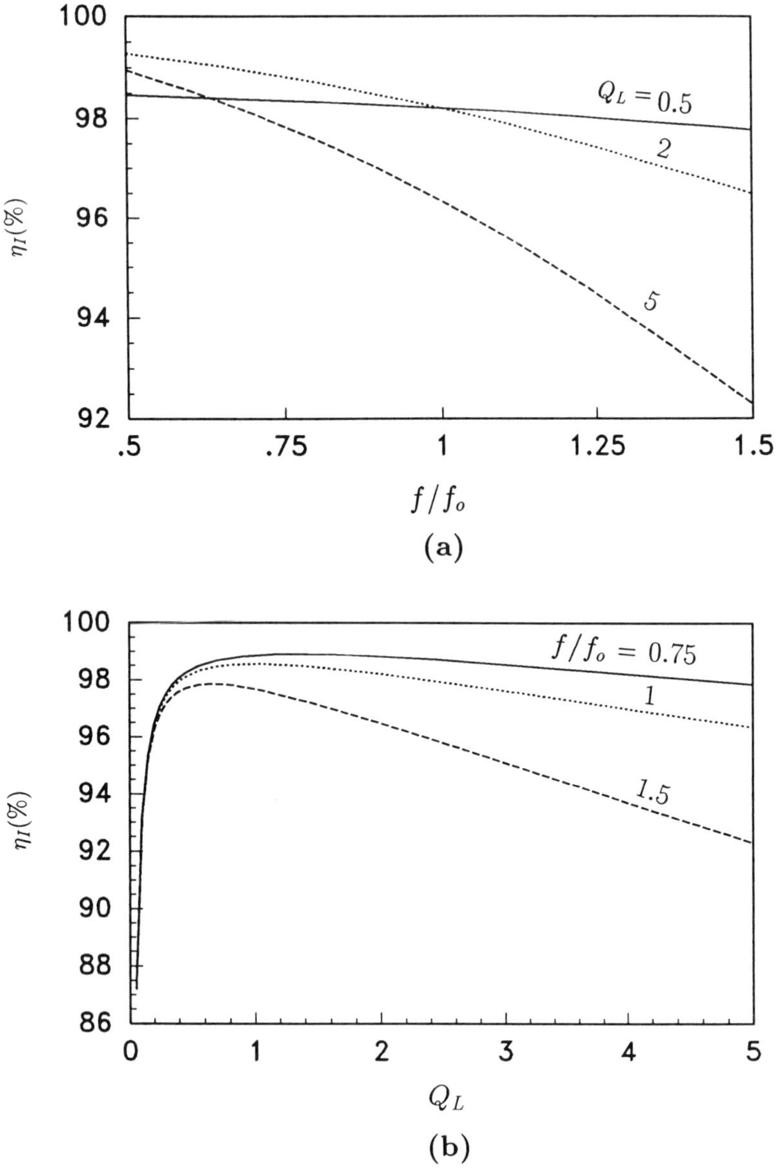

Figure 7.23: Efficiency η_I as a function of f/f_o and Q_L. (a) η_I as a function of f/f_o at various values of Q_L. (b) η_I as a function of Q_L at various values of f/f_o.

$$I_m = \frac{V_m}{r} = \frac{2V_I}{\pi r}, \tag{7.70}$$

and the amplitudes of the voltages across C and L are

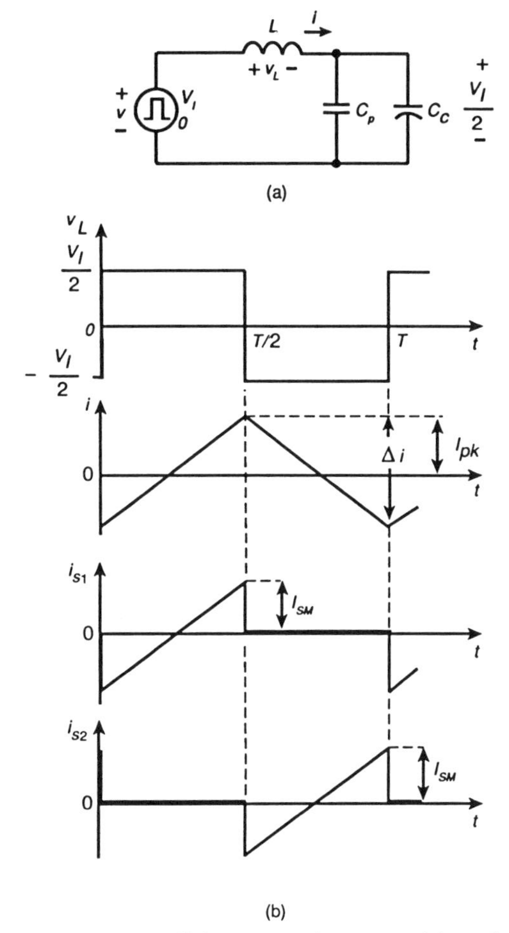

Figure 7.24: Operation of the parallel resonant inverter with a short circuit at the output. (a) Equivalent circuit. (b) Waveforms.

$$V_{Cm} = V_{Lm} = \frac{I_m}{\omega_o C} = \omega_o L I_m = I_m Z_o = \frac{2V_I Z_o}{\pi r}. \tag{7.71}$$

For instance, for $V_I = 200$ V, $Z_o = 250\ \Omega$, and $r = 1\ \Omega$, $I_m = 127$ A, and $V_{Cm} = V_{Lm} = 31.75$ kV! Thus, the transistors and the resonant components would be destroyed. Therefore, the operation at $f = f_o$ should be avoided in most applications. In the case of electronic ballasts, this feature can be utilized to start the lamp. After the ballast is turned on, a control circuit containing a voltage-controlled oscillator (VCO) can increase the operating frequency f close to the corner frequency f_o. This causes V_{Cm} to increase to several hundred volts, starting the lamp and protecting the inverter. Then R_i decreases and f is reduced by the VCO, ensuring safe operation. Resistance

r is much lower than the reactance of the L-C circuit at f sufficiently lower or higher than f_o and (7.69) becomes

$$I_m = \frac{V_m}{Z} = \frac{2V_I}{\pi Z_o \left(\frac{\omega}{\omega_o} - \frac{\omega_o}{\omega} \right)}. \tag{7.72}$$

Thus, I_m can be reduced to a safe level by making f sufficiently different from f_o.

In contrast, the Class D inverter with a series-resonant circuit cannot operate safely under short-circuit conditions, but can operate safely with an open circuit at the output.

7.5 DESIGN EXAMPLE

Example 7.1

Design a Class D inverter of Fig. 7.1(a) with the following specifications: $V_I = 200$ V, $P_{Ri} = 75$ W, and $f = 100$ kHz. Assume the loaded quality factor at full load to be $Q_L = 2.5$. Estimate the efficiency of the designed inverter, switching losses, and gate-drive power loss if MTP5N40 MOSFETs (Motorola) are employed.

Solution: Assume a typical value of the inverter efficiency $\eta_I = 95\%$. Thus, the dc supply power is

$$P_I = \frac{P_{Ri}}{\eta_I} = \frac{75}{0.95} = 78.95 \text{ W}, \tag{7.73}$$

and the dc supply current is

$$I_I = \frac{P_I}{V_I} = \frac{78.95}{200} = 394.7 \text{ mA}. \tag{7.74}$$

Assuming that $f = f_r = 100$ kHz at full power and using (7.15), the corner frequency is

$$f_o = \frac{f_r}{\sqrt{1 - \frac{1}{Q_L^2}}} = \frac{100 \times 10^3}{\sqrt{1 - \frac{1}{2.5^2}}} = 109.1 \text{ kHz}. \tag{7.75}$$

Using (7.45) and (7.46), the ac load resistance of the inverter is

$$R_i = \frac{V_{Ri}^2}{P_{Ri}} = \frac{2V_I^2\eta_I^2}{\pi^2 P_{Ri}\{[1-(\frac{\omega}{\omega_o})^2]^2 + [\frac{1}{Q_L}(\frac{\omega}{\omega_o})]^2\}}$$

$$= \frac{2 \times 200^2 \times 0.95^2}{\pi^2 \times 75[(1-\frac{100^2}{109.1^2})^2 + (\frac{100}{2.5\times 109.1})^2]} = 609.7\ \Omega. \qquad (7.76)$$

From (7.5), the characteristic impedance can be obtained as

$$Z_o = \frac{R_i}{Q_L} = \frac{609.7}{2.5} = 243.9\ \Omega. \qquad (7.77)$$

Using (7.4), the elements of the resonant circuit are

$$L = \frac{Z_o}{\omega_o} = \frac{243.9}{2\pi \times 109.1 \times 10^3} = 355.7\ \mu\text{H} \qquad (7.78)$$

and

$$C = \frac{1}{\omega_o Z_o} = \frac{1}{2\pi \times 109.1 \times 10^3 \times 243.9} = 5.98\ \text{nF}; \quad \text{let } C = 6\ \text{nF}. \qquad (7.79)$$

From (7.50), the maximum value of the switch peak current is

$$I_{m(max)} = I_{SM(max)} = \frac{2V_I\sqrt{Q_L^2+1}}{\pi Z_o} = \frac{2 \times 200\sqrt{2.5^2+1}}{\pi \times 243.9} = 1.41\ \text{A}. \qquad (7.80)$$

From (7.52) and (7.54), the voltage stresses of the resonant components are

$$V_{Cm(max)} = \frac{2V_I Q_L}{\pi} = \frac{2 \times 200 \times 2.5}{\pi} = 318.3\ \text{V} \qquad (7.81)$$

and

$$V_{Lm(max)} = \frac{2V_I\sqrt{Q_L^2+1}}{\pi} = \frac{2 \times 200 \times \sqrt{2.5^2+1}}{\pi} = 342.8\ \text{V}. \qquad (7.82)$$

The MTP5N40 transistor has the following data: $r_{DS} = 1\ \Omega$ and $Q_g = 27$ nC. Assuming that the equivalent series resistance of the inductor is $r_L = 1.4\ \Omega$, the parasitic resistance of the inverter is $r = r_{DS} + r_L = 2.4\ \Omega$. Using (7.15), (7.48), and keeping in mind that $f = f_r$,

$$I_m = I_{SM} = \frac{2V_I}{\pi Z_o}\sqrt{\frac{1+Q_L^2(\frac{\omega}{\omega_o})^2}{Q_L^2[1-(\frac{\omega}{\omega_o})^2]^2 + (\frac{\omega}{\omega_o})^2}} = \frac{2V_I Q_L}{\pi Z_o}$$

$$= \frac{2 \times 200 \times 2.5}{\pi \times 243.9} = 1.305\ \text{A}. \qquad (7.83)$$

The conduction loss per MOSFET is

$$P_{rDS} = \frac{I_m^2 r_{DS}}{4} = \frac{1.305^2 \times 1}{4} = 0.426 \text{ W}, \tag{7.84}$$

and the power loss in the resonant inductor is

$$P_{rL} = \frac{I_m^2 r_L}{2} = \frac{1.305^2 \times 1.4}{2} = 1.192 \text{ W}. \tag{7.85}$$

The amplitude of the current through the resonant capacitor C is

$$I_{Cm} = \frac{I_m R_i}{R_i + | X_C |} = \frac{1.305 \times 609.7}{609.7 + 265} = 0.91 \text{ A}. \tag{7.86}$$

Assuming that the ESR of the resonant capacitor C is $r_C = 100$ mΩ at 100 kHz, the conduction power loss in this capacitor is

$$P_{rC} = \frac{I_{Cm}^2 r_C}{2} = \frac{0.91^2 \times 0.1}{2} = 41 \text{ mW}. \tag{7.87}$$

The conduction loss is

$$P_r = 2P_{rDS} + P_{rL} + P_{rC} = 2 \times 0.426 + 1.192 + 0.041 = 2.085 \text{ W}. \tag{7.88}$$

Equation (7.60) gives the efficiency of the inverter associated with the conduction losses only at full power

$$\eta_I = \frac{P_{Ri}}{P_{Ri} + P_r} = \frac{75}{75 + 2.085} = 97.29\%. \tag{7.89}$$

Assuming $V_{GSpp} = 15$ V and using (2.117), the gate-drive power loss for two MOSFETs is

$$2P_G = 2fQ_g V_{GSpp} = 2 \times 100 \times 10^3 \times 27 \times 10^{-9} \times 15 = 0.081 \text{ W}. \tag{7.90}$$

The sum of the conduction loss and the gate-drive power loss is

$$P_{LS} = P_r + 2P_G = 2.085 + 0.081 = 2.166 \text{ W}. \tag{7.91}$$

The turn-on switching loss is zero because the input impedance of the resonant circuit is inductive. The inverter efficiency is

$$\eta_I = \frac{P_{Ri}}{P_{Ri} + P_{LS}} = \frac{75}{75 + 2.166} = 96.19\%. \tag{7.92}$$

7.6 FULL-BRIDGE PARALLEL RESONANT INVERTER

7.6.1 Voltage Transfer Function

A full-bridge parallel resonant resonant inverter is shown in Fig. 7.25. The input voltage of the resonant circuit v is a square wave of magnitude V_I, given by

$$v = \begin{cases} V_I, & \text{for } 0 < \omega t \leq \pi \\ -V_I, & \text{for } \pi < \omega t \leq 2\pi. \end{cases} \tag{7.93}$$

The fundamental component of this voltage is

$$v_{i1} = V_m sin\omega t, \tag{7.94}$$

where the amplitude of v_{i1} can be found from Fourier analysis as

$$V_m = \frac{4}{\pi} V_I = 1.273 V_I. \tag{7.95}$$

Hence, one obtains the rms value of v_{i1}

$$V_{rms} = \frac{V_m}{\sqrt{2}} = \frac{2\sqrt{2}}{\pi} V_I = 0.9 V_I, \tag{7.96}$$

which leads to the voltage transfer function from V_I to the fundamental component at the input of the resonant circuit

$$M_{Vs} \equiv \frac{V_{rms}}{V_I} = \frac{2\sqrt{2}}{\pi} = 0.9. \tag{7.97}$$

The magnitude of the voltage transfer function of the resonant circuit is given by (7.35). The magnitude of the dc-to-ac voltage transfer function of the Class D inverter is obtained from (7.35) and (7.97)

$$M_{VI} \equiv \frac{V_{Ri}}{V_I} = M_{Vs} M_{Vr} = \frac{2\sqrt{2}}{\pi\sqrt{[1 - (\frac{\omega}{\omega_o})^2]^2 + [\frac{1}{Q_L}(\frac{\omega}{\omega_o})]^2}}. \tag{7.98}$$

The range of M_{VI} is from zero to ∞.

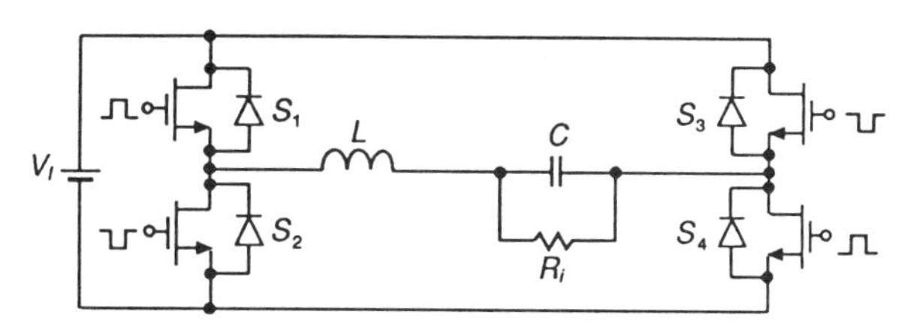

Figure 7.25: Circuit of the Class D full-bridge parallel resonant inverter.

7.6.2 Currents, Voltages, and Powers

The current through resonant inductor L is given by

$$i = I_m sin(\omega t - \psi), \tag{7.99}$$

where I_m is the amplitude of the inductor current equal to the peak value of the switch current I_{SM}. Assuming that $r << R_s$ and using (7.9) and (7.95),

$$I_m = I_{SM} = \frac{V_m}{Z} = \frac{4V_I}{\pi Z} = \frac{4V_I}{\pi Z_o}\sqrt{\frac{1 + (Q_L\frac{\omega}{\omega_o})^2}{Q_L^2[1 - (\frac{\omega}{\omega_o})^2]^2 + (\frac{\omega}{\omega_o})^2}}. \tag{7.100}$$

From (7.5) and (7.9), (7.100) simplifies for $f = f_o$ to the form

$$I_m = \frac{4V_I\sqrt{Q_L^2 + 1}}{\pi Z_o} \approx \frac{4V_I R_i}{\pi Z_o^2}, \quad \text{for} \quad Q_L^2 >> 1. \tag{7.101}$$

Thus, I_m is directly proportional to R_i for $f = f_o$ and $Q_L^2 >> 1$. If f is increased from f_o, Z also increases [Fig. 7.4(a)], reducing the valuc of I_m.

The amplitude of the output current is

$$I_{om} = \frac{4V_I}{\pi Z_o\sqrt{Q_L^2[1 - (\frac{\omega}{\omega_o})^2]^2 + (\frac{\omega}{\omega_o})^2}}. \tag{7.102}$$

At $f = f_o$,

$$I_{om} = \frac{4V_I}{\pi Z_o} = \frac{4V_I}{\pi\omega_o L} = \frac{4V_I\omega_o C}{\pi}. \tag{7.103}$$

From (7.45), the amplitude of the voltage across the resonant capacitor C is

$$V_{Cm} = \frac{4V_I}{\pi\sqrt{[1 - (\frac{\omega}{\omega_o})^2]^2 + [\frac{1}{Q_L}(\frac{\omega}{\omega_o})]^2}}. \tag{7.104}$$

The maximum value of the capacitor peak voltage occurs at $f = f_o$

$$V_{Cm(max)} = \frac{4V_I Q_L}{\pi}. \tag{7.105}$$

From (7.100), the amplitude of the voltage across the resonant inductor L is found as

$$V_{Lm} = \omega L I_m = \left(\frac{\omega}{\omega_o}\right) Z_o I_m = \frac{4V_I}{\pi}\left(\frac{\omega}{\omega_o}\right)\sqrt{\frac{1+(Q_L\frac{\omega}{\omega_o})^2}{Q_L^2[1-(\frac{\omega}{\omega_o})^2]^2+(\frac{\omega}{\omega_o})^2}}. \tag{7.106}$$

The maximum value of the inductor peak voltage occurs at $f = f_o$

$$V_{Lm(max)} = \frac{4V_I\sqrt{Q_L^2+1}}{\pi}. \tag{7.107}$$

The output power is obtained from (7.102)

$$P_{Ri} = \frac{R_i I_{om}^2}{2} = \frac{8V_I^2 R_i}{\pi^2 Z_o^2\{Q_L^2[1-(\frac{\omega}{\omega_o})^2]^2+(\frac{\omega}{\omega_o})^2\}} \tag{7.108}$$

$$= \frac{8V_I^2 Q_L}{\pi^2 Z_o\{Q_L^2[1-(\frac{\omega}{\omega_o})^2]^2+(\frac{\omega}{\omega_o})^2\}}. \tag{7.109}$$

The output power at $f = f_o$ becomes

$$P_{Ri} = \frac{8V_I^2 R_i}{\pi^2 Z_o^2}. \tag{7.110}$$

Thus, P_{Ri} increases linearly with R_i at $f = f_r$. Therefore, the inverter can operate safely for load resistances R_i ranging from a short circuit to a maximum value of R_i limited by a maximum value of $I_{SM} = I_m$.

7.6.3 Efficiency

The conduction power loss in the MOSFETs and the inductor is

$$P_r = \frac{r I_m^2}{2} = \frac{8rV_I^2[1+(Q_L\frac{\omega}{\omega_o})^2]}{\pi^2 Z_o^2\{Q_L^2[1-(\frac{\omega}{\omega_o})^2]^2+(\frac{\omega}{\omega_o})^2\}}, \tag{7.111}$$

where

$$r = 2r_{DS} + r_L. \tag{7.112}$$

Neglecting switching losses, the conduction loss in the resonant capacitor and the coupling capacitor, and the gate-drive loss, $P_I = P_{Ri} + P_r$. Hence, from (7.110) one arrives at the efficiency of the inverter

$$\eta_I = \frac{P_{Ri}}{P_I} = \frac{P_{Ri}}{P_{Ri}+P_r} = \frac{1}{1+\frac{r}{Z_o Q_L}[1+(\frac{\omega}{\omega_o})^2 Q_L^2]} = \frac{1}{1+\frac{r}{R_i}[1+(\frac{\omega}{\omega_o})^2(\frac{R_i}{Z_o})^2]}. \tag{7.113}$$

This expression can also be obtained from (7.108) and (7.110). For a given r, the maximum value of the efficiency occurs at a critical value of the loaded factor $Q_{Lcr} = 1/(\omega/\omega_o)$, which corresponds to $R_{icr} = 1/(\omega C) = Z_o/(\omega/\omega_o)$. Hence, the maximum efficiency is given by

$$\eta_{Imax} = \frac{1}{1 + 2r\omega C} = \frac{1}{1 + \frac{2r}{R_{icr}}} = \frac{1}{1 + 2(\frac{\omega}{\omega_o})(\frac{r}{Z_o})}. \tag{7.114}$$

7.6.4 Short-Circuit and Open-Circuit Operation

If $R_i = 0$, the voltage across L is a square wave

$$v_L = \begin{cases} V_I, & \text{for } 0 < t \le \frac{T}{2} \\ -V_I, & \text{for } \frac{T}{2} < t \le T. \end{cases} \tag{7.115}$$

Hence, the inductor current is

$$i = \begin{cases} V_I t/L + i(0), & \text{for } 0 < t \le \frac{T}{2} \\ -V_I t/L + i(\frac{T}{2}), & \text{for } \frac{T}{2} < t \le T, \end{cases} \tag{7.116}$$

where $i(0)$ and $i(T/2)$ are the boundary conditions and $T = 1/f$. The peak-to-peak inductor current is

$$\Delta i = i(T/2) - i(0) = \frac{V_I}{2Lf}, \tag{7.117}$$

and the peak value of the inductor current I_{pk}, which is equal to the peak values of the switch currents I_{SM}, is

$$I_{pk} = I_{SM} = \frac{\Delta i}{2} = \frac{V_I}{4Lf}. \tag{7.118}$$

The inductor limits the current of the switches, providing inherent short-circuit protection.

An *open circuit* at the output ($R_i = \infty$) can cause 1) an excessive current through the resonant circuit and the transistors, and 2) a very high voltage across the resonant capacitor C and the resonant inductor L. In this case, the transistors are loaded by a series-resonant circuit L-C-r. Hence, the resonant frequency f_r becomes equal to the corner frequency f_o. The input impedance of the L-C-r resonant circuit is given by (7.66)–(7.68). The amplitude I_m of current i is

$$I_m = \frac{V_m}{Z} = \frac{4V_I}{\pi Z} = \frac{4V_I}{\pi\sqrt{r^2 + Z_o^2(\frac{\omega}{\omega_o} - \frac{\omega_o}{\omega})^2}}. \tag{7.119}$$

The amplitude I_m depends on f/f_o. At $f = f_o$, $Z_i = r$, and I_m is

$$I_m = \frac{V_m}{r} = \frac{4V_I}{\pi r}, \tag{7.120}$$

and the amplitudes of the voltages across C and L are

$$V_{Cm} = V_{Lm} = \frac{I_m}{\omega_o C} = \omega_o L I_m = I_m Z_o = \frac{4V_I Z_o}{\pi r}. \tag{7.121}$$

7.7 SUMMARY

- The corner frequency f_o of the resonant circuit is independent of the load resistance.
- The resonant frequency f_r does not exist for $Q_L \leq 1$ and depends on the load resistance for $Q_L > 1$, making the boundary between the capacitive and inductive load dependent on the load resistance.
- The resonant circuit acts as an impedance inverter at the resonant frequency f_r. It transforms high load resistances to low input resistances. The input resistance is inversely proportional to the load resistance. The peak transistor currents and the output power increase with increasing load resistance at operating frequencies close to the resonant frequency.
- The part-load efficiency of the PRI is low.
- The inverter offers an inherent short-circuit protection at any operating frequency. The transistors under short-circuit conditions are loaded by the resonant inductor which limits the current.
- The inverter cannot operate safely at the resonant frequency with an open circuit at the output because of excessive transistor current and resonant capacitor voltage.
- At the resonant frequency, the peak current in the resonant circuit and the transistor peak currents increase as the load resistance increases.
- The inverter can operate under safe conditions from a short circuit to an open circuit if the operating frequency is higher than the resonant frequency.
- The resonant frequency increases with the load resistance. If the inverter operates slightly above the resonant frequency at a given load resistance, and then the load resistance is increased while maintaining a constant operating frequency, the inverter can operate below the resonant frequency. Hence, the operating frequency should be selected so that it is sufficiently higher than the resonant frequency at the maximum value of the load resistance.

- The operation of the inverter without external fast diodes is not recommended below the resonant frequency.
- Because the dc voltage source V_I and the switches form an ideal square-wave voltage source, many loads can be connected between the two switches and ground and operated without mutual interactions.

7.8 REFERENCES

1. R. L. Steigerwald, "A comparison of half-bridge resonant converter topologies," *IEEE Trans. Ind. Electron.*, vol. IE-35, pp. 174–182, Apr. 1988.
2. M. Kazimierczuk, "Class D voltage-switching MOSFET power inverter," *Proc. Inst. Electr. Eng., Pt. B, Electric Power Appl.*, vol. 138, pp. 285–296, Nov. 1991.
3. M. K. Kazimierczuk, W. Szaraniec, and S. Wang, "Analysis of parallel resonant converter at high Q_L," *IEEE Trans. Aerospace and Electronic Systems*, vol. AES-27, pp. 35–50, Jan. 1992.
4. M. K. Kazimierczuk and W. Szaraniec, "Electronic ballast for fluorescent lamps," *IEEE Trans. Power Electronics*, vol. PE-7, pp. 386–395, October 1994.
5. M. K. Kazimierczuk, D. Czarkowski, and N. Thirunarayan, "A new phase-controlled parallel resonant converter," *IEEE Trans. Industrial Electronics*, vol. PE-40, pp. 542–552, December 1993.
6. M. K. Kazimierczuk, and K. Puczko, "Impedance inverter for Class E dc/dc converters," *29th Midwest Symposium on Circuits and Systems*, August 10–12, 1986, Lincoln, NE, pp. 707–710.

7.9 REVIEW QUESTIONS

7.1 Does the boundary between the capacitive and inductive load depend on the load resistance in Class D parallel resonant inverter?

7.2 Is the operation with capacitive or inductive load preferred?

7.3 Is the voltage transfer function of the PRI dependent of the load?

7.4 Is the part-load efficiency of the PRI high?

7.5 Does the output power increase or decrease while increasing the load resistance at the corner frequency f_o?

7.6 Is operation with an open circuit at the output safe?

7.7 Is operation with a short circuit at the output safe?

7.8 Does the voltage stress across the resonant capacitor increase with increasing Q_L?

7.9 Is it possible to match any load impedance in the PRI?

7.10 PROBLEMS

7.1 The resonant circuit of Fig. P7.1 has the following parameters: $L = 200\ \mu H$, $C = 4.7$ nF, and $R_i = 500\ \Omega$. The circuit is driven by a variable frequency voltage source. Find the boundary frequency between the inductive and capacitive load for that source.

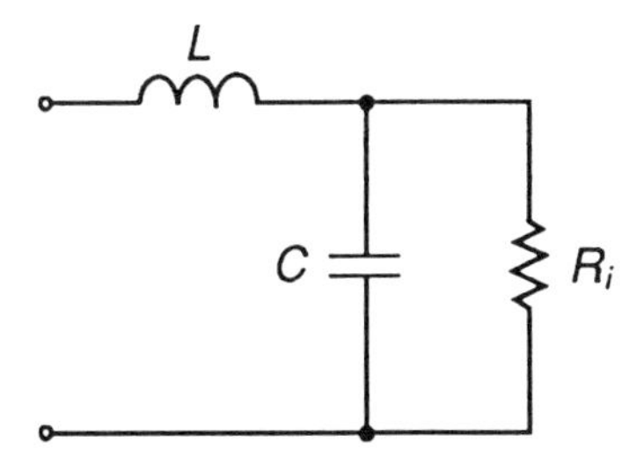

Figure P7.1: Resonant circuit of the Class D parallel resonant inverter.

7.2 The resonant circuit of Fig. P7.1 has the following parameters: $L = 400\ \mu H$, $C = 10$ nF, and $R_i = 600\ \Omega$. The circuit is driven by a variable frequency voltage source $v = 100 sin\omega t$ (V). Calculate exactly the maximum voltage stresses for the resonant components. Compare the results with voltages across the inductor and the capacitor at the corner frequency.

7.3 Derive equations (7.40) and (7.41).

7.4 Show that $R_{icr} = 1/(\omega C)$ in (7.61).

7.5 Design a full-bridge Class D parallel resonant inverter to meet the following specifications: $V_I = 200$ V, $P_{Ri} = 75$ W, and $f = 100$ kHz. Assume the loaded quality factor at full load to be $Q_L = 2.5$ and that the inverter efficiency $\eta_I = 92\%$.

CHAPTER 8

CLASS D SERIES-PARALLEL RESONANT INVERTER

8.1 INTRODUCTION

In this chapter the circuit and major characteristics of a series-parallel resonant inverter (SPRI) [1]–[11] are presented. The topology of this inverter is the same as that of the parallel resonant inverter (PRI), except for an additional capacitor in series with the resonant inductor, or is the same as that of the series resonant inverter (SRI), except for an additional capacitor in parallel with the load. As a result, the inverter exhibits the characteristics that are intermediate between those of the SRI and the PRI. In particular, it has high part-load efficiency.

8.2 PRINCIPLE OF OPERATION

A circuit of the Class D series-parallel resonant inverter [1]–[11] is shown in Fig. 8.1. The inverter is composed of two bidirectional two-quadrant switches S_1 and S_2 and a resonant circuit L-C_1-C_2-R_i, where R_i is the ac load resistance. Capacitor C_1 is connected in series with resonant inductor L as in the SRI, and capacitor C_2 is connected in parallel with the load as in the PRI. In the case of the transformer inverter, the parallel capacitor C_2 can be placed either on the primary or the secondary side of the transformer. If it is placed on the secondary, the transformer leakage inductance is absorbed into the resonant inductance L. However, for a step-down transformer with a high turns ratio, a large current flows through the parallel capacitor, causing a large conduction loss. Therefore, a capacitor with a low ESR should be used. The switches consist of MOSFETs and their body diodes. Each switch can conduct

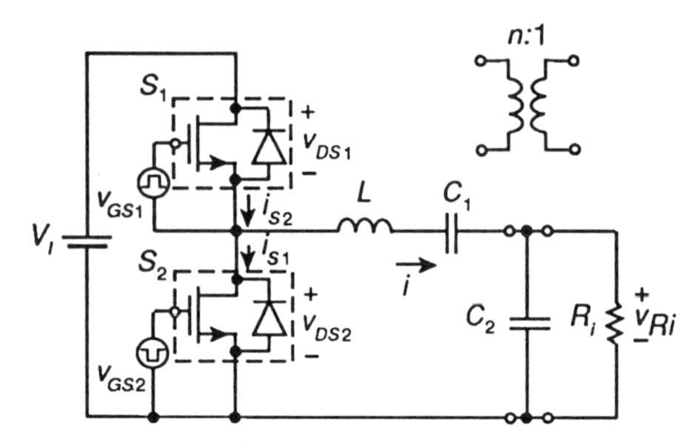

Figure 8.1: Series-parallel resonant inverter.

a positive or a negative current. The transistors are driven by rectangular-wave voltage sources v_{GS1} and v_{GS2}. Switches S_1 and S_2 are alternately turned ON and OFF at the switching frequency $f = \omega/2\pi$ with a duty cycle of 50%. If capacitance C_1 becomes very large (i.e., capacitor C_1 is replaced by a dc-blocking capacitor), the SPRI becomes the PRI. If capacitance C_2 becomes zero (i.e., capacitor C_2 is removed from the circuit), the SPRI becomes the SRI. In fact, the transformer version of the Class D SRI is the same as that of the SPRI because of the transformer stray capacitance, which is in parallel with the transformer windings. For the reasons given earlier, the SPRI exhibits intermediate characteristics between those of the SRI and the PRI.

Figure 8.2 shows equivalent circuits of the inverter. The MOSFETs are modeled by switches whose on-resistances are r_{DS1} and r_{DS2}. If the output voltage of the inverter v_{Ri} is sinusoidal, the input power of the rectifier contains only the fundamental component. The parallel R_i-C_2 circuit of Fig. 8.2(a) is converted into a series R_s-C_s circuit of Fig. 8.2(b). In Fig. 8.2(c), the dc voltage source V_I and the switches S_1 and S_2 are replaced by a square-wave voltage source with a low level value of zero and a high level value of V_I. Resistance r is the total parasitic resistance of the inverter. Basic waveforms in the series-parallel resonant inverter are similar to those in the parallel resonant inverter, shown in Fig. 7.2.

8.3 ANALYSIS

8.3.1 Assumptions

The analysis of the Class D SPRI inverter shown in Fig. 8.1 is based on the following assumptions:

1) The switches are modeled by the on-resistances r_{DS1} and r_{DS2}.

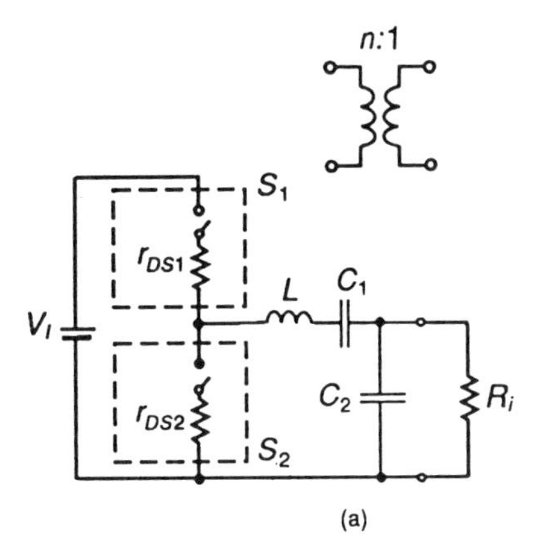

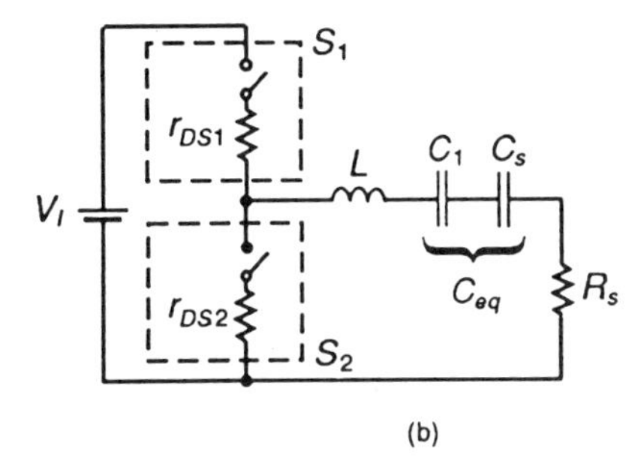

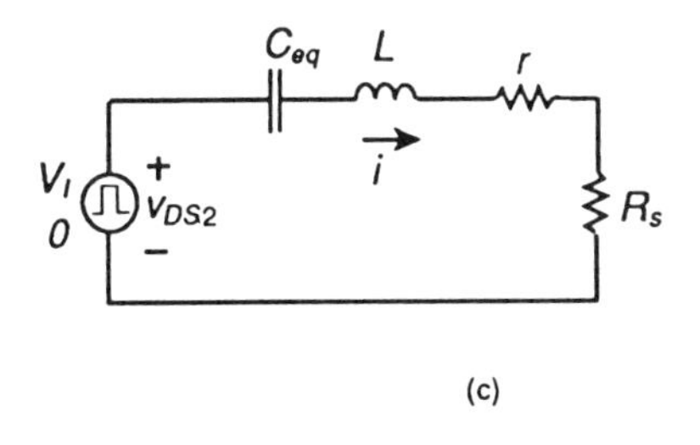

Figure 8.2: Equivalent circuits of the series-parallel resonant inverter. (a) The MOSFETs are modeled by switches with on-resistances r_{DS1} and r_{DS2}. (b) The parallel R_i-C_2 circuit is converted into the R_s-C_s circuit. (c) The dc input source V_I and the transistors are replaced by a square-wave voltage source.

2) Switching losses are neglected.
3) The current through the resonant inductor is nearly sinusoidal (i.e., $Q_r > 2.5$).

8.3.2 Resonant Circuit

The resonant circuit in the inverter of Fig. 8.1 is a third-order low-pass filter and can be described by the following normalized parameters:

- the ratio of the capacitances

$$A = \frac{C_2}{C_1} \tag{8.1}$$

- the equivalent capacitance of C_1 and C_2 connected in series

$$C = \frac{C_1 C_2}{C_1 + C_2} = \frac{C_2}{1 + A} = \frac{C_1}{1 + 1/A} \tag{8.2}$$

- the corner frequency (or the undamped natural frequency)

$$\omega_o = \frac{1}{\sqrt{LC}} = \sqrt{\frac{C_1 + C_2}{LC_1 C_2}} \tag{8.3}$$

- the characteristic impedance

$$Z_o = \omega_o L = \frac{1}{\omega_o C} = \sqrt{\frac{L}{C}} \tag{8.4}$$

- the loaded quality factor at the corner frequency f_o

$$Q_L = \omega_o C R_i = \frac{R_i}{\omega_o L} = \frac{R_i}{Z_o} \tag{8.5}$$

- the equivalent capacitance of C_1 and C_s connected in series

$$C_{eq} = \frac{C_1 C_s}{C_1 + C_s} \tag{8.6}$$

- the resonant frequency

$$\omega_r = \frac{1}{\sqrt{LC_{eq}}} = \sqrt{\frac{C_1 + C_s}{LC_1 C_s}} \tag{8.7}$$

- the loaded quality factor at the resonant frequency f_r

$$Q_r = \frac{\omega_r L}{R_s} = \frac{C_1 + C_s}{\omega_r R_s C_1 C_s}. \tag{8.8}$$

The input impedance of the resonant circuit shown in Fig. 8.1 is

$$\mathbf{Z} = \frac{R_i\{(1+A)[1 - (\frac{\omega}{\omega_o})^2] + j\frac{1}{Q_L}(\frac{\omega}{\omega_o} - \frac{\omega_o}{\omega}\frac{A}{A+1})\}}{1 + jQ_L(\frac{\omega}{\omega_o})(1+A)} = Ze^{j\psi} = R_s + jX_s, \tag{8.9}$$

where

$$Z = Z_o Q_L \sqrt{\frac{(1+A)^2[1-(\frac{\omega}{\omega_o})^2]^2 + \frac{1}{Q_L^2}(\frac{\omega}{\omega_o} - \frac{\omega_o}{\omega}\frac{A}{A+1})^2}{1+[Q_L(\frac{\omega}{\omega_o})(1+A)]^2}} \quad (8.10)$$

$$\psi = arctan\left\{\frac{1}{Q_L}\left(\frac{\omega}{\omega_o} - \frac{\omega_o}{\omega}\frac{A}{A+1}\right) - Q_L(1+A)^2\left(\frac{\omega}{\omega_o}\right)\left[1-\left(\frac{\omega}{\omega_o}\right)^2\right]\right\} \quad (8.11)$$

$$R_s = Z cos\psi \quad (8.12)$$

$$X_s = Z sin\psi. \quad (8.13)$$

At $f/f_o = 1$,

$$Z = \frac{Z_o}{(1+A)\sqrt{1+Q_L^2(1+A)^2}} \approx \frac{Z_o^2}{R_i(1+A)^2} \quad \text{for} \quad Q_L^2(1+A)^2 >> 1. \quad (8.14)$$

Thus, Z decreases with increasing A and R_i at $f = f_o$. For $A > 0$, $Z \to \infty$ as $f = 0$ or $f \to \infty$. In Fig. 8.3, Z/Z_o and ψ are plotted as functions of f/f_o at fixed values of Q_L for $A = 1$. At $f = f_o$,

$$\psi = arctan\left[\frac{1}{Q_L(1+A)}\right]. \quad (8.15)$$

Since $\psi > 0$, the resonant circuit always represents an inductive load for the switches at $f = f_o$.

The resonant frequency f_r is defined as the frequency at which the phase shift ψ is equal to zero [Fig. 8.3(b)]. This frequency forms the boundary between capacitive and inductive loads. For $f < f_r$, ψ is less than zero and the resonant circuit represents a capacitive load. Therefore, the current through the resonant inductor i leads the fundamental component of the voltage v_{DS2}. The operation in this frequency range is not recommended because the antiparallel diodes of the MOSFETs turn off at high di/dt, generating high reverse-recovery current spikes. For $f > f_r$, $\psi > 0$ and the resonant circuit represents an inductive load. Consequently, the inductor current i lags behind the fundamental component of the voltage v_{DS2}. The antiparallel diodes turn off at low di/dt and do not generate reverse-recovery current spikes. Hence, the operation in this frequency range is recommended for practical applications. Setting ψ given in (8.11) to zero yields

$$\frac{f_r}{f_o} = \sqrt{\frac{Q_L^2(1+A)^2 - 1 + \sqrt{[Q_L^2(1+A)^2-1]^2 + 4Q_L^2 A(1+A)}}{2Q_L^2(1+A)^2}}. \quad (8.16)$$

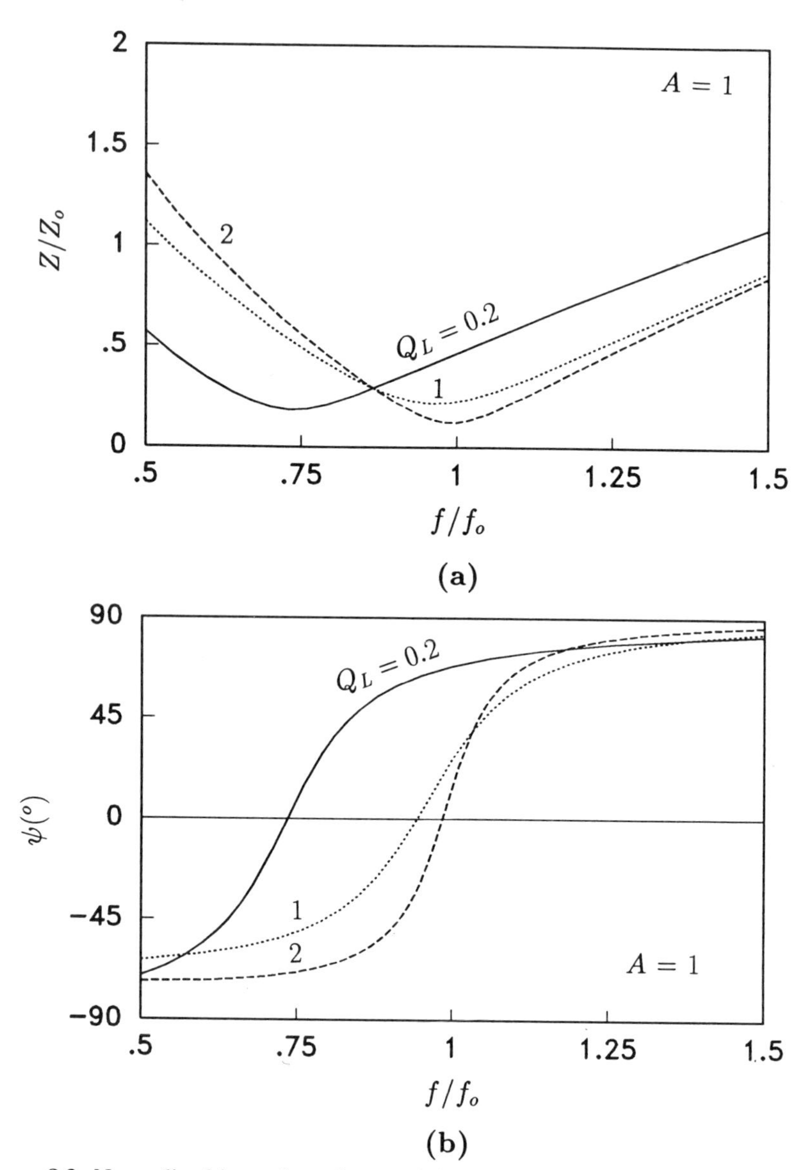

Figure 8.3: Normalized input impedance of the resonant circuit Z/Z_o and phase of the input impedance of the resonant circuit ψ as functions of f/f_o at constant values of Q_L and $A = 1$. (a) Z/Z_o versus f/f_o. (b) ψ versus f/f_o.

The resonant frequency f_r depends on Q_L and A. As $Q_L \to 0$, $f_r/f_o \to 1/\sqrt{1 + 1/A}$. Figure 8.4 shows f_r/f_o versus Q_L at selected values of A.

The R_i–C_2 parallel two-terminal network of Fig. 8.2(a) can be converted into the R_s–C_s series two-terminal network of Fig. 8.2(b). This results in the basic topology of the Class D series resonant inverter. The reactance factor at the resonant angular frequency $\omega_r = 1/\sqrt{LC_{eq}}$ is defined as

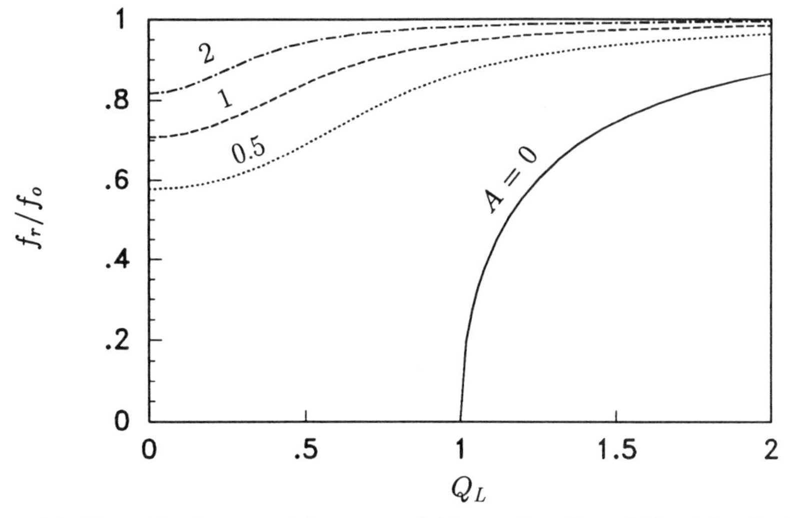

Figure 8.4: Normalized resonant frequency f_r/f_o as a function of Q_L at fixed values of A.

$$q_r = \frac{1}{\omega_r C_s R_s} = \omega_r C_2 R_i = Q_L(1+A)\left(\frac{\omega_r}{\omega_o}\right), \tag{8.17}$$

where ω_r/ω_o is given by (8.16). Using the equivalent two-terminal networks method and (8.17) yields the relationships among R_i, R_s, C_2, and C_s at $f = f_r$

$$R_s = \text{Re}\{R_i||C_2\} = \frac{R_i}{1+q_r^2} = \frac{R_i}{1+(\omega_r C_2 R_i)^2} \tag{8.18}$$

$$X_{Cs} = \text{Im}\{R_i||C_2\} = \frac{X_{C_2}}{1+\frac{1}{q_r^2}} \tag{8.19}$$

$$C_s = C_2\left(1+\frac{1}{q_r^2}\right) = C_2\left[1+\frac{1}{(\omega_r C_2 R_i)^2}\right] \tag{8.20}$$

where $X_{C_2} = 1/(\omega_r C_2)$ and $X_{Cs} = 1/(\omega_r C_s)$. The maximum value of $R_s = R_i/2$ occurs at $q_r = 1$. Capacitance C_s decreases from ∞ to C_2 as q_r is increased from zero to ∞. For $q_r^2 >> 1$, $C_s \approx C_2$, $C \approx C_{eq}$, and $f_r \approx f_o$. For $Q_L^2(1+A)^2 >> 1$, $Q_r = \omega_r L/R_s \approx Q_L(1+A)^2$.

8.3.3 Voltage Transfer Function

The input voltage of the resonant circuit v_{DS2} is a square wave of magnitude V_I given by

$$v_{DS2} = \begin{cases} V_I, & \text{for} \quad 0 < \omega t \le \pi \\ 0, & \text{for} \quad \pi < \omega t \le 2\pi. \end{cases} \tag{8.21}$$

Its fundamental component is $v_{i1} = V_m sin\omega t$, where

$$V_m = \frac{2}{\pi}V_I = 0.6366V_I. \tag{8.22}$$

The rms value of v_{i1} is $V_{rms} = V_m/\sqrt{2} = \sqrt{2}V_I/\pi = 0.4502V_I$. The voltage transfer function from V_I to the fundamental component at the input of the resonant circuit is

$$M_{Vs} \equiv \frac{V_{rms}}{V_I} = \frac{\sqrt{2}}{\pi} = 0.4502. \tag{8.23}$$

Referring to Fig. 8.2(b), the voltage transfer function of the resonant circuit is

$$\mathbf{M_{Vr}} = \frac{1}{(1+A)[1-(\frac{\omega}{\omega_o})^2] + j\frac{1}{Q_L}(\frac{\omega}{\omega_o} - \frac{\omega_o}{\omega}\frac{A}{A+1})} = M_{Vr}e^{j\varphi}, \tag{8.24}$$

where

$$M_{Vr} \equiv \frac{V_{Ri}}{V_{rms}} = \frac{1}{\sqrt{(1+A)^2[1-(\frac{\omega}{\omega_o})^2]^2 + \frac{1}{Q_L^2}(\frac{\omega}{\omega_o} - \frac{\omega_o}{\omega}\frac{A}{A+1})^2}} \tag{8.25}$$

$$\varphi = -arctan\left\{\frac{\frac{1}{Q_L}(\frac{\omega}{\omega_o} - \frac{\omega_o}{\omega}\frac{A}{A+1})}{(1+A)[1-(\frac{\omega}{\omega_o})^2]}\right\}, \tag{8.26}$$

and V_{Ri} is the rms value of the voltage across R_i. Figure 8.5 shows M_{Vr} versus f/f_o and Q_L at $A = 0$, 1, and 2. In Fig. 8.6, M_{Vr} is plotted as a function of f/f_o at selected values of Q_L for $A = 0$, 0.5, 1, and 2. Equation (8.25) is illustrated in Fig. (8.25) for $A = 1$. From (8.25), $M_{Vr} = 1$ at $f/f_o = f_{rs}/f_o = 1/\sqrt{1+1/A} = 1/\sqrt{1+C_1/C_2}$, where $f_{rs} = 1/(2\pi\sqrt{LC_1})$ is the resonant frequency of the L-C_1 series-resonant circuit. The reactance of this circuit at $f = f_{rs}$ is zero, and therefore the fundamental component of v_{DS2} appears across C_2 and R_i, making M_{Vr} independent of the load. As A is increased from 0 to ∞, f_{rs} increases from 0 to f_o. Note that the input impedance of the entire resonant circuit is capacitive at $f = f_{sr}$.

Using (8.23) and (8.25), one obtains the magnitude of the dc-to-ac voltage transfer function of the Class D inverter without losses

$$M_{VI} \equiv \frac{V_{Ri}}{V_I} = M_{Vs}M_{Vr} = \frac{\sqrt{2}}{\pi\sqrt{(1+A)^2[1-(\frac{\omega}{\omega_o})^2]^2 + \frac{1}{Q_L^2}(\frac{\omega}{\omega_o} - \frac{\omega_o}{\omega}\frac{A}{A+1})^2}}. \tag{8.27}$$

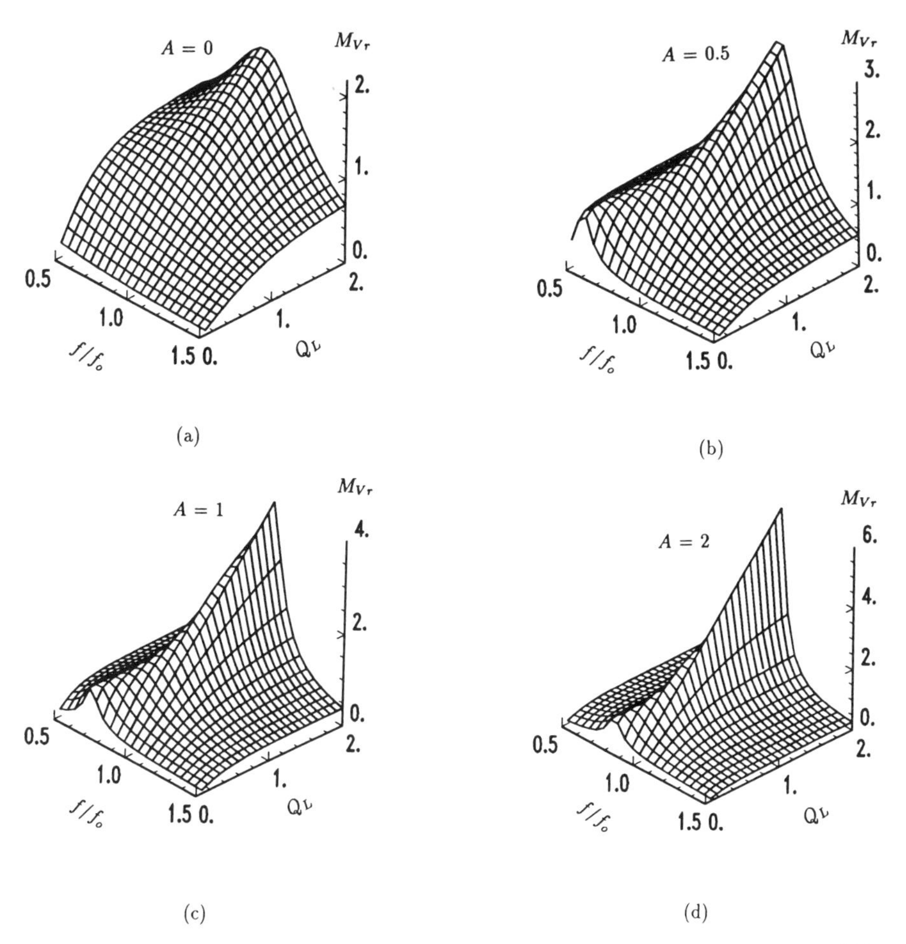

Figure 8.5: Voltage transfer function of the resonant circuit M_{Vr} as a function of f/f_o and Q_L. (a) $A = 0$. (b) $A = 0.5$. (c) $A = 1$. (d) $A = 2$.

The dc-to-ac voltage transfer function of the actual inverter can be estimated as

$$\mathbf{M_{VIa}} = \eta_I \mathbf{M_{VI}} \tag{8.28}$$

where η_I is the efficiency of the inverter, discussed in the next section.

8.3.4 Energy Parameters

The current through the resonant inductor L is

$$i = I_m sin(\omega t - \psi), \tag{8.29}$$

where (8.10) gives the amplitude of i

$$I_m = \frac{V_m}{Z} = \frac{2V_I}{\pi Z} = \frac{2V_I}{\pi Z_o Q_L} \sqrt{\frac{1 + [Q_L(1+A)(\frac{\omega}{\omega_o})]^2}{(1+A)^2[1-(\frac{\omega}{\omega_o})^2]^2 + \frac{1}{Q_L^2}(\frac{\omega}{\omega_o} - \frac{\omega_o}{\omega}\frac{A}{A+1})^2}}. \tag{8.30}$$

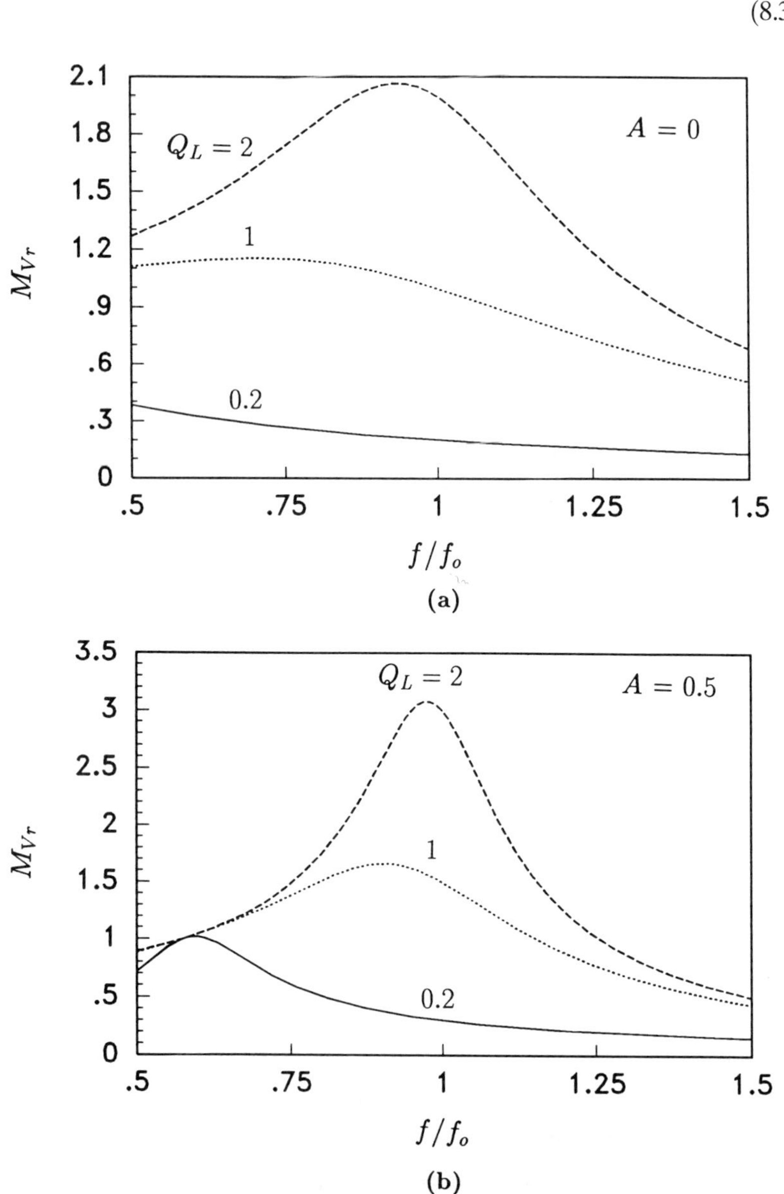

Figure 8.6: Voltage transfer function of the resonant circuit M_{Vr} as a function of f/f_o at constant values of Q_L. (a) $A = 0$ (PRC). (b) $A = 0.5$. (c) $A = 1$. (d) $A = 2$.

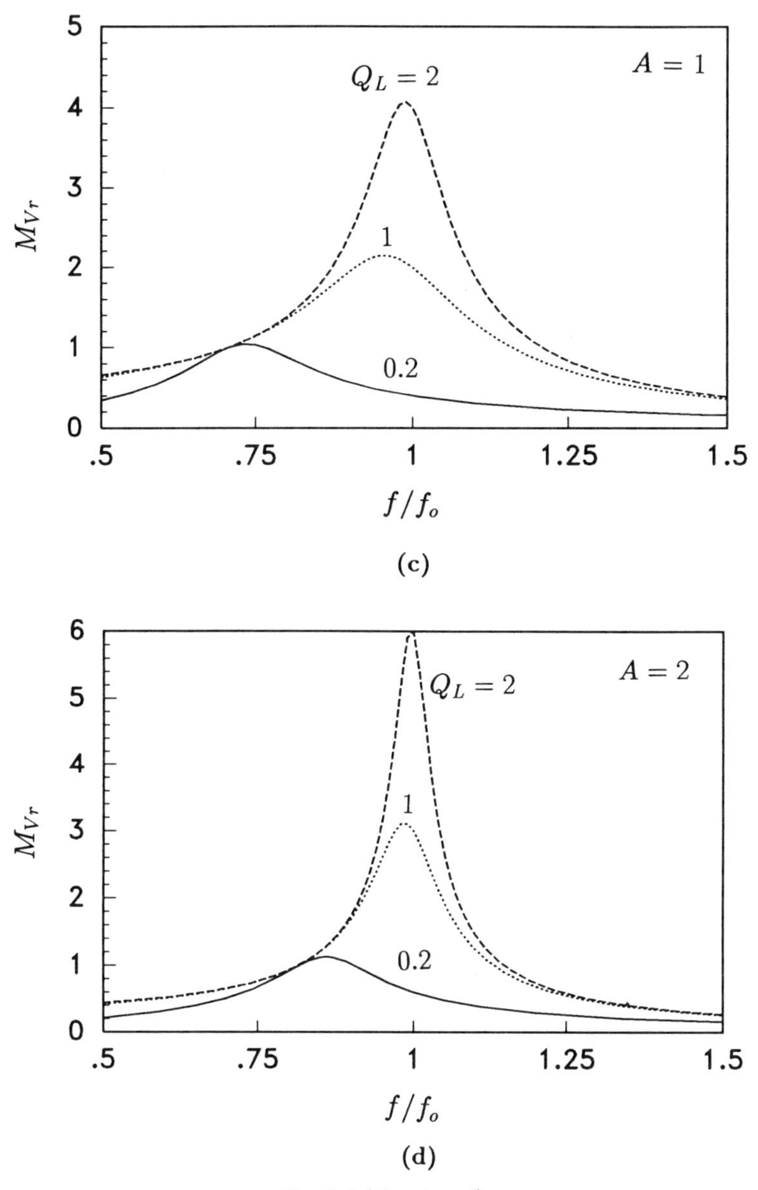

Fig. 8.6 (*Continued*)

Using (8.25), the amplitude of i at a fixed value of M_{Vr} becomes

$$I_m = \frac{2V_I M_{Vr}}{\pi Z_o Q_L}\sqrt{1 + \left[Q_L(1+A)\left(\frac{\omega}{\omega_o}\right)\right]^2}. \quad (8.31)$$

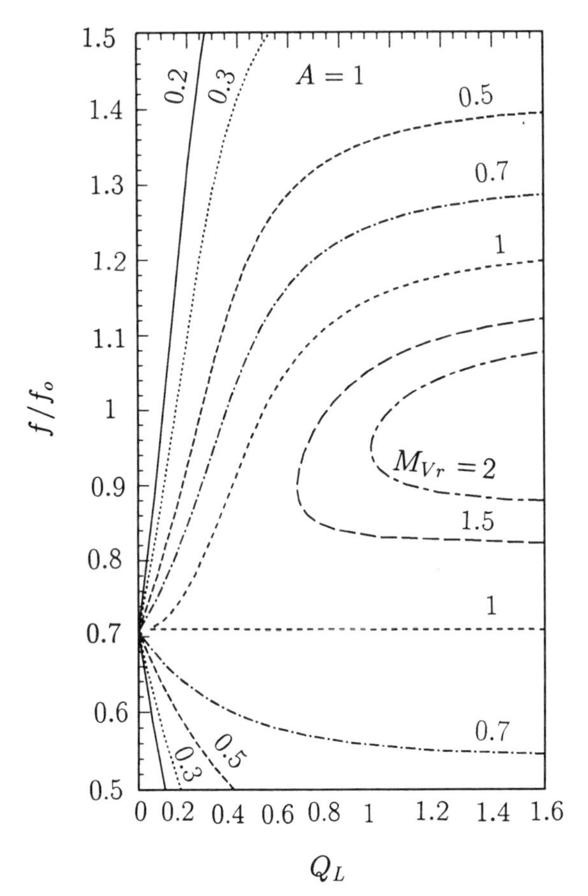

Figure 8.7: Plots of f/f_o against Q_L at fixed values of M_{Vr} for $A = 1$.

Figure 8.8 shows $I_m Z_o/V_I$ versus f/f_o and Q_L at $A = 0$, 0.5, 1, and 2. It can be seen that high values of I_m occur at the resonant frequency f_r. Therefore, if full load occurs at a low value of Q_L, I_m decreases with increasing Q_L, reducing conduction loss $P_r = rI_m^2/2$ and maintaining high efficiency at partial load. If, however, full load occurs at a high value of Q_L, I_m is almost independent of the load, reducing part-load efficiency. Plots of $I_m Z_o/V_I$ against f/f_o at fixed values of M_{Vr} for $A = 1$ are depicted in Fig. 8.9.

The magnitude of the voltage across the resonant inductor is

$$V_{Lm} = \omega L I_m = \frac{2V_I \frac{\omega}{\omega_o}}{\pi Q_L} \sqrt{\frac{1 + [Q_L(1+A)(\frac{\omega}{\omega_o})]^2}{(1+A)^2[1-(\frac{\omega}{\omega_o})^2]^2 + \frac{1}{Q_L^2}(\frac{\omega}{\omega_o} - \frac{\omega_o}{\omega}\frac{A}{A+1})^2}}. \quad (8.32)$$

Plots of V_{Lm} against f/f_o at fixed values of Q_L for $A = 1$ are depicted in Fig. 8.10.

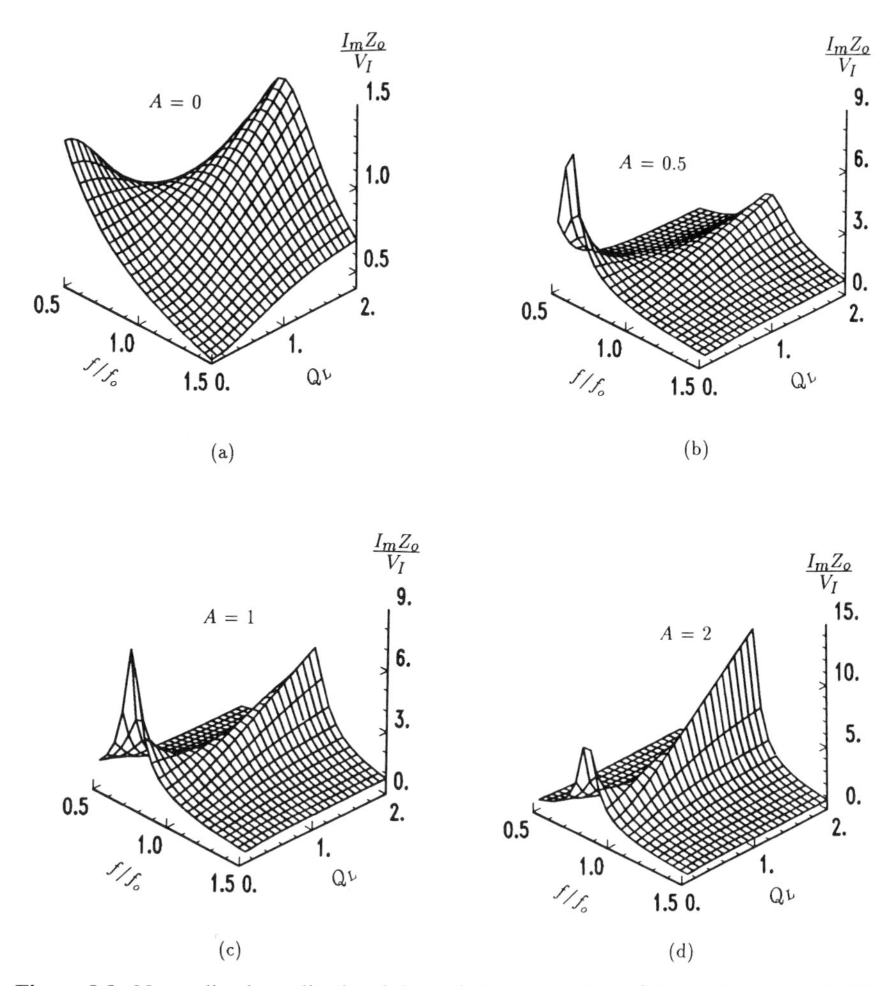

Figure 8.8: Normalized amplitude of the switch current $I_m Z_o / V_I$ as a function of f/f_o and Q_L. (a) $A = 0$. (b) $A = 0.5$. (c) $A = 1$. (d) $A = 2$.

The magnitude of the voltage across the resonant capacitor C_1 is

$$V_{C1m} = \frac{I_m}{\omega C_1} = \frac{2V_I A \frac{\omega_o}{\omega}}{\pi Q_L (1+A)} \sqrt{\frac{1 + [Q_L(1+A)(\frac{\omega}{\omega_o})]^2}{(1+A)^2[1-(\frac{\omega}{\omega_o})^2]^2 + \frac{1}{Q_L^2}(\frac{\omega}{\omega_o} - \frac{\omega_o}{\omega}\frac{A}{A+1})^2}}. \tag{8.33}$$

Plots of V_{C1m} against f/f_o at fixed values of Q_L for $A = 1$ are depicted in Fig. 8.11.

The voltage across the resonant capacitor C_2 is equal to the output voltage of the inverter. The magnitude of the voltage across capacitor C_2 is

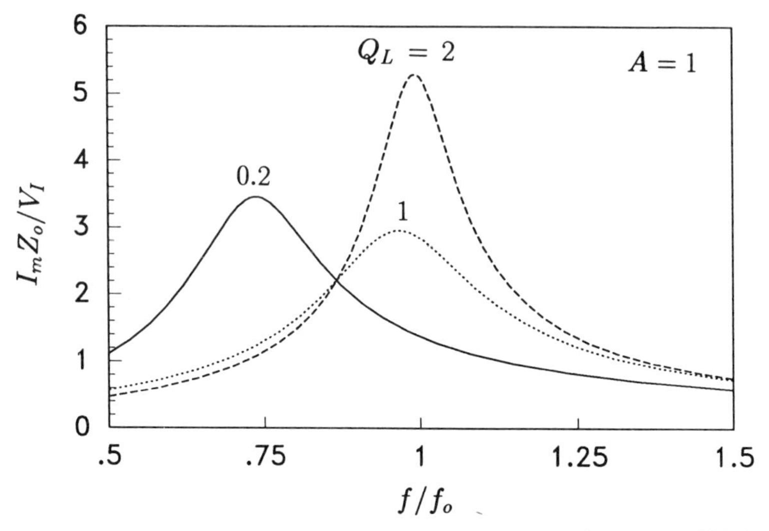

Figure 8.9: Normalized amplitude of the switch current $I_m Z_o/V_I$ versus f/f_o for various values of Q_L and $A = 1$.

$$V_{C2m} = \sqrt{2}V_{Ri} = \sqrt{2}M_{VI}V_I = \frac{2V_I}{\pi\sqrt{(1+A)^2[1-(\frac{\omega}{\omega_o})^2]^2 + [\frac{1}{Q_L}(\frac{\omega}{\omega_o} - \frac{\omega_o}{\omega}\frac{A}{A+1})]^2}}. \tag{8.34}$$

The magnitude of the output current of the inverter is

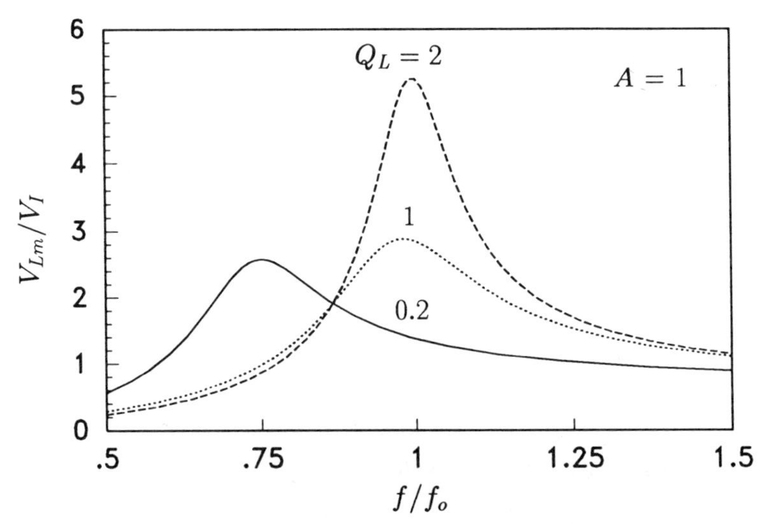

Figure 8.10: Amplitude of the voltage across the resonant inductor V_{Lm} versus f/f_o for various values of Q_L and $A = 1$.

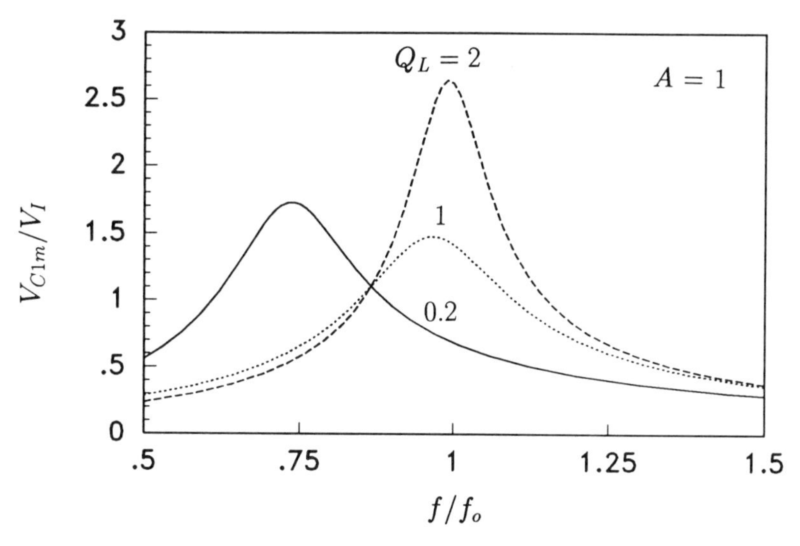

Figure 8.11: Amplitude of the voltage across the resonant capacitor V_{C1m} versus f/f_o for various values of Q_L and $A = 1$.

$$I_{om} = \frac{\sqrt{2}V_{Ri}}{R_i} = \frac{2V_I}{\pi R_i \sqrt{(1+A)^2[1-(\frac{\omega}{\omega_o})^2]^2 + [\frac{1}{Q_L}(\frac{\omega}{\omega_o} - \frac{\omega_o}{\omega}\frac{A}{A+1})]^2}}$$
$$= \frac{2V_I}{\pi Z_o Q_L \sqrt{(1+A)^2[1-(\frac{\omega}{\omega_o})^2]^2 + [\frac{1}{Q_L}(\frac{\omega}{\omega_o} - \frac{\omega_o}{\omega}\frac{A}{A+1})]^2}}. \quad (8.35)$$

Figure 8.12 shows plots of $I_{om}Z_o/V_I$ versus f/f_o at fixed values of Q_L for $A = 1$.

Using (8.27), the output power can be found as

$$P_{Ri} = \frac{V_{Ri}^2}{R_i} = \frac{2V_I^2}{\pi^2 Z_o Q_L[(1+A)^2[1-(\frac{\omega}{\omega_o})^2]^2 + \frac{1}{Q_L^2}(\frac{\omega}{\omega_o} - \frac{\omega_o}{\omega}\frac{A}{A+1})^2]}. \quad (8.36)$$

Figure 8.13 shows the normalized output power $P_{Ri}Z_o/V_I^2$ versus f/f_o and Q_L at $A = 0, 0.5, 1,$ and 2. Plots of $P_{Ri}Z_o/V_I^2$ against f/f_o at fixed values of Q_L for $A = 1$ are depicted in Fig. 8.14. The output power P_{Ri} at $f = f_r$ is obtained

$$P_{Ri} = \frac{2V_I^2}{\pi^2 R_s} = \frac{2V_I^2 R_i}{\pi^2 Z_o^2} \approx 0.2026\frac{V_I^2}{R_s} \approx \frac{V_{DD}^2}{5R_s} = \frac{V_I^2 R_i}{5Z_o^2}. \quad (8.37)$$

From (8.30), the conduction loss is

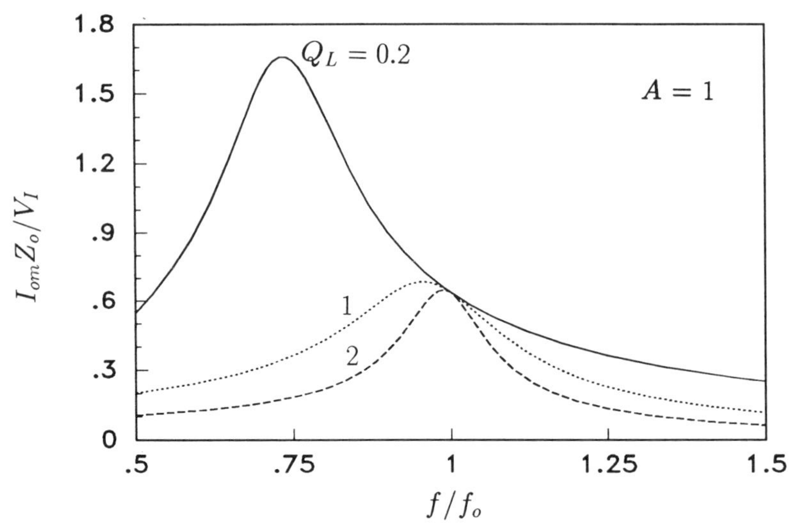

Figure 8.12: Normalized amplitude of the output current $I_{om}Z_o/V_I$ as a function of f/f_o for various values of Q_L and $A = 1$.

$$P_r = \frac{rI_m^2}{2} = \frac{2V_I^2 r}{\pi^2 Z_o^2 Q_L^2} \frac{1 + [Q_L(1+A)(\frac{\omega}{\omega_o})]^2}{(1+A)^2[1-(\frac{\omega}{\omega_o})^2]^2 + \frac{1}{Q_L^2}(\frac{\omega}{\omega_o} - \frac{\omega_o}{\omega}\frac{A}{A+1})^2}$$
$$= \frac{2rV_I^2 M_{Vr}^2 \{1 + [Q_L(1+A)(\frac{\omega}{\omega_o})]^2\}}{\pi^2 Z_o^2 Q_L^2}, \qquad (8.38)$$

where $I_{rms} = I_m/\sqrt{2}$. The total parasitic resistance of the inverter is

$$r = r_{DS} + r_L + r_{C1} + \frac{r_{C2}}{1 + (\frac{1}{\omega C_2 R_i})^2}, \qquad (8.39)$$

where $r_{DS} = (r_{DS1} + r_{DS2})/2$ is the averaged on-resistance of the MOSFETs, r_L is the ac ESR of the inductor, and r_{C1} and r_{C2} are the ESRs of the capacitors C_1 and C_2, respectively. Neglecting switching losses, $P_I = P_{Ri} + P_r$. Hence, one obtains the efficiency of the inverter

$$\eta_I = \frac{P_{Ri}}{P_I} = \frac{P_{Ri}}{P_{Ri} + P_r} = \frac{1}{1 + \frac{r}{R_i}\{1 + [\frac{R_i}{Z_o}(\frac{\omega}{\omega_o})(1+A)]^2\}}$$
$$= \frac{1}{1 + \frac{r}{Z_o Q_L}\{1 + [Q_L(\frac{\omega}{\omega_o})(1+A)]^2\}}. \qquad (8.40)$$

Figure 8.15 shows the inverter efficiency η_I versus f/f_o and Q_L for $A = 1$ and $r/Z_o = 0.006779$ (i.e., $r = 3.42\ \Omega$ and $Z_o = 504.5\ \Omega$). The maximum efficiency occurs if the following condition is met:

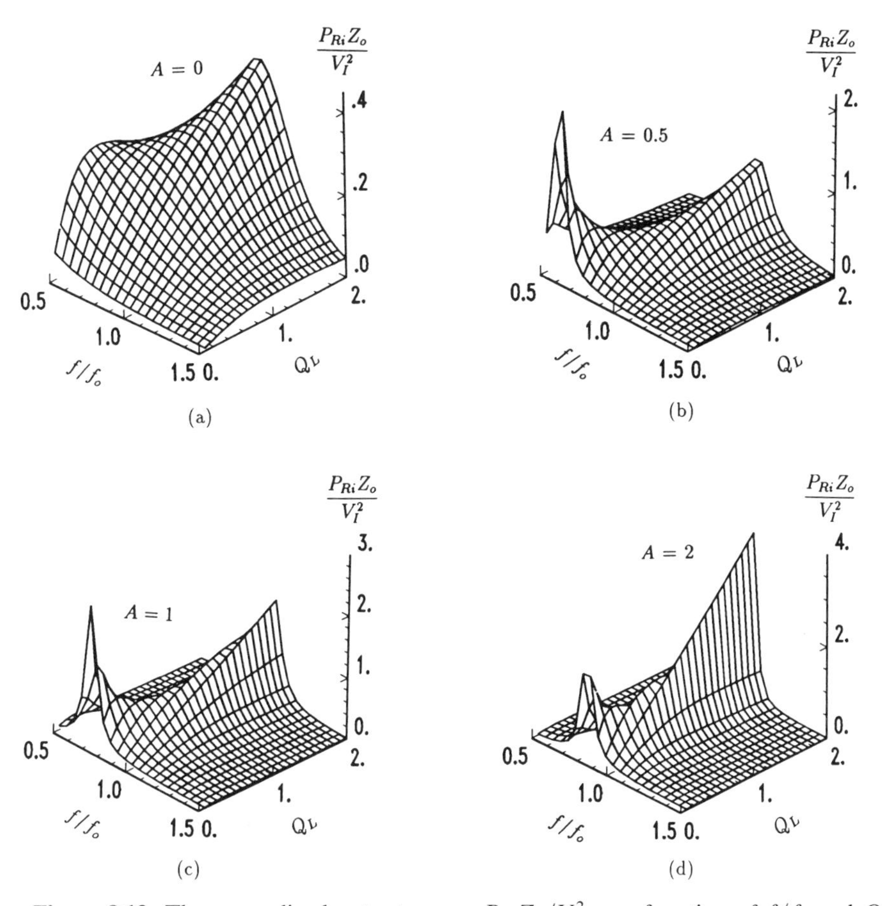

Figure 8.13: The normalized output power $P_{Ri}Z_o/V_I^2$ as a function of f/f_o and Q_L. (a) $A = 0$. (b) $A = 0.5$. (c) $A = 1$. (d) $A = 2$.

$$Q_L = \frac{1}{(1 + A)\left(\frac{\omega}{\omega_o}\right)}. \tag{8.41}$$

The turn-on switching loss is zero for operation above resonance (i.e., for inductive loads for which $f/f_r > 1$). The turn-off power loss can be reduced by adding a capacitor in parallel with one of the transistors.

8.3.5 Short-Circuit and Open-Circuit Operation

The inverter is not safe under the short-circuit and the open-circuit conditions. At $R_i = 0$, the capacitor C_2 is shorted-circuited and the resonant circuit consists of L and C_1. If the switching frequency f is equal to the resonant

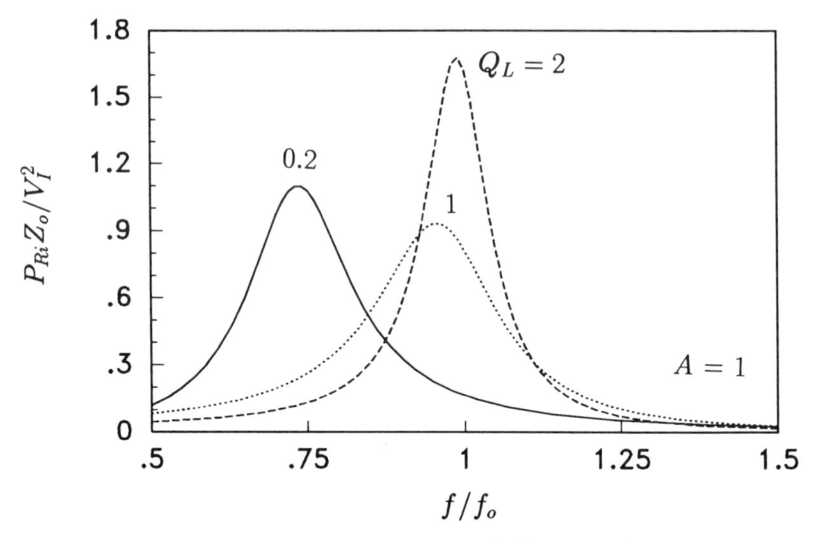

Figure 8.14: The normalized output power $P_{Ri}Z_o/V_I^2$ versus f/f_o for various values of Q_L and $A = 1$.

frequency of this circuit $f_{rs} = 1/(2\pi\sqrt{LC_1})$, the magnitude of the current through the switches and the L-C_1 resonant circuit is $I_m \approx 2V_I/(\pi r)$. This current may become excessive and may destroy the circuit. If f is far from f_{rs}, I_m is limited by the reactance of the resonant circuit. It is observed that $f_{rs} < f_o$. Therefore, the inverter is safe above f_o. At $R_i = \infty$, the resonant circuit is comprised of L and the series combination of C_1 and C_2. Consequently, its resonant frequency is equal to f_o, $I_m \approx 2V_I/(\pi r)$, and the inverter is not safe at or close to this frequency.

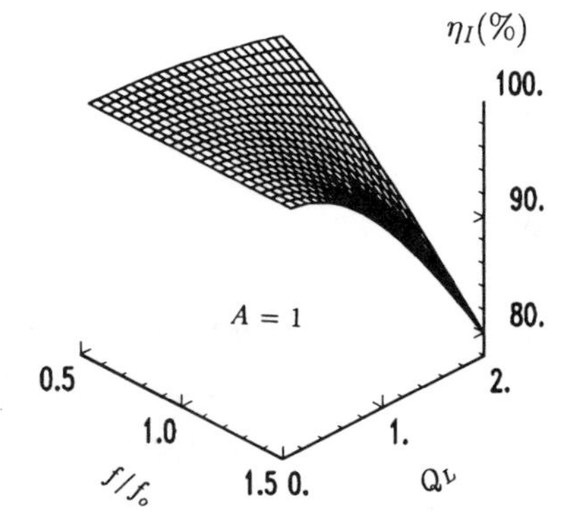

Figure 8.15: Inverter efficiency η_I versus f/f_o and Q_L at $A = 1$, $r = 3.42\ \Omega$, and $Z_o = 504.5\ \Omega$.

8.4 DESIGN EXAMPLE

Example 8.1

Design a Class D series-parallel inverter of Fig. 8.1 to meet the following specifications: $V_I = 250$ V, $f_o = 100$ kHz, $R_{imin} = 100\ \Omega$, and $P_{Rimax} = 85$ W. Assume the inverter efficiency $\eta_I = 90\%$, $A = 1$, and $Q_{Lmin} = 0.2$.

Solution: The maximum dc input power is

$$P_{Imax} = \frac{P_{Rimax}}{\eta_I} = \frac{85}{0.9} = 94.4\ \text{W}, \tag{8.42}$$

and the maximum value of the dc input current is

$$I_{Imax} = \frac{P_{Imax}}{V_I} = \frac{94.4}{250} = 0.38\ \text{A}. \tag{8.43}$$

The voltage transfer function of the inverter is

$$M_{VIa} = \frac{V_{Ri}}{V_I} = \frac{\sqrt{P_{Ri}R_i}}{V_I} = \frac{\sqrt{85 \times 100}}{250} = 0.369. \tag{8.44}$$

Substituting (8.27) into (8.28) and solving numerically the resulting equation with respect to ω/ω_o, one obtains the normalized switching frequency $f/f_o = 0.7926$ and $f = 79.26$ kHz. The component values of the resonant circuit are

$$L = \frac{R_{imin}}{\omega_o Q_L} = \frac{100}{2 \times \pi \times 100 \times 10^3 \times 0.2} = 796\ \mu\text{H} \tag{8.45}$$

$$C = \frac{Q_L}{\omega_o R_{imin}} = \frac{0.2}{2 \times \pi \times 100 \times 10^3 \times 100} = 3.2\ \text{nF} \tag{8.46}$$

$$C_1 = C\left(1 + \frac{1}{A}\right) = 2C = 6.4\ \text{nF} \tag{8.47}$$

$$C_2 = C(1 + A) = 2C = 6.4\ \text{nF}. \tag{8.48}$$

The characteristic impedance of the resonant circuit is $Z_o = R_{imin}/Q_L = 500$ Ω.

From (8.16), the resonant frequency is

$$f_r = f_o\sqrt{\frac{Q_L^2(1+A)^2 - 1 + \sqrt{[Q_L^2(1+A)^2 - 1]^2 + 4Q_L^2 A(1+A)}}{2Q_L^2(1+A)^2}}$$
$$= 100 \times 10^3\sqrt{\frac{4 \times 0.2^2 - 1 + \sqrt{(4 \times 0.2^2 - 1)^2 + 8 \times 0.2^2}}{8 \times 0.2^2}} = 73.5\ \text{kHz}. \tag{8.49}$$

Since $f > f_r$, the switches are loaded inductively, which is a desired feature.

The amplitude of the current through the resonant circuit can be calculated using (8.5), (8.27), (8.31), and (8.45) as

$$I_m = \sqrt{\frac{2P_{Ri}\{1 + [Q_L(1 + A)(\frac{\omega}{\omega_o})]^2\}}{R_i}}$$
$$= \sqrt{\frac{2 \times 85\{1 + [0.2 \times (1 + 1) \times 0.7926]^2\}}{100}} = 1.37 \text{ A}. \quad (8.50)$$

Hence, the maximum value of the current through the switches is $I_{SM} = I_m = 1.37$ A.

The peak values of the voltages across the reactive components can be obtained from (8.32), (8.33), and (8.34) as

$$V_{Lm} = \omega L I_m = 2\pi \times 79.26 \times 10^3 \times 796 \times 10^{-6} \times 1.37 = 543.1 \text{ V}, \quad (8.51)$$

$$V_{C1m} = \frac{I_m}{\omega C_1} = \frac{1.37}{2\pi \times 79.26 \times 10^3 \times 6.4 \times 10^{-9}} = 429.8 \text{ V}, \quad (8.52)$$

and

$$V_{C2m} = \sqrt{2}V_{Ri} = \sqrt{2P_{Ri}R_i} = \sqrt{2 \times 85 \times 100} = 130.4 \text{ V}. \quad (8.53)$$

Assume $r_{DS} = r_L = 1\ \Omega$ and $r_{C1} = r_{C2} = 0.1\ \Omega$. The amplitude of the current through C_2 at full power is

$$I_{C2m} = \frac{I_m R_i}{R_i + \frac{1}{\omega C_2}} = \frac{1.37 \times 100}{100 + 313.75} = 0.331 \text{ A}. \quad (8.54)$$

The conduction losses are

$$P_{rDS} = \frac{I_m^2 r_{DS}}{4} = \frac{1.37^2 \times 1}{4} = 0.469 \text{ W} \quad (8.55)$$

$$P_{rL} = \frac{I_m^2 r_L}{2} = \frac{1.37^2 \times 1}{2} = 0.938 \text{ W} \quad (8.56)$$

$$P_{rC1} = \frac{I_m^2 r_{C1}}{2} = \frac{1.37^2 \times 0.1}{2} = 0.094 \text{ W} \quad (8.57)$$

$$P_{rC2} = \frac{I_{C2m}^2 r_{C2}}{2} = \frac{0.331^2 \times 0.1}{2} = 0.005 \text{ W}. \quad (8.58)$$

Hence, the total conduction loss is

$$P_r = 2P_{rDS} + P_{rL} + P_{rC1} + P_{rC2}$$
$$= 2 \times 0.469 + 0.938 + 0.094 + 0.005 = 1.975 \text{ W}, \quad (8.59)$$

and the inverter efficiency associated with the conduction loss is

$$\eta_I = \frac{P_{Ri}}{P_{Ri} + P_r} = \frac{85}{85 + 1.975} = 97.73\%. \tag{8.60}$$

Assuming $Q_g = 27$ nC and $V_{GSpp} = 15$ V, the gate-drive power is

$$P_G = fQ_g V_{GSpp} = 79.26 \times 10^3 \times 27 \times 10^{-9} \times 15 = 0.032 \text{ W}. \tag{8.61}$$

The turn-on switching power loss is zero because the input impedance of the resonant circuit is inductive. The total power loss is

$$P_{LS} = 2P_G + P_r = 2 \times 0.032 + 1.975 = 2.039 \text{ W}, \tag{8.62}$$

resulting in the inverter efficiency

$$\eta_I = \frac{P_{Ri}}{P_{Ri} + P_{LS}} = \frac{85}{85 + 2.039} = 97.66\%. \tag{8.63}$$

8.5 FULL-BRIDGE SERIES-PARALLEL RESONANT INVERTER

The full-bridge series-parallel resonant inverter is depicted in Fig. 8.16. As will be shown shortly, the full-bridge configuration of the switches results, at the same input voltage, in two times higher amplitude of the fundamental component of the voltage at the input of the resonant circuit than that for the half-bridge configuration. Thus, all the parameters of the inverter that are directly proportional to this amplitude are doubled compared to the half-bridge inverter. This section focuses on presenting the expressions for the parameters of the full-bridge inverter. The operation of the inverter is similar to that of the half-bridge inverter and, to avoid unnecessary repetitions, is not given here.

8.5.1 Voltage Transfer Function

Referring to Fig. 8.16, the input voltage of the resonant circuit v_{DS2} is a square wave of magnitude $2V_I$ given by

$$v_{DS2} = \begin{cases} V_I, & \text{for } 0 < \omega t \leq \pi \\ -V_I, & \text{for } \pi < \omega t \leq 2\pi. \end{cases} \tag{8.64}$$

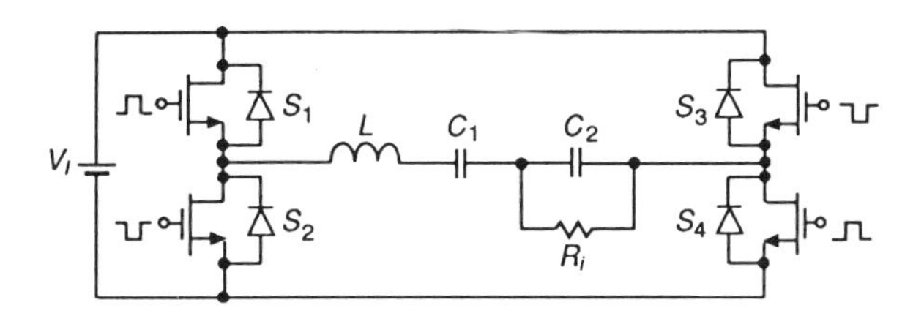

Figure 8.16: Full-bridge series-parallel resonant inverter.

Its fundamental component is $v_{i1} = V_m sin\omega t$, where

$$V_m = \frac{4}{\pi} V_I = 1.273 V_I. \tag{8.65}$$

Thus, the rms value of v_{i1} is $V_{rms} = V_m/\sqrt{2} = 2\sqrt{2}V_I/\pi = 0.9V_I$. The voltage transfer function from V_I to the fundamental component at the input of the resonant circuit is

$$M_{Vs} \equiv \frac{V_{rms}}{V_I} = \frac{2\sqrt{2}}{\pi} = 0.9. \tag{8.66}$$

Using (8.25) and (8.66), one obtains the magnitude of the dc-to-ac voltage transfer function of the Class D full-bridge lossless inverter

$$\begin{aligned} M_{VI} &\equiv \frac{V_{Ri}}{V_I} = M_{Vs} M_{Vr} \\ &= \frac{2\sqrt{2}}{\pi\sqrt{(1+A)^2[1-(\frac{\omega}{\omega_o})^2]^2 + [\frac{1}{Q_L}(\frac{\omega}{\omega_o} - \frac{\omega_o}{\omega}\frac{A}{A+1})]^2}}. \end{aligned} \tag{8.67}$$

The dc-to-ac voltage transfer function of the lossy inverter is given by (8.28).

8.5.2 Currents and Voltages

The current through inductance L is $i = I_m sin(\omega t - \psi)$, where (8.10) gives

$$I_m = \frac{V_m}{Z} = \frac{4V_I}{\pi Z} = \frac{4V_I}{\pi Z_o Q_L}\sqrt{\frac{1 + [Q_L(1+A)(\frac{\omega}{\omega_o})]^2}{(1+A)^2[1-(\frac{\omega}{\omega_o})^2]^2 + \frac{1}{Q_L^2}(\frac{\omega}{\omega_o} - \frac{\omega_o}{\omega}\frac{A}{A+1})^2}}. \tag{8.68}$$

Using (8.25), the amplitude of i at a fixed value of M_{Vr} becomes

$$I_m = \frac{4V_I M_{Vr}}{\pi Z_o Q_L}\sqrt{1 + \left[Q_L(1+A)\left(\frac{\omega}{\omega_o}\right)\right]^2}. \tag{8.69}$$

The magnitude of the voltage across the resonant inductor is

$$V_{Lm} = \omega L I_m = \frac{4V_I \frac{\omega}{\omega_o}}{\pi Q_L}\sqrt{\frac{1 + [Q_L(1+A)(\frac{\omega}{\omega_o})]^2}{(1+A)^2[1-(\frac{\omega}{\omega_o})^2]^2 + \frac{1}{Q_L^2}(\frac{\omega}{\omega_o} - \frac{\omega_o}{\omega}\frac{A}{A+1})^2}}. \tag{8.70}$$

The magnitude of the voltage across the resonant capacitor C_1 is

$$V_{C1m} = \frac{I_m}{\omega C_1} = \frac{4V_I A \frac{\omega_o}{\omega}}{\pi Q_L(1+A)} \sqrt{\frac{1 + [Q_L(1+A)(\frac{\omega}{\omega_o})]^2}{(1+A)^2[1 - (\frac{\omega}{\omega_o})^2]^2 + \frac{1}{Q_L^2}(\frac{\omega}{\omega_o} - \frac{\omega_o}{\omega}\frac{A}{A+1})^2}}. \tag{8.71}$$

The voltage across the resonant capacitor C_2 is equal to the output voltage of the inverter. From (8.67), the magnitude of the voltage across capacitor C_2 is

$$V_{C2m} = \sqrt{2} V_{Ri} = \sqrt{2} M_{VI} V_I = \frac{4V_I}{\pi\sqrt{(1+A)^2[1 - (\frac{\omega}{\omega_o})^2]^2 + [\frac{1}{Q_L}(\frac{\omega}{\omega_o} - \frac{\omega_o}{\omega}\frac{A}{A+1})]^2}}. \tag{8.72}$$

The magnitude of the output current of the inverter is

$$I_{om} = \frac{\sqrt{2} V_{Ri}}{R_i} = \frac{4V_I}{\pi R_i \sqrt{(1+A)^2[1 - (\frac{\omega}{\omega_o})^2]^2 + [\frac{1}{Q_L}(\frac{\omega}{\omega_o} - \frac{\omega_o}{\omega}\frac{A}{A+1})]^2}}. \tag{8.73}$$

8.5.3 Powers and Efficiency

Using (8.67), the output power can be found as

$$P_{Ri} = \frac{V_{Ri}^2}{R_i} = \frac{8V_I^2}{\pi^2 Z_o Q_L[(1+A)^2[1 - (\frac{\omega}{\omega_o})^2]^2 + \frac{1}{Q_L^2}(\frac{\omega}{\omega_o} - \frac{\omega_o}{\omega}\frac{A}{A+1})^2]}. \tag{8.74}$$

From (8.68), the conduction loss is

$$\begin{aligned} P_r = \frac{r I_m^2}{2} &= \frac{8V_I^2 r}{\pi^2 Z_o^2 Q_L^2} \frac{1 + [Q_L(1+A)(\frac{\omega}{\omega_o})]^2}{(1+A)^2[1 - (\frac{\omega}{\omega_o})^2]^2 + \frac{1}{Q_L^2}(\frac{\omega}{\omega_o} - \frac{\omega_o}{\omega}\frac{A}{A+1})^2} \\ &= \frac{8rV_I^2 M_{Vr}^2\{1 + [Q_L(1+A)\frac{\omega}{\omega_o})]^2\}}{\pi^2 Z_o^2 Q_L^2}, \end{aligned} \tag{8.75}$$

where the total parasitic resistance of the inverter is

$$r = 2r_{DS} + r_L + r_{C1} + \frac{r_{C2}}{1 + (\frac{1}{\omega C_2 R_i})^2} \tag{8.76}$$

Neglecting switching losses, $P_I = P_{Ri} + P_r$. Hence, one obtains the efficiency of the inverter

$$\begin{aligned} \eta_I &= \frac{P_{Ri}}{P_I} = \frac{P_{Ri}}{P_{Ri} + P_r} = \frac{1}{1 + \frac{r}{R_i}\{1 + [\frac{R_i}{Z_o}(1+A)(\frac{\omega}{\omega_o})]^2\}} \\ &= \frac{1}{1 + \frac{r}{Z_o Q_L}\{1 + [Q_L(1+A)(\frac{\omega}{\omega_o})]^2\}}. \end{aligned} \tag{8.77}$$

8.6 SUMMARY

- In the transformer series-parallel resonant inverter, the capacitor C_2 can be placed on the secondary side of the transformer. In this case, the transformer leakage inductance is included in the resonant inductance L.
- The voltage transfer function of the resonant circuit is independent of the load at the resonant frequency $f_{rs} = 1/(2\pi\sqrt{LC_1})$ of the L-C_1 resonant circuit. The whole resonant circuit represents a capacitive load to the transistors at f_{rs} because $f_{rs} < f_r$.
- If full load occurs at a low value of Q_L, the magnitude of the current through the switches and the resonant inductor decreases with increasing load resistance, reducing the conduction loss and maintaining high part-load efficiency. However, beyond a certain value of Q_L, the amplitude of the current becomes essentially constant, reducing the efficiency at light loads.
- If full load occurs at a high value of Q_L, the magnitude of the switch current is almost independent of the load, keeping a constant conduction loss and reducing efficiency at part load (as for the PRC).
- The resonant frequency f_r, which forms the boundary between a capacitive and an inductive load, is dependent on the load.
- The inverter cannot operate safely with an open circuit at the output at frequencies close to the corner frequency f_o.
- The inverter cannot operate safely with a short circuit at the output at frequencies close to the resonant frequency f_r.
- The sensitivity of the dc voltage transfer function to the load decreases with increasing C_1/C_2 for high values of Q_L.
- As A is increased, low values of M_{Vr} can be achieved with f/f_o close to 1.

8.7 REFERENCES

1. A. K. S. Bhat and S. B. Dewan, "Analysis and design of a high frequency resonant converter using LCC-type commutation," *Proc. IEEE Industry Applications Society Annual Meeting,* 1986, pp. 657–663; reprinted in *IEEE Trans. Power Electronics*, vol. PE-2, pp. 291–301, Oct. 1987.
2. A. K. S. Bhat and S. B. Dewan, "Steady-state analysis of a LCC-type commutated high-frequency inverter," *IEEE Power Electronics Specialists Conference Record*, Kyoto, Japan, Apr. 11–14, 1988, pp. 1220–1227.
3. R. L. Steigerwald, "A comparison of half-bridge resonant converter topologies," *IEEE Trans. Power Electronics*, vol. PE-3, pp. 174–182, Apr. 1988.
4. I. Batarseh, R. Liu, and C. Q. Lee, "Design of parallel resonant converter with LCC-type commutation," *Electronics Letters*, vol. 24, no. 3, pp. 177–179, Feb. 1988.

5. I. Batarseh, R. Liu, C. Q. Lee, and A. K. Upadhyay, "150 watts and 140 kHz multi-output LCC-type parallel resonant converter," in *IEEE Applied Power Electronics Conference*, 1989, pp. 221–230.
6. I. Batarseh and C. Q. Lee, "High-frequency high-order parallel resonant converter," *IEEE Industrial Electronics*, vol. IE-36, pp. 485–498, Nov. 1989.
7. A. K. S. Bhat, "Analysis, optimization and design of a series-parallel resonant converter," in *IEEE Applied Power Electronics Conference*, Los Angeles, CA, Mar. 11–19, 1990, pp. 155–164.
8. S. Shah and A. K. Upadhyay, "Analysis and design of a half-bridge series-parallel resonant converter operating in the discontinuous conduction mode," in *IEEE Applied Power Electronics Conference*, Los Angeles, CA, Mar. 11–19, 1990, pp. 165–174.
9. I. Batarseh, R. Liu, C. Q. Lee, and A. K. Upadhyay, "Theoretical and experimental studies of the LCC-type parallel resonant converter," *IEEE Trans. Power Electronics*, vol. PE-5, pp. 140–150, Apr. 1990.
10. M. K. Kazimierczuk, N. Thirunarayan, and S. Wang, "Analysis of series-parallel resonant converter," *IEEE Trans. Aerospace and Electronic Systems*, vol. AES-29, pp. 88–99, Jan. 1993.
11. D. Czarkowski and M. K. Kazimierczuk, "Phase-controlled series-parallel resonant converter," *IEEE Trans. Power Electronics*, vol. PE-8, pp. 309–319, July 1993.

8.8 REVIEW QUESTIONS

8.1 Is the transformer leakage inductance absorbed into the topology of the series-parallel resonant inverter?

8.2 What are the advantages and disadvantages of placing the capacitor C_2 on the secondary side of the transformer?

8.3 Is the resonant frequency, which is the boundary between the capacitive and inductive load, dependent on the load in the SPRI?

8.4 Is the voltage transfer function always dependent on the load in the SPRI?

8.5 What is the value of f/f_o at which the voltage transfer function of the SPRI is independent of the load? Is the load inductive in this case?

8.6 Is the part-load efficiency of the SPRI high?

8.7 What is the condition needed to obtain a high part-load efficiency of the SPRI?

8.8 How does the output power change when the load resistance is increased?

8.9 Is the SPRI open-circuit proof?

8.10 Is the SPRI short-circuit proof?

8.9 PROBLEMS

8.1 Derive, step by step, the input impedance of the resonant circuit of Fig. 8.1. Compare your result to (8.9).

8.2 The resonant circuit of the inverter of Fig. 8.1 with $L = 500\ \mu$H, $C_1 = C_2 = 4.7$ nF, and $R_i = 600\ \Omega$ is driven by a sinusoidal voltage source $v = 100 sin(840 \times 10^3 t)$. What is the amplitude of the voltage across the ac load R_i in this circuit?

8.3 Calculate the voltage stresses for the resonant components of the circuit from Problem 8.2.

8.4 Derive equation (8.41).

8.5 Design a full-bridge Class D series-parallel inverter of Fig. 8.16 to meet the following specifications: $V_I = 250$ V, $f_o = 100$ kHz, $R_{imin} = 200\ \Omega$, and $P_{Rimax} = 85$ W. Assume the inverter efficiency $\eta_I = 90\%$, $A = 1$, and $f/f_o = 0.85$.

CHAPTER 9

CLASS D CLL RESONANT INVERTER

9.1 INTRODUCTION

In this chapter a steady-state analysis is given for a Class D resonant inverter in which impedance transformation is achieved by tapping the resonant inductor. As a result, the resonant circuit of the inverter contains one resonant capacitor and two resonant inductors. The inverter is also called a CLL resonant inverter [1], [2]. The analysis of the inverter is carried out in the frequency domain using Fourier series techniques. Design equations describing the steady-state operation are derived. The dc-to-ac voltage transfer function of the inverter is almost *independent* of the load variations at a switching frequency higher than the resonant frequency. An important advantage of the inverter is that the load presented by the resonant circuit to the switches is inductive at this frequency. In addition, the circuit has high efficiency over a wide range of load resistance.

9.2 PRINCIPLE OF OPERATION

The CLL resonant inverter [1] shown in Fig. 9.1 is composed of two bidirectional two-quadrant switches S_1 and S_2 and a resonant circuit C-L_1-L_2. The resonant capacitor C is connected in series with the tapped inductor L_1-L_2. The load is connected in parallel with the inductor L_2. The switches consist of MOSFETs and their body diodes, and are driven by rectangular-wave voltage sources v_{GS1} and v_{GS2}. Each switch can conduct a positive or a negative current. The transistors are driven by rectangular-wave voltage sources v_{GS1} and v_{GS2}. Switches S_1 and S_2 are alternately turned ON and OFF at the

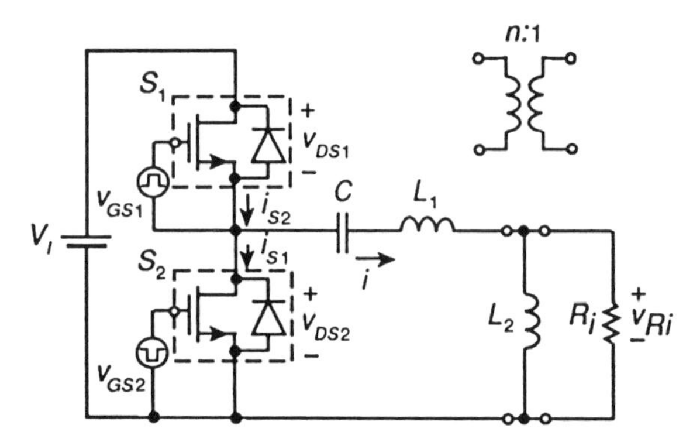

Figure 9.1: Class D CLL resonant dc-ac inverter.

switching frequency $f = \omega/2\pi$ with a duty cycle of 50%. The inductor L_2 can be replaced by a transformer to obtain isolation or the desired amplitude of the ac output voltage. In this case, the transformer magnetizing inductance can be used as inductance L_2. Gapped cores are usually used to obtain low magnetizing inductances. Note that the transformer versions of the series resonant inverter and the CLL inverter are almost the same. The only difference is that the magnetizing inductance in the series resonant inverter is large, whereas the magnetizing inductance in the CLL inverter is small. The leakage inductance of the transformer is absorbed into inductance L_1.

The parallel R_i-L_2 circuit of Fig. 9.2(a) can be converted into a series R_s-L_s circuit of Fig. 9.2(b) at a given frequency. The total series equivalent inductance is $L_{eq} = L_1 + L_s$. In Fig. 9.2(c), the dc voltage source V_I and the switches S_1 and S_2 are modeled by a square-wave voltage source, where the low level of the square wave is zero and the high level is V_I. The equivalent on-resistance of the MOSFETs is $r_{DS} \approx (r_{DS1} + r_{DS2})/2$. The parasitic resistance r of the inverter is composed of the resistance of the switch r_{DS}, the equivalent series resistance (ESR) of the capacitor r_{Cr}, and the ESRs of the inductors r_{L1} and r_{L2}.

Waveforms in the CLL inverter for $f > f_r = 1/(2\pi\sqrt{L_{eq}C})$ are the same as in the parallel resonant inverter shown in Fig. 7.2. The operation of the inverter above resonance is preferred because the reverse recovery of the MOSFET body diodes does not affect adversely the circuit operation. The input voltage v_{DS2} of the resonant circuit is a square wave. Assuming that loaded quality factor Q_r at the resonant frequency f_r is high, the capacitor current i is nearly sinusoidal and flows alternately through switches S_1 or S_2.

If $R_i << X_{L2} = \omega L_2$, most of the capacitor current i flows through the load resistance, and therefore I_m is inversely proportional to the load resistance, resulting in high part-load efficiency. When the load resistance R_i becomes greater than X_{L2}, most of the capacitor current i flows through the resonant inductor L_2, making I_m independent of R_i. Therefore, the efficiency is low at part loads.

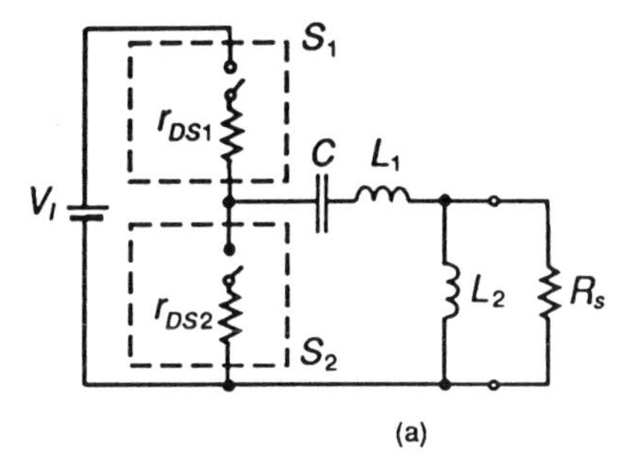

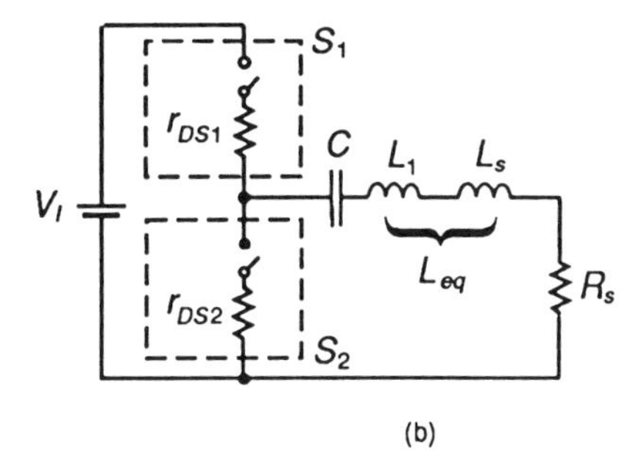

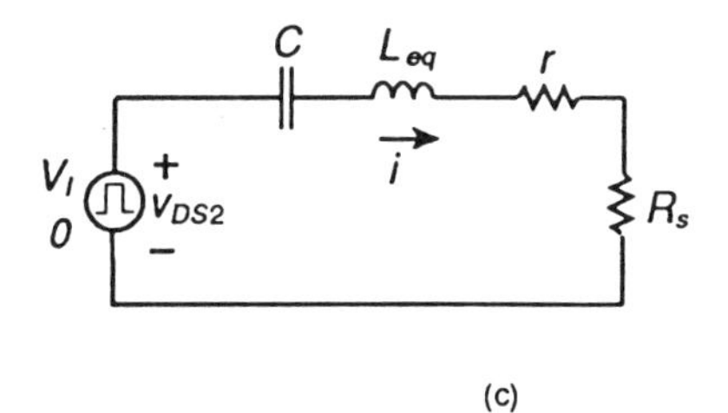

Figure 9.2: Equivalent circuits of the CLL resonant inverter.

9.3 ANALYSIS

9.3.1 Assumptions

The analysis of the Class D inverter of Fig. 9.1 is carried out under the following assumptions:

1) The MOSFETs are modeled by switches with on-resistances r_{DS1} and r_{DS2}.
2) Switching losses are ignored.
3) The current i through the resonant capacitor is nearly sinusoidal.

9.3.2 Boundary Between Capacitive and Inductive Load

The resonant circuit in the inverter of Fig. 9.1 can be described by the following normalized parameters:

- the ratio of the inductances

$$A = \frac{L_1}{L_2} \tag{9.1}$$

- the equivalent inductance of L_1 and L_2 connected in series

$$L = L_1 + L_2 = L_2(1 + A) = L_1\left(1 + \frac{1}{A}\right) \tag{9.2}$$

- the corner frequency (or the undamped natural frequency)

$$\omega_o = \frac{1}{\sqrt{LC}} = \frac{1}{\sqrt{(L_1 + L_2)C}} \tag{9.3}$$

- the characteristic impedance

$$Z_o = \omega_o L = \frac{1}{\omega_o C} = \sqrt{\frac{L}{C}} \tag{9.4}$$

- the loaded quality factor at the corner frequency f_o

$$Q_L = \omega_o C R_i = \frac{R_i}{\omega_o L} = \frac{R_i}{Z_o} \tag{9.5}$$

- the equivalent inductance of L_1 and L_s connected in series

$$L_{eq} = L_1 + L_s \tag{9.6}$$

- the resonant frequency

$$\omega_r = \frac{1}{\sqrt{L_{eq}C}} = \frac{1}{\sqrt{(L_1 + L_s)C}} \tag{9.7}$$

- the loaded quality factor at the resonant frequency f_r

$$Q_r = \frac{1}{\omega_r C R_s} = \frac{\omega_r(L_1 + L_s)}{R_s}. \tag{9.8}$$

The boundary between capacitive and inductive load is determined by the resonant frequency f_r. If the MOSFETs in the inverter are loaded by inductive loads, high efficiency is obtained.

The input impedance of the resonant circuit is given by

$$\mathbf{Z} = \frac{R_i\{(1+A)[1-(\frac{\omega_o}{\omega})^2] + j\frac{1}{Q_L}(\frac{\omega}{\omega_o}\frac{A}{A+1} - \frac{\omega_o}{\omega})\}}{1 - jQ_L(\frac{\omega_o}{\omega})(1+A)} = Ze^{j\psi}$$
$$= R_s + jX_s, \tag{9.9}$$

where

$$\frac{Z}{Z_o} = Q_L\sqrt{\frac{(1+A)^2[1-(\frac{\omega_o}{\omega})^2]^2 + \frac{1}{Q_L^2}(\frac{\omega}{\omega_o}\frac{A}{A+1} - \frac{\omega_o}{\omega})^2}{1 + [Q_L(\frac{\omega_o}{\omega})(1+A)]^2}} \tag{9.10}$$

$$\psi = arctan\left\{\frac{1}{Q_L}\left(\frac{\omega}{\omega_o}\frac{A}{A+1} - \frac{\omega_o}{\omega}\right) + Q_L\left(\frac{\omega_o}{\omega}\right)(1+A)^2\left[1-\left(\frac{\omega_o}{\omega}\right)^2\right]\right\} \tag{9.11}$$

$$R_s = Zcos\psi \tag{9.12}$$

$$X_s = Zsin\psi. \tag{9.13}$$

Figure 9.3 shows normalized input impedance Z/Z_o and phase ψ as functions of f/f_o at fixed values of Q_L for $A = 0.5$. At $f = f_o$,

$$\psi = -arctan\left[\frac{1}{Q_L(1+A)}\right] < 0. \tag{9.14}$$

Thus, the resonant circuit represents a capacitive load at $f = f_o$.

The resonant frequency f_r is defined as the frequency at which the phase shift ψ is equal to zero [Fig. 9.3(b)]. This frequency constitutes the boundary between capacitive and inductive loads. For $f < f_r$, $\psi < 0$; this means that the resonant circuit represents a capacitive load. Therefore, the current through the resonant capacitor i leads the fundamental component of the voltage v_{DS2}. The operation in this frequency range is not recommended because the antiparallel diodes of the MOSFETs turn off at high di/dt, generating high reverse-recovery current spikes. For $f > f_r$, $\psi > 0$ and the resonant circuit represents an inductive load. Consequently, the capacitor current i lags behind the fundamental component of the voltage v_{DS2}. The antiparallel diodes turn off at low di/dt and do not generate reverse-recovery current spikes. Hence, the operation in this frequency range is recommended for practical applications. The boundary between inductive and capacitive loads occurs when the phase $\psi = 0$. Hence, from (9.11) the resonant frequency is found to be

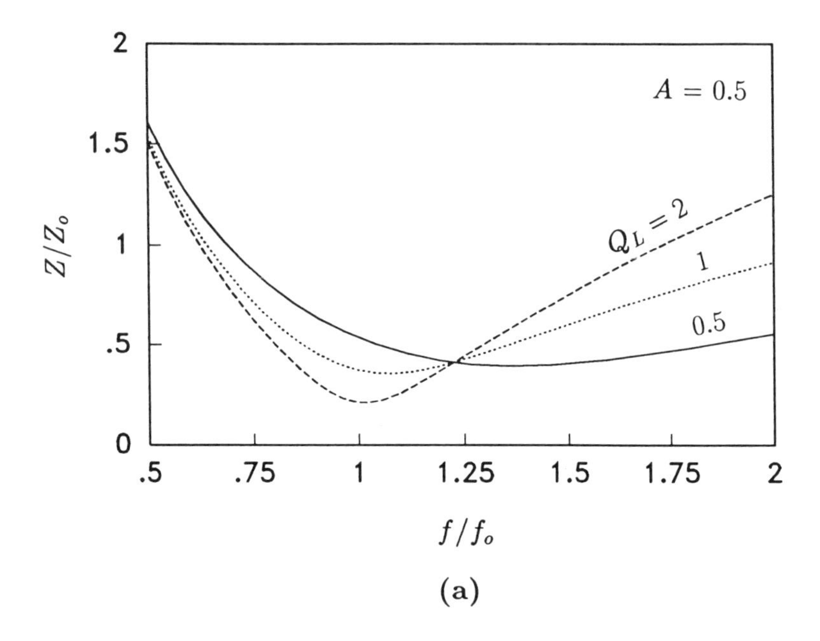

(a)

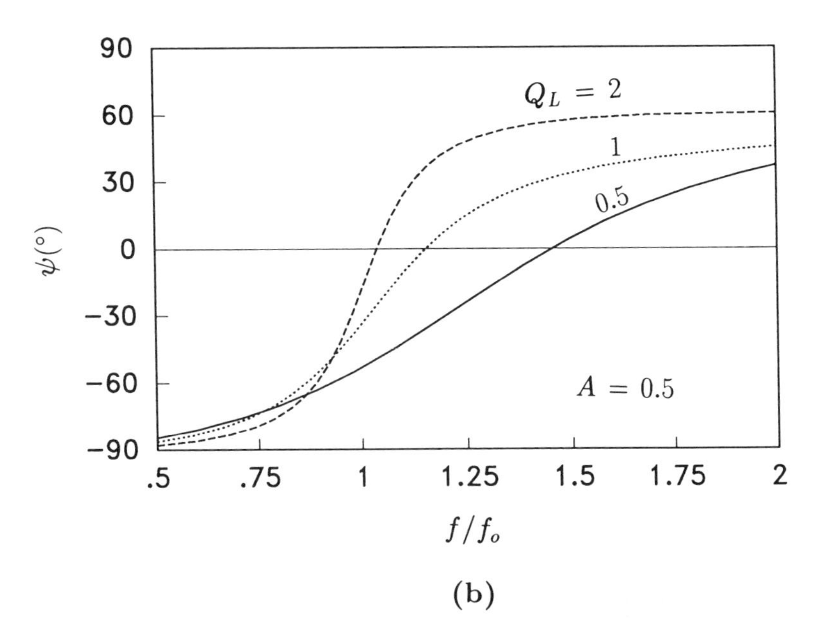

(b)

Figure 9.3: Normalized input impedance Z/Z_o and phase ψ of the resonant circuit as functions of f/f_o at constant values of Q_L for $A = 0.5$. (a) Z/Z_o versus f/f_o. (b) ψ versus f/f_o.

$$\frac{f_r}{f_o} = \sqrt{\frac{\{1 - Q_L^2(1+A)^2 + \sqrt{[Q_L^2(1+A)^2 - 1]^2 + 4Q_L^2 A(1+A)}\}(1+A)}{2A}}. \tag{9.15}$$

The resonant frequency f_r depends on Q_L and A. As $Q_L \to 0$, $f_r/f_o \to \sqrt{(1+A)/A}$. Figure 9.4 shows f_r/f_o versus Q_L at fixed values of A.

The R_i-L_2 parallel two-terminal network of Fig. 9.2(a) can be converted into the R_s-L_s series two-terminal network of Fig. 9.2(b). This leads to basic topology of the Class D series resonant inverter, where $L_{eq} = L_1 + L_s$. The reactance factor at the resonant frequency $\omega_r = 1/\sqrt{L_{eq}C}$ is defined as

$$q_r = \frac{\omega_r L_s}{R_s} = \frac{R_i}{\omega_r L_2} = Q_L(1+A)\left(\frac{\omega_o}{\omega_r}\right), \tag{9.16}$$

where ω_r/ω_o is given by (9.15). The equivalent two-terminal networks method and (9.16) yields the relationships among R_i, R_s, L_2, and L_s at $f = f_r$

$$R_s = \text{Re}\{R_i||X_{L_2}\} = \frac{R_i}{1+q_r^2} \tag{9.17}$$

$$X_{Ls} = \text{Im}\{R_i||L_2\} = \frac{X_{L_2}}{1+\frac{1}{q_r^2}} = \frac{X_{L_2}}{1+(\frac{\omega_r L_2}{R_i})^2}, \tag{9.18}$$

where $X_{L_2} = \omega_r L_2$ and $X_{Ls} = \omega_r L_s$. The maximum value of $R_s = R_i/2$ occurs at $q_r = 1$. Inductance L_s increases from 0 to L_2 as q_r is increased from 0 to ∞. For $q_r^2 >> 1$, $L_s \approx L_2$, $L \approx L_{eq}$, and $f_r \approx f_o$. If Q_r is high enough (e.g., $Q_r > 2.5$), the current through the resonant capacitor i is nearly sinusoidal.

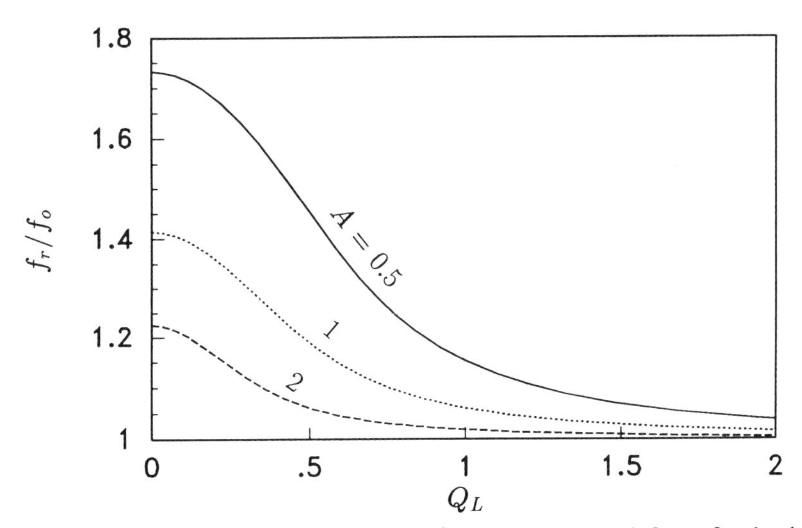

Figure 9.4: Normalized resonant frequency f_r/f_o as a function of Q_L at fixed values of A.

9.3.3 Voltage Transfer Function

The input voltage of the resonant circuit v_{DS2} is a square wave and is given by

$$v_{DS2} = \begin{cases} V_I, & \text{for } 0 < \omega t \le \pi \\ 0, & \text{for } \pi < \omega t \le 2\pi. \end{cases} \tag{9.19}$$

The fundamental component of this voltage is $v_{i1} = V_m sin\omega t$, where

$$V_m = \frac{2}{\pi} V_I, \tag{9.20}$$

resulting in its rms value $V_{rms} = V_m/\sqrt{2} = \sqrt{2}V_I/\pi$. The dc-to-ac voltage transfer function from the dc input voltage V_I to the fundamental component of the input voltage of the resonant circuit is given by

$$M_{Vs} \equiv \frac{V_{rms}}{V_I} = \frac{\sqrt{2}}{\pi}. \tag{9.21}$$

Referring to Fig. 9.2(a), one arrives at the ac-to-ac voltage transfer function of the resonant circuit

$$\mathbf{M_{Vr}} \equiv \frac{V_{Ri}}{V_{rms}} = \frac{1}{(1+A)[1-(\frac{\omega_o}{\omega})^2] + j\frac{1}{Q_L}(\frac{\omega}{\omega_o}\frac{A}{A+1} - \frac{\omega_o}{\omega})} = M_{Vr}e^{j\varphi}, \tag{9.22}$$

where

$$M_{Vr} = \frac{1}{\sqrt{(1+A)^2[1-(\frac{\omega_o}{\omega})^2]^2 + \frac{1}{Q_L^2}(\frac{\omega}{\omega_o}\frac{A}{A+1} - \frac{\omega_o}{\omega})^2}}, \tag{9.23}$$

$$\varphi = -arctan\left\{\frac{\frac{1}{Q_L}(\frac{\omega}{\omega_o}\frac{A}{A+1} - \frac{\omega_o}{\omega})}{(1+A)[1-(\frac{\omega_o}{\omega})^2]}\right\}, \tag{9.24}$$

and V_{Ri} is the rms value of the voltage across R_i. In Fig. 9.5, M_{Vr} is plotted as a function of f/f_o and Q_L at fixed values of A in three-dimensional space. Figure 9.6 shows plots of M_{Vr} versus f/f_o at fixed values of Q_L for $A = 0$, 0.5, 1, and 2. From (9.23), $M_{Vr} = 1$ at a normalized critical frequency given by

$$\frac{f_{rs}}{f_o} = \sqrt{1 + \frac{1}{A}} = \sqrt{1 + \frac{L_2}{L_1}}, \tag{9.25}$$

where $f_{rs} = 1/(2\pi\sqrt{L_1 C})$ is the resonant frequency of the C-L_1 series-resonant circuit. For $A = 1$, $f_{rs}/f_o = \sqrt{2}$. As A is increased from 0 to ∞, f_{rs} decreases from ∞ to f_o. If A is small, the critical frequency is far from the

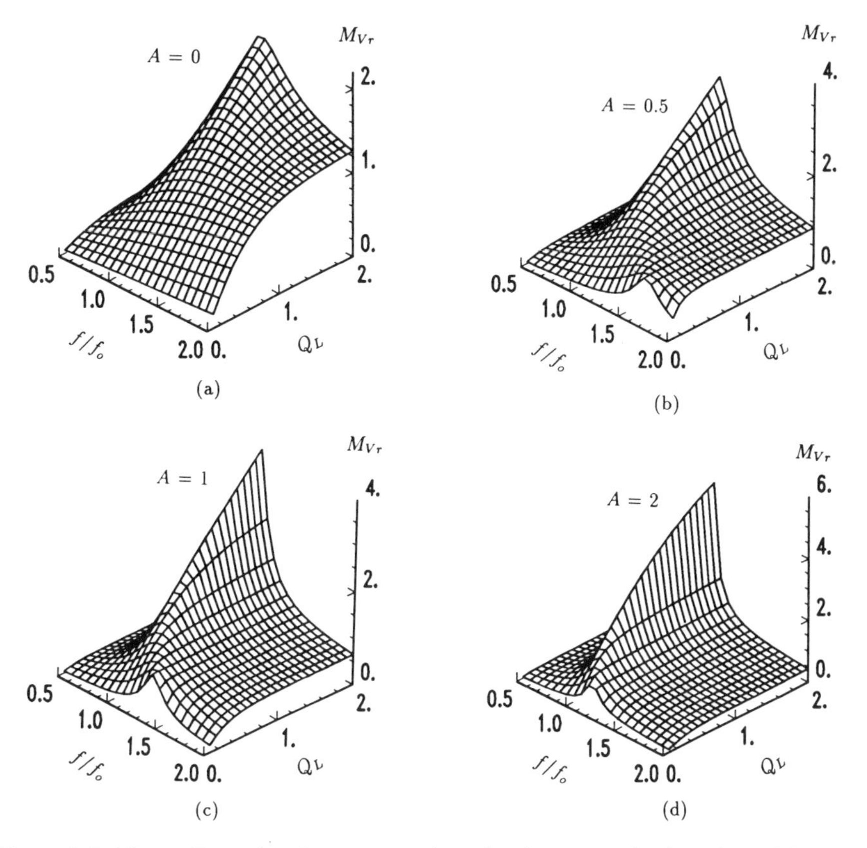

Figure 9.5: Three-dimensional representation of voltage transfer function of the resonant circuit M_{Vr} as a function of f/f_o and Q_L at fixed values of A. (a) $A = 0$. (b) $A = 0.5$. (c) $A = 1$. (d) $A = 2$.

corner frequency. On the other hand, if A is high, the critical frequency is very close to the resonant frequency and the transfer function becomes very sensitive to frequency variations. Note that the reactance of the series-resonant circuit C-L_1 at $f = f_{rs}$ is zero and therefore the fundamental component of v_{DS2} appears directly across L_2, making M_{Vr} independent of load and equal to 1. Equation (9.23) was solved numerically for f/f_o as a function of Q_L at fixed values of M_{Vr}. The results are illustrated in Fig. 9.7.

The product of (9.21) and (9.23) yields the magnitude of the dc-to-ac voltage transfer function of the Class D inverter

$$M_{VI} \equiv \frac{V_{Ri}}{V_I} = M_{Vs}M_{Vr} = \frac{\sqrt{2}}{\pi\sqrt{(1+A)^2[1-(\frac{\omega_o}{\omega})^2]^2 + [\frac{1}{Q_L}(\frac{\omega}{\omega_o}\frac{A}{A+1} - \frac{\omega_o}{\omega})]^2}}. \tag{9.26}$$

The dc-to-ac voltage transfer function of the actual inverter can be estimated as

$$\mathbf{M_{VIa}} = \eta_I \mathbf{M_{VI}}, \tag{9.27}$$

where η_I is the efficiency of the inverter and is derived in the next section.

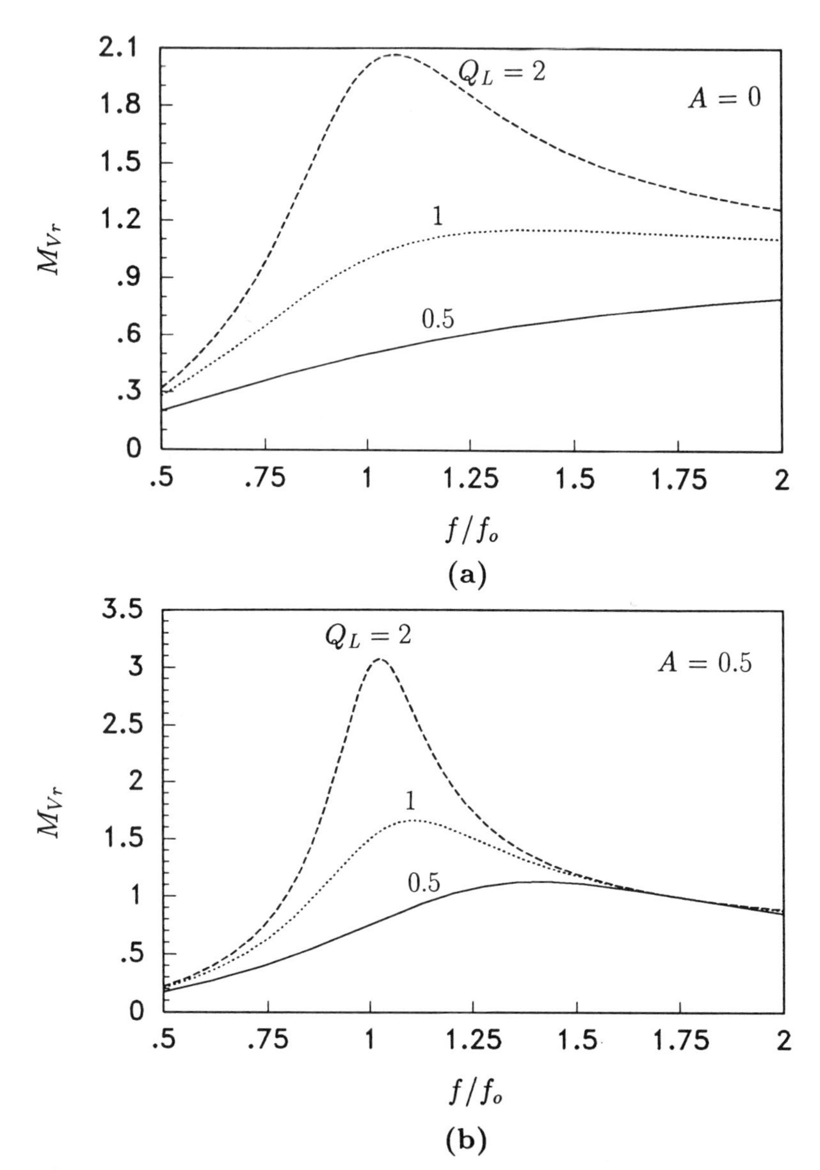

Figure 9.6: Voltage transfer function of the resonant circuit M_{Vr} as a function of f/f_o at constant values of Q_L. (a) $A = 0$. (b) $A = 0.5$. (c) $A = 1$. (d) $A = 2$.

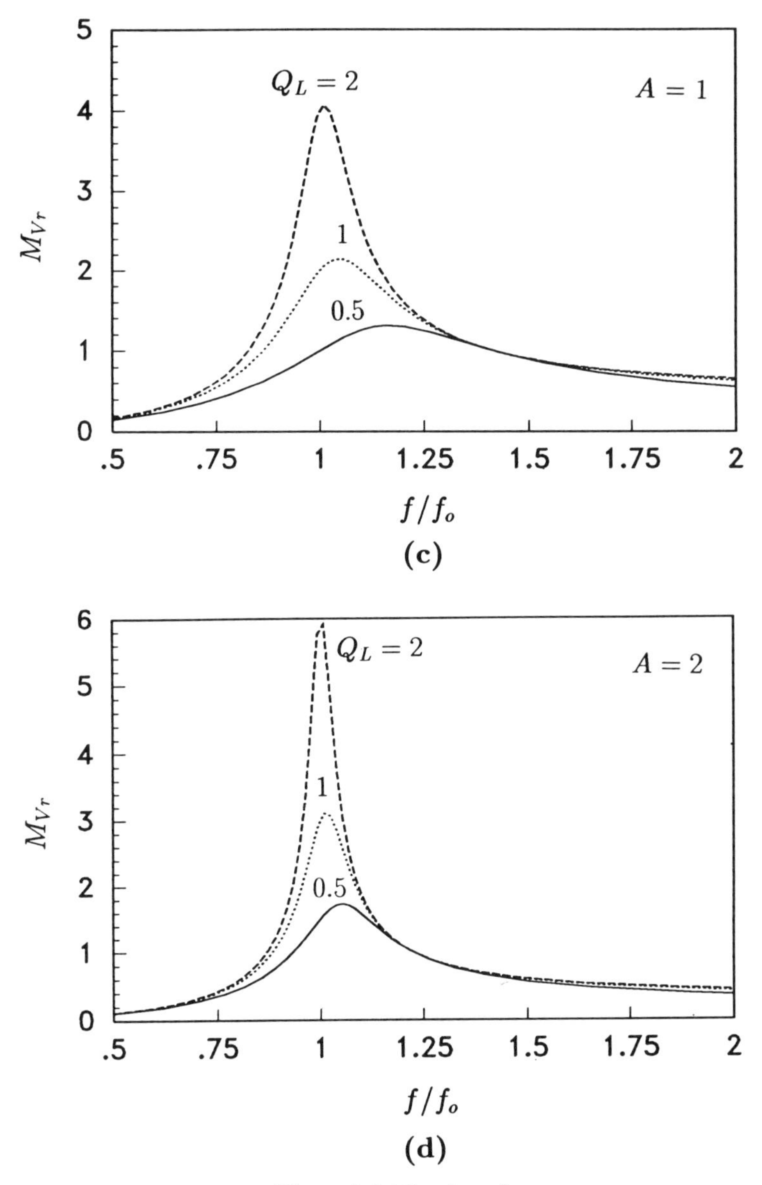

Figure 9.6 (*Continued*)

9.3.4 Energy Parameters

The current through resonant capacitor C is approximately sinusoidal, given by

$$i = I_m sin(\omega t - \psi). \tag{9.28}$$

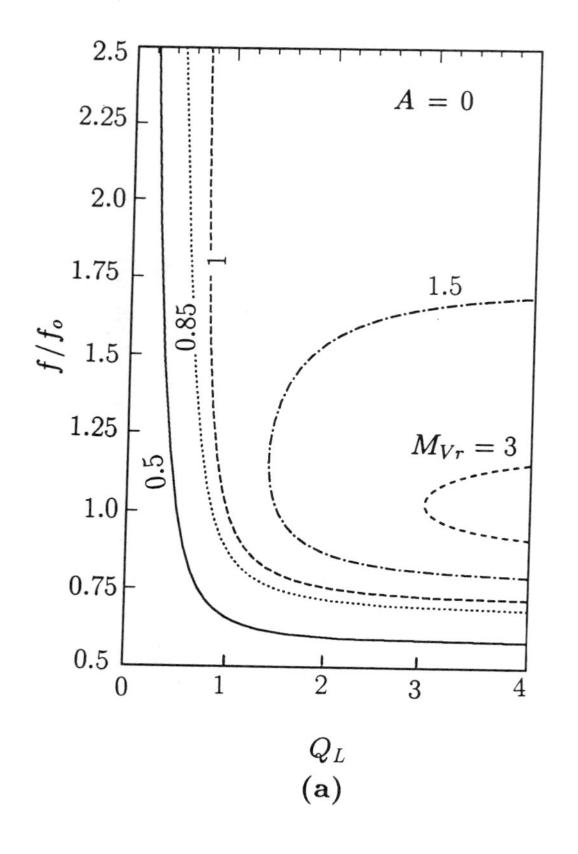
A = 0
f/fo
2.5
2.25
2.0
1.75
1.5
1.25
1.0
0.75
0.5
0
1
2
3
4
0.5
0.85
1
1.5
M_Vr = 3
Q_L
(a)

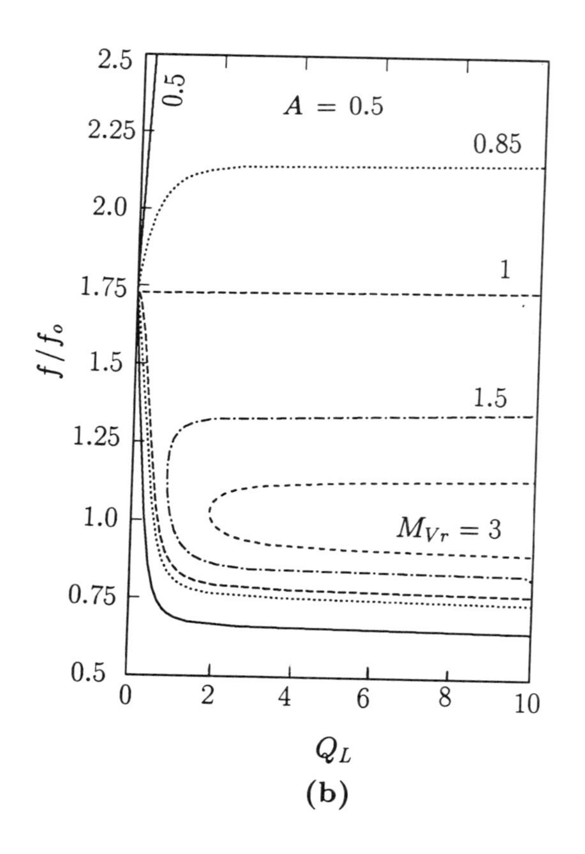
A = 0.5
f/fo
2.5
2.25
2.0
1.75
1.5
1.25
1.0
0.75
0.5
0
2
4
6
8
10
0.5
0.85
1
1.5
M_Vr = 3
Q_L
(b)

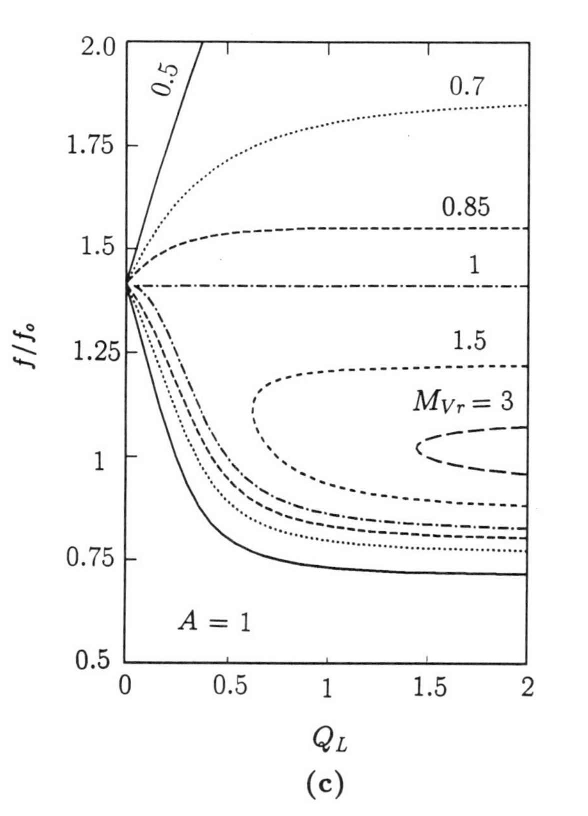

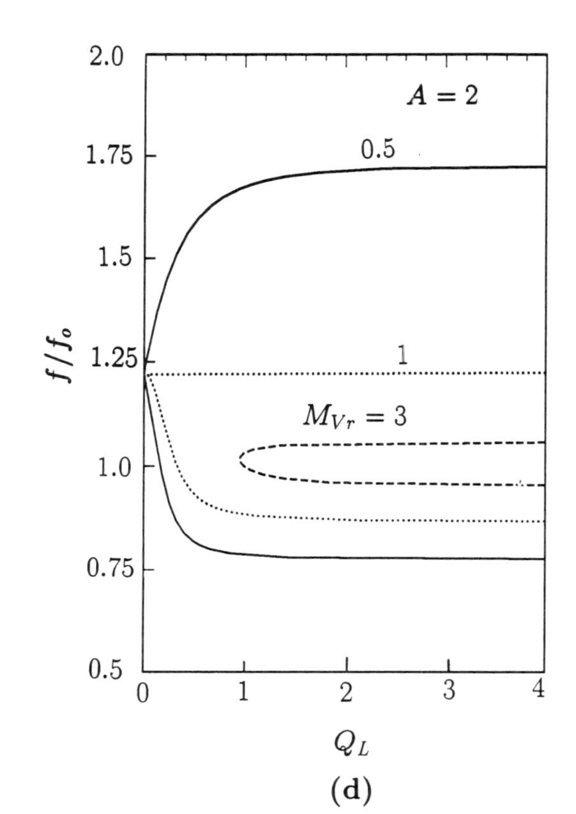

Figure 9.7: Plots of f/f_o against Q_L at fixed values of M_{Vr}. (a) $A = 0$. (b) $A = 0.5$. (c) $A = 1$. (d) $A = 2$.

From (9.10), (9.20), and (9.23), one obtains the amplitude of the capacitor current I_m, which is equal to the peak value of the switch current I_{SM}

$$I_m = I_{SM} = \frac{V_m}{Z} = \frac{2V_I}{\pi Z}$$
$$= \frac{2V_I}{\pi Z_o Q_L}\sqrt{\frac{1 + [Q_L(\frac{\omega_o}{\omega})(1+A)]^2}{(1+A)^2[1-(\frac{\omega_o}{\omega})^2]^2 + \frac{1}{Q_L^2}(\frac{\omega}{\omega_o}\frac{A}{A+1} - \frac{\omega_o}{\omega})^2}}.$$
$$= \frac{2V_I M_{Vr}}{\pi Z_o Q_L}\sqrt{1 + \left[Q_L\left(\frac{\omega_o}{\omega}\right)(1+A)\right]^2}. \tag{9.29}$$

Figure 9.8 shows $I_m Z_o / V_I$ versus f/f_o and Q_L at $A = 0$, 0.5, 1, and 2. The normalized amplitude of the switch current $I_m Z_o/V_I$ as a function of f/f_o at selected values of Q_L and A is depicted in Fig. 9.9. It can be seen that high values of I_m occur at the resonant frequency f_r. Therefore, if full load occurs at a low value of Q_L, I_m decreases with increasing Q_L, reducing conduction loss $P_r = rI_m^2/2$ and maintaining high efficiency at partial load. If, however, full load occurs at a high value of Q_L, I_m is almost independent of load, reducing part-load efficiency. Note that I_m increases with A.

Using (9.26), the amplitude of the output current of the inverter can be expressed as

$$I_{om} = \frac{\sqrt{2}V_{Ri}}{R_i} = \frac{2V_I}{\pi Z_o Q_L\sqrt{(1+A)^2[1-(\frac{\omega_o}{\omega})^2]^2 + \frac{1}{Q_L^2}(\frac{\omega}{\omega_o}\frac{A}{A+1} - \frac{\omega_o}{\omega})^2}}. \tag{9.30}$$

Figure 9.10 shows the normalized amplitude of the output current $I_{om}Z_o/V_I$ as a function of f/f_o at fixed values of Q_L for $A = 0.5$.

From (9.26), the output power of the inverter is obtained as

$$P_{Ri} = \frac{V_{Ri}^2}{R_i} = \frac{M_{VI}^2 V_I^2}{R_i}$$
$$= \frac{2V_I^2}{\pi^2 Z_o Q_L\{(1+A)^2[1-(\frac{\omega_o}{\omega})^2]^2 + \frac{1}{Q_L^2}(\frac{\omega}{\omega_o}\frac{A}{A+1} - \frac{\omega_o}{\omega})^2\}}. \tag{9.31}$$

The normalized output power $P_{Ri}Z_o/V_I^2$ is illustrated in three- and two-dimensional space at $A = 0$, 0.5, 1, and 2 in Figs. 9.11 and 9.12.

From (9.29), the conduction loss is

$$P_r = \frac{rI_m^2}{2} = \frac{2rV_I^2 M_{Vr}^2\{1 + [Q_L(\frac{\omega_o}{\omega})(1+A)]^2\}}{\pi^2 Z_o^2 Q_L^2}, \tag{9.32}$$

where

$$r = r_{DS} + r_{Cr} + r_{L1} + \frac{r_{L2}}{1 + (\frac{\omega L_2}{R_i})^2}$$
$$= r_{DS} + r_{Cr} + r_{L1} + \frac{r_{L2}}{1 + (\frac{\omega}{\omega_o})^2(\frac{Z_o}{R_i})^2\frac{1}{(1+A)^2}}. \tag{9.33}$$

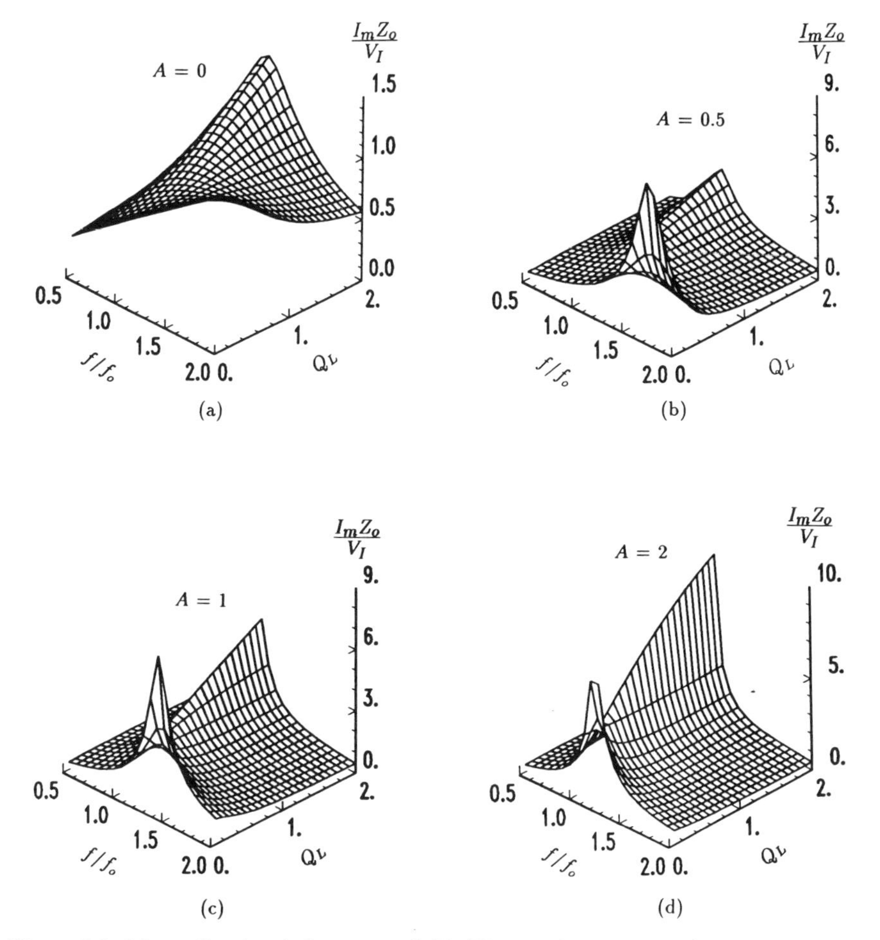

Figure 9.8: Normalized switch current $I_m Z_o / V_I$ as a function of f/f_o and Q_L. (a) $A = 0$. (b) $A = 0.5$. (c) $A = 1$. (d) $A = 2$.

Neglecting switching losses, $P_I = P_{Ri} + P_r$. Hence, from (9.23), (9.29), and (9.31), one obtains the efficiency of the inverter

$$\eta_I = \frac{P_{Ri}}{P_{Ri} + P_r} = \frac{1}{1 + \frac{r}{R_i}\{1 + [(\frac{R_i}{Z_o})(\frac{\omega_o}{\omega})(1 + A)]^2\}}. \tag{9.34}$$

Assuming that r is constant, the maximum efficiency occurs at

$$Q_L = \frac{\frac{\omega}{\omega_o}}{1 + A}. \tag{9.35}$$

Figure 9.13 shows efficiency η_I versus f/f_o and Q_L for $r_{DS} = 0.5\ \Omega$, $r_{Cr} = 0.08\ \Omega$, $r_{L1} = r_{L2} = 0.8\ \Omega$, and $Z_o = 212\ \Omega$. Efficiency η_I is illustrated as

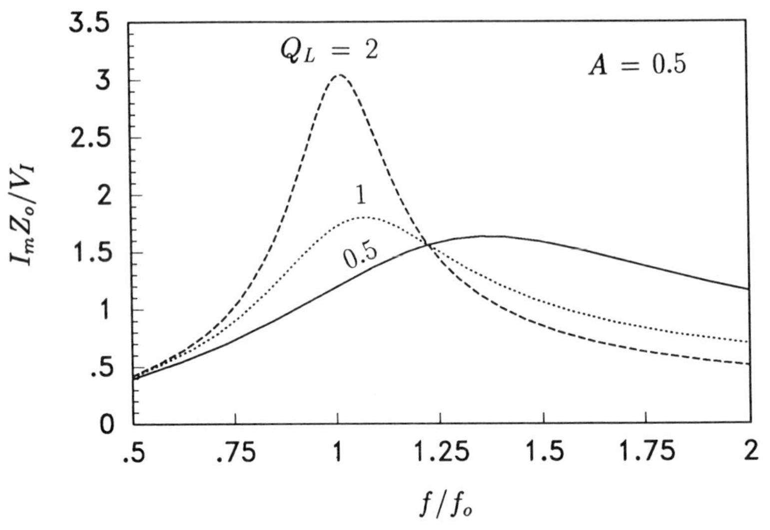

Figure 9.9: Normalized switch current $I_m Z_o / V_I$ as a function of f/f_o at fixed values of Q_L for $A = 0.5$.

a function of f/f_o at selected values of Q_L for $r_{DS} = 0.5\ \Omega$, $r_{Cr} = 0.08\ \Omega$, $r_{L1} = r_{L2} = 0.8\ \Omega$, and $Z_o = 212\ \Omega$ in Fig. 9.14.

The turn-on switching loss is zero for inductive loads, that is, for $f/f_r > 1$. The turn-off power loss can be reduced by adding a capacitor in parallel with one of the transistors, as shown in Chapter 10.

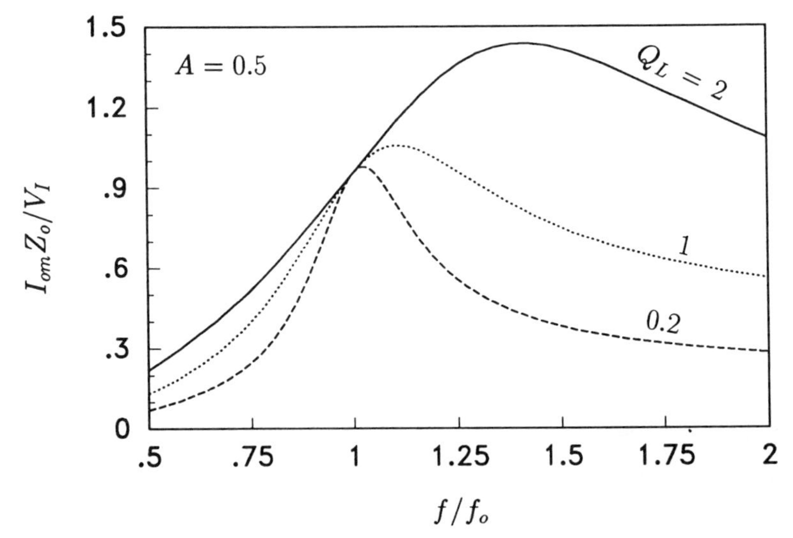

Figure 9.10: Normalized amplitude of the output current $I_{om} Z_o / V_I$ as a function of f/f_o at various values of Q_L for $A = 0.5$.

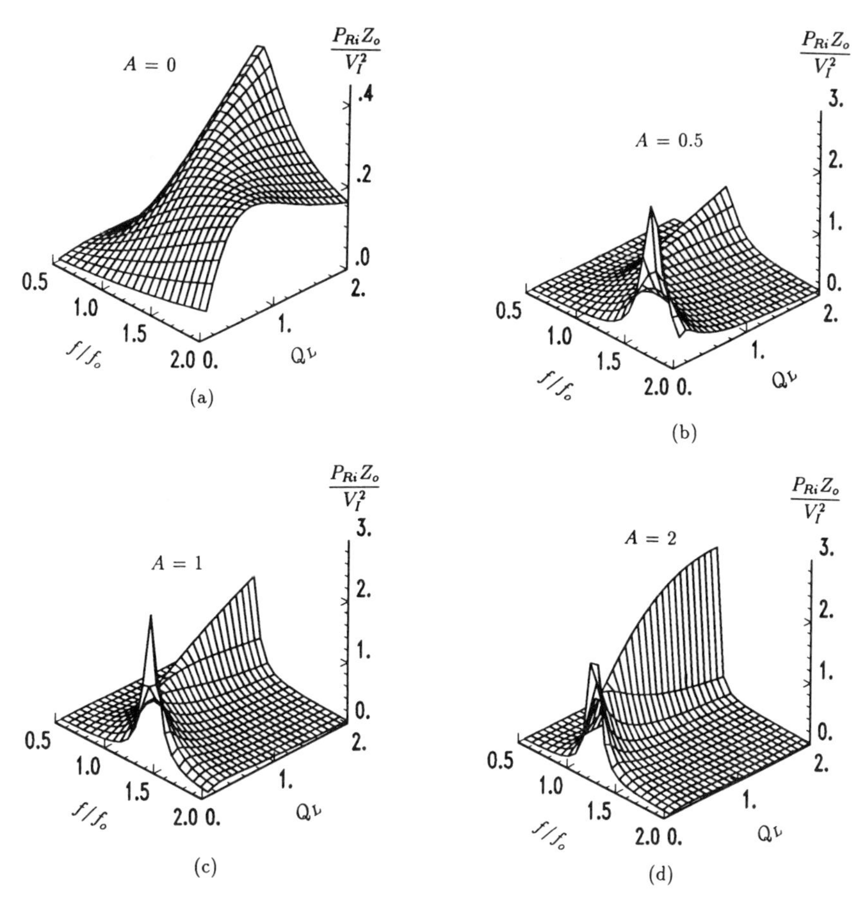

Figure 9.11: Normalized output power $P_{Ri}Z_o/V_I^2$ as a function of f/f_o and Q_L. (a) $A = 0$. (b) $A = 0.5$. (c) $A = 1$. (d) $A = 2$.

9.3.5 Short-Circuit and Open-Circuit Operation

The inverter is not safe under short-circuit and open-circuit conditions. At $R_i = 0$, the inductor L_2 is shorted-circuited and the resonant circuit consists of L_1 and C. If the switching frequency f is equal to the resonant frequency of the C-L_1 circuit $f_{rs} = 1/(2\pi\sqrt{L_1 C})$, the magnitude of the current through the switches and the C-L_1 resonant circuit is $I_m \approx 2V_I/(\pi r)$. This current may become excessive and may destroy the circuit. If f is far from f_{rs}, I_m is limited by the reactance of the resonant circuit. At $R_i = \infty$, the resonant circuit is comprised of C and the series combination of L_1 and L_2. Consequently, its resonant frequency is equal to f_o, $I_m \approx 2V_I/(\pi r)$, and the inverter is not safe at or close to this frequency. The peak values of the voltages across the reactive components can be calculated by considering the full load flowing through the resonant circuit when the load resistance is mini-

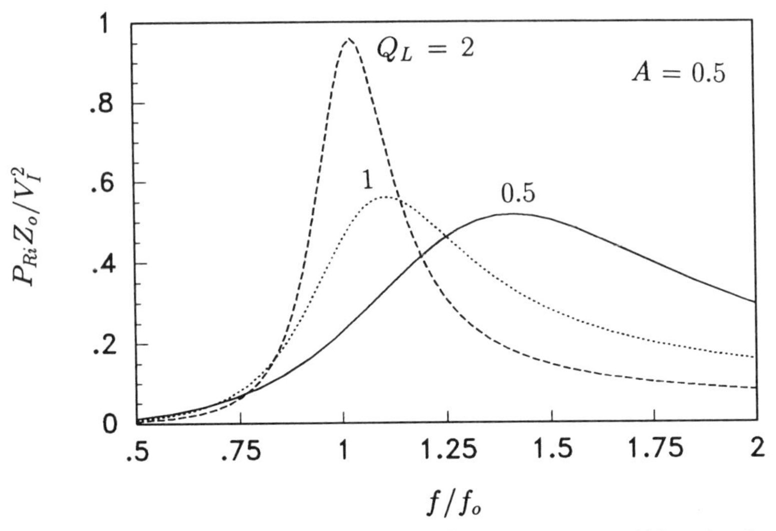

Figure 9.12: Normalized output power $P_{Ri}Z_o/V_I^2$ as a function of f/f_o at fixed values of Q_L for $A = 0.5$.

mum. The reactance of the inductors can be calculated as follows: $X_L = \omega L$, $L_1 = L/(1 + 1/A)$, and $L_2 = L/(1 + A)$. For $A = 1$, the inductor values L_1 and L_2 become equal and the reactances of the inductors are equal, that is, $X_{L1} = X_{L2} = \omega L_1 = \omega L_2$. The reactance of the capacitor is $X_C = 1/(\omega C)$. From (9.26) and (9.29), the peak values of the voltage across the various reactive components can be found as

$$V_{L1m} = X_{L1}I_m = (\omega L_1)\left(\frac{V_m}{Z}\right) = \left(\frac{\omega}{\omega_o}\right)(\omega_o L_1)\frac{2V_I}{\pi Z}$$
$$= \left(\frac{\omega}{\omega_o}\right)\left(\frac{A}{1+A}\right)\left(\frac{2V_I M_{Vr}}{\pi Q_L}\right)\sqrt{1 + [Q_L(\frac{\omega_o}{\omega})(1+A)]^2} \quad (9.36)$$

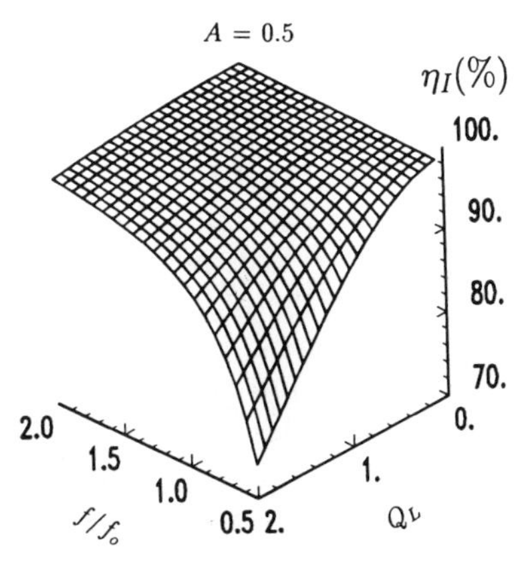

Figure 9.13: Inverter efficiency η_I versus f/f_o and Q_L for $r_{DS} = 0.5\ \Omega$, $r_{Cr} = 0.08\ \Omega$, $r_{L1} = r_{L2} = 0.8\ \Omega$, and $Z_o = 212\ \Omega$ for $A = 0.5$.

$$
\begin{aligned}
V_{L2m} &= \sqrt{2} M_{Vs} M_{Vr} V_I \\
&= \frac{2V_I}{\pi\sqrt{(1+A)^2[1-(\frac{\omega_o}{\omega})^2]^2 + [\frac{1}{Q_L}(\frac{\omega}{\omega_o}\frac{A}{A+1} - \frac{\omega_o}{\omega})]^2}}
\end{aligned}
\tag{9.37}
$$

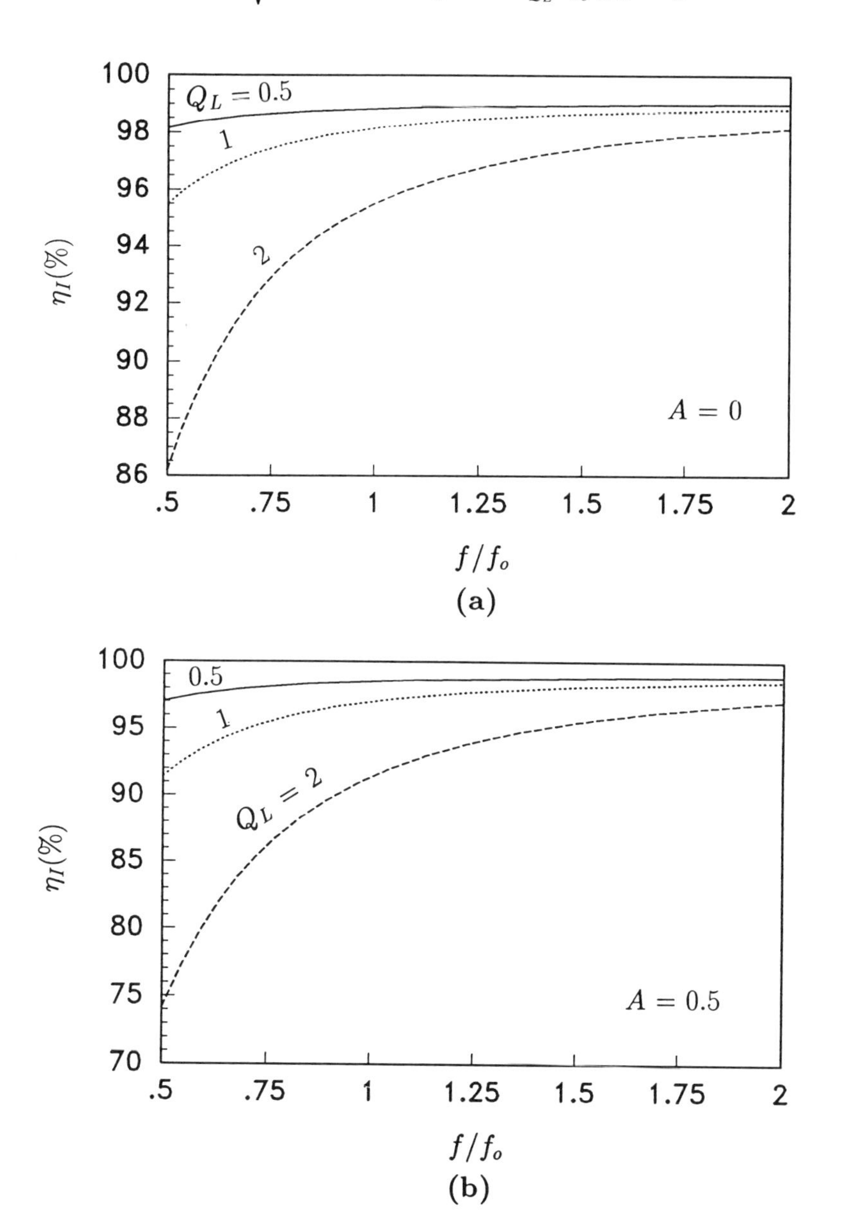

Figure 9.14: Inverter efficiency η_I versus f/f_o at fixed values of Q_L for $r_{DS} = 0.5\ \Omega$, $r_{Cr} = 0.08\ \Omega$, $r_{L1} = r_{L2} = 0.8\ \Omega$, and $Z_o = 212\ \Omega$. (a) $A = 0$. (b) $A = 0.5$. (c) $A = 1$. (d) $A = 2$.

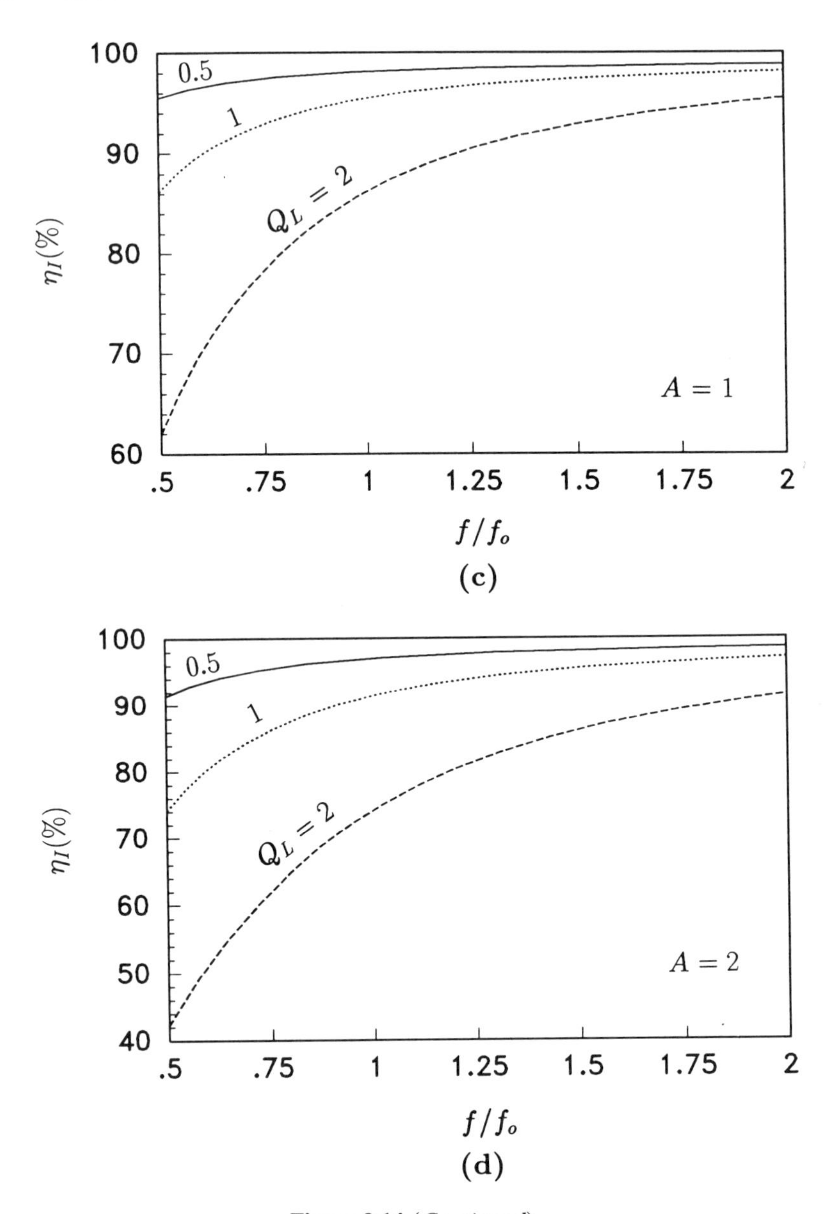

Figure 9.14 (*Continued*)

$$V_{Cm} = X_C I_m = \left(\frac{1}{\omega C}\right)\left(\frac{V_m}{Z}\right) = \left(\frac{\omega_o}{\omega}\right)\left(\frac{1}{\omega_o C}\right)\frac{2V_I}{\pi Z}$$

$$= \left(\frac{\omega_o}{\omega}\right)\left(\frac{2V_I M_{Vr}}{\pi Q_L}\right)\sqrt{1 + [Q_L(\frac{\omega_o}{\omega})(1+A)]^2}. \tag{9.38}$$

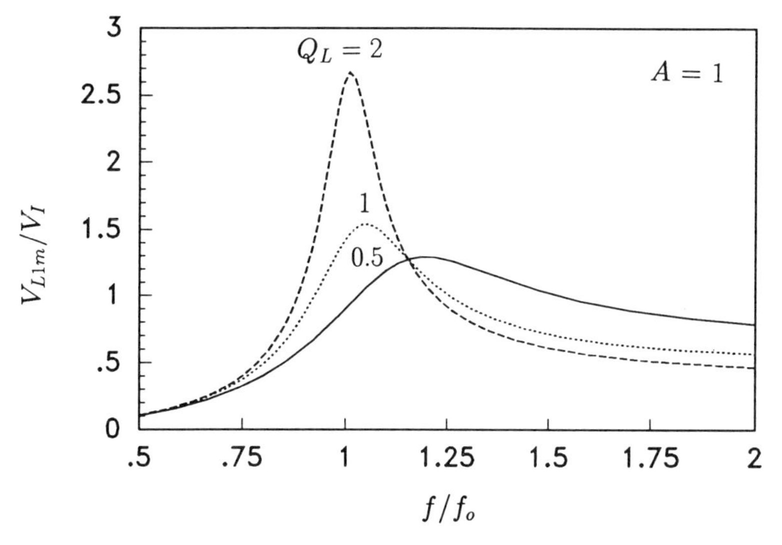

Figure 9.15: Normalized amplitude V_{L1m}/V_I of the voltage across the resonance inductor L_1 as a function of f/f_o at fixed values of Q_L for $A = 0.5$.

The normalized amplitude V_{L1m}/V_I of the voltage across the resonance inductor L_1 is shown in Fig. 9.15 as a function of f/f_o at selected values of Q_L and $A = 0.5$. The amplitude of the voltage V_{L2m} across the inductor L_2 is proportional to the voltage transfer function of the resonant circuit M_{Vr} illustrated in Figs. 9.5 and 9.6. Plots of V_{Cm}/V_I against f/f_o at constant values of Q_L and $A = 0.5$ are displayed in Fig. 9.16.

9.4 DESIGN EXAMPLE

Example 9.1

Design a CLL inverter shown in Fig. 9.1 to meet the following specifications: $V_I = 250$ V, $R_{imin} = 150\ \Omega$, and $P_{Rimax} = 84$ W. Assume $f_o = 100$ kHz and the total efficiency of the inverter $\eta_I = 90\%$.

Solution: The maximum dc input power is

$$P_{Imax} = \frac{P_{Rimax}}{\eta_I} = \frac{84}{0.9} = 93.3 \text{ W} \tag{9.39}$$

and the maximum value of the dc input current

$$I_{Imax} = \frac{P_{Imax}}{V_I} = \frac{93.3}{250} = 0.373 \text{ A}. \tag{9.40}$$

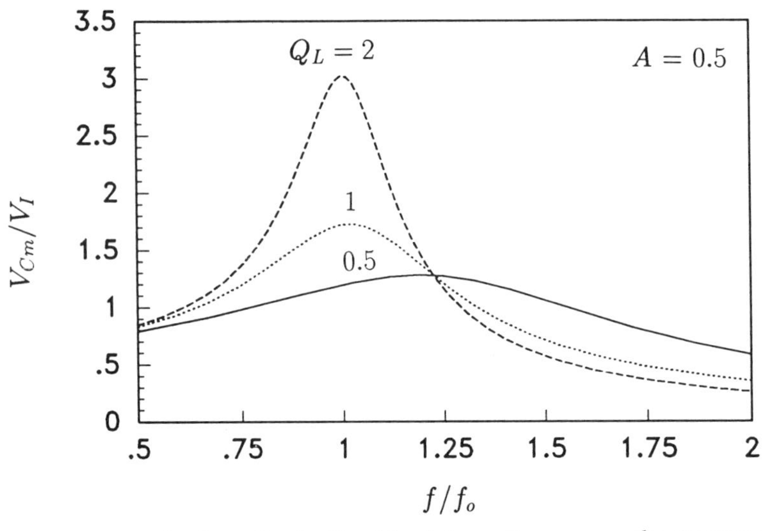

Figure 9.16: Normalized amplitude V_{Cm}/V_I of the voltage across the resonance capacitor C as a function of f/f_o at fixed values of Q_L for $A = 0.5$.

To obtain a required voltage transfer function and a high full-load efficiency, a system of equations consisting of (9.26) and (9.35) should be solved with respect to A and Q_L. Assuming $f/f_o = 1.414$, the results are $A = 1$ and $Q_L = 1/\sqrt{2}$. It should be noticed that the designed inverter has a very desirable feature, namely, the voltage transfer function is independent of the load (see Fig. 9.6). This is because $f = f_{rs}$, and consequently $M_{Vr} = 1$, as described in comments to (9.25).

The component values of the resonant circuit are

$$C = \frac{Q_L}{\omega_o R_{imin}} = \frac{0.707}{2 \times \pi \times 100 \times 10^3 \times 150} = 7.5 \text{ nF} \tag{9.41}$$

$$L = \frac{R_{imin}}{\omega_o Q_L} = \frac{150}{2 \times \pi \times 100 \times 10^3 \times 0.707} = 338 \ \mu\text{H} \tag{9.42}$$

$$L_1 = \frac{L}{1 + \frac{1}{A}} = 169 \ \mu\text{H} \tag{9.43}$$

$$L_2 = \frac{L}{1 + A} = 169 \ \mu\text{H}. \tag{9.44}$$

The characteristic impedance of the resonant circuit is $Z_o = R_{imin}/Q_L = 150\sqrt{2} = 212 \ \Omega$.

From (9.29), the peak value of the current through the resonant circuit and the switches is

$$I_m = I_{SM} = \frac{2V_I M_{Vr}}{\pi Z_o Q_L}\sqrt{1 + [Q_L(\frac{\omega_o}{\omega})(1+A)]^2}$$
$$= \frac{2 \times 250 \times 1}{\pi \times 212 \times 0.707}\sqrt{1 + [0.707 \times 0.707(1+1)]^2} = 1.5 \text{ A}. \quad (9.45)$$

The peak values of the voltages across the reactive components can be calculated using (9.36) through (9.38):

$$V_{L1m} = \omega L_1 I_m = 1.414 \times 2 \times \pi \times 100 \times 10^3 \times 169 \times 10^{-6} \times 1.5 = 225.3 \text{ V}, \quad (9.46)$$

$$V_{L2m} = \sqrt{2} M_{Vs} M_{Vr} V_I = \sqrt{2} \times \frac{\sqrt{2}}{\pi} \times 1 \times 250 = 159.2 \text{ V}, \quad (9.47)$$

and

$$V_{Cm} = \frac{I_m}{\omega C} = \frac{1.5}{1.414 \times 2 \times \pi \times 100 \times 10^3 \times 7.5 \times 10^{-9}} = 225.1 \text{ V}. \quad (9.48)$$

From (9.8), (9.15), (9.16), and (9.18), one obtains $f_r = 111$ kHz, $q_r = 1.272$, $R_s = 57.3\ \Omega$, and $Q_r = \omega_r L/R_s = 4.12$. The high value of Q_r results in the sinusoidal waveforms of the capacitor current i and the output voltage of the inverter v_{Ri}.

Let us select MTP5N40 power MOSFETs whose $r_{DS} = 1\ \Omega$ and $Q_g = 27$ nC. The conduction loss per transistor is

$$P_{rDS} = \frac{I_m^2 r_{DS}}{4} = \frac{1.5^2 \times 1}{4} = 0.5625 \text{ W}. \quad (9.49)$$

Assume that both resonant inductors L_1 and L_2 have the unloaded quality factor $Q_{Lo} = 300$ at $f = 141$ kHz. The ESRs of these inductors at $f = 141$ kHz are

$$r_{L1} = r_{L2} = \frac{\omega L_1}{Q_{Lo}} = \frac{2 \times \pi \times 141 \times 10^3 \times 169 \times 10^{-6}}{300} = 0.5\ \Omega. \quad (9.50)$$

The amplitude of the current through L_2 is

$$I_{L2m} = \frac{I_m R_i}{R_i + \omega L_2} = \frac{1.5 \times 150}{150 + 150} = 0.75 \text{ A}. \quad (9.51)$$

Hence, one obtains the conduction loss in r_{L1}

$$P_{rL1} = \frac{I_m^2 r_{L1}}{2} = \frac{1.5^2 \times 0.5}{2} = 0.5625 \text{ W}, \quad (9.52)$$

and in r_{L2}

$$P_{rL2} = \frac{I_{L2m}^2 r_{L2}}{2} = \frac{0.75^2 \times 0.5}{2} = 0.141 \text{ W}. \quad (9.53)$$

Assume that the ESR of the resonant capacitor is $r_{Cr} = 50$ mΩ. The conduction loss in r_C is

$$P_{rCr} = \frac{I_m^2 r_C}{2} = \frac{1.5^2 \times 0.05}{2} = 0.056 \text{ W}. \tag{9.54}$$

The total conduction loss is

$$\begin{aligned} P_r &= 2P_{rDS} + P_{rL1} + P_{rL2} + P_{rCr} \\ &= 2 \times 0.5625 + 0.5625 + 0.141 + 0.056 = 1.884 \text{ W}. \end{aligned} \tag{9.55}$$

Thus, the inverter efficiency associated with the conduction loss at full power is

$$\eta_I = \frac{P_{Ri}}{P_{Ri} + P_r} = \frac{84}{84 + 1.884} = 97.81\%. \tag{9.56}$$

Assuming the peak-to-peak gate-source voltage $V_{GSpp} = 15$ V, the gate-drive power per MOSFETs is

$$P_G = fQ_g V_{GSpp} = 141 \times 10^3 \times 27 \times 10^{-9} \times 15 = 0.057 \text{ W}. \tag{9.57}$$

The sum of the conduction loss and gate-drive power loss is

$$P_{LS} = 2P_G + P_r = 2 \times 0.057 + 1.884 = 1.998 \text{ W}. \tag{9.58}$$

The turn-on switching loss is zero because the input impedance of the resonant circuit is inductive. The inverter efficiency is

$$\eta_I = \frac{P_{Ri}}{P_{Ri} + P_{LS}} = \frac{84}{84 + 1.998} = 97.68\%. \tag{9.59}$$

9.5 FULL-BRIDGE CLL RESONANT INVERTER

The full-bridge CLL resonant inverter is depicted in Fig. 9.17. In the full-bridge configuration, the amplitude of the fundamental component of the voltage at the input of the resonant circuit is two times higher than that

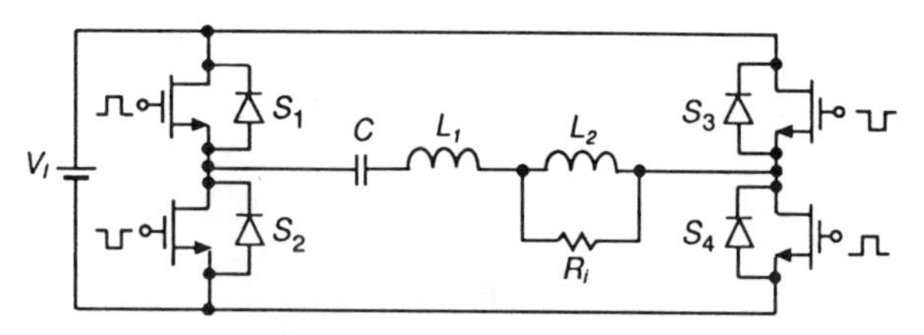

Figure 9.17: Full-bridge CLL resonant inverter.

for the half-bridge configuration at the same dc input voltage. Thus, all the parameters of the inverter that are directly proportional to this amplitude are doubled comparing the half-bridge inverter. This section focuses on presenting the expressions for the parameters of the full-bridge inverter. The operation of the inverter is similar to that of the half-bridge inverter and, to avoid unnecessary repetitions, is not given here.

9.5.1 Voltage Transfer Function

Referring to Fig. 9.17, the input voltage of the resonant circuit v_{DS2} is a square wave of magnitude $2V_I$ given by

$$v_{DS2} = \begin{cases} V_I, & \text{for } 0 < \omega t \leq \pi \\ -V_I, & \text{for } \pi < \omega t \leq 2\pi. \end{cases} \tag{9.60}$$

Its fundamental component is $v_{i1} = V_m sin\omega t$, where

$$V_m = \frac{4}{\pi} V_I = 1.273 V_I. \tag{9.61}$$

Thus, the rms value of v_{i1} is $V_{rms} = V_m/\sqrt{2} = 2\sqrt{2}V_I/\pi = 0.9V_I$. The voltage transfer function from V_I to the fundamental component at the input of the resonant circuit is

$$M_{Vs} \equiv \frac{V_{rms}}{V_I} = \frac{2\sqrt{2}}{\pi} = 0.9. \tag{9.62}$$

Using (9.62) and (9.23), one obtains the magnitude of the dc-to-ac voltage transfer function of the Class D lossless inverter

$$\begin{aligned} M_{VI} &\equiv \frac{V_{Ri}}{V_I} = M_{Vs} M_{Vr} \\ &= \frac{2\sqrt{2}}{\pi\sqrt{(1+A)^2[1-(\frac{\omega_o}{\omega})^2]^2 + [\frac{1}{Q_L}(\frac{\omega}{\omega_o}\frac{A}{A+1} - \frac{\omega_o}{\omega})]^2}}. \end{aligned} \tag{9.63}$$

The dc-to-ac voltage transfer function of the lossy inverter is given by (9.27).

9.5.2 Currents and Voltages

The current through the capacitor C is $i = I_m sin(\omega t - \psi)$, where (9.10) gives

$$I_m = \frac{V_m}{Z} = \frac{4V_I}{\pi Z} = \frac{4V_I}{\pi Z_o Q_L}\sqrt{\frac{1 + [Q_L(\frac{\omega_o}{\omega})(1+A)]^2}{(1+A)^2[1-(\frac{\omega_o}{\omega})^2]^2 + \frac{1}{Q_L^2}(\frac{\omega}{\omega_o}\frac{A}{A+1} - \frac{\omega_o}{\omega})^2}}. \tag{9.64}$$

Using (9.23), the amplitude of i at a fixed value of M_{Vr} becomes

$$I_m = \frac{4V_I M_{Vr}}{\pi Z_o Q_L}\sqrt{1+\left[Q_L(\frac{\omega_o}{\omega})(1+A)\right]^2}. \tag{9.65}$$

From (9.63), the amplitude of the output current of the full-bridge inverter is

$$I_{om} = \frac{\sqrt{2}V_{Ri}}{R_i} = \frac{4V_I}{\pi Z_o Q_L\sqrt{(1+A)^2[1-(\frac{\omega_o}{\omega})^2]^2+[\frac{1}{Q_L}(\frac{\omega}{\omega_o}\frac{A}{A+1}-\frac{\omega_o}{\omega})]^2}}. \tag{9.66}$$

The magnitude of the voltage across the resonant capacitor is

$$V_{Cm} = X_C I_m = \left(\frac{1}{\omega C}\right)\left(\frac{V_m}{Z}\right) = \left(\frac{\omega_o}{\omega}\right)\left(\frac{1}{\omega_o C}\right)\frac{4V_I}{\pi Z}$$
$$= \left(\frac{\omega_o}{\omega}\right)\left(\frac{4V_I M_{Vr}}{\pi Q_L}\right)\sqrt{1+\left[Q_L(\frac{\omega_o}{\omega})(1+A)\right]^2}. \tag{9.67}$$

The magnitude of the voltage across the resonant inductor L_1 is

$$V_{L1m} = X_{L1} I_m = (\omega L_1)\left(\frac{V_m}{Z}\right) = \left(\frac{\omega}{\omega_o}\right)(\omega_o L_1)\frac{4V_I}{\pi Z}$$
$$= \left(\frac{\omega}{\omega_o}\right)\left(\frac{A}{1+A}\right)\left(\frac{4V_I M_{Vr}}{\pi Q_L}\right)\sqrt{1+\left[Q_L(\frac{\omega_o}{\omega})(1+A)\right]^2}. \tag{9.68}$$

The voltage across the resonant inductor L_2 is equal to the output voltage of the inverter. The magnitude of the voltage across inductor L_2 is

$$V_{L2m} = \sqrt{2}V_{Ri} = \sqrt{2}M_{VI}V_I$$
$$= \frac{4V_I}{\pi\sqrt{(1+A)^2[1-(\frac{\omega_o}{\omega})^2]^2+[\frac{1}{Q_L}(\frac{\omega}{\omega_o}\frac{A}{A+1}-\frac{\omega_o}{\omega})]^2}}. \tag{9.69}$$

9.5.3 Powers and Efficiency

Using (9.63), the output power can be found as

$$P_{Ri} = \frac{V_{Ri}^2}{R_i} = \frac{8V_I^2}{\pi^2 Z_o Q_L\{(1+A)^2[1-(\frac{\omega_o}{\omega})^2]^2+\frac{1}{Q_L^2}(\frac{\omega}{\omega_o}\frac{A}{A+1}-\frac{\omega_o}{\omega})^2\}}. \tag{9.70}$$

From (9.64), the conduction loss is

$$P_r = \frac{rI_m^2}{2} = \frac{8rV_I^2 M_{Vr}^2\{1+[Q_L(\frac{\omega_o}{\omega})(1+A)]^2\}}{\pi^2 Z_o^2 Q_L^2}, \tag{9.71}$$

where the total parasitic resistance of the inverter is

$$r = 2r_{DS} + r_{Cr} + r_{L1} + \frac{r_{L2}}{1 + (\frac{\omega L_2}{R_i})^2} \tag{9.72}$$

Neglecting switching losses, $P_I = P_{Ri} + P_r$. Hence, one obtains the efficiency of the inverter

$$\eta_I = \frac{P_{Ri}}{P_I} = \frac{P_{Ri}}{P_{Ri} + P_r} = \frac{1}{1 + \frac{r}{R_i}\{1 + [(\frac{R_i}{Z_o})(\frac{\omega_o}{\omega})(1 + A)]^2\}}. \tag{9.73}$$

9.6 SUMMARY

- In the transformer version of the CLL inverter, the inductance L_2 can be made of the magnetizing inductance of the transformer. A gapped core is usually needed to obtain a low value of the magnetizing inductance.
- The transformer leakage inductance is absorbed into inductance L_1.
- The dc voltage transfer function of the CLL inverter is *independent* of the load resistance for the normalized switching frequency $f/f_o = f_{sr}/f_o = \sqrt{1 + L_2/L_1}$. This occurs at inductive loads of the switches, which is a very desirable feature if power MOSFETs are used as switches.
- The maximum efficiency of the CLL inverter occurs at $Q_L = (f/f_o)/(1 + A) = (f/f_o)/(1 + L_1/L_2)$.
- The efficiency of the CLL inverter decreases with increasing $A = L_1/L_2$ at fixed values of Q_L and f/f_o.
- The efficiency of the CLL inverter decreases with increasing Q_L at fixed values of A and f/f_o for light loads.
- If $f/f_o = f_{sr}/f_o$ at which the voltage transfer function is independent of the load, the maximum efficiency occurs at $Q_{Lopt} = 1/[A(1 + A)] = 1/[(L_1/L_2)(1 + L_1/L_2)]$.
- The CLL inverter cannot operate safely with a short circuit at frequencies close to the resonant frequency f_r because of the excessive peak value of the current through the resonant capacitor and switches.
- The inverter cannot operate safely with an open circuit at frequencies close to the corner frequency f_o.

9.7 REFERENCES

1. M. K. Kazimierczuk and N. Thirunarayan, "Class D voltage-switching inverter with tapped resonant inductor," *Proc. IEE, Pt. B, Electric Power Applications*, vol. 140, pp. 177–185, May 1993.
2. D. Czarkowski and M. K. Kazimierczuk, "Phase-controlled CLL resonant converter," *IEEE Applied Power Electronics Conf.*, San Diego, CA, March 7–11, 1993, pp. 432–438.

9.8 REVIEW QUESTIONS

9.1 Does the boundary between the capacitive and inductive loads depend on the load in the CLL inverter?

9.2 Is the voltage transfer function of the CLL inverter always dependent on the load?

9.3 What are the conditions for the voltage transfer function of the CLL inverter to be independent of the load? Is the load of the switches inductive or capacitive in this case?

9.4 How many magnetic components are required to build a transformer CLL inverter?

9.5 Is the transformer leakage inductance included in the topology of the CLL inverter?

9.6 Is the transformer magnetizing inductance included in the topology of the CLL inverter?

9.7 Is the part-load efficiency of the CLL inverter high?

9.8 Can the CLL inverter operate safely under short-circuit conditions at the output?

9.9 Can the CLL inverter operate safely under open-circuit conditions at the output?

9.9 PROBLEMS

9.1 Derive step by step the input impedance of the resonant circuit of Fig. 9.1. Compare your result to (9.9).

9.2 Show that the argument of the input impedance of the resonant circuit of Fig. 9.1 is given by (9.11).

9.3 The resonant circuit of the inverter of Fig. 9.1 has the following parameters: $L_1 = 300\ \mu$H, $L_2 = 200\ \mu$H, $C = 2$ nF, and $R_i = 600\ \Omega$. The circuit is driven by a sinusoidal voltage source $v = 100 sin(1.41 \times 10^6)t$. What is the amplitude of the voltage across the ac load R_i in this circuit?

9.4 For the circuit of Problem 9.3, find the voltage stress across the resonant capacitor.

9.5 Design a full-bridge CLL resonant inverter shown in Fig. 9.17 to meet the following specifications: $V_I = 250$ V and $P_{Rimax} = 85$ W. Assume the corner frequency $f_o = 100$ kHz, the normalized switching frequency $\omega/\omega_o = 1.5$, and the total efficiency of the inverter $\eta_I = 90\%$. Make the voltage transfer function of the inverter independent of the load.

CHAPTER 10

CLASS D ZERO-VOLTAGE-SWITCHING RESONANT INVERTERS

10.1 INTRODUCTION

The major factors limiting the performance of Class D inverters are switching losses and switching noise. Switching losses increase with the square of the supply voltage and linearly with the operating frequency. Therefore, the operating frequencies of Class D inverters are limited to about 500 kHz. For operation below the resonant frequency, the transistors turn off at zero voltage, reducing the turn-off switching loss to nearly zero. However, the transistors turn on at high voltage (equal to the dc input voltage), causing a turn-on switching loss. In addition, the antiparallel diodes turn off at a high di/dt, generating high spikes in the switch current waveforms because of the diode reverse recovery. For operation above resonance, the transistors are turned on at zero voltage, eliminating the turn-on switching loss. On the other hand, during the turn-off transition, the drain current and the drain-source voltage waveforms are overlapping, causing a turn-off switching loss and reducing efficiency. Moreover, the diodes turn off at low di/dt and therefore do not generate current spikes.

To reduce switching losses, various soft-switching techniques have been reported [1]–[18]. This chapter will discuss Class D zero-voltage switching (ZVS) inverters in which a single capacitor is connected in parallel with one of the switches and the drive voltages have a dead time so that the switches are off during the transitions.

10.2 PRINCIPLE OF OPERATION

A circuit of a Class D ZVS inverter is shown in Fig. 10.1. It consists of two bidirectional switches S_1 and S_2, a series-resonant circuit L-C-R_i, and a shunt

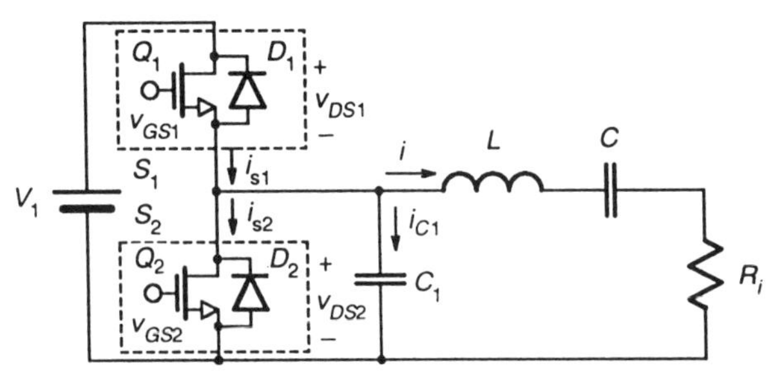

Figure 10.1: Circuit of the Class D ZVS resonant inverter with zero turn-on and turn-off switching losses.

capacitor C_1. The switches S_1 and S_2 consist of power MOSFETs Q_1 and Q_2 and antiparallel diodes D_1 and D_2. The body-to-drain *pn* junction diodes may be utilized as the antiparallel diodes in the case of inductive loads [9]. Other switching devices can also be used, such as MOS-controlled thyristors (MCTs), bipolar junction transistors (BJTs), and insulated gate bipolar transistors (IGBTs). The shunt capacitor C_1 can be connected in parallel with the switch S_1 as well as with the switch S_2, or even divided into two parts connected in parallel with the switches. Therefore, the parasitic capacitances of both switches (i.e., transistor output capacitances and transistor-heat sink capacitances) are absorbed into C_1. This is because the dc voltage source V_I represents a short circuit for the ac current. As a result, all the capacitances connected in parallel with the transistors are placed between the input of the resonant circuit and the negative terminal of V_I (usually connected to ground) for the ac component. The switches are driven alternately by the rectangular voltages v_{GS1} and v_{GS2} with a sufficiently long dead time, that is, the ON duty cycle of each drive voltage is less then 0.5. Figure 10.2 depicts steady-state current and voltage waveforms in the inverter for the operating frequency f higher than the resonant frequency $f_o = 1/(2\pi\sqrt{LC})$. The series-resonant circuit forces a sinusoidal current if the loaded quality factor $Q_L = \omega_o L/R_i$ is sufficiently high (i.e., $Q_L > 3$).

Consider the charging process of the shunt capacitor C_1. Prior to time $t = 0$, the drive voltage v_{GS1} is low and the drive voltage v_{GS2} is high. Therefore, transistor Q_2 is ON, while diode D_2, transistor Q_1, and diode D_1 are OFF. The voltage across the upper switch v_{DS1} is nearly V_I, and the voltage across the bottom switch v_{DS2} is approximately equal to zero. The current through the shunt capacitor $i_{C1} = C_1 dv_{DS2}/dt$ is zero because the capacitor voltage is constant and equal to zero.

At time $t = 0$, transistor Q_2 is turned off by the drive voltage v_{GS2}. Transistor Q_1 is still maintained OFF by the drive voltage v_{GS1} because of the dead time t_D. Both diodes are reverse biased and remain OFF. Thus, all the semiconductor devices are simultaneously off and therefore both switches are simultaneously off. The current waveform in the series-resonant circuit is given by

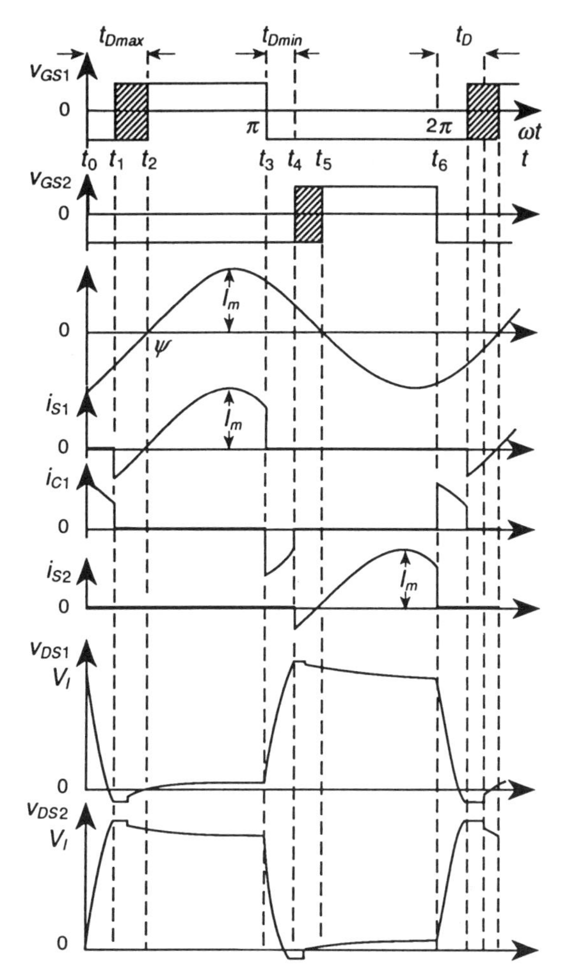

Figure 10.2: Waveforms in the Class D ZVS inverter of Fig. 10.1 for $f > f_o$.

$$i = I_m sin(\omega t - \psi), \tag{10.1}$$

where I_m and ψ are the amplitude and the initial phase, respectively. The waveform of the current i is a continuous function because this current flows through the resonant inductor L. When transistor Q_2 is turned off at time $t = 0$, the current i is diverted from transistor Q_2 to shunt capacitor C_1. Thus, the current through the shunt capacitor is

$$i_{C1} = -i = -I_m sin(\omega t - \psi), \quad \text{for} \quad 0 < t \leq t_1. \tag{10.2}$$

Because the current i_{C1} is positive, the shunt capacitor C_1 is charged by the L-C resonant circuit. Consequently, the voltage across the capacitor and switch v_{DS2} gradually increases from zero to V_I, in accordance with the equation:

$i_{C1} = C_1 dv_{DS2}/dt$. The increase in the voltage v_{DS2} causes the voltage $v_{DS1} = V_I - v_{DS2}$ to decrease from V_I to about -1 V. Assuming that the transistor output capacitance is linear and absorbed into C_1, the voltage waveform of the shunt capacitor and the bottom switch can be found as

$$v_{DS2} = \frac{1}{\omega C_1}\int_0^{\omega t} i_{C1} d(\omega t) = \frac{1}{\omega C_1}\int_0^{\omega t} [-I_m sin(\omega t - \psi)] d(\omega t)$$

$$= \frac{I_m}{\omega C_1}\left[cos(\omega t - \psi) - cos\psi\right], \quad \text{for} \quad 0 < t \leq t_1. \tag{10.3}$$

The current i_{S2} through transistor Q_2 falls to zero when the voltage v_{DS2} across transistor Q_2 increases gradually and still remains close to zero. The product of the transistor current and voltage waveforms is very low, and therefore the turn-off switching loss in the transistor Q_2 is approximately zero.

At time t_1, the voltage v_{DS1} reaches -0.7 V, the diode D_1 turns on, and the current i is diverted from the shunt capacitor C_1 to the diode D_1. During the interval from t_1 to t_2 when the switch current i_{S1} is negative, the drive voltage v_{GS1} turns the upper transistor on. Before turn-on, the diode D_1 conducts the negative switch current and the transistor voltage is kept by this diode at a low value of about -1 V. At turn-on, the change of the transistor voltage v_{DS1} is small, typically from -1 to -0.3 V. Consequently, the turn-on switching loss in the transistor Q_1 is nearly zero. After the transistor turns on, the diode begins to turn off, and its voltage changes from -1 to -0.3 V. The diode reverse-recovery current is a portion of a sine wave, and therefore the diode turns off at a very low di/dt. During the second part of the reverse-recovery time interval, the diode voltage is normally high and the diode current decreases to zero. In the inverter of Fig. 10.1, the diode voltage is kept at a low voltage on the order of 1 V by the transistor Q_2. Since the product of the diode current and voltage waveforms is low, the turn-off switching loss of the diode is negligible. The diode reverse-recovery current flows through the transistor on-resistance r_{DS1} and does not flow outside the diode-transistor combination. The shaded regions of the drive voltages v_{GS1} and v_{GS2} indicate the range during which the transistors should be turned on.

At time t_3, the drive voltage v_{GS1} turns transistor Q_1 off. Since transistor Q_2 still remains OFF, the current of the resonant circuit discharges the shunt capacitor C_1, decreasing v_{DS2} and thereby increasing v_{DS1}. The amount of charge delivered by the resonant circuit to the shunt capacitor during the charging interval is equal to that removed from the capacitor by the resonant circuit during the discharging interval. Therefore, the amount of energy $W = (1/2)C_1 V_I^2$ delivered by the resonant circuit to the shunt capacitor C_1 during the charging interval is equal to the amount of energy delivered by the capacitor to the resonant circuit during the discharging interval. Thus, both the turn-on and the turn-off switching losses are zero. The discharging process of C_1 is taking place during the time interval from t_3 to t_4. Miller's effect at both the turn-on and the turn-off is reduced to almost zero because the transistors are off during the voltage transitions. When the gate voltage changes, the drain voltage is constant, and vice versa. Notice that the turn-off transitions of the switches are forced, whereas the turn-on transitions are automatic (natural). It should be emphasized that zero-voltage switching at both transitions can be accomplished only when the operating frequency is higher than the resonant frequency. In other words, the resonant circuit must represent an inductive load to the switches so that the input current of the resonant circuit i lags behind the fundamental component of the input voltage of the resonant circuit v_{DS2}. Below the resonant frequency (i.e., for capacitive loads), zero-voltage switching cannot be achieved because the direction of the resonant current and the derivative of the voltage across the capacitor would be of the opposite sign.

Conventional Class D inverters generate a large amount of electromagnetic interference (EMI) because of switching noise (the high frequency ringing at the switching instants) and rapid changes in the switch voltages. On the other hand, soft switching in the improved inverter reduces switching noise. Also, since the switch voltage waveforms are trapezoidal in the improved Class D inverters, they contain a smaller amount of harmonics than the square waves encountered in conventional Class D inverters. Consequently, the EMI and noise level are inherently reduced in the improved inverters, generating a smaller amount of "electromagnetic pollution" and thereby reducing the environmental damage. Other Class D inverters — both half-bridge and full-bridge — can be improved in the same way.

10.3 DEAD TIME

The charging and discharging times of the shunt capacitor C_1 can be obtained by solving equation (10.3). However, this equation can only be solved numerically. Therefore, a simpler analysis is given below. Referring to Fig. 10.2, the charging process of the shunt capacitor C_1 begins at $t = 0$. If the charging time t_1 is much shorter than the cycle $T = 1/f$, the charging current can be assumed to be constant and given by

$$I = i_{C1}(0) = I_m sin\psi. \tag{10.4}$$

Thus

$$i_{C1} = C_1 \frac{dv_{DS2}}{dt} = I. \tag{10.5}$$

Solving this equation, one obtains

$$v_{DS2} = \frac{I}{C_1} t. \tag{10.6}$$

The charging process of the capacitor ends when its voltage reaches V_I, that is, $v_{DS2}(t_1) = V_I$. Hence, the charging time of the shunt capacitor is obtained as

$$t_1 = \frac{C_1 V_I}{I} = \frac{C_1 V_I}{I_m sin\psi} \leq t_D. \tag{10.7}$$

The minimum value of the dead time is then

$$t_{Dmin} = \frac{C_1 V_I}{I} = \frac{C_1 V_I}{I_m sin\psi}. \tag{10.8}$$

Thus, the minimum dead time should be increased as C_1 and V_I are increased, and I_m and ψ are decreased.

The shunt capacitor C_1 is short-circuited during most of the cycle, and its influence on the input impedance of the resonant circuit is assumed to be negligible. Let us define the resonant frequency

$$\omega_o = \frac{1}{\sqrt{LC}}, \tag{10.9}$$

the characteristic impedance

$$Z_o = \sqrt{\frac{L}{C}}, \tag{10.10}$$

and the loaded quality factor

$$Q_L = \frac{\omega_o L}{R_i} = \frac{1}{\omega_o C R_i} = \frac{Z_o}{R_i}. \tag{10.11}$$

The input impedance of the resonant circuit is

$$\begin{aligned} \mathbf{Z} &= R_i + j\left(\omega L - \frac{1}{\omega C}\right) = R_i\left[1 + jQ_L\left(\frac{\omega}{\omega_o} - \frac{\omega_o}{\omega}\right)\right] \\ &= Z_o\left[\frac{R_i}{Z_o} + j\left(\frac{\omega}{\omega_o} - \frac{\omega_o}{\omega}\right)\right] = R_i + jX = Ze^{j\psi}, \end{aligned} \tag{10.12}$$

where

$$Z = R_i\sqrt{1 + Q_L^2\left(\frac{\omega}{\omega_o} - \frac{\omega_o}{\omega}\right)^2} = Z_o\sqrt{\left(\frac{R_i}{Z_o}\right)^2 + \left(\frac{\omega}{\omega_o} - \frac{\omega_o}{\omega}\right)^2}, \tag{10.13}$$

$$tan\psi = Q_L\left(\frac{\omega}{\omega_o} - \frac{\omega_o}{\omega}\right), \tag{10.14}$$

$$R_i = Zcos\psi, \tag{10.15}$$

$$X = Zsin\psi. \tag{10.16}$$

From (10.14) and trigonometric relationships,

$$sin\psi = \frac{Q_L(\frac{\omega}{\omega_o} - \frac{\omega_o}{\omega})}{\sqrt{1 + Q_L^2(\frac{\omega}{\omega_o} - \frac{\omega_o}{\omega})^2}} \tag{10.17}$$

$$cos\psi = \frac{1}{\sqrt{1 + Q_L^2(\frac{\omega}{\omega_o} - \frac{\omega_o}{\omega})^2}}. \tag{10.18}$$

The amplitude of the fundamental component of the voltage at the input of the resonant circuit is

$$V_m = \frac{2V_I}{\pi}. \tag{10.19}$$

From (10.15) and (10.19), one obtains the amplitude of the current through the resonant circuit

$$I_m = \frac{V_m}{Z} = \frac{2V_I}{\pi Z} = \frac{2V_I cos\psi}{\pi R_i}. \tag{10.20}$$

Using (10.4), (10.17), (10.18), and (10.20) yields

$$I = I_m sin\psi = \frac{2V_I sin\psi cos\psi}{\pi R_i} = \frac{2V_I Q_L(\frac{\omega}{\omega_o} - \frac{\omega_o}{\omega})}{\pi R_i[1 + Q_L^2(\frac{\omega}{\omega_o} - \frac{\omega_o}{\omega})^2]}. \tag{10.21}$$

Hence, the charging time is

$$t_1 = \frac{C_1 V_I}{I} = \frac{\pi C_1 Z_o[(\frac{R_i}{Z_o})^2 + (\frac{\omega}{\omega_o} - \frac{\omega_o}{\omega})^2]}{2(\frac{\omega}{\omega_o} - \frac{\omega_o}{\omega})}. \tag{10.22}$$

The discharging time is the same as the charging time and is given by (10.22). A three-dimensional representation of the normalized charging time $t_1/(C_1 Z_o)$ is depicted in Fig. 10.3. Figure 10.4 shows the normalized charging

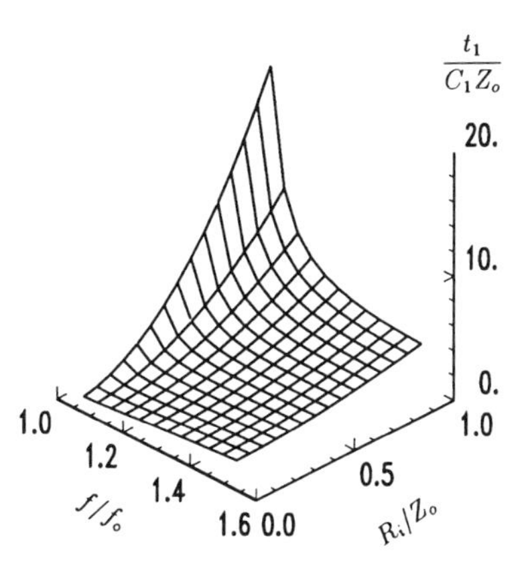

Figure 10.3: Normalized charging time $t_1/(C_1 Z_o)$ as functions of f/f_o and R_i/Z_o for the SRI.

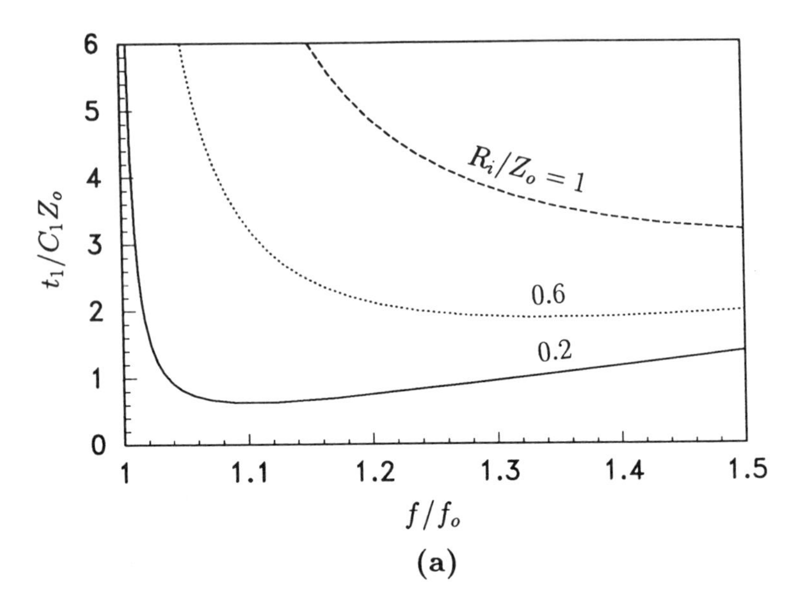

(a)

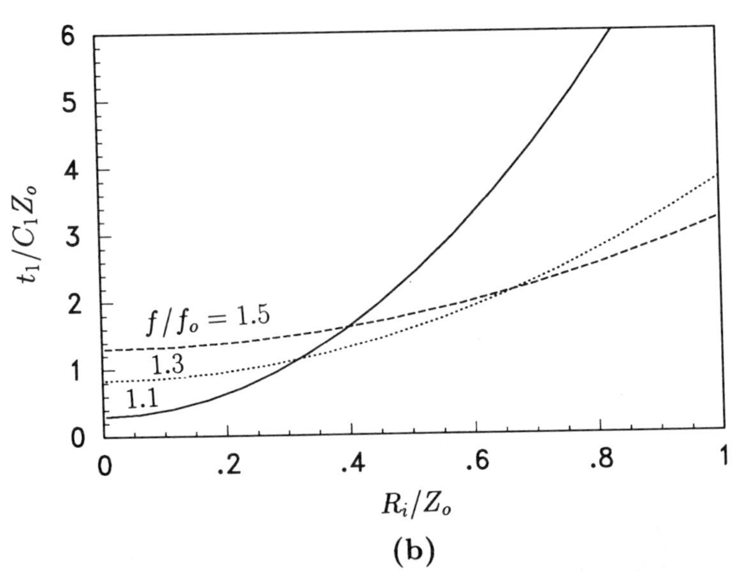

(b)

Figure 10.4: Normalized charging time $t_1/(C_1Z_o)$ as functions of the normalized frequency f/f_o and the normalized load resistance R_i/Z_o for the SRI. (a) $t_1/(C_1Z_o)$ against f/f_o at various values of R_i/Z_o. (b) $t_1/(C_1Z_o)$ against R_i/Z_o at various values of f/f_o.

time $t_1/(C_1 Z_o)$ as functions of the normalized operating frequency f/f_o and the normalized load resistance R_i/Z_o. It can be seen that $t_1/(C_1 Z_o)$ increases with increasing R_i/Z_o. This is because the amplitude I_m of the current in the series-resonant circuit decreases with increasing load resistance R_i.

To obtain zero-voltage switching, the dead time t_D should be longer than the charging or discharging time t_1. Consider the discharging process of C_1. Initially, the capacitor C_1 is discharged by the resonant circuit. If the dead time t_D is shorter than the discharging time t_1, the transistor Q_2 will be turned on too early when the increasing voltage v_{DS1} is still less than V_I, and therefore the decreasing voltage $v_{DS2} = V_I - v_{DS1}$ is still greater than zero. The energy stored in the capacitor at $t = t_D$ is

$$W_C = \frac{1}{2} C_1 v_{DS2}^2(t_D), \tag{10.23}$$

where $v_{DS2}(t_D)$ is the voltage across the shunt capacitor C_1 when Q_2 is turned on. The capacitor will be discharged by the transistor Q_2. Therefore, the energy stored in the capacitor will be dissipated in the transistor Q_2 as heat. Thus, the turn-on switching loss in Q_2 is

$$P_{ton} = \frac{W_C}{T} = f W_C = \frac{1}{2} f C_1 v_{DS2}^2(t_D). \tag{10.24}$$

The turn-on switching loss in Q_1 during charging the shunt capacitor is the same.

Close examination of the parallel resonant inverter shows that the charging (or discharging) time is

$$t_1 = \frac{\pi C_1 Z_o}{2} \sqrt{\frac{[1 - (\frac{\omega}{\omega_o})^2]^2 + (\frac{1}{Q_L}\frac{\omega}{\omega_o})^2}{1 + (Q_L \frac{\omega}{\omega_o})^2}} \times \frac{\sqrt{1 + \{Q_L(\frac{\omega}{\omega_o})[(\frac{\omega}{\omega_o})^2 + \frac{1}{Q_L^2} - 1]\}^2}}{(\frac{\omega}{\omega_o})[(\frac{\omega}{\omega_o})^2 + \frac{1}{Q_L^2} - 1]}, \tag{10.25}$$

where $\omega_o = 1/\sqrt{LC}$ is the corner frequency, $Z_o = \sqrt{L/C}$, and $Q_L = R_i/Z_o$. This expression is illustrated in Figs. 10.5 and 10.6. It can be seen that $t_1/(C_1 Z_o)$ decreases with increasing Q_L.

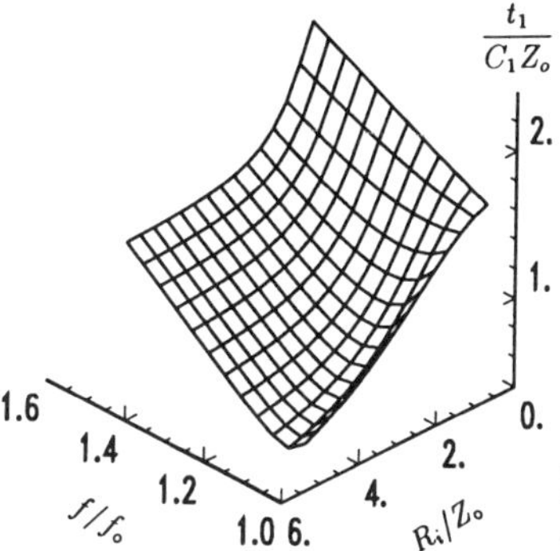

Figure 10.5: Normalized charging time $t_1/(C_1 Z_o)$ as functions of f/f_o and $Q_L = R_i/Z_o$ for the PRI.

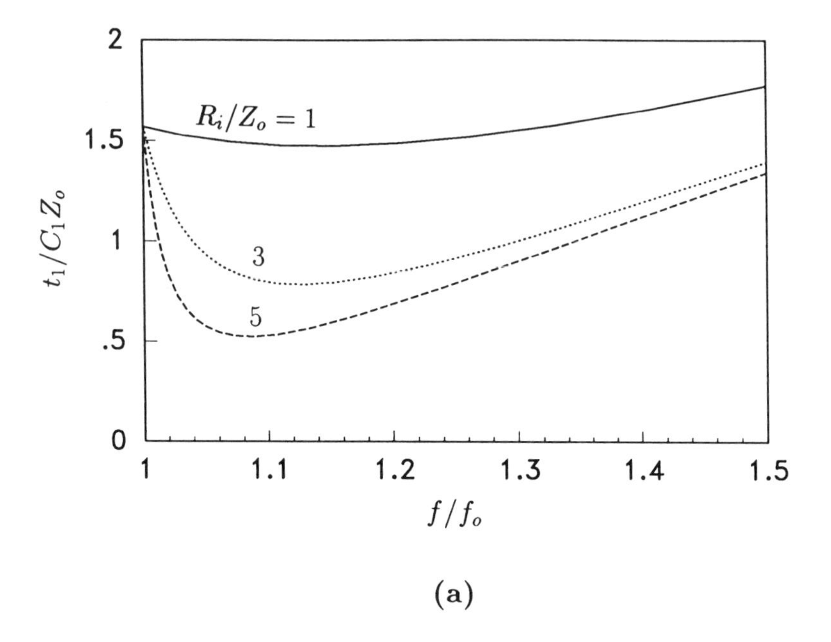

(a)

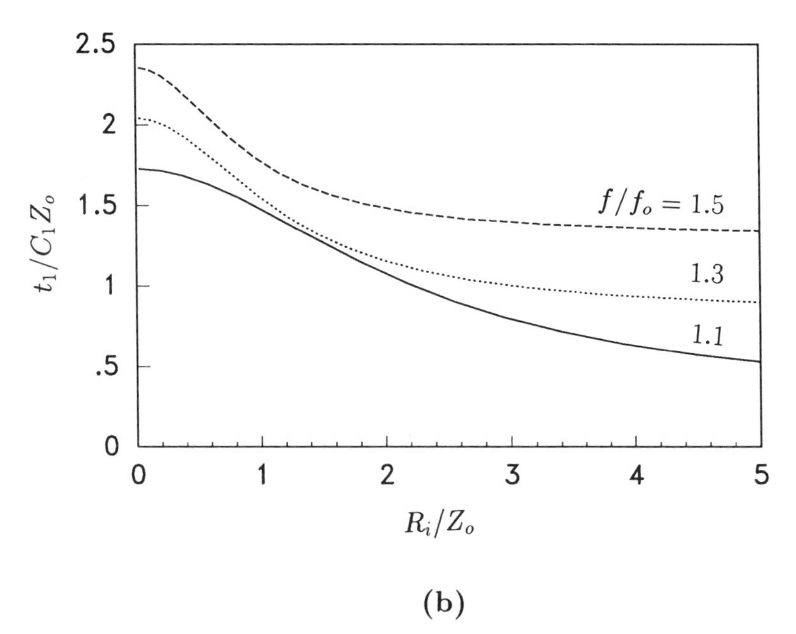

(b)

Figure 10.6: Normalized charging time $t_1/(C_1Z_o)$ as functions of the normalized frequency f/f_o and the normalized load resistance Q_L. (a) $t_1/(C_1Z_o)$ against f/f_o at various values of Q_L. (b) $t_1/(C_1Z_o)$ against Q_L at various values of f/f_o.

Assuming that the waveform of voltage v_{DS2} is trapezoidal, the amplitude of the fundamental component of v_{DS2} is expressed by

$$V_{m1} = \frac{2}{\pi} V_I \frac{sin(\frac{\omega t_1}{2})}{(\frac{\omega t_1}{2})}. \tag{10.26}$$

All other equations given in Chapter 6 remain valid.

10.3.1 Sinusoidal Drive

The dead time can be obtained using a driver whose output voltages are sinusoidal. The principle of operation is explained in Fig. 10.7. The gate-source voltages are sinusoidal and 180° out of phase

$$v_{GS1} = V_{GSm} sin\omega t \tag{10.27}$$

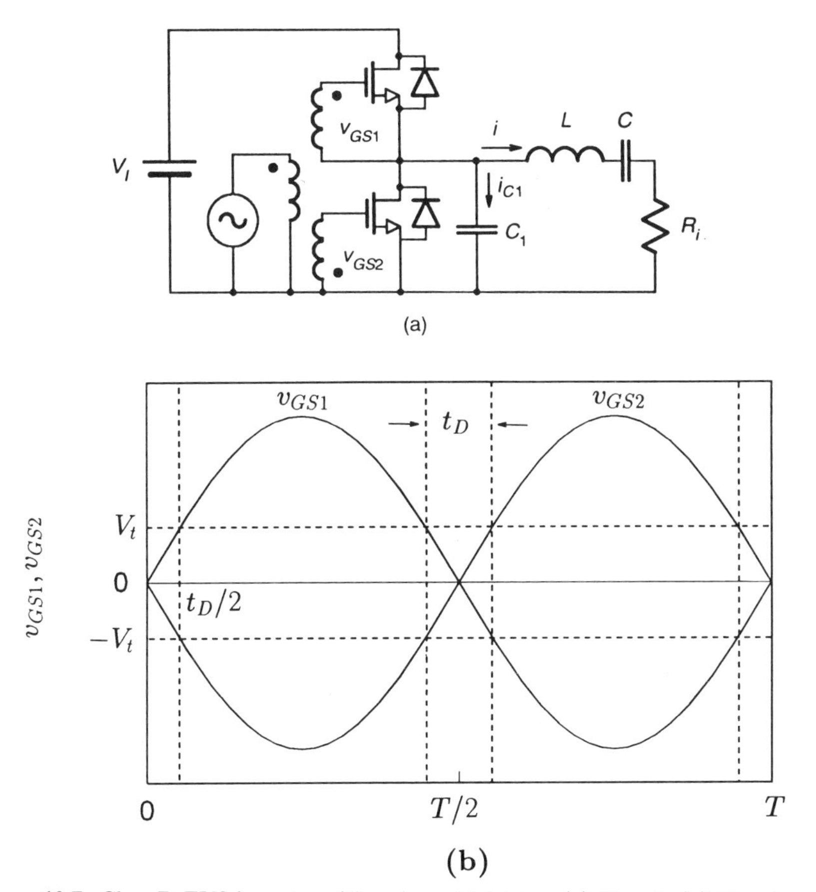

Figure 10.7: Class D ZVS inverter with a sinusoidal drive. (a) Circuit. (b) Waveforms.

and

$$v_{GS2} = -V_{GSm} sin \omega t. \tag{10.28}$$

The voltage v_{GS1} reaches the MOSFET threshold voltage V_t at time $t_D/2$. Hence,

$$V_t = v_{GS1}\left(\frac{t_D}{2}\right) = V_{GSm} sin\left(\frac{\omega t_D}{2}\right), \tag{10.29}$$

which results in the dead time

$$t_D = \frac{1}{\pi f} arcsin\left(\frac{V_t}{V_{GSm}}\right). \tag{10.30}$$

The desired value of the dead time can be accomplished by selecting an appropriate amplitude V_{GSm}. This amplitude must be lower than the gate-source breakdown voltage.

10.4 SUMMARY

- The ZVS operation of all Class D half-bridge and full-bridge inverters can be achieved by adding a capacitor in parallel with one of the switching devices (or both) and maintaining both switching devices in the off-state during the transitions.
- ZVS operation can be accomplished for inductive loads only, that is, for operation above resonance.
- The output capacitances of the switching devices are absorbed into the parallel capacitor.
- The switching devices can be kept in the off-state during the transitions by using a dead time in the rectangular drive voltages. Another method is to use sinusoidal drive voltages.
- ZVS operation cannot be accomplished for capacitive loads (i.e., below resonance). In fact, switching losses are increased if an external capacitor is added.
- ZVS operation can be achieved over a narrow load range in the SRI. This is because the amplitude of the current through the resonant circuit is inversely proportional to the load.
- ZVS operation can be achieved over a wide load range in the PRI. This is because amplitude of the current through the resonant inductor is almost independent of the load.

10.5 REFERENCES

1. S. A. Zhukov and V. B. Kozyrev, “Double-ended switching generator without commutating (switching) loss,” *Poluprovodnikovye Pribory v Tekhnike Elektrosvyazi*, vol. 15, Moscow, pp. 95–107, 1975.

2. F. Goldfarb, "A new non-dissipative load-line shaping technique elements switching stress in bridge converters," *Proc. Powercon 8*, paper D-4. pp. 1–8, 1981.
3. M. Boidin, H. Foch, and P. Proudlock, "The design, construction and evaluation of a new generation h.f. 40 kW d.c. converter," *Proc. Power Conversion International Conf.*, pp. 124–133, 1984.
4. H. Foch and J. Roux, "Static semiconductor electrical energy converter," U.S. Patent no. 4,330,819, May 18, 1982.
5. B. Carsten, "Fast, accurate measurement of core loss at high frequencies," *Power Conversion and Intelligent Motion*, vol. 12, pp. 29–33, Mar. 1986.
6. B. Carsten, "A hybrid series-parallel converter for high frequencies and power levels," in *Proc. High Frequency Power Conversion Conf.*, Washington, DC, Apr. 1987, pp. 41–47.
7. C. P. Henze, H. C. Martin, and D. W. Parsley, 'Zero-voltage switching in high frequency power converters using pulse width modulation,' *Proceedings of IEEE Applied Power Electronics Conference*, New Orleans, LA, USA, 1–5 February 1988, pp. 33–40.
8. R. L. Steigerwald, "A comparison of half-bridge resonant converter topologies," *IEEE Trans. Ind. Electron.*, vol. IE-3, pp. 174–182, Apr. 1988.
9. F. M. Magalhaes, F. T. Dickens, G. R. Westerman, and N. G. Ziesse, "Zero-voltage-switched resonant half-bridge high-voltage dc-dc converter," in *High Frequency Power Conversion Conference*, 1988, pp. 332–343.
10. R. L. Steigerwald and K. D. T. Ngo, "Full-bridge lossless switching converter", U.S. Patent no. 4,864,479, Sept. 5, 1989.
11. R. Redl, N. O. Sokal, and L. Balogh, "A novel soft-switching full-bridge dc/dc converter: analysis, design, and experimental results at 1.5 kW, 100 kHz," *IEEE Power Electron.*, vol. PE-6, pp. 408–419, July 1991.
12. L. H. Mweene, C. A. Wright, and M. F. Schlecht, "A 1 kW 500 kHz front-end converter for a distributed power supply system," *IEEE Trans. Power Electron.*, vol. PE-6, pp. 398–407, July 1991.
13. S. Hamada, Y. Ogino, and M. Nakaoka, "Saturable reactor assisted soft-switching full-bridge dc-dc convertors," *Proc. IEE, Pt. B, Electric Power Appl.*, vol. 138, pp. 95–103, Mar. 1991.
14. A. El-Hamamasy and G. Jernakoff, "Driver for a high-efficiency, high-frequency Class-D power amplifier," U.S. Patent no. 5,023,566, June 11, 1991.
15. M. K. Kazimierczuk, W. Szaraniec, and S. Wang, "Analysis and design of parallel resonant converter at high Q_L," *IEEE Trans. Aerospace Electron. Syst.*, vol. AES-27, pp. 35–50, Jan. 1992.
16. M. K. Kazimierczuk and S. Wang, "Frequency-domain analysis of series resonant converter for continuous conduction mode," *IEEE Trans. Power Electron.*, vol. PE-7, pp. 270–279, Mar. 1992.
17. M. K. Kazimierczuk and W. Szaraniec, "Class D voltage-switching inverter with only one shunt capacitor," *IEE Proc., Pt.B, Electric Power Appl.*, vol. 139, pp. 449–456, Sept. 1992.
18. D. Czarkowski and M. K. Kazimierczuk, "Simulation and experimental results for Class D ZVS series resonant inverter," *Proceedings of the IEEE International Telecommunications Energy Conference (INTELEC'92)*, Washington, DC, October 4–8, 1992, pp. 153–157.

10.6 REVIEW QUESTIONS

10.1 Explain what is meant by the term *soft switching*.

10.2 Explain how ZVS operation of Class D resonant inverters is achieved.

10.3 Is it possible to achieve ZVS operation of Class D half-bridge using two capacitors, each in parallel with one switch?

10.4 Is it possible to achieve ZVS operation of Class D half-bridge using only one capacitor in parallel with one of the switches?

10.5 Is it better to connect the single capacitor in parallel with the upper or the bottom switch?

10.6 Are the output capacitances absorbed into the parallel capacitor?

10.7 Explain how the switching devices can be maintained in the off-state during the transitions.

10.8 Is it possible to obtain ZVS operation for capacitive loads?

10.9 Is it possible to obtain ZVS operation over a wide load range in the SRI?

10.10 Is it possible to obtain ZVS operation over a wide load range in the PRI?

10.7 PROBLEMS

10.1 A Class D series resonant inverter of Fig. 10.1 has the following parameters: $L = 220\ \mu\text{H}$, $C = 3.3$ nF, $Q_L = 4$, and $f = 200$ kHz. Calculate the minimum dead time in the drive voltages that is required for achieving zero-voltage-switching operation if $C_1 = 500$ pF.

10.2 The circuit of Problem 10.1 has sinusoidal drive voltages of power MOSFETs. Select the amplitude of the drive signals that ensures that shunt capacitor C_1 is charged and discharged completely. Assume that MOSFETs with a threshold voltage $V_T = 2$ V are used.

10.3 Derive equation (10.25).

CHAPTER 11

CLASS D CURRENT-SOURCE RESONANT INVERTER

11.1 INTRODUCTION

Class D voltage-source resonant dc-ac inverters, studied in the preceding chapters, employ either a series-resonant circuit or resonant circuits that are derived from the series-resonant circuit. In voltage-source resonant inverters, the current drawn from the dc voltage supply is a pulse current. On the other hand, in current-source resonant inverters [1]–[7], the current drawn from the dc voltage supply is constant and continuous. The objectives of this chapter are to 1) introduce the current-source parallel-resonant inverter, 2) present a comprehensive frequency-domain analysis of this current-source inverter for steady-state operation, and 3) give a design procedure.

One of the major advantages of a current-source parallel-resonant inverter is a simple gate drive circuit. Both transistors are driven with respect to ground and, therefore, the drivers do not require isolation transformers or optical couplers. In a typical voltage-source series resonant inverter, the output voltage of the inverter increases as the frequency approaches the resonant frequency. In the current-source parallel-resonant inverter, the output voltage decreases as the frequency is closer to the resonant frequency.

11.2 PRINCIPLE OF OPERATION

A circuit of a current-source inverter with a parallel-resonant circuit is shown in Fig. 11.1(a). It consists of a large choke inductor L_{fi}, two switches S_1 and S_2, and an R_iLC parallel-resonant circuit. The circuit requires unidirectional switches for the current and bidirectional switches for the voltage.

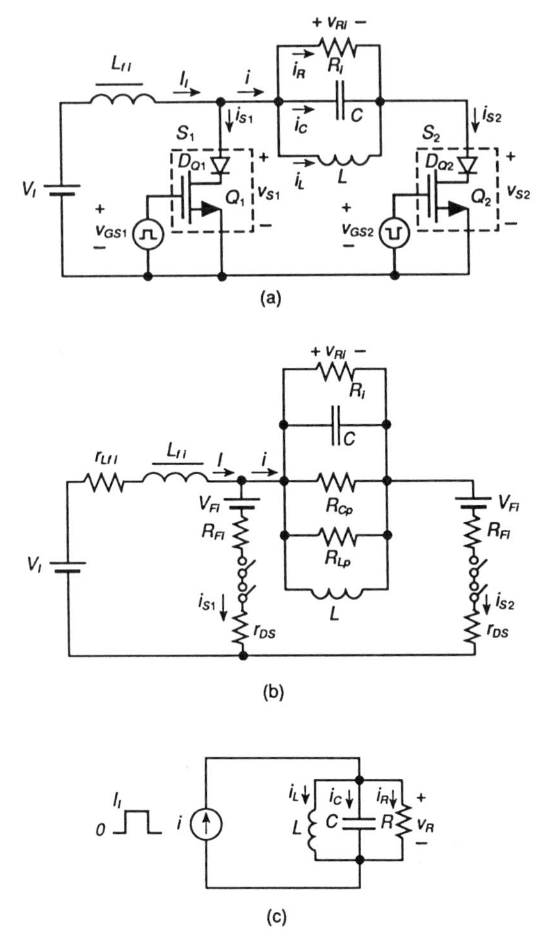

Figure 11.1: Class D current-source inverter with a parallel-resonant circuit. (a) Circuit. (b) Equivalent circuit with parasitic resistances and offset voltage sources. (c) Simplified model.

Each switch consists of a MOSFET in series with a diode. The intrinsic body-drain *pn* junction diode of the MOSFET is disabled by the series diode. As a result, the switch can conduct only a positive current and can block either positive or negative voltage. The MOSFETs are driven by rectangular gate-to-source voltages v_{GS1} and v_{GS2} at the operating frequency $f = 1/T$ and with an on-duty cycle of slightly greater than 50%. To provide the path for the dc input current source I_I, either one or both switches should be ON. Therefore, slightly overlapping gate-to-source voltages should be used to obtain the simultaneous conduction of the power MOSFETs. The parallel-resonant circuit can be modified to achieve impedance transformation.

Figure 11.1(b) shows an equivalent circuit of the inverter with parasitic resistances and offset voltage sources, where r_{Lfi} is the series equivalent resistance (ESR) of the input inductor L_{fi}, r_{DS} is the MOSFET on-resistance, R_{Fi} is the diode forward resistance, V_{Fi} is the diode offset voltage, R_{Lp} is the equivalent parallel resistance (EPR) of the resonant inductor L, and R_{Cp} is the EPR of the resonant capacitor C. When switch S_1 is OFF and switch S_2 is ON, the dc input current I_I flows into the resonant circuit and the energy is transferred from the dc input source to the resonant circuit. When S_1 is ON and S_2 is OFF, the dc input current I_I flows through switch S_1 and the energy stored in the resonant circuit is partially discharged into the load. Assuming that the input inductor and the switches are ideal, the circuit composed of V_I, L_{fi}, S_1, and S_2 can be modeled by a square-wave current source $i = i_{S2}$, as shown in Fig. 11.1(c).

The principle of operation of the current-source inverter is explained by the current and voltage waveforms, depicted in Fig. 11.2. Figure 11.2(a), (b), and (c) shows the waveforms for $f < f_o$, $f = f_o$, and $f > f_o$, respectively,

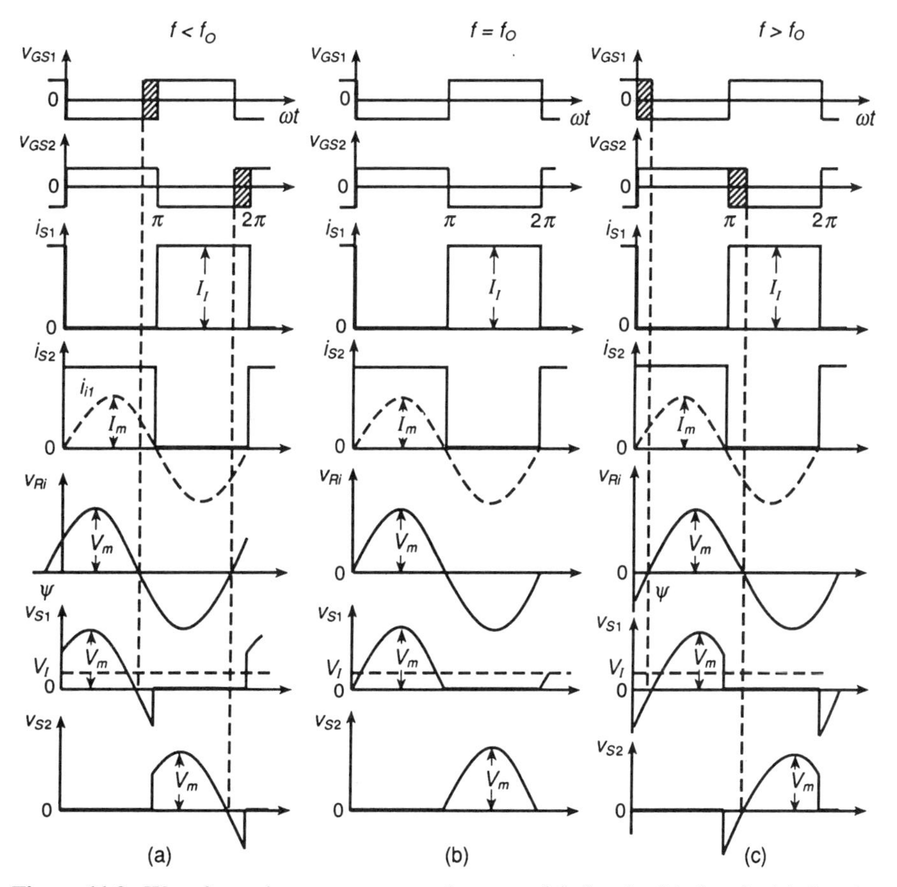

Figure 11.2: Waveforms in current-source inverter. (a) $f < f_o$. (b) $f = f_o$. (c) $f > f_o$.

where $f_o = 1/2\pi\sqrt{LC}$ is the resonant frequency. The input current of the resonant circuit is a square wave of magnitude I_I. When the switch voltage is positive and higher than the diode threshold voltage, the series diode is ON and, therefore, the MOSFET must be OFF to block the voltage. When the switch voltage is negative, the series diode is OFF and the transistor can be either ON or OFF because the diode can block the switch voltage. For this reason, the duty cycle D_t of the MOSFET should be greater than or equal to the duty cycle of the entire switch D. Thus, the range of D_t is $D \le D_t \le D_{t(max)}$, as indicated by the cross hatched areas in Fig. 11.2(a) and (c).

For $f < f_o$, the parallel-resonant circuit represents an inductive load. Therefore, the voltage across the resonant circuit v_{Ri} leads the fundamental component i_{i1} of the current through the resonant circuit i by the phase angle $|\psi|$. The switch begins the off-state with a positive voltage and ends with a negative voltage. Let us consider the operation of switch S_1. Prior to $\omega t = 0$, switch S_1 is ON, its current is I_I, and its voltage is approximately zero. At $\omega t = 0$, voltage v_{GS1} is reduced from high to low and the MOSFET Q_1 is turned off. During the interval $0 < \omega t \le \pi$, voltage v_{S1} is positive, diode D_{Q_1} is ON, and transistor Q_1 must be OFF to support the switch voltage. When the switch voltage increases, the diode conducts a small current that charges the transistor output capacitance. Thus, the voltage across the series diode D_{Q_1} is 0.7 V, and the switch voltage across transistor Q_1 is equal to $v_{S1} - 0.7$ V. When v_{S1} is decreased below 0.7 V, the series diode D_{Q_1} is turned off and transistor Q_1 can be either ON or OFF. The diode turns off at low dv/dt and at zero di/dt. The derivative dv/dt is limited by the sinusoidal voltage across the parallel-resonant circuit. The MOSFET prevents the current flow prior to diode turn-off. Consequently, the turn-off switching loss in the diode is zero. The diode supports the switch voltage when the switch voltage is negative if the transistor is ON. If the transistor is OFF, both the diode and the transistor block the switch voltage. At $\omega t = \pi$, voltage v_{GS2} is reduced from a high to a low level and transistor Q_2 is turned off. At this time, the inductor current I_I causes the series diode D_{Q_1} to turn on. During the interval $\pi < \omega t \le 2\pi$, Q_1 is already ON and, therefore, switch S_1 is turned on and stays ON until the beginning of the next period. When the switch voltage increases, the MOSFET output capacitance C_{out} is charged via diode D_{Q_1} to the peak value of the switch voltage $V_{SM} = V_m$ and then remains at that voltage until the transistor turns on. At this time, the capacitance C_{out} is discharged through the transistor, resulting in a turn-on switching loss in the MOSFET $P_D \approx fC_{out}V_{SM}^2/2$. Only the turn-off transition of the switch is directly controllable by the driver. The turn-on switch transition is caused by the turn-off transition of the other switch.

For $f = f_o$, the parallel-resonant circuit represents a purely resistive load and, therefore, the current i_{i1} and voltage v_{Ri} are in phase, as shown in Fig. 11.2(b). In this case, the MOSFETs turn on and off at zero voltage, resulting in zero-voltage switching, zero switching losses, and high efficiency. Since the switch voltages are never negative, the series diodes are not required and can be removed, reducing conduction losses. In this case, frequency control of the output voltage cannot be used.

For $f > f_o$, the parallel-resonant circuit represents a capacitive load. Hence, the voltage across the resonant circuit v_{Ri} lags behind the fundamental component i_{i1} of the current i by the phase angle ψ. The switch begins the off-state with a negative voltage and ends with a positive voltage. Let us consider the operation of switch S_1. Prior to $\omega t = 0$, switch S_1 is ON, its current is I_I, and its voltage is approximately zero. At $\omega t = 0$, voltage v_{GS2} is increased from a low to a high level and the MOSFET Q_2 is turned on. As a result, the dc input current I_I is diverted from switch S_1 to switch S_2, which causes the series diode D_{Q_1} to turn off. Diode D_{Q1} turns off at high di/dt, causing a reverse-recovery switching loss. When the switch voltage v_{S1} is negative, the series diode D_{Q_1} is OFF and the MOSFET Q_1 can be either ON or OFF. Transistor Q_1 must be turned off when the voltage v_{S1} is negative, as indicated by the shaded area shown in Fig. 11.2(c). As the switch voltage v_{S1} crosses zero, diode D_{Q_1} turns on at low dv/dt and zero di/dt while transistor Q_1 is already OFF and supports the switch voltage. When the switch voltage increases, a small current flows through diode D_{Q1} to charge the output capacitance of transistor Q_1. Hence, the voltage across D_{Q_1} is approximately 0.7 V and the voltage across Q_1 is $v_{S1} - 0.7$ V. At $\omega t = \pi$, voltage v_{GS1} is increased from a low to a high value, which causes Q_1 to turn on. During the interval $\pi < \omega t \leq 2\pi$, switch S_1 stays ON until the beginning of the next period. Just before turn-on, the switch voltage V_{ton} is greater than zero. Therefore, the transistor turn-on switching loss is $P_{turn\text{-}on} = fC_{out}V_{ton}^2/2$. Notice that only the turn-on transition of the switch is directly controllable by the driver. The turn-off switch transition is caused by the turn-on transition of the other switch. In summary, for $f > f_o$, the MOSFETs experience soft switching and zero turn-off switching loss and the series diodes experience zero turn-on switching loss. However, there is turn-on switching loss in each MOSFET and reverse-recovery turn-off loss in each series diode. For these two reasons, the efficiency above resonance is less than that below resonance. Other topologies of Class D current-source resonant inverters are given in [1]–[7].

To design the value of the choke inductance, let us consider the operation at the switching frequency equal to the resonant frequency. In such a case, the current in the choke inductor increases when switch S_1 is ON and decreases when switch S_2 is ON. When switch S_1 is ON, the voltage across the choke inductor is V_I. Hence, the peak-to-peak value of the ripple current in the choke inductor can be expressed as

$$I_r = \frac{V_I}{L_{fi}} \frac{T}{2}, \tag{11.1}$$

from which the minimum value of the filter inductance for a given maximum allowable ripple current is

$$L_{fi(min)} = \frac{V_I}{2fI_{r(max)}}. \tag{11.2}$$

11.3 ANALYSIS OF THE PARALLEL-RESONANT CIRCUIT

The current-source parallel-resonant circuit is shown in Fig. 11.1. The parallel-resonant circuit contains a capacitor C, an inductor L, and a resistor R. The resistor R is given by

$$R = \frac{1}{G} = \frac{R_i R_d}{R_i + R_d}, \tag{11.3}$$

where R_i is the ac load resistance and R_d is the parasitic resistance of the resonant circuit. The parasitic resistance R_d is given by

$$R_d = \frac{R_{Lp} R_{Cp}}{R_{Lp} + R_{Cp}}, \tag{11.4}$$

where R_{Lp} is the *equivalent parallel resistance* (EPR) of L and R_{Cp} is the EPR of C. The relationships between the ESRs and EPRs are

$$R_{Lp} = r_L(1 + Q_{Lo}^2) \approx r_L Q_{Lo}^2 \tag{11.5}$$

$$R_{Cp} = r_C(1 + Q_{Co}^2) \approx r_C Q_{Co}^2, \tag{11.6}$$

where $Q_{Lo} = \omega L / r_L = R_{Lp}/(\omega L)$ and $Q_{Co} = 1/(\omega C r_C) = \omega C R_{Cp}$ are the unloaded quality factors of L and C, respectively.

The parallel-resonant circuit can be characterized by the following normalized parameters:

- the resonant frequency

$$\omega_o = \frac{1}{\sqrt{LC}}, \tag{11.7}$$

- the characteristic impedance

$$Z_o = \omega_o L = \frac{1}{\omega_o C} = \sqrt{\frac{L}{C}}, \tag{11.8}$$

- the loaded quality factor at the corner frequency f_o

$$Q_L = \omega_o C R = \frac{R}{\omega_o L} = \frac{R}{Z_o} = \frac{1}{G Z_o}. \tag{11.9}$$

The input admittance of the parallel-resonant circuit is given by

$$\mathbf{Y} = G + j\omega C + \frac{1}{j\omega L} = G + j\omega_o C\left(\frac{\omega}{\omega_o} - \frac{\omega_o}{\omega}\right)$$
$$= G + j\frac{1}{Z_o}\left(\frac{\omega}{\omega_o} - \frac{\omega_o}{\omega}\right) = G\left[1 + jQ_L\left(\frac{\omega}{\omega_o} - \frac{\omega_o}{\omega}\right)\right]$$
$$= \frac{1}{Z_o}\left[\frac{1}{Q_L} + j\left(\frac{\omega}{\omega_o} - \frac{\omega_o}{\omega}\right)\right] = Ye^{j\psi} = G + jB, \qquad (11.10)$$

where $G = Y cos\psi$ and $B = Y sin\psi$. It follows from trigonometric relationships that

$$cos\psi = \frac{1}{\sqrt{1 + [Q_L(\frac{\omega}{\omega_o} - \frac{\omega_o}{\omega})]^2}}. \qquad (11.11)$$

The magnitude of the admittance is

$$Y = \frac{I_{rms}}{V_{Ri}} = \frac{1}{Z_o}\sqrt{\frac{1}{Q_L^2} + \left(\frac{\omega}{\omega_o} - \frac{\omega_o}{\omega}\right)^2}, \qquad (11.12)$$

where I_{rms} is the rms value of the fundamental component of the resonant circuit input current i defined in (11.18) and V_{Ri} is the rms value of the ac output voltage of the inverter. The phase of the admittance Y is

$$\psi = arctan\left[Q_L\left(\frac{\omega}{\omega_o} - \frac{\omega_o}{\omega}\right)\right]. \qquad (11.13)$$

The two-dimensional representation of the normalized input admittance YZ_o and the phase ψ against f/f_o are shown in Fig. 11.3.

The input power of the resonant circuit is $P_R = V_{Ri}^2/R$, the output power of the resonant circuit is $P_{Ri} = V_{Ri}^2/R_i$, and the power dissipated in the resonant circuit is $P_{R_d} = V_{Ri}^2/R_d$. Hence, one obtains the efficiency of the resonant circuit

$$\eta_{rc} \equiv \frac{P_{Ri}}{P_R} = \frac{R}{R_i} = \frac{R_d}{R_i + R_d}. \qquad (11.14)$$

11.4 ANALYSIS OF THE INVERTER

11.4.1 Voltage Transfer Function

Referring to Fig. 11.1(c), the input current of the resonant circuit i is a square wave of magnitude I_I expressed by

$$i = \begin{cases} I_I, & \text{for } 0 < \omega t \leq \pi \\ 0, & \text{for } \pi < \omega t \leq 2\pi. \end{cases} \qquad (11.15)$$

Its fundamental component is

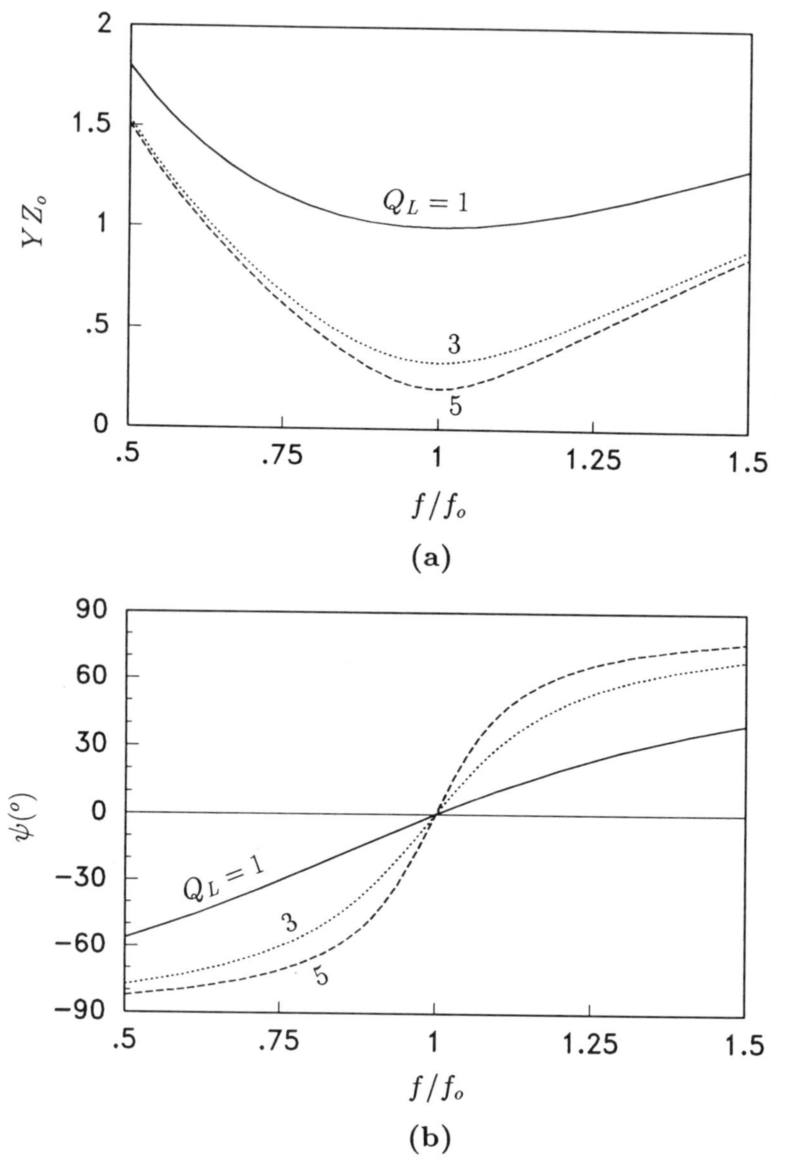

Figure 11.3: Two-dimensional representation of the input admittance of the resonant circuit. (a) YZ_o as a function of f/f_o at various values of Q_L. (b) ψ as a function of f/f_o at various values of Q_L.

$$i_{i1} = I_m sin\omega t, \tag{11.16}$$

where

$$I_m = \frac{2}{\pi} I_I = 0.6366 I_I. \tag{11.17}$$

Hence, the rms value of i_{i1} is

$$I_{rms} = \frac{I_m}{\sqrt{2}} = \frac{\sqrt{2}I_I}{\pi} = 0.4502 I_I, \tag{11.18}$$

which leads to the current transfer function from the dc input current source I_I to the fundamental component at the input of the resonant circuit

$$M_{Is} \equiv \frac{I_{rms}}{I_I} = \frac{\sqrt{2}}{\pi} = 0.4502. \tag{11.19}$$

The magnitude of the transfer function of the dc current to the ac voltage of the inverter is obtained from (11.12) and (11.19)

$$\begin{aligned} M_{Rr} &\equiv \frac{V_{Ri}}{I_I} = \frac{V_{Ri}}{I_{rms}} \times \frac{I_{rms}}{I_I} = \frac{M_{Is}}{Y} \\ &= \frac{\sqrt{2}Z_o}{\pi\sqrt{\frac{1}{Q_L^2} + \left(\frac{\omega}{\omega_o} - \frac{\omega_o}{\omega}\right)^2}} = \frac{\sqrt{2}R}{\pi\sqrt{1 + Q_L^2\left(\frac{\omega}{\omega_o} - \frac{\omega_o}{\omega}\right)^2}}. \end{aligned} \tag{11.20}$$

The voltage transfer function M_{VI} from the dc input voltage V_I to output of the inverter V_{Ri} can be determined as follows. The dc input power of the inverter is

$$P_I = V_I I_I, \tag{11.21}$$

and the ac output power of the inverter is

$$P_{Ri} = \frac{V_{Ri}^2}{R_i}. \tag{11.22}$$

The efficiency of the inverter η_I is the ratio of the output power to the input power. From (11.20), (11.21), and (11.22), one obtains

$$\eta_I = \frac{P_{Ri}}{P_I} = \frac{V_{Ri}^2}{V_I I_I R_i} = \frac{V_{Ri}}{V_I}\frac{M_{Rr}}{R_i}. \tag{11.23}$$

Substitution of (11.20) into (11.23) yields an expression for the dc-to-ac voltage transfer function of the inverter

$$M_{VI} \equiv \frac{V_{Ri}}{V_I} = \frac{\pi\eta_I R_i\sqrt{\frac{1}{Q_L^2} + \left(\frac{\omega}{\omega_o} - \frac{\omega_o}{\omega}\right)^2}}{\sqrt{2}Z_o} = \frac{\pi\eta_I\sqrt{1 + Q_L^2\left(\frac{\omega}{\omega_o} - \frac{\omega_o}{\omega}\right)^2}}{\sqrt{2}\eta_{rc}}. \tag{11.24}$$

Ideally, the range of M_{VI} is from $\pi/\sqrt{2} = 2.22$ to ∞. The voltage transfer function M_{VI} has the same phase as the admittance Y and is given by (11.13).

A three-dimensional representation of M_{VI} as a function of f/f_o and Q_L is shown in Fig. 11.4 for $\eta_I = 0.95$ and $\eta_{rc} = R/R_i = 0.98$. In Fig. 11.5(a), M_{VI} is plotted as a function of f/f_o at different values of Q_L. If (11.24) is inverted, a function of f/f_o in terms of M_{VI} and Q_L can be obtained as

$$\frac{f}{f_o} = \frac{-\sqrt{a} + \sqrt{a+4}}{2}, \qquad \text{for } \frac{f}{f_o} \leq 1 \tag{11.25}$$

and

$$\frac{f}{f_o} = \frac{\sqrt{a} + \sqrt{a+4}}{2}, \qquad \text{for } \frac{f}{f_o} \geq 1, \tag{11.26}$$

where

$$a = \frac{2M_{VI}^2 R^2}{\pi^2 \eta_I^2 R_i^2 Q_L^2} - \frac{1}{Q_L^2} \tag{11.27}$$

and a must be greater than zero. Equation (11.25) corresponds to operation below resonance and equation (11.26) corresponds to operation above resonance, and their graphs are shown in Fig. 11.5(b). Figure 11.5(c) and (d) shows M_{VI} versus Q_L at various values of f/f_o for $f/f_o < 1$ and $f/f_o > 1$, respectively. From (11.24), the peak value of the switch voltage and the voltage across the resonant circuit is

$$V_{SM} = V_{Rim} = \sqrt{2}V_{Ri} = \sqrt{2}M_{VI}V_I = \frac{\pi\eta_I V_I \sqrt{1 + Q_L^2\left(\frac{\omega}{\omega_o} - \frac{\omega_o}{\omega}\right)^2}}{\eta_{rc}}. \tag{11.28}$$

Hence, assuming $\eta_I = 1$ and $\eta_{rc} = 1$, the minimum value of the switch peak voltage occurs at $f = f_o$ and is given by

$$V_{Rim} = V_{SMmin} = \pi V_I. \tag{11.29}$$

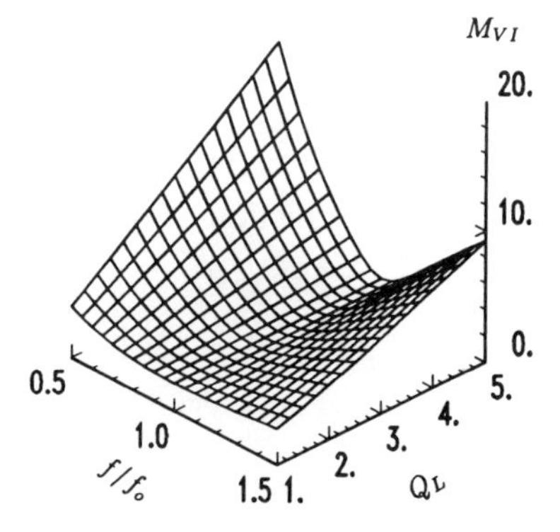

Figure 11.4: Three-dimensional representation of the dc-to-ac voltage transfer function M_{VI} as a function of f/f_o and Q_L.

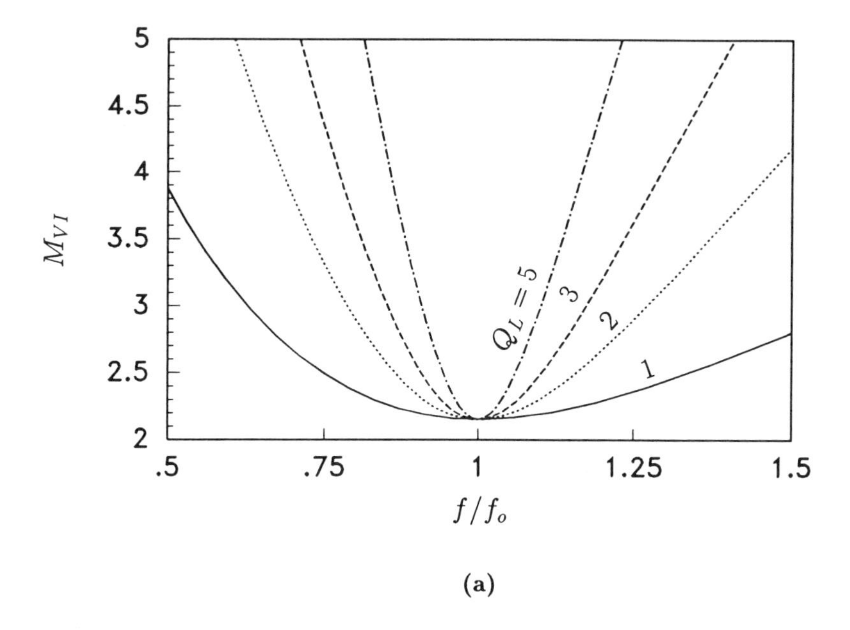

(a)

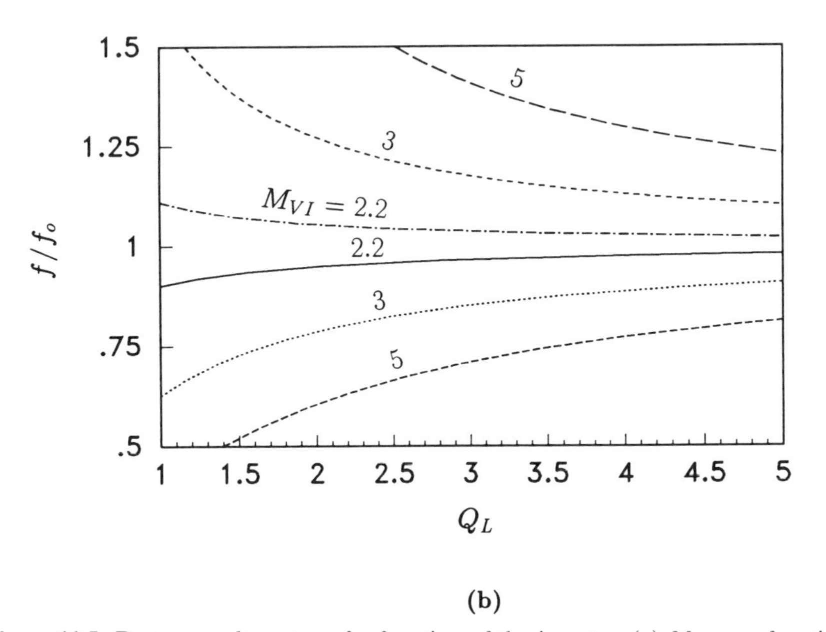

(b)

Figure 11.5: Dc-to-ac voltage transfer function of the inverter. (a) M_{VI} as a function of f/f_o at various values of Q_L. for $f/f_o \leq 1$. (b) f/f_o versus Q_L at fixed values of M_{VI}. (c) M_{VI} as a function of Q_L for $f/f_o \leq 1$. (d) M_{VI} versus Q_L for $f/f_o \geq 1$.

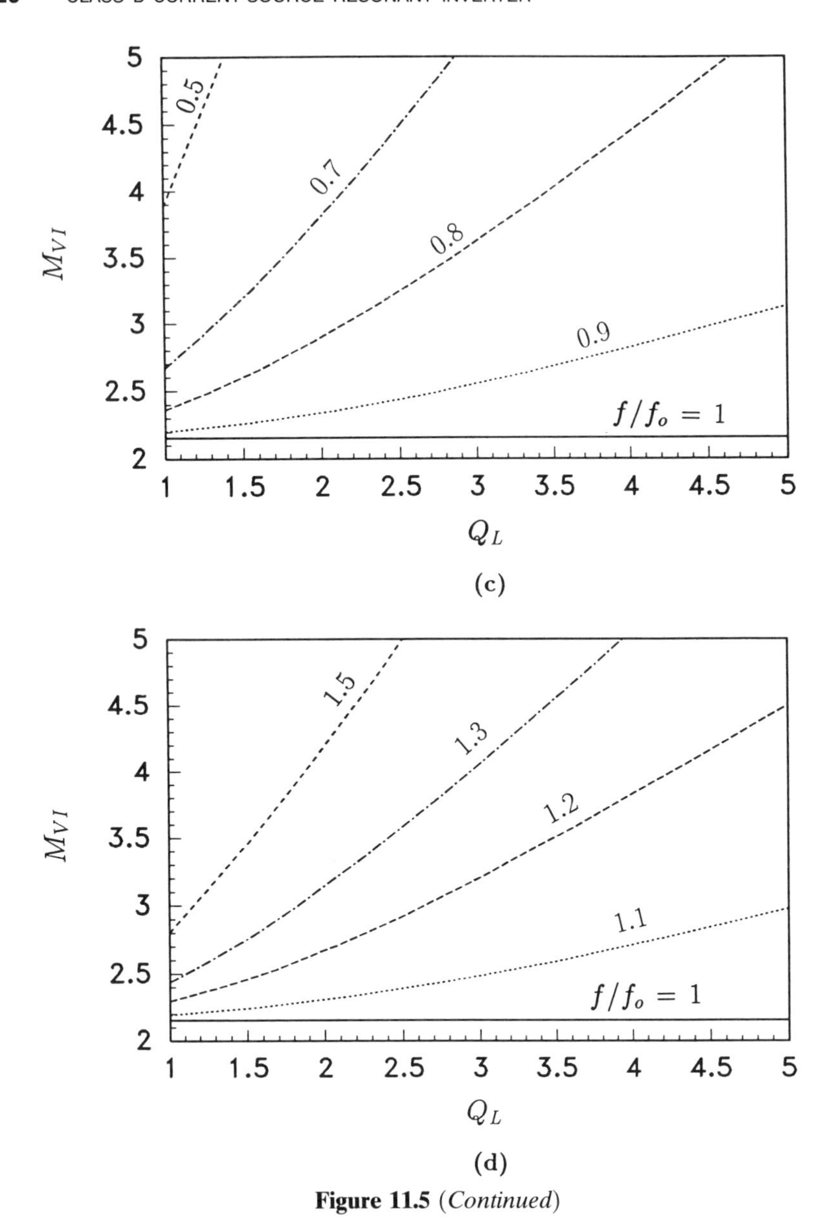

Figure 11.5 *(Continued)*

11.4.2 Output Power

The voltage across the resonant circuit is

$$v_{Ri} = V_{Rim} sin(\omega t + \psi). \tag{11.30}$$

From (11.24), the ac output power of the inverter is

$$P_{Ri} = \frac{V_{Ri}^2}{R_i} = \frac{M_{VI}^2 V_I^2}{R_i} = \frac{\pi^2 \eta_I^2 V_I^2 \{1 + [Q_L(\frac{\omega}{\omega_o} - \frac{\omega_o}{\omega})]^2\}}{2R_i \eta_{rc}^2}$$
$$= \frac{\pi^2 \eta_I^2 V_I^2 \{1 + [Q_L(\frac{\omega}{\omega_o} - \frac{\omega_o}{\omega})]^2\}}{2Q_L Z_o \eta_{rc}}. \quad (11.31)$$

Figure 11.6 shows the normalized output power $P_{Ri}Z_o/V_I^2$ versus f/f_o and Q_L at $\eta_I = 0.95$ and $\eta_{rc} = 0.98$. Plots of $P_{Ri}Z_o/V_I^2$ against f/f_o at fixed values of Q_L for $\eta_I = 0.95$ and $\eta_{rc} = 0.98$ are depicted in Fig. 11.7. From (11.29), the output power at the resonant frequency $f = f_o$ is

$$P_{Ri} = \frac{V_{Rim}^2}{2R_i} = \frac{\pi^2 V_I^2}{2R_i}. \quad (11.32)$$

The maximum values of the amplitudes of the currents through the resonant inductor L and the resonant capacitor C occur at the resonant frequency and are given by

$$I_{Lm} = I_{Cm} = Q_L I_m = \frac{2Q_L I_{Imax}}{\pi}. \quad (11.33)$$

The peak value of the switch current is

$$I_{SM} = I_I. \quad (11.34)$$

11.4.3 Conduction Power Loss

Figure 11.1(b) shows an equivalent circuit of the inverter with parasitic resistances. The conduction power loss in the inverter circuit is mainly present in four major parts. From (11.20),

$$I_I^2 = \frac{\pi^2 V_{Ri}^2 [1 + Q_L^2(\frac{\omega}{\omega_o} - \frac{\omega_o}{\omega})^2]}{2R^2} = \frac{\pi^2 R_i [1 + Q_L^2(\frac{\omega}{\omega_o} - \frac{\omega_o}{\omega})^2]}{2R^2} P_{Ri}. \quad (11.35)$$

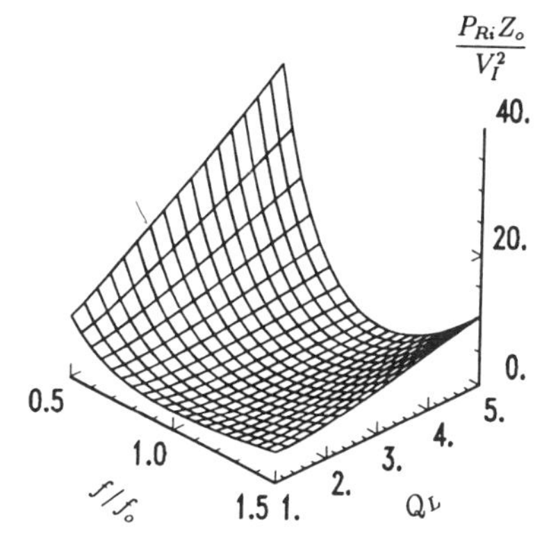

Figure 11.6: Normalized output power $P_{Ri}Z_o/V_I^2$ as a function of f/f_o and Q_L at $\eta_I = 0.95$ and $\eta_{rc} = 0.98$.

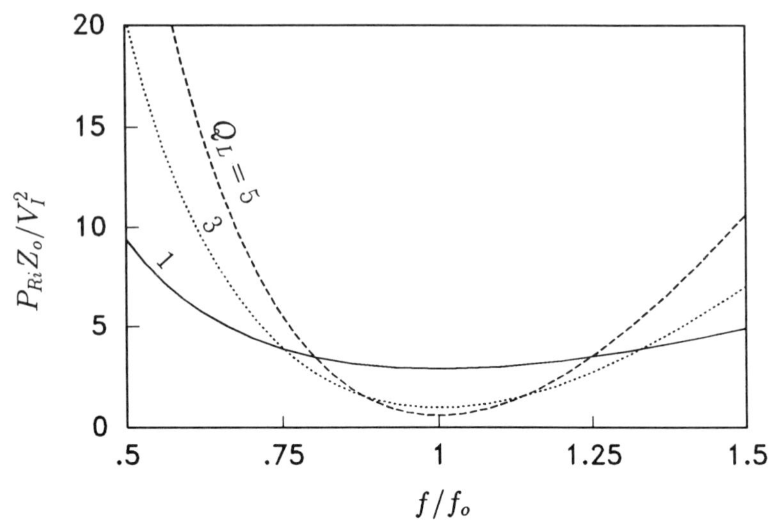

Figure 11.7: Normalized output power $P_{Ri}Z_o/V_I^2$ versus f/f_o for various values of Q_L, $\eta_I = 0.95$, and $\eta_{rc} = 0.98$.

Power Loss in the Inductor L_{fi}

Neglecting the ripple current in L_{fi}, only the dc current I_I flows through the inductor L_{fi}. Using (11.35), the power loss in the dc series equivalent resistance (ESR) r_{Lfi} of L_{fi} is

$$P_{Lfi} = r_{Lfi}I_I^2 = \frac{\pi^2 r_{Lfi}R_i[1 + Q_L^2(\frac{\omega}{\omega_o} - \frac{\omega_o}{\omega})^2]}{2R^2}P_{Ri}. \tag{11.36}$$

Power Loss in the MOSFETs

The MOSFETs are modeled by switches whose on-resistances are r_{DS1} and r_{DS2}. Assume that $r_{DS} = r_{DS1} = r_{DS2}$. From (11.15), the rms value of the current through the switch can be found as

$$I_{S(rms)} = \sqrt{\frac{1}{2\pi}\int_0^{2\pi} i_{S1}^2 d(\omega t)} = \sqrt{\frac{1}{2\pi}\int_\pi^{2\pi} I_I^2 d(\omega t)} = \frac{I_I}{\sqrt{2}}. \tag{11.37}$$

Hence, using (11.35) the conduction power loss in each MOSFET is

$$P_{rDS} = r_{DS}I_{S(rms)}^2 = \frac{r_{DS}I_I^2}{2} = \frac{\pi^2 r_{DS}R_i[1 + Q_L^2(\frac{\omega}{\omega_o} - \frac{\omega_o}{\omega})^2]}{4R^2}P_{Ri}. \tag{11.38}$$

Power Loss in the Series Diodes

The diode is modeled by a voltage source V_{Fi} and an on-resistance R_{Fi}. Let us assume that the forward resistances of the series diodes are identical and equal to R_{Fi}. The average current through the diode is

$$I_{S(AV)} = \frac{1}{2\pi}\int_0^{2\pi} i_{S1} d(\omega t) = \frac{1}{2\pi}\int_{\pi}^{2\pi} I_I d(\omega t) = \frac{I_I}{2}, \tag{11.39}$$

and the power loss associated with V_{Fi} is

$$P_{VF} = V_{Fi} I_{S(AV)} = \frac{V_{Fi} I_I}{2} = \frac{V_{Fi} V_I I_I}{2V_I} = \frac{V_{Fi} P_I}{2V_I} = \frac{V_{Fi} P_{Ri}}{2V_I \eta_I} \approx \frac{V_{Fi}}{2V_I} P_{Ri}. \tag{11.40}$$

From (11.35) and (11.37), the power loss in R_{Fi} is found as

$$P_{rF} = R_{Fi} I_{Srms}^2 = \frac{R_{Fi} I_I^2}{2} = \frac{\pi^2 R_{Fi} R_i [1 + Q_L^2 (\frac{\omega}{\omega_o} - \frac{\omega_o}{\omega})^2]}{4R^2} P_{Ri}. \tag{11.41}$$

Therefore, from (11.40) and (11.41) the total conduction power loss in each diode can be written as

$$P_D = P_{VF} + P_{rF} = \frac{V_{Fi} I_I}{2} + \frac{I_I^2 R_{Fi}}{2} = \frac{I_I^2}{2}\left(\frac{V_{Fi}}{I_I} + R_{Fi}\right)$$

$$= \left\{ \frac{\pi^2 R_i R_{Fi} [1 + Q_L^2 (\frac{\omega}{\omega_o} - \frac{\omega_o}{\omega})^2]}{4R^2} + \frac{V_{Fi}}{2V_I} \right\} P_{Ri}. \tag{11.42}$$

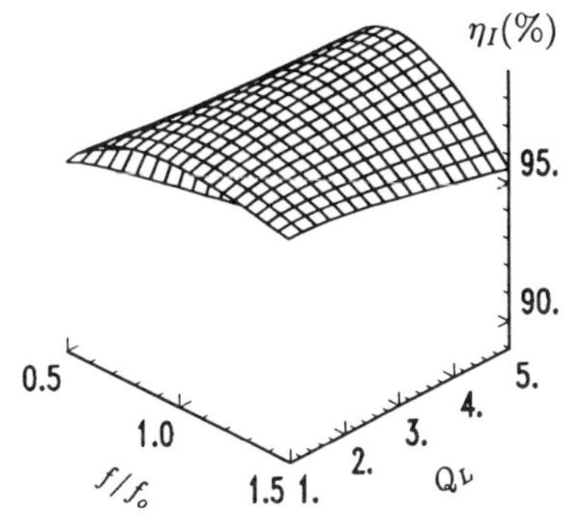

Figure 11.8: Three-dimensional representation of the efficiency η_I as a function of Q_L and f/f_o at $V_{Fi} = 0.7$ V, $r = 1.4\ \Omega$, $R_i = 2.918$ kΩ, $R_d = 318$ kΩ, and $V_I = 120$ V.

Power Loss in the Resonant Circuit

The resistance R_d was defined as the parallel combination of the parallel equivalent resistance R_{Lp} of L and the parallel equivalent resistance R_{Cp} of C. The total conduction power loss in the resonant circuit P_{Rd} can be obtained as

$$P_{Rd} = \frac{V_{Ri}^2}{R_d} = \frac{V_{Ri}^2}{R_i}\frac{R_i}{R_d} = \frac{R_i}{R_d}P_{Ri}. \tag{11.43}$$

11.4.4 Efficiency

The total conduction power loss P_r in the current-source inverter is obtained by using (11.36), (11.38), (11.42), and (11.43)

$$\begin{aligned} P_r &= P_{Lfi} + 2P_{rDS} + 2P_D + P_{Rd} \\ &= \left\{ (r_{Lfi} + r_{DS} + R_{Fi})\frac{\pi^2 R_i[1 + Q_L^2(\frac{\omega}{\omega_o} - \frac{\omega_o}{\omega})^2]}{2R^2} + \frac{V_{Fi}}{V_I} + \frac{R_i}{R_d} \right\} P_{Ri} \\ &= \left\{ \frac{\pi^2 r R_i[1 + Q_L^2(\frac{\omega}{\omega_o} - \frac{\omega_o}{\omega})^2]}{2R^2} + \frac{V_{Fi}}{V_I} + \frac{R_i}{R_d} \right\} P_{Ri}, \end{aligned} \tag{11.44}$$

where

$$r = r_{Lfi} + r_{DS} + R_{Fi}. \tag{11.45}$$

Neglecting switching losses, the dc input power of the inverter is

$$P_I = P_{Ri} + P_r, \tag{11.46}$$

and the efficiency of the inverter can be found using (11.31), (11.44), and (11.46)

$$\begin{aligned} \eta_I &= \frac{P_{Ri}}{P_I} = \frac{P_{Ri}}{P_{Ri} + P_r} = \frac{1}{1 + \frac{P_r}{P_{Ri}}} = \frac{1}{1 + \left\{ \frac{\pi^2 r R_i[1+Q_L^2(\frac{\omega}{\omega_o} - \frac{\omega_o}{\omega})^2]}{2R^2} + \frac{V_{Fi}}{V_I} + \frac{R_i}{R_d} \right\}} \\ &= \frac{1}{1 + \left\{ \frac{\pi^2 r[1+Q_L^2(\frac{\omega}{\omega_o} - \frac{\omega_o}{\omega})^2]}{2\eta_{rc} Q_L Z_o} + \frac{V_{Fi}}{V_I} + \frac{R_i}{R_d} \right\}}. \end{aligned} \tag{11.47}$$

Figures 10.8 and 11.9 show the efficiency η_I as a function of f/f_o and Q_L at $V_{Fi} = 0.7$ V, $r = 1.4\ \Omega$, $R_i = 2.918$ kΩ, $R_d = 318$ kΩ, and $V_I = 120$ V.

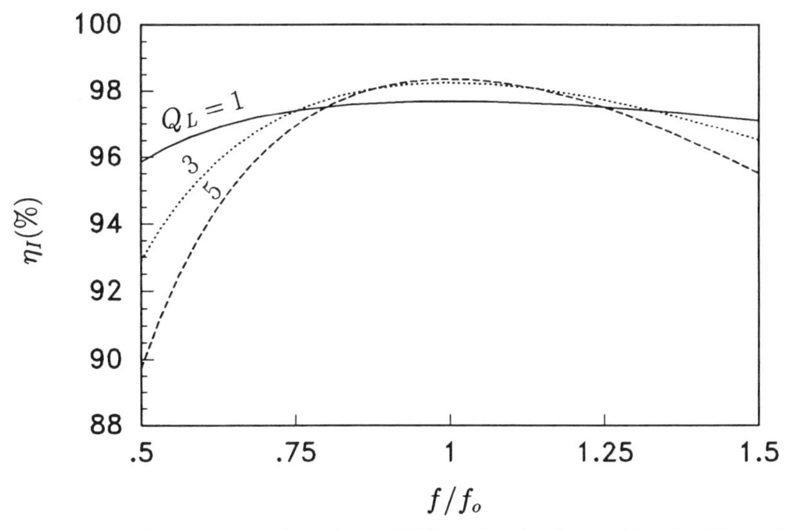

Figure 11.9: Efficiency η_I as a function of f/f_o at fixed values of Q_L for $V_{Fi} = 0.7$ V, $r = 1.4\ \Omega$, $R_i = 2.918$ kΩ, $R_d = 318$ kΩ, and $V_I = 120$ V.

11.5 DESIGN EXAMPLE

Example 11.1

Design a current-source inverter shown in Fig. 11.1(a) to meet the following specifications: $V_I = 120$ V, $R_{imin} = 3000\ \Omega$, and $P_{Rimax} = 25$ W. Assume $f_o = 100$ kHz, the total efficiency of the inverter $\eta_I = 95\%$, and the ratio $R/R_i = 0.98$.

Solution: The maximum dc input power is

$$P_{Imax} = \frac{P_{Rimax}}{\eta_I} = \frac{25}{0.95} = 26.3\text{W}, \tag{11.48}$$

and the maximum value of the dc input current and the switch current is

$$I_{Imax} = I_{SMmax} = \frac{P_{Imax}}{V_I} = \frac{26.3}{120} = 0.22\text{ A}. \tag{11.49}$$

The total resistance of the inverter is $R = R_i(R/R_i) = 3000 \times 0.98 = 2940\ \Omega$. Using (11.24), relationship $V_{Ri} = \sqrt{P_{Ri}R_i}$ and assuming $\omega/\omega_o = 0.95$, the operating frequency is $f = 95$ kHz and the loaded quality factor can be calculated as

$$Q_L = \frac{\sqrt{\frac{2R^2 P_{Ri}}{\pi^2 \eta_I^2 R_i V_I^2} - 1}}{\mid \frac{\omega}{\omega_o} - \frac{\omega_o}{\omega} \mid} = \frac{\sqrt{\frac{2\times 2940^2 \times 25}{\pi^2 \times 0.95^2 \times 3000 \times 120^2} - 1}}{\mid 0.95 - \frac{1}{0.95} \mid} = 3.42. \quad (11.50)$$

This value of Q_L ensures an almost sinusoidal voltage across the resonant circuit at full load, which justifies the use of fundamental component method in the design.

The component values of the resonant circuit are

$$L = \frac{R}{\omega_o Q_L} = \frac{2940}{2 \times \pi \times 100 \times 10^3 \times 3.42} = 1.368 \text{ mH} \quad (11.51)$$

and

$$C = \frac{Q_L}{\omega_o R} = \frac{3.42}{2 \times \pi \times 100 \times 10^3 \times 2940} = 1.851 \text{ nF}. \quad (11.52)$$

The characteristic impedance is $Z_o = \sqrt{L/C} = 860\ \Omega$. Let us assume that the peak-to-peak value of the the ripple current in the choke inductor cannot exceed 0.1 A. The minimum value of the choke inductance can be obtained using (11.2) as

$$L_{fi(min)} = \frac{V_I}{2f I_{r(max)}} = \frac{120}{2 \times 95 \times 10^3 \times 0.1} = 6.3 \text{ mH}. \quad (11.53)$$

From (11.33), the maximum value of the current through the resonant inductor L and the resonant capacitor C is

$$I_{Lm} = I_{Cm} = \frac{2Q_L I_{Imax}}{\pi} = \frac{2 \times 3.42 \times 0.22}{\pi} = 0.479 \text{ A}, \quad (11.54)$$

and the maximum value of the switch voltage and the voltage across the resonant circuit is

$$V_{SM} = V_{Rim} = \sqrt{2 P_{Rimax} R_{imin}} = \sqrt{2 \times 25 \times 3000} = 387.3 \text{ V}. \quad (11.55)$$

Using (11.36) and assuming that the dc resistance of the choke is $r_{Lfi} = 0.5\ \Omega$, the conduction power loss in r_{Lfi} is

$$P_{Lfi} = r_{Lfi} I_I^2 = 0.5 \times 0.22^2 = 0.024 \text{ W}. \quad (11.56)$$

From (11.37), the rms value of the switch current is

$$I_{Srms} = \frac{I_I}{\sqrt{2}} = \frac{0.22}{\sqrt{2}} = 0.156 \text{ A}. \quad (11.57)$$

If $r_{DS} = 1\ \Omega$, the conduction power loss in each transistor is

$$P_{rDS} = \frac{r_{DS} I_I^2}{2} = \frac{1 \times 0.22^2}{2} = 0.024\ \text{W}. \quad (11.58)$$

The typical parameters for the series diode are $V_{Fi} = 0.7$ V and $R_{Fi} = 0.025$ mΩ. Hence, the conduction loss in each series diode is

$$P_D = \frac{V_{Fi} I_I}{2} + \frac{I_I^2 R_{Fi}}{2} = \frac{0.7 \times 0.22}{2} + \frac{0.22^2 \times 0.025}{2} = 0.077\ \text{W}. \quad (11.59)$$

Assume that the unloaded quality factors of the resonant inductor and the resonant capacitor at $f = 95$ kHz are $Q_{Lo} = R_{Lp}/\omega L = 275$ and $Q_{Co} = R_{Cp}\omega C = 1800$, respectively. Hence,

$$R_{Lp} = \omega L Q_{Lo} = 2 \times \pi \times 95 \times 10^3 \times 1.368 \times 10^{-3} \times 275 = 225\ \text{k}\Omega, \quad (11.60)$$

$$R_{Cp} = \frac{Q_{Co}}{\omega C} = \frac{1800}{2 \times \pi \times 95 \times 10^3 \times 1.851 \times 10^{-9}} = 1.629\ \text{M}\Omega, \quad (11.61)$$

and

$$R_d = \frac{R_{Lp} R_{Cp}}{R_{Lp} + R_{Cp}} = \frac{225 \times 1629}{225 + 1629} = 198\ \text{k}\Omega. \quad (11.62)$$

Thus, the power loss in the resonant circuit is

$$P_{Rd} = \frac{V_{Rim}^2}{2R_d} = \frac{387.3^2}{2 \times 198 \times 10^3} = 0.379\ \text{W}. \quad (11.63)$$

The total conduction loss is

$$\begin{aligned} P_r &= 2P_{rDS} + 2P_D + P_{Rd} + P_{Lfi} \\ &= 2 \times 0.024 + 2 \times 0.077 + 0.379 + 0.024 = 0.605\ \text{W}, \end{aligned} \quad (11.64)$$

resulting in the inverter efficiency associated with the conduction loss

$$\eta_I = \frac{P_{Ri}}{P_{Ri} + P_r} = \frac{25}{25 + 0.605} = 97.64\%. \quad (11.65)$$

11.6 SUMMARY

- The current-source inverter has a nonpulsating input current with a very small ac ripple.
- It is easy to drive because both gates are referenced to ground.
- The inverter efficiency decreases while increasing the ac load resistance R_i because the ratio R_i/R_d increases with increasing R_i.

- Operation closer to the resonant frequency is desirable because the inverter does not draw a high current at this frequency.
- Above resonance, the series diodes of the inverter turn off at a very high di/dt, causing high reverse-recovery turn-off switching losses. In addition, the transistors turn on at nonzero voltage, resulting in turn-on switching losses.
- Below resonance, the series diodes of the inverter turn off at zero di/dt, yielding zero reverse-recovery losses. The MOSFETs turn on at a low voltage, resulting in low switching losses. Therefore, the efficiency below resonance is higher than that above resonance.

11.7 REFERENCES

1. P. J. Baxandall, "Transistor sine-wave *LC* oscillators, some general considerations and new developments," *Proc. IEE*, vol. 106, pt. B, pp. 748–758, May 1959.
2. M. R. Osborne, "Design of tuned transistor power inverters," *Electron. Eng.*, vol. 40, no. 486, pp. 436–443, 1968.
3. W. J. Chudobiak and D. F. Page, "Frequency and power limitations of Class-D transistor inverter," *IEEE J. Solid-State Circuits*, vol. SC–4, pp. 25–37, Feb. 1969.
4. H. L. Krauss, C. W. Bostian, and F. H. Raab, *Solid State Radio Engineering*, New York: John Wiley & Sons, ch. 14.1, pp. 439–441, 1980.
5. J. G. Kassakian, "A new current mode sine wave inverter," *IEEE Trans. Industry Applications*, vol. IA–18, pp. 273–278, May/June 1982.
6. R. L. Steigerwald, "High-frequency resonant transistor dc-dc converters," *IEEE Trans. Industrial Electronics*, vol. IE–31, pp. 181–191, May 1984.
7. J. G. Kassakian, M. F. Schlecht, and G. C. Verghese, *Principles of Power Electronics*, Reading, MA: Addison-Wesley, 1991, ch. 9.3, pp. 212–217.

11.8 REVIEW QUESTIONS

11.1 Is it difficult to drive the Class D current-source resonant inverter?

11.2 Is the input current pulsating in the current-source inverter?

11.3 Is the operation of the Class D current-source inverter below or above resonance preferred?

11.4 What is the voltage stress of the switches?

11.5 What is the current stress of the switches?

11.6 What is the behavior of the transfer function in the Class D current-source resonant inverter versus frequency?

11.7 Is the operation of the inverter safe under short-circuit conditions?

11.8 Is the operation of the inverter safe under open-circuit conditions?

11.9 PROBLEMS

11.1 In the parallel resonant circuit of Fig. 11.1, the load resistance is $R_i = 500\ \Omega$ and the efficiency is $\eta_{rc} = 99\%$. What is the EPR of the resonant inductor if the EPR of the resonant capacitor is $R_{Cp} = 100\ \text{k}\Omega$?

11.2 A parallel resonant circuit with $R = 300\ \Omega$, $L = 500\ \mu\text{H}$, and $C = 2$ nF is driven by a sinusoidal current source $i = 2sin(800 \times 10^3 t)$. Calculate the power dissipated in this circuit.

11.3 Assume that the frequency of the current source of Problem 11.2 can be adjusted. At what frequency are the current and voltage stresses of the resonant components maximum? Calculate those stresses.

11.4 Derive equation (11.25).

11.5 Design a current-source inverter of Fig. 11.1. The following specifications should be satisfied: $V_I = 200$ V, $R_{imin} = 1000\ \Omega$, and $P_{Rimax} = 200$ W. Assume the resonant frequency $f_o = 100$ kHz, the normalized switching frequency $f/f_o = 0.95$, the total efficiency of the inverter $\eta_I = 95\%$, and the ratio $R/R_i = 0.99$.

CHAPTER 12

PHASE-CONTROLLED RESONANT INVERTERS

12.1 INTRODUCTION

The ac output current, voltage, or power in most resonant dc-ac inverters and the dc output voltage in most resonant dc-dc converters is controlled by varying the switching frequency f. This method is called frequency modulation (FM) control. The FM control implies a variable-frequency operation, which has several disadvantages: 1) a wide and unpredictable noise spectrum, causing difficulty in EMI control, 2) more complex filtering of the output-voltage ripple, and 3) poor utilization of magnetic components. Noise generated by pulse-width modulated and FM-controlled power processors is a major problem in industrial controls, where computer systems are being used more each day for control of high-power systems. The problem of EMI pollution is very serious for computer systems. It can interfere with communications, system operations, and can even damage components if currents or voltages are too large. The engineer is faced with two choices to guard the computer systems in such an environment. The first option is to electromagnetically isolate the computer system. This is unlikely because the system should be integrated with the rest of the manufacturing equipment. The second choice is to remove or reduce the source of noise. In addition to just good design practice, EMI control for industrial plants is required by various regulations (e.g., ANSI Std C63.12-1987, IEC555-3, VDE0160, FCC, and FAA). Lately, these laws are becoming more strict and engineers not only should be concerned with a particular piece of equipment but the combined effect of all equipment used in the facility.

The problems mentioned above can be overcome by using fixed-frequency phase-controlled (PC), also called "outphasing modulation" dc-ac inverters.

Phase-controlled power systems were first developed during the 1930s by Chireix [1] to improve the efficiency of AM radio transmitters.

Since then, phase-controlled power processors have been studied by many researchers [2]–[26]. The principle of operation of phase-controlled inverters is as follows. Two power resonant dc-ac inverters synchronized to the same switching frequency drive the same ac load. The phase shift between the output currents or voltages of the dc-ac inverters is varied to control the ac output current, voltage, or power. The same concept can be used in resonant dc-dc converters to regulate the dc output voltage against load and line variations. The phase-controlled power processors have a fast dynamic response. The objective of this chapter is to present one of many possible topologies of phase-controlled resonant inverters as an example. A single-capacitor phase-controlled series resonant inverter (SC PC SRI) has been selected for that purpose. Two other topologies of phase-controlled resonant inverters are analyzed in Problems at the end of the chapter.

12.2 ANALYSIS OF SC PC SRI

12.2.1 Circuit Description

A single-capacitor fixed-frequency phase-controlled full-bridge Class D inverter is shown in Fig. 12.1(a). It consists of a dc input voltage source V_I, two switching legs, two resonant inductors L, one resonant capacitor C, a

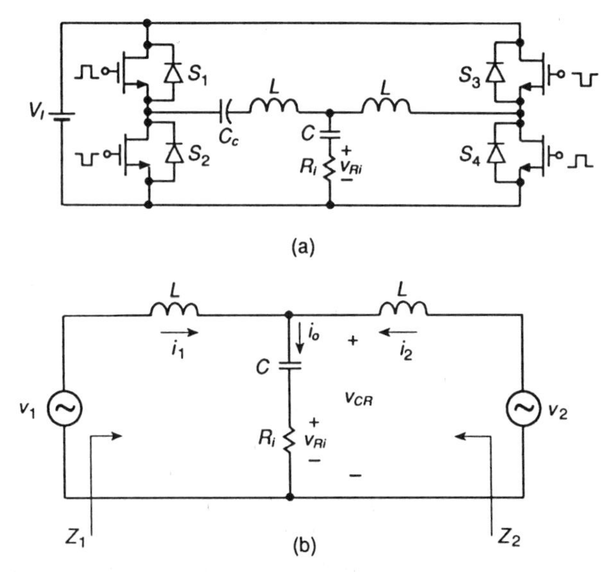

Figure 12.1: Single-capacitor fixed-frequency phase-controlled full-bridge Class D series resonant inverter. (a) Circuit. (b) Equivalent circuit for the fundamental components.

coupling capacitor C_C, and an ac load R_i. Each switching leg comprises two switches with antiparallel diodes. The switches in both legs are turned on and off alternately by rectangular voltage sources at a frequency $f = \omega/(2\pi)$ with a duty cycle of slightly less than 50%. The gate-drive voltages of either switching leg have a dead time to achieve zero-voltage-switching turn-on for all the transistors over a wide range of the line voltage and from full load to no load. When the drain-source voltages decrease from V_I to zero, the transistors are off. Therefore, the transistor output capacitances, the stray capacitances, and snubbing external capacitances that can be added in parallel with one of the transistors in each switching leg are discharged by inductor currents. The transistors are turned on when their voltage is zero, yielding zero turn-on switching loss. However, zero-voltage-switching turn-on of all the transistors can only be achieved for inductive loads of both switching legs. In such a case, the inductor currents are negative during the dead time. Consequently, they can discharge the transistor shunt capacitances and bring the transistor voltages to zero before the turn-on drive signal. In addition, the amplitudes of the inductor currents should be high at light loads to completely discharge the transistor shunt capacitances during the dead time. In most full-bridge inverters with a single resonant circuit, the load of one switching leg is inductive and the other capacitive. Therefore, the MOSFETs in the switching leg with a capacitive load have large turn-on switching losses, resulting in a poor efficiency.

12.2.2 Assumptions

The analysis of the single-capacitor phase-controlled Class D inverter of Fig. 12.1(a) is performed under the following simplifying assumptions:

1) The loaded quality factor Q_L of the resonant circuit is high enough (e.g., $Q_L > 3$) that the currents through the resonant inductors i_1 and i_2 are sinusoidal.
2) The power MOSFETs are modeled by switches with a linear ON-resistance r_{DS}.
3) The reactive components of the resonant circuit are linear, time-invariant, and the operating frequency is much lower than the self-resonant frequencies of the reactive components.
4) Both resonant inductors are identical.

12.2.3 Voltage Transfer Function

In the inverter shown in Fig. 12.1(a), the switching legs and the dc input voltage V_I form square-wave voltage sources. Since the currents i_1 and i_2 through the resonant inductors are sinusoidal, only the power of the fundamental component of each input voltage source is transferred to the output. Therefore, the square wave voltage sources can be replaced by sinusoidal voltage sources representing the fundamental components as shown in Fig. 12.1(b). These components are

$$v_1 = V_m cos\left(\omega t + \frac{\phi}{2}\right) \tag{12.1}$$

and

$$v_2 = V_m cos\left(\omega t - \frac{\phi}{2}\right), \tag{12.2}$$

where

$$V_m = \frac{2}{\pi}V_I \tag{12.3}$$

and ϕ is the phase shift between v_1 and v_2. The voltages at the inputs of the resonant circuits are expressed in the complex domain by

$$\mathbf{V_1} = V_m e^{j(\phi/2)} \tag{12.4}$$

and

$$\mathbf{V_2} = V_m e^{-j(\phi/2)}. \tag{12.5}$$

The voltages across the branch C-R_i caused by voltage sources $\mathbf{V_1}$ and $\mathbf{V_2}$ separately, that is, with the other voltage source shorted, are expressed as

$$\mathbf{V_{CR1}} = \frac{\mathbf{V_1}[(R_i + \frac{1}{j\omega C}) \parallel j\omega L]}{j\omega L + [(R_i + \frac{1}{j\omega C}) \parallel j\omega L]} \tag{12.6}$$

and

$$\mathbf{V_{CR2}} = \frac{\mathbf{V_2}[(R_i + \frac{1}{j\omega C}) \parallel j\omega L]}{j\omega L + [(R_i + \frac{1}{j\omega C}) \parallel j\omega L]}, \tag{12.7}$$

respectively. Using the principle of superposition, one obtains the voltage across the branch C-R_i

$$\mathbf{V_{CR}} = \mathbf{V_{CR1}} + \mathbf{V_{CR2}} = \frac{V_m cos(\frac{\phi}{2})(1 - jQ_L\frac{\omega_o}{\omega})}{1 + jQ_L(\frac{\omega}{\omega_o} - \frac{\omega_o}{\omega})} \tag{12.8}$$

and the output voltage of the inverter

$$\mathbf{V_{Ri}} = \frac{\mathbf{V_{CR}}R_i}{R_i + \frac{1}{j\omega C}} = \frac{V_m cos(\frac{\phi}{2})}{1 + jQ_L(\frac{\omega}{\omega_o} - \frac{\omega_o}{\omega})} = \frac{2V_I cos(\frac{\phi}{2})}{\pi[1 + jQ_L(\frac{\omega}{\omega_o} - \frac{\omega_o}{\omega})]}, \tag{12.9}$$

where

$$\omega_o = \sqrt{\frac{2}{LC}} \tag{12.10}$$

is the resonant frequency,

$$Q_L \equiv \frac{\omega_o L}{2R_i} = \frac{Z_o}{2R_i} \tag{12.11}$$

is the loaded quality factor, and

$$Z_o = \sqrt{\frac{2L}{C}} \tag{12.12}$$

is the characteristic impedance of the resonant circuit. The factor 2 in those definitions arises from the fact that the series connection C-R_i can be divided into two identical branches $C/2$-$2R_i$ at $\phi = 0$. Thus, one can imagine the SC PC SRC as two identical conventional series resonant inverters: inverter 1 and inverter 2, in which the resonant capacitors and the ac loads are connected in parallel. Rearrangement of (12.9) gives the dc-to-ac voltage transfer function of the phase-controlled Class D inverter

$$\mathbf{M_{VI}} \equiv \frac{\mathbf{V_{Ri}}}{\sqrt{2}V_I} = \frac{\sqrt{2}cos(\frac{\phi}{2})}{\pi[1 + jQ_L(\frac{\omega}{\omega_o} - \frac{\omega_o}{\omega})]}. \tag{12.13}$$

Hence, the magnitude of $\mathbf{M_{VI}}$ is

$$M_{VI} = \frac{V_{Ri(rms)}}{V_I} = \frac{\sqrt{2}cos(\frac{\phi}{2})}{\pi\sqrt{1 + Q_L^2(\frac{\omega}{\omega_o} - \frac{\omega_o}{\omega})^2}}. \tag{12.14}$$

Figure 12.2 illustrates M_{VI} as a function of loaded quality factor Q_L and phase shift ϕ at $f/f_o = 1.33$. The dc-to-ac voltage transfer function of the actual inverter is

$$\mathbf{M_{VIa}} = \eta_I \mathbf{M_{VI}}, \tag{12.15}$$

where η_I is the efficiency of the inverter, derived in Section 12.2.6.

12.2.4 Currents

The currents through the resonant inductors L of inverter 1 and inverter 2 are given by

$$\mathbf{I_1} = \frac{\mathbf{V_1} - \mathbf{V_{CR}}}{j\omega L} = \frac{2V_I\{\frac{\omega_o}{\omega}sin(\frac{\phi}{2}) + Q_L cos(\frac{\phi}{2}) - jQ_L sin(\frac{\phi}{2})[(\frac{\omega_o}{\omega})^2 - 1]\}}{\pi Z_o[1 + jQ_L(\frac{\omega}{\omega_o} - \frac{\omega_o}{\omega})]} \tag{12.16}$$

and

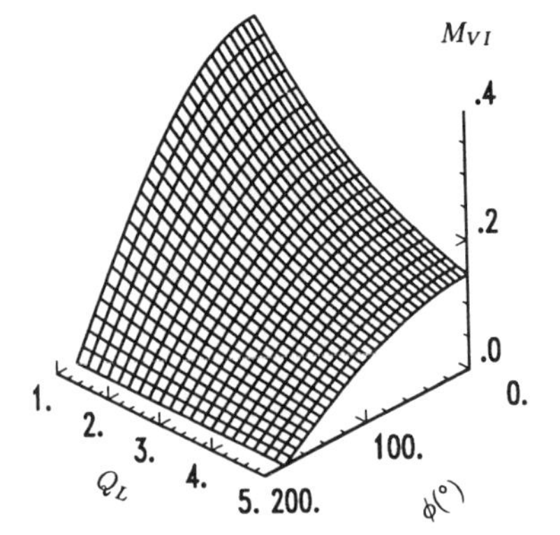

Figure 12.2: Three-dimensional representation of the magnitude of the dc-to-ac transfer function of the phase-controlled Class D inverter M_{VI} as a function of Q_L and ϕ at $f/f_o = 1.33$.

$$\mathbf{I_2} = \frac{\mathbf{V_2} - \mathbf{V_{CR}}}{j\omega L} = \frac{2V_I\{-\frac{\omega_o}{\omega}sin(\frac{\phi}{2}) + Q_L cos(\frac{\phi}{2}) + jQ_L sin(\frac{\phi}{2})[(\frac{\omega_o}{\omega})^2 - 1]\}}{\pi Z_o[1 + jQ_L(\frac{\omega}{\omega_o} - \frac{\omega_o}{\omega})]}. \tag{12.17}$$

Hence, one arrives at the amplitude of the current through the resonant inductor of inverter 1

$$I_{m1} = \frac{2V_I}{\pi Z_o}\sqrt{\frac{[\frac{\omega_o}{\omega}sin(\frac{\phi}{2}) + Q_L cos(\frac{\phi}{2})]^2 + Q_L^2 sin^2(\frac{\phi}{2})[(\frac{\omega_o}{\omega})^2 - 1]^2}{1 + Q_L^2(\frac{\omega}{\omega_o} - \frac{\omega_o}{\omega})^2}} \tag{12.18}$$

and the amplitude of the current through the resonant inductor of inverter 2

$$I_{m2} = \frac{2V_I}{\pi Z_o}\sqrt{\frac{[\frac{\omega_o}{\omega}sin(\frac{\phi}{2}) - Q_L cos(\frac{\phi}{2})]^2 + Q_L^2 sin^2(\frac{\phi}{2})[(\frac{\omega_o}{\omega})^2 - 1]^2}{1 + Q_L^2(\frac{\omega}{\omega_o} - \frac{\omega_o}{\omega})^2}}. \tag{12.19}$$

Figure 12.3 shows normalized amplitudes $I_{m1}Z_o/V_I$ and $I_{m2}Z_o/V_I$ as functions of ϕ at $f/f_o = 1.33$ and $Q_L = 1$, 3, and 5. The output current is

$$\mathbf{I_o} = \mathbf{I_1} + \mathbf{I_2} = \frac{4V_I Q_L cos(\frac{\phi}{2})}{\pi Z_o[1 + jQ_L(\frac{\omega}{\omega_o} - \frac{\omega_o}{\omega})]}. \tag{12.20}$$

12.2.5 Boundary Between Capacitive and Inductive Load

To determine whether the switches are loaded capacitively or inductively, the impedances seen by the switching legs at the fundamental frequency are calculated and their angles are examined. From (12.4) and (12.16), the impedance seen by the voltage source v_1 is

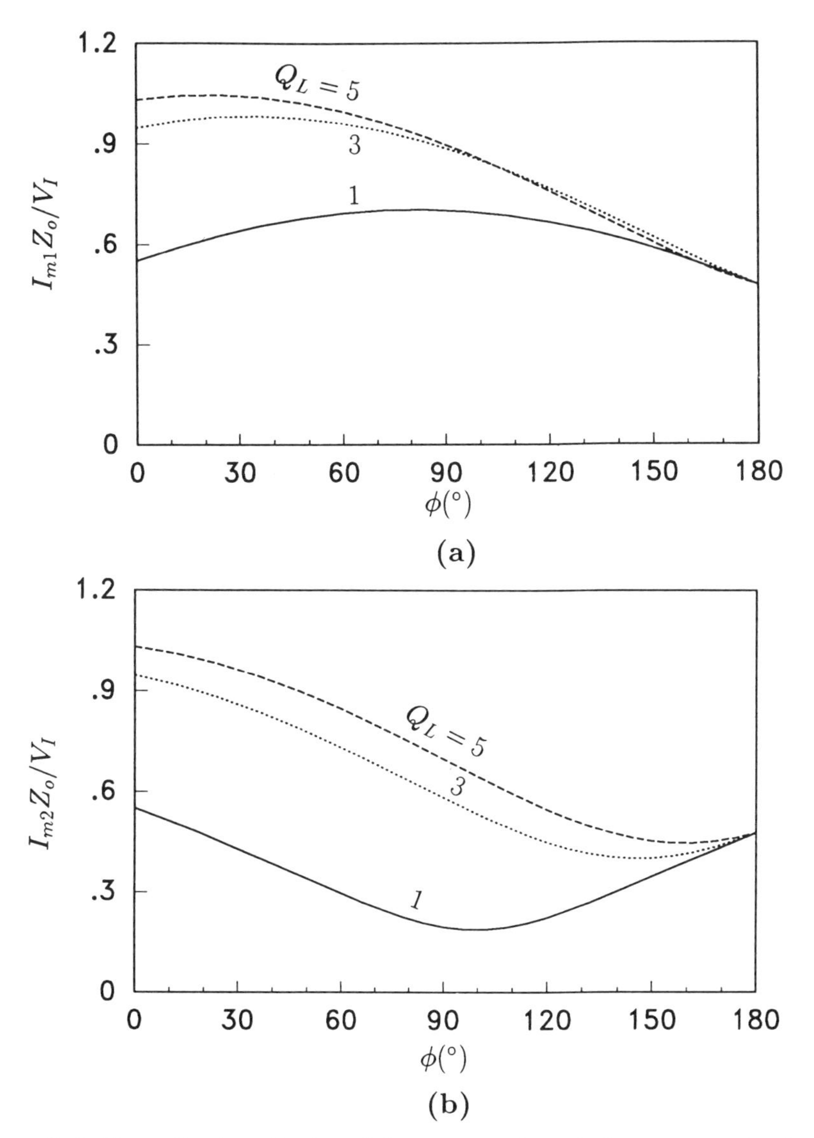

Figure 12.3: Normalized amplitudes of the currents through the resonant circuits as functions of ϕ at $f/f_o = 1.33$ and $Q_L = 1$, 3, and 5. (a) $I_{m1}Z_o/V_I$ versus ϕ. (b) $I_{m2}Z_o/V_I$ versus ϕ.

$$\mathbf{Z_1} \equiv \frac{\mathbf{V_1}}{\mathbf{I_1}} = Z_o\left(\frac{\omega}{\omega_o}\right)$$
$$\times \frac{cos(\frac{\phi}{2})Q_L(\frac{\omega}{\omega_o} - \frac{\omega_o}{\omega}) + sin(\frac{\phi}{2}) + j[sin(\frac{\phi}{2})Q_L(\frac{\omega}{\omega_o} - \frac{\omega_o}{\omega}) - cos(\frac{\phi}{2})]}{sin(\frac{\phi}{2})Q_L(\frac{\omega}{\omega_o} - \frac{\omega_o}{\omega}) - j[sin(\frac{\phi}{2}) + Q_L\frac{\omega}{\omega_o}cos(\frac{\phi}{2})]}$$
$$= Z_1 e^{j\psi_1}, \tag{12.21}$$

and from (12.5) and (12.17) the impedance seen by the voltage source v_2 is

$$\mathbf{Z_2} \equiv \frac{\mathbf{V_2}}{\mathbf{I_2}} = Z_o\left(\frac{\omega}{\omega_o}\right) \times \frac{cos(\frac{\phi}{2})Q_L(\frac{\omega}{\omega_o} - \frac{\omega_o}{\omega}) - sin(\frac{\phi}{2}) - j[sin(\frac{\phi}{2})Q_L(\frac{\omega}{\omega_o} - \frac{\omega_o}{\omega}) + cos(\frac{\phi}{2})]}{-sin(\frac{\phi}{2})Q_L(\frac{\omega}{\omega_o} - \frac{\omega_o}{\omega}) + j[sin(\frac{\phi}{2}) - Q_L\frac{\omega}{\omega_o}cos(\frac{\phi}{2})]} = Z_2 e^{j\psi_2}. \tag{12.22}$$

Figure 12.4 depicts principal arguments ψ_1 and ψ_2 as functions of ϕ at $f/f_o = 1.33$ and $Q_L = 1, 3$, and 5. Numerical calculations and experimental evidence show that ψ_1 and ψ_2 are positive for $f/f_o > 1.15$ at any load and any phase shift. This indicates that both the inverter 1 and the inverter 2 are loaded by inductive loads for $f/f_o > 1.15$.

The maximum value of the amplitude of the current through the resonant inductor $I_{m(max)}$ can be found from (12.18) because I_{m1} is always greater than I_{m2}. Hence, one obtains the maximum amplitude of the voltage across the resonant inductor L as

$$V_{Lm} = \omega L I_{m(max)}. \tag{12.23}$$

The maximum value of the voltage across the resonant capacitor C occurs at a full load and is

$$V_{Cm} = \frac{I_{om(max)}}{\omega C}, \tag{12.24}$$

where $I_{om(max)}$ is the maximum value of the amplitude of the output current of the inverter and can be calculated using (12.20).

12.2.6 Efficiency

From (12.14), the output power of the phase-controlled Class D inverter is obtained as

$$P_{Ri} = \frac{V^2_{Ri(rms)}}{R_i} = \frac{2V_I^2 cos^2(\frac{\phi}{2})}{\pi^2 R_i[1 + Q_L^2(\frac{\omega}{\omega_o} - \frac{\omega_o}{\omega})^2]}. \tag{12.25}$$

The parasitic resistance of each switching leg with resonant inductor can be estimated as $r = r_{DS} + r_L + r_{C_c}/2$, where $r_{DS} = (r_{DS1} + r_{DS2})/2$ is the average resistance of the on-resistances of the MOSFETs, r_L is the ESR of the resonant inductor, and r_{C_c} is the ESR of the coupling capacitor C_c. Hence, one obtains the conduction power loss in the switching legs and the inductors of inverter 1 and inverter 2 as $P_{r1} = rI^2_{m1}/2$ and $P_{r2} = rI^2_{m2}/2$, respectively. Substituting (12.18) and (12.19) into these equations, one obtains the conduction

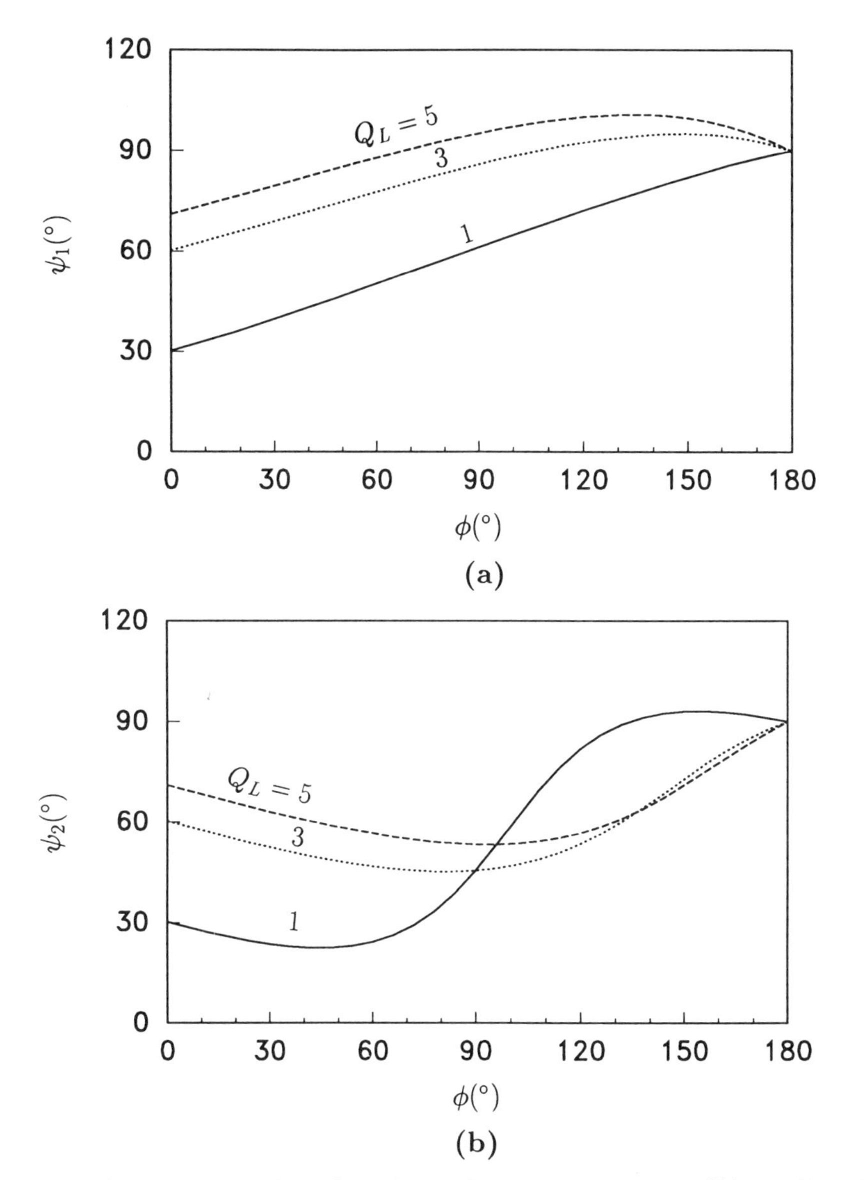

Figure 12.4: Phases of the input impedances Z_1 and Z_2 versus ϕ at $f/f_o = 1.33$ and $Q_L = 1$, 3, and 5. (a) ψ_1 versus ϕ. (b) ψ_2 versus ϕ.

loss in four MOSFETs, two inductors L, and coupling capacitor C_c

$$P_{rs} = P_{r1} + P_{r2} = \frac{r(I_{m1}^2 + I_{m2}^2)}{2}$$
$$= \frac{4V_I^2 r\{Q_L^2 + sin^2(\frac{\phi}{2})(\frac{\omega_o}{\omega})^2[Q_L^2(\frac{\omega_o}{\omega})^2 - 2Q_L^2 + 1]\}}{\pi^2 Z_o^2[1 + Q_L^2(\frac{\omega}{\omega_o} - \frac{\omega_o}{\omega})^2]}. \quad (12.26)$$

The conduction loss in the resonant capacitor C is found as

$$P_C = I_{o(rms)}^2 r_C = \frac{8V_I^2 Q_L^2 cos^2(\frac{\phi}{2})}{\pi^2 Z_o^2[1 + Q_L^2(\frac{\omega}{\omega_o} - \frac{\omega_o}{\omega})^2]} r_C, \quad (12.27)$$

where $I_{o(rms)}$ is the rms value of the output current given by (12.20) and r_C is the ESR of the resonant capacitor C. The total conduction loss in the inverter is

$$P_r = P_{rs} + P_C$$
$$= 4V_I^2 \frac{r\{Q_L^2 + sin^2(\frac{\phi}{2})(\frac{\omega_o}{\omega})^2[Q_L^2(\frac{\omega_o}{\omega})^2 - 2Q_L^2 + 1]\} + 2r_C Q_L^2 cos^2(\frac{\phi}{2})}{\pi^2 Z_o^2[1 + Q_L^2(\frac{\omega}{\omega_o} - \frac{\omega_o}{\omega})^2]}. \quad (12.28)$$

Neglecting switching losses, drive power, and second-order effects (such as nonlinear interactions) and using (12.25) and (12.28), one arrives at the efficiency of the phase-controlled inverter

$$\eta_I \equiv \frac{P_{Ri}}{P_{Ri} + P_r} = \frac{1}{1 + \frac{P_r}{P_{Ri}}}$$
$$= \frac{1}{1 + \frac{1}{Z_o Q_L cos^2(\frac{\phi}{2})}\{r[Q_L^2 + sin^2(\frac{\phi}{2})(\frac{\omega_o}{\omega})^2(Q_L^2 \frac{\omega_o^2}{\omega^2} - 2Q_L^2 + 1)] + 2r_C Q_L^2 cos^2(\frac{\phi}{2})\}}. \quad (12.29)$$

Figure 12.5 shows the efficiency of the inverter as a function of phase shift ϕ and loaded quality factor Q_L for $f/f_o = 1.33$, $r = 2\ \Omega$, $r_C = 0.2\ \Omega$, and $Z_o = 226\ \Omega$. It can be seen that the inverter has an excellent efficiency at both full and part loads.

12.3 DESIGN EXAMPLE

Example 12.1

Design a single-capacitor phase-controlled series resonant inverter of Fig. 12.1 to meet the following specifications: the input voltage $V_I = 270$ to 300 V, the maximum output power $P_{Ri} = 80$ W, and the full load ac resistance $R_i = 50\ \Omega$. Assume $f_o = 150$ kHz, $f/f_o = 1.33$, and the inverter efficiency $\eta_I = 95\%$.

Solution:
Consider the case for full power. The rms value of the output voltage is

$$V_{Ri(rms)} = \sqrt{P_{Ri} R_i} = \sqrt{80 \times 50} = 63.25 \text{ V}. \quad (12.30)$$

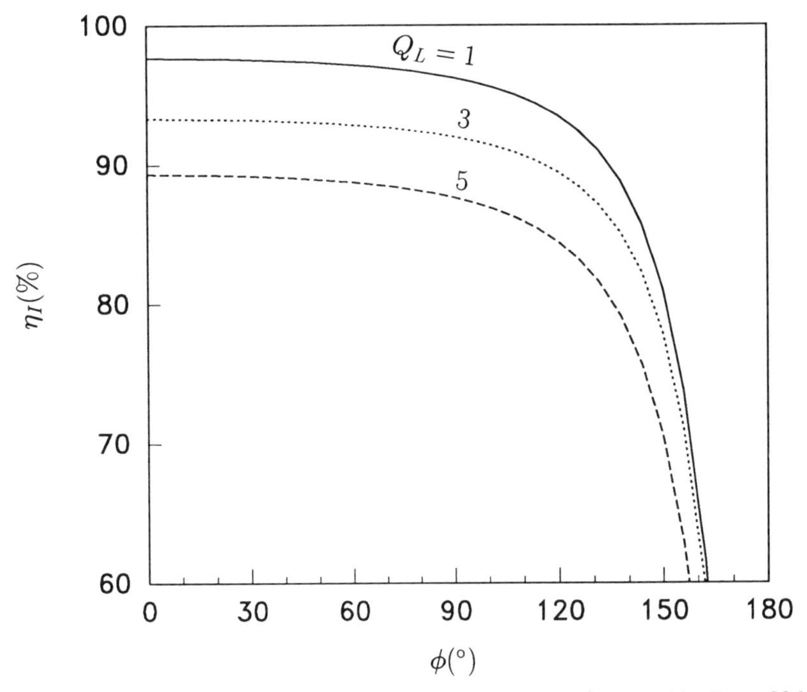

Figure 12.5: Inverter efficiency η_I as a function of ϕ at $f/f_o = 1.33$, $Z_o = 226\ \Omega$, $r = 2\ \Omega$, $r_C = 0.2\ \Omega$, and $Q_L = 1$, 3, and 5.

Assume that for the minimum output voltage, $cos(\phi/2) = 0.9$. From (12.14) and (12.15), the quality factor at full load can be calculated as

$$Q_L = \frac{1}{\mid \frac{\omega}{\omega_o} - \frac{\omega_o}{\omega} \mid}\sqrt{\frac{2\eta_I^2 V_I^2 cos^2(\frac{\phi}{2})}{\pi^2 V_{Ri(rms)}^2} - 1}$$

$$= \frac{1}{\mid 1.33 - \frac{1}{1.33} \mid}\sqrt{\frac{2 \times 0.95^2 \times 270^2 \times 0.9^2}{63.25^2 \pi^2} - 1} = 2.25. \quad (12.31)$$

Hence,

$$L = \frac{2R_i Q_L}{\omega_o} = \frac{2 \times 50 \times 2.23}{2\pi \times 150 \times 10^3} = 238.7\ \mu\text{H} \quad (12.32)$$

and

$$C = \frac{2}{\omega_o^2 L} = \frac{2}{(2\pi \times 150 \times 10^3)^2 \times 236.6 \times 10^{-6}} = 9.43\ \text{nF}. \quad (12.33)$$

12.4 SUMMARY

- Phase-controlled resonant inverters can control the output voltage, the output current, or the output power against load and line variations by varying the phase shift between the drive voltages of the two inverters while maintaining a fixed operating frequency.
- Operation at a fixed frequency allows the magnetic and filter components to be optimized at a specific frequency. In addition, the EMI control is simple.
- For the SC PC SRI, both inverters are loaded by inductive loads for $f/f_o > 1.15$.
- The efficiency of the inverter is high at light loads because the rectifier input resistance R_i and, thereby, the ratio R_i/r increases with increasing R_L.
- The inverter is inherently short-circuit and open-circuit protected by the impedances of the resonant circuits.
- Operation at a constant frequency with inductive loads for both switching legs is achievable at the expense of a second resonant inductor, whereas the simplest frequency-controlled full-bridge inverters such as the series resonant inverter and the parallel resonant inverter employ only one inductor.
- Operation of the phase-controlled inverters is not symmetrical.

12.5 REFERENCES

1. H. Chireix, "High power outphasing modulation," *Proc. IRE*, vol. 23, pp. 1370–1392, Nov. 1935.
2. L. F. Gaudernack, "A phase-opposition system of amplitude modulation," *Proc. IRE*, vol. 26, pp. 983–1008, Aug. 1938.
3. F. H. Raab, "Efficiency of outphasing RF power-amplifier systems," *IEEE Trans. Commun.*, vol. COM-33, pp. 1094–1099, Oct. 1985.
4. P. Savary, M. Nakaoka, and T. Maruhashi, "Resonant vector control base high frequency inverter," *IEEE Power Electronics Specialists Conf. Rec.*, 1985, pp. 204–213.
5. I. J. Pitel, "Phase-modulated resonant power conversion techniques for high-frequency link inverters," *Proc. IEEE Industry Applications Society Annual Meeting*, 1985; reprinted in *IEEE Trans. Ind. Appl.*, vol. IA-22, pp. 1044–1051, Nov./Dec. 1986.
6. F. S. Tsai and F. C. Y. Lee, " Constant-frequency, phase-controlled resonant power processor," in *Proc. IEEE Industry Applications Society Annual Meeting*, Denver, Colorado, Sept. 28–Oct. 3, 1986, pp. 617–622.
7. F. S. Tsai, P. Materu, and F. C. Y. Lee, "Constant-frequency clamped-mode resonant converter," *IEEE Power Electronics Specialists Conference Record*, 1987, pp. 557–566.

8. F. S. Tsai and F. C. Y. Lee, "A complete DC characterization of a constant-frequency, clamped-mode, series-resonant converter," *IEEE Power Electronics Specialists Conference Record*, Kyoto, Japan, April 11–14, 1988, pp. 987–996.
9. F. S. Tsai, Y. Chin, and F. C. Y. Lee, "State-plane analysis of a constant-frequency clamped-mode parallel-resonant converter," *IEEE Trans. Power Electron.*, vol. PE-3, pp. 364–378, July 1988.
10. Y. Chin and F. C. Y. Lee, "Constant-frequency parallel-resonant converter," *IEEE Trans. Ind. Appl.*, vol. 25, no. 1, pp. 133–142, Jan./Feb. 1989.
11. R. J. King, "A design procedure for the phase-controlled parallel-resonant inverter," *IEEE Trans. Aerospace Electron. Syst.*, vol. AES-25, pp. 497–507, July 1989.
12. R. Redl, N. O. Sokal, and L. Balogh, "A novel soft-switching full-bridge dc-dc converter: analysis, design considerations, and experimental results at 1.5 KW, 100 kHz," *IEEE Trans. Power Electron.*, vol. 6, pp. 408–418, July 1991.
13. P. Jain, "A constant frequency resonant dc-dc converter with zero switching losses," *IEEE Industry Applications Society Annual Meeting*, Oct. 1991, pp. 1067–1073.
14. M. K. Kazimierczuk, "Synthesis of phase-modulated resonant DC/AC inverters and DC/DC convertors," *IEE Proc., Pt. B, Electric Power Appl.*, vol. 139, pp. 387–394, July 1992.
15. P. Jain, D. Bannard, and M. Cardella, "A phase-shift modulated double tuned resonant dc/dc converter: analysis and experimental results," *IEEE Applied Power Electronics Conference*, 1992, pp. 90–97.
16. P. Jain and H. Soin, "A constant frequency parallel tuned resonant dc/dc converter for high voltage applications," *IEEE Power Electronics Specialists Conference Record*, 1992, pp. 71–76.
17. D. Czarkowski and M. K. Kazimierczuk, "Phase-controlled CLL resonant converter," *IEEE Applied Power Electronics Conference (APEC'93)*, San Diego, CA, March 7–11, 1993, pp. 432–438.
18. M. K. Kazimierczuk and D. Czarkowski, "Phase-controlled series resonant converter," *IEEE Power Electronics Specialists Conf. (PESC'93)*, Seattle, WA, June 20–24, 1993, pp. 1002–1008.
19. D. Czarkowski and M. K. Kazimierczuk, "Single-capacitor phase-controlled series resonant converter," *IEEE Trans. Circuits Syst.*, vol. CAS-40, pp. 381–391, June 1993.
20. D. Czarkowski and M. K. Kazimierczuk, "Phase-controlled series-parallel resonant converter," *IEEE Trans. Power Electronics*, vol. PE-8, pp. 309–319, July 1993.
21. M. K. Kazimierczuk and M. K. Jutty, "Phase-modulated series-parallel resonant converter with a series load," *IEE Proc., Pt. B, Electric Power Applications*, vol. 140, pp. 297–306, September 1993.
22. M. K. Kazimierczuk, D. Czarkowski, and N. Thirunarayan, "A new phase-controlled parallel resonant converter," *IEEE Trans. Industrial Electronics*, vol. IE-40, pp. 542–552, Dec. 1993.
23. R. Bonert and P. Blanchard, "Design of a resonant inverter with variable voltage and constant frequency," *IEEE IAS Conf.*, 1988, pp. 1003–1008.

24. R. L. Steigerwald and K. D. T. Ngo, "Half-bridge lossless switching converter, U.S. Patent no. 4,864,479, Sept. 5, 1989.
25. F. Goodenough, "Phase-modulation cuts large switching losses," *Electronic Design*, pp. 39–44, Apr. 25, 1991.
26. F. S. Tsai, J. Sabate, and F. C. Y. Lee, "Constant-frequency zero-voltage-switched parallel-resonant converter," *IEEE INTELEC Conf.*, Florence, Italy, Oct. 15–18, 1991, Paper 16.4, pp. 1–7.

12.6 REVIEW QUESTIONS

12.1 Is it possible to control the output voltage, output current, or output power at a constant operating frequency in phase-controlled resonant inverters?

12.2 Is it necessary to maintain a constant operating frequency in phase-controlled resonant inverters?

12.3 What are the advantages of inverters operated at a fixed frequency?

12.4 Is it possible to obtain inductive loads for both switching legs in the single-capacitor phase-controlled resonant inverter?

12.5 What is the lowest normalized switching frequency at which both switching legs are still inductively loaded at any load?

12.6 Are the full-load and part-load efficiencies high in the single-capacitor phase-controlled resonant inverter?

12.7 Is the single-capacitor phase-controlled resonant inverter short-circuit proof?

12.8 Is the single-capacitor phase-controlled resonant inverter open-circuit proof?

12.9 Is the operation of the phase-controlled inverters symmetrical?

12.7 PROBLEMS

12.1 A single-capacitor phase-controlled series resonant inverter operates at a normalized switching frequency $f/f_o = 1.25$. The phase shift at the full load is $\phi = 20°$, and the maximum value of the loaded quality factor is $Q_L = 3$. What is the phase shift at 50% of the full load?

12.2 A single-capacitor phase-controlled series resonant inverter operating at a switching frequency $f = 200$ kHz has the following parameters: $V_I = 180$ V, $L = 400$ μH, $C = 4.7$ nF, $\phi = 25°$, and $R_i = 50$ Ω. Calculate the output power of the inverter.

12.3 An equivalent circuit for the fundamental component of the phase-controlled Class D series resonant inverter [18] is depicted in Fig. P12.1,

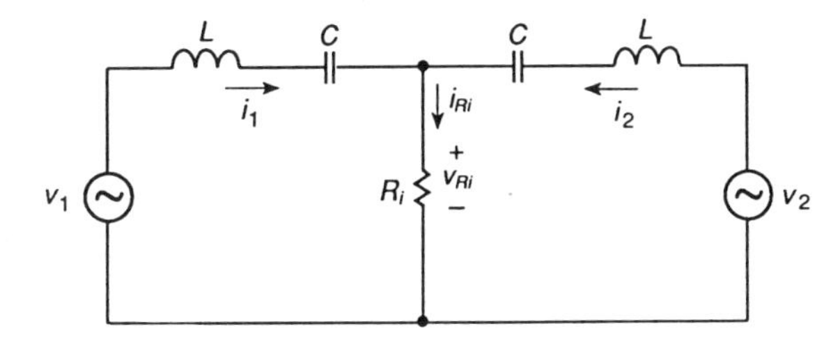

Figure P12.1: Equivalent circuit of the phase-controlled Class D series resonant inverter of Problem 12.3 for the fundamental component.

where $v_1 = V_m cos(\omega t + \phi/2)$ and $v_2 = V_m cos(\omega t - \phi/2)$. Find the voltage transfer function $M_{VI} = V_{Ri}/V_m$ of the inverter in terms of the phase shift ϕ, normalized switching frequency ω/ω_o, where $\omega_o = 1/\sqrt{LC}$, and loaded quality factor $Q_L = \omega_o L/(2R_i)$. Compare your result to (12.13).

12.4 Figure P12.2 shows an equivalent circuit for the fundamental component of the phase-controlled Class D parallel resonant inverter [22], where $v_1 = V_m cos(\omega t + \phi/2)$ and $v_2 = V_m cos(\omega t - \phi/2)$. Derive an expression for the voltage transfer function $M_{VI} = V_{Ri}/V_m$ of the inverter in terms of phase shift ϕ, normalized switching frequency ω/ω_o, where $\omega_o = \sqrt{2/LC}$, and loaded quality factor $Q_L = 2R_i/(\omega_o L)$.

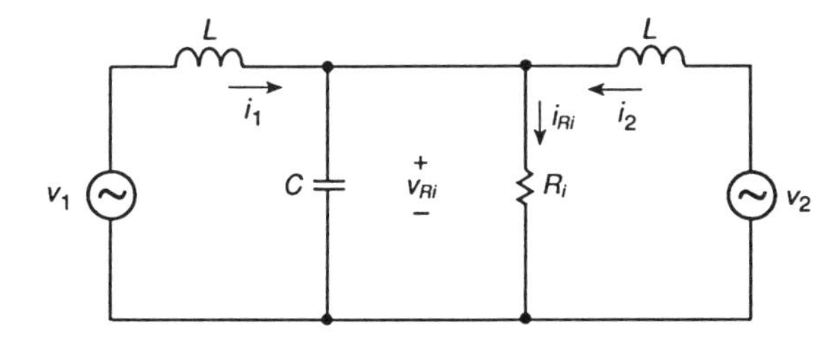

Figure P12.2: Equivalent circuit of the phase-controlled Class D parallel resonant inverter of Problem 12.4 for the fundamental component.

12.5 Design a single-capacitor phase-controlled series resonant inverter of Fig. 12.1 that delivers 100 W power to 25 Ω load resistance. The input voltage of the inverter is $V_I = 180$ V. Assume the resonant frequency $f_o = 100$ kHz, the normalized switching frequency $f/f_o = 1.25$, $cos(\phi/2) = 0.9$, and the inverter efficiency $\eta_I = 94\%$.

CHAPTER 13

CLASS E ZERO-VOLTAGE-SWITCHING RESONANT INVERTER

13.1 INTRODUCTION

There are two types of Class E dc-ac inverters: Class E zero-voltage-switching (ZVS) inverters, which are the subject of this chapter, and Class E zero-current-switching (ZCS) inverters, which we shall study in Chapter 14. Class E ZVS inverters [1]–[35] are the most efficient inverters known so far. The current and voltage waveforms of the switch are displaced with respect to time to produce a very high efficiency of the inverter. In particular, the switch turns on at zero voltage if the component values of the resonant circuit are properly chosen. Since the switch current and voltage waveforms do not overlap during the switching time intervals, switching losses are virtually zero, yielding high efficiency.

We shall start by presenting a simple qualitative description of the operation of the Class E ZVS inverter. Though simple, this description provides considerable insight into the performance of the inverter as a basic power cell. Next, we shall quickly move to a quantitative description of the inverter. Finally, we will present matching resonant circuits and give a design procedure of the inverter. By the end of the chapter, the reader will be able to perform rapid first-order analysis as well as design a single-stage Class E ZVS inverter.

13.2 PRINCIPLE OF OPERATION

13.2.1 Circuit Description

The basic circuit of the Class E ZVS inverter is shown in Fig. 13.1(a). It consists of a power MOSFET operating as a switch, a L-C-R_i series-resonant

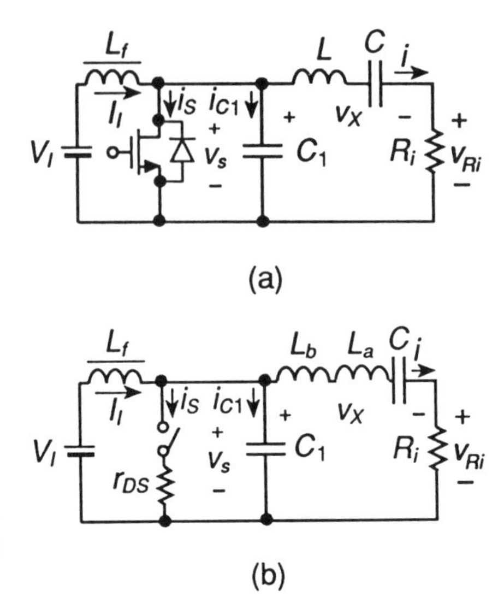

Figure 13.1: Class E zero-voltage-switching inverter. (a) Circuit. (b) Equivalent circuit for operation above resonance.

circuit, a shunt capacitor C_1, and a choke inductor L_f. The switch turns on and off at the operating frequency $f = \omega/(2\pi)$ determined by a driver. The transistor output capacitance, the choke parasitic capacitance, and stray capacitances are included in the shunt capacitance C_1. For high operating frequencies, all of capacitance C_1 can be supplied by the overall shunt parasitic capacitance. The resistor R_i is an ac load. The choke inductance L_f is assumed to be high enough so that the ac ripple on the dc supply current I_I can be neglected. A small inductance with a large current ripple is also possible [22], but the consideration of this case is beyond the scope of this text. When the switch is ON, the resonant circuit consists of L, C, and R_i because the capacitance C_1 is short-circuited by the switch. However, when the switch is OFF, the resonant circuit consists of C_1, L, C, and R_i connected in series. Because C_1 and C are connected in series, the equivalent capacitance $C_{eq} = CC_1/(C + C_1)$ is lower than C and C_1. The load network is characterized by two resonant frequencies and two loaded quality factors. When the switch is ON, $f_{o1} = 1/(2\pi\sqrt{LC})$ and $Q_{L1} = \omega_{o1}L/R_i = 1/(\omega_{o1}CR_i)$. When the switch is OFF, $f_{o2} = 1/[2\pi\sqrt{LCC_1/(C + C_1)}]$ and $Q_{L2} = \omega_{o2}L/R_i = 1/[\omega_{o2}LCC_1/(C + C_1)]$. Note that $f_{o1}/f_{o2} = Q_{L1}/Q_{L2} = \sqrt{C_1/(C_1 + C)}$. Figure 13.1(b) shows an equivalent circuit of the inverter for operation above resonance. If the operating frequency f is greater than the resonant frequency f_{o1}, the L-C-R_i series-resonant circuit represents an inductive load at the operating frequency f. Therefore, the inductance L can be divided into two inductances, L_a and L_b, connected in series such that $L = L_a + L_b$ and L_a resonates with C at the operating frequency f, that is,

$$\omega = \frac{1}{\sqrt{L_a C}}. \tag{13.1}$$

The loaded quality factor defined at the operating frequency is

$$Q_L = \frac{\omega L}{R_i} = \frac{\omega(L_a + L_b)}{R_i} = \frac{1}{\omega C R_i} + \frac{\omega L_b}{R_i}. \tag{13.2}$$

13.2.2 Circuit Operation

Figure 13.2 shows current and voltage waveforms in the Class E ZVS inverter for three cases: $dv_S(\omega t)/d(\omega t) = 0$, $dv_S(\omega t)/d(\omega t) < 0$, and $dv_S(\omega t)/d(\omega t) > 0$ at $\omega t = 2\pi$ when the switch turns on. In all three cases, the voltage v_S

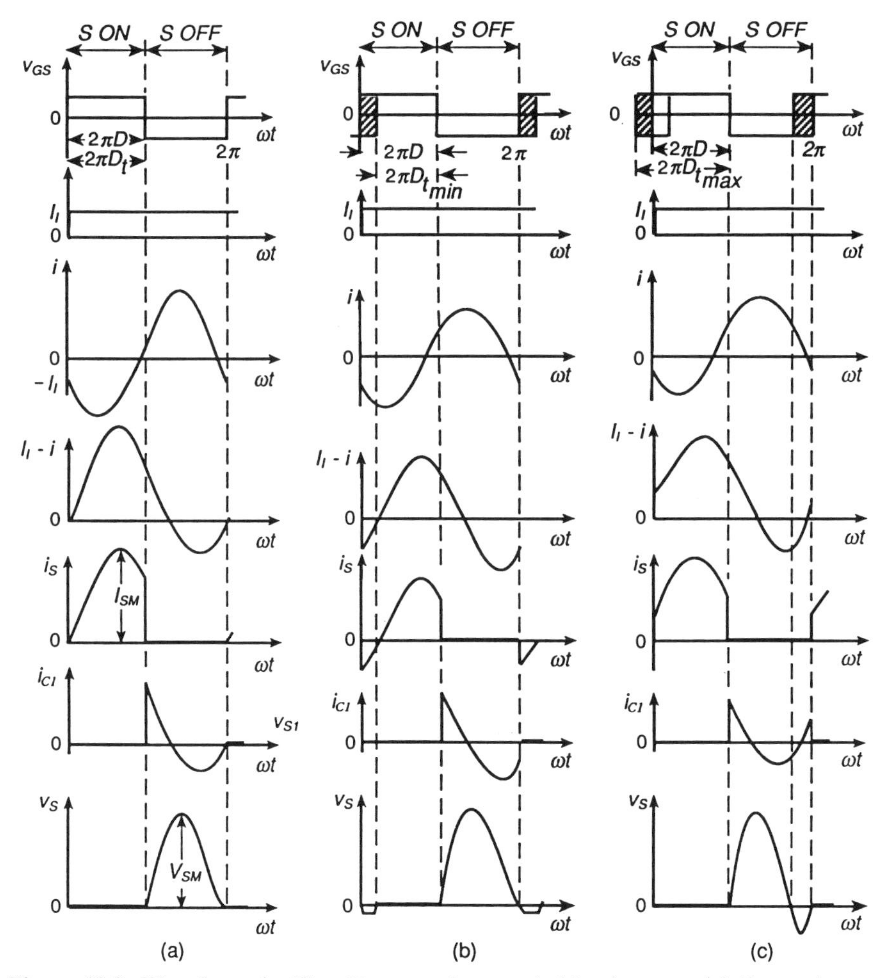

Figure 13.2: Waveforms in Class E zero-voltage-switching inverter. (a) For optimum operation. (b) For suboptimum operation with $dv_S(\omega t)/d(\omega t) < 0$ at $\omega t = 2\pi$. (c) For suboptimum operation with $dv_S(\omega t)/d(\omega t) > 0$ at $\omega t = 2\pi$.

across the switch and the shunt capacitance C_1 is zero when the switch turns on. Therefore, the energy stored in the shunt capacitance C_1 is zero when the switch turns on, yielding zero turn-on switching loss. Thus, the ZVS condition is expressed by

$$v_S(2\pi) = 0. \tag{13.3}$$

The choke inductor L_f forces a dc current I_I. To achieve zero-voltage switching turn-on of the switch, the operating frequency $f = \omega/(2\pi)$ should be greater than the resonant frequency $f_{o1} = 1/(2\pi\sqrt{LC})$, that is, $f > f_{o1}$. However, the operating frequency is usually lower than $f_{o2} = 1/(2\pi\sqrt{LC_{eq}})$, that is, $f < f_{o2}$. The shape of the waveform of the current i depends on the loaded quality factor. If Q_L is high (i.e., $Q_L \geq 2.5$), the shape of the waveform of current i is approximately sinusoidal. If Q_L is low, the shape of the waveform of current i becomes close to an exponential function [19], [22]. The combination of the choke inductor and the L-C-R_i series-resonant circuit acts as a current source whose current is $I_I - i$. When the switch is ON, the current $I_I - i$ flows through the switch. When the switch is OFF, the current $I_I - i$ flows through the capacitor C_1, producing the voltage across the shunt capacitor C_1 and the switch. Therefore, the shunt capacitor C_1 shapes the voltage across the switch.

13.2.3 Optimum Operation

Figure 13.2(a) shows current and voltage waveforms for optimum operation. In this case, both the switch voltage v_S and its derivative dv_S/dt are zero when the switch turns on. The second condition is given by

$$\frac{dv_S(\omega t)}{d(\omega t)}\Big|_{\omega t = 2\pi} = 0. \tag{13.4}$$

Because the derivative of v_S is zero at the time the switch turns on, the switch current i_S increases gradually from zero after the switch is closed. Note that both the switch voltage and the switch current are positive for optimum operation; therefore, there is no need to add any diode to the switch.

Close relationships among C_1, L_b, R_i, f, and D must be satisfied to achieve optimum operation [22]. Therefore, optimum operation can be achieved only at an optimum load resistance $R_i = R_{opt}$. If $R_i > R_{opt}$, the amplitude I_m of the current i through the L-C-R_i series-resonant circuit is lower than that for optimum operation, the voltage drop across the shunt capacitor C_1 decreases, and the switch voltage v_S is greater than zero at turn-on. On the other hand, if $R_i < R_{opt}$, the amplitude I_m is higher than that for optimum operation, the voltage drop across the shunt capacitor C_1 increases, and the switch voltage v_S is less than zero at turn-on. In both cases, assuming a linear capacitance C_1, the energy stored in C_1 just before turn-on of the switch is $W(2\pi-) = 0.5C_1v_S^2(2\pi-)$. This energy is dissipated in the transistor as heat after the switch is turned on, resulting in a turn-on switching loss. To obtain ZVS operation at a wider load range, an antiparallel or a series diode can be added to the transistor. This improvement ensures that the switch automatically turns on at zero voltage for $R_i \leq R_{opt}$.

13.2.4 Suboptimum Operation

In many applications, the load resistance varies over a certain range. The turn-on of the switch at zero voltage can be achieved for suboptimum operation for $0 \leq R_i \leq R_{opt}$. For suboptimum operation, $v_S(2\pi) = 0$ and either $dv_S(\omega t)/d(\omega t) < 0$ or $dv_S(\omega t)/d(\omega t) > 0$. Figure 13.2(b) shows current and voltage waveforms for the case when $v_S(2\pi) = 0$ and $dv_S(\omega t)/d(\omega t) < 0$ at $\omega t = 2\pi$. Power MOSFETs are bidirectional switches because their current can flow in both directions, but their voltage can be only greater than –0.7 V. When the switch voltage reaches –0.7 V, the antiparallel diode turns on and therefore the switch automatically turns on. The diode accelerates the time at which the switch turns on. This time is no longer determined by the gate-to-source voltage. Since the switch turns on at zero voltage, the turn-on switching loss is zero, yielding high efficiency. Such an operation can be achieved for $0 \leq R_i \leq R_{opt}$. In addition, if $R_i < R_{opt}$, the operating frequency f and the transistor ON switch duty cycle D_t can vary in bounded ranges. When the switch current is negative, the antiparallel diode is ON, but the transistor can be either ON or OFF. Therefore, the transistor ON switch duty cycle D_t is less than or equal to the ON switch duty cycle of the entire switch D. When the switch current is positive, the diode is OFF and the transistor must be ON. Hence, the range of D_t is $D_{t\ min} \leq D_t \leq D$, as indicated in Fig. 13.2(b) by the shaded area.

Figure 13.2(c) depicts current and voltage waveforms for the case when $v_S(2\pi) = 0$ and $dv_S(\omega t)/d(\omega t) > 0$ at $\omega t = 2\pi$. Notice that the switch current i_S is always positive, but the switch voltage v_S has positive and negative values. Therefore, a unidirectional switch for current and bidirectional for voltage is needed. Such a switch can be obtained by adding a diode in series with a MOSFET. When the switch voltage v_S is negative the diode is OFF and supports the switch voltage, regardless of the state of the MOSFET. The MOSFET is turned on during the time interval when the switch voltage is negative. Once the switch voltage reaches 0.7 V with a positive derivative, the diode turns on, turning the switch on. The series diode delays the time at which the switch turns on. The range of D_t is $D \leq D_t \leq D_{max}$, as shown in Fig. 13.2(c) by the shaded area. One disadvantage of the switch with a series diode is a higher on-voltage and a higher conduction loss. Another disadvantage is associated with the transistor output capacitance. When the switch voltage increases, the transistor output capacitance is charged via a series diode to the peak value of the switch voltage and then remains at this voltage until the transistor turns on because the diode is OFF. At this time, the transistor output capacitance is discharged through the MOSFET on-resistance, dissipating the stored energy.

13.3 ANALYSIS

13.3.1 Assumptions

The analysis of the Class E ZVS inverter of Fig. 13.1(a) is carried out under the following assumptions:

1) The transistor and diode form an ideal switch whose on-resistance is zero, off-resistance is infinity, and switching times are zero.
2) The choke inductance is high enough so that its ac component is much lower than the dc component of the input current.
3) The loaded quality factor Q_L of the LCR_i series-resonant circuit is high enough so that the current i through the resonant circuit is sinusoidal.

13.3.2 Current and Voltage Waveforms

The current through the series-resonant circuit is sinusoidal and given by

$$i = I_m sin(\omega t + \phi), \tag{13.5}$$

where I_m is the amplitude and ϕ is the initial phase of current i. According to Fig. 13.1(a),

$$i_S + i_{C1} = I_I - i = I_I - I_m sin(\omega t + \phi). \tag{13.6}$$

For the time interval $0 < \omega t \le 2\pi D$, the switch is ON and therefore $i_{C1} = 0$. Consequently, the current through the MOSFET is given by

$$i_S = \begin{cases} I_I - I_m sin(\omega t + \phi), & \text{for} \quad 0 < \omega t \le 2\pi D, \\ 0, & \text{for} \quad 2\pi D < \omega t \le 2\pi. \end{cases} \tag{13.7}$$

For the time interval $2\pi D < \omega t \le 2\pi$, the switch is OFF, which implies that $i_S = 0$. Hence, the current through the shunt capacitor C_1 is given by

$$i_{C1} = \begin{cases} 0, & \text{for} \quad 0 < \omega t \le 2\pi D, \\ I_I - I_m sin(\omega t + \phi), & \text{for} \quad 2\pi D < \omega t \le 2\pi. \end{cases} \tag{13.8}$$

The voltage across the shunt capacitor and the switch is found as

$$\begin{aligned} v_S &= \frac{1}{\omega C_1} \int_{2\pi D}^{\omega t} i_{C1} d(\omega t) \\ &= \begin{cases} 0, & \text{for} \quad 0 < \omega t \le 2\pi D, \\ \frac{1}{\omega C_1}\{I_I(\omega t - 2\pi D) \\ +I_m[cos(\omega t + \phi) - cos(2\pi D + \phi)]\}, & \text{for} \quad 2\pi D < \omega t \le 2\pi. \end{cases} \end{aligned} \tag{13.9}$$

Substitution of the condition $v_S(2\pi) = 0$ into (13.9) yields the relationship among I_I, I_m, D, and ϕ

$$I_m = I_I \frac{2\pi(1-D)}{cos(2\pi D + \phi) - cos\phi}. \tag{13.10}$$

Figure 13.3(a) shows a plot of I_m/I_I versus D.

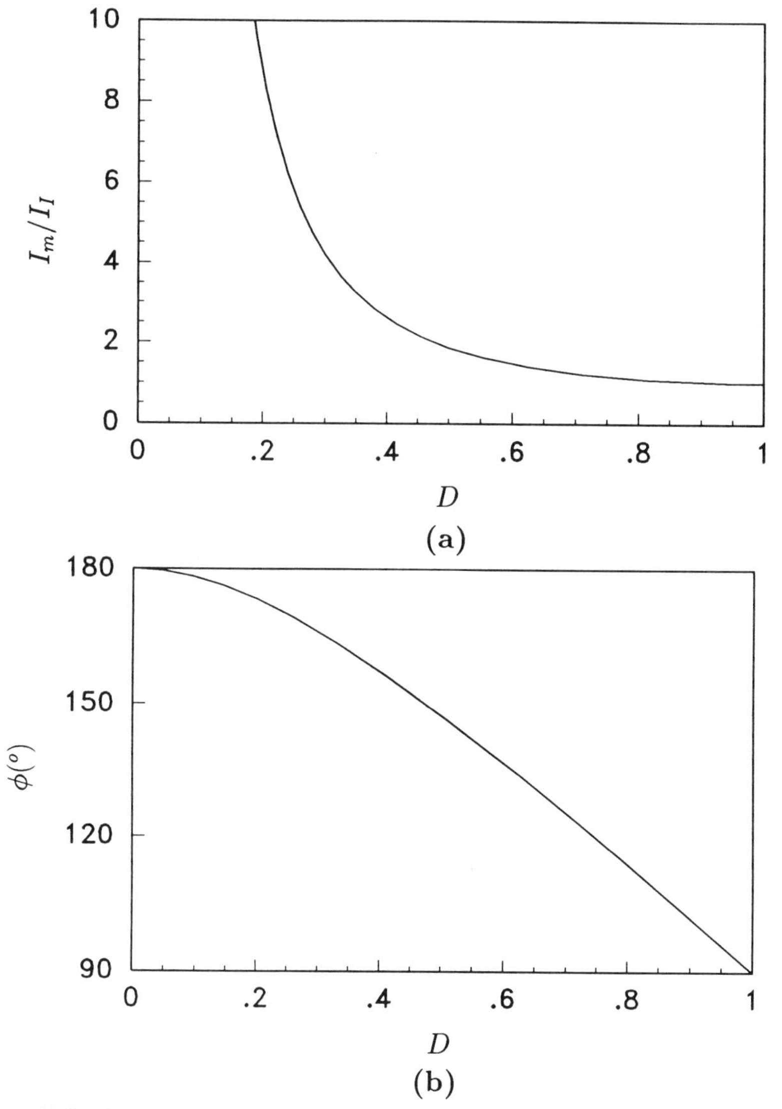

Figure 13.3: Normalized amplitude and phase of the current i through the series-resonant circuit as functions of the duty cycle D. (a) I_m/I_I versus D. (b) ϕ versus D.

Substitution of (13.10) into (13.7) yields the switch current

$$\frac{i_S}{I_I} = \begin{cases} 1 - \frac{2\pi(1-D)sin(\omega t+\phi)}{cos(2\pi D+\phi)-cos\phi}, & \text{for} \quad 0 < \omega t \leq 2\pi D, \\ 0, & \text{for} \quad 2\pi D < \omega t \leq 2\pi. \end{cases} \tag{13.11}$$

Likewise, substituting (13.10) into (13.8), one obtains the current through the shunt capacitor

$$\frac{i_{C1}}{I_I} = \begin{cases} 0, & \text{for} \quad 0 < \omega t \leq 2\pi D, \\ 1 - \frac{2\pi(1-D)sin(\omega t+\phi)}{cos(2\pi D+\phi)-cos\phi}, & \text{for} \quad 2\pi D < \omega t \leq 2\pi. \end{cases} \tag{13.12}$$

From (13.10), (13.9) becomes

$$v_S = \begin{cases} 0, & \text{for } 0 < \omega t \leq 2\pi D, \\ \frac{I_I}{\omega C_1}\left\{\omega t - 2\pi D + \frac{2\pi(1-D)[cos(\omega t+\phi)-cos(2\pi D+\phi)]}{cos(2\pi D+\phi)-cos\phi}\right\}, & \text{for } 2\pi D < \omega t \leq 2\pi. \end{cases} \tag{13.13}$$

Using the condition $dv_S/d(\omega t) = 0$ at $\omega t = 2\pi$, one obtains the relationship between phase ϕ and duty cycle D

$$tan\phi = \frac{cos2\pi D - 1}{2\pi(1-D) + sin2\pi D}, \tag{13.14}$$

from which

$$\phi = \pi + arctan\left\{\frac{cos2\pi D - 1}{2\pi(1-D) + sin2\pi D}\right\}. \tag{13.15}$$

Figure 13.3(b) shows a plot of the initial phase ϕ as a function of the duty cycle D.

From (13.13), the dc input voltage is found as

$$V_I = \frac{1}{2\pi}\int_{2\pi D}^{2\pi} v_S d(\omega t) = \frac{I_I}{\omega C_1}\left\{\frac{(1-D)[\pi(1-D)cos\pi D + sin\pi D]}{tan(\pi D + \phi)sin\pi D}\right\}. \tag{13.16}$$

Rearrangement of this produces the dc input resistance of the Class E inverter

$$R_{DC} \equiv \frac{V_I}{I_I} = \frac{(1-D)[\pi(1-D)cos\pi D + sin\pi D]}{\omega C_1 tan(\pi D + \phi)sin\pi D}. \tag{13.17}$$

Figure 13.4 shows a plot of the normalized dc input resistance $\omega C_1 R_{DC}$ as a function of the duty cycle D. From (13.13) and (13.17), one arrives at the

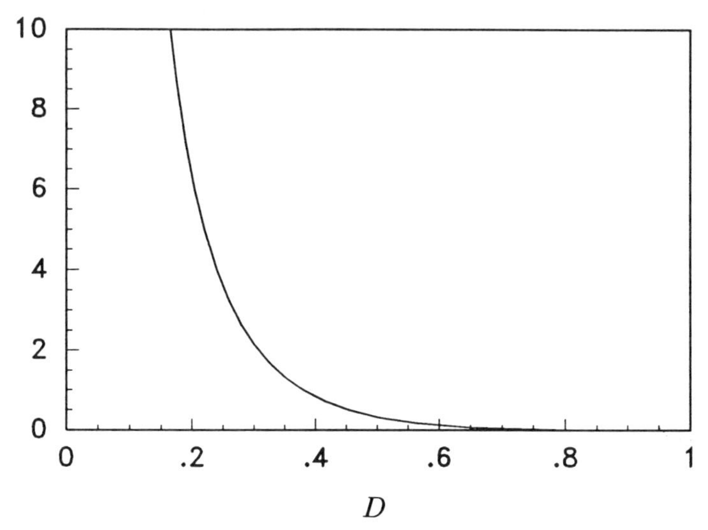

Figure 13.4: Normalized dc input resistance $\omega C_1 R_{DC}$ of the Class E ZVS inverter as a function of the duty cycle D.

normalized switch voltage waveform

$$\frac{v_S}{V_I} = \begin{cases} 0, & \text{for } 0 < \omega t \leq 2\pi D, \\ \frac{tan(\pi D+\phi)sin\pi D}{(1-D)[\pi(1-D)cos\pi D+sin\pi D]}\{\omega t - 2\pi D + \\ \frac{2\pi(1-D)}{cos(2\pi D+\phi)-cos\phi}[cos(\omega t+\phi) - cos(2\pi D+\phi)]\}, & \text{for } 2\pi D < \omega t \leq 2\pi. \end{cases} \tag{13.18}$$

13.3.3 Voltage and Current Stresses

Differentiating (13.11), one obtains the value of ωt at which the peak value of the switch current occurs

$$\omega t_{im} = \frac{3\pi}{2} - \phi. \tag{13.19}$$

Substitution of this into (13.11) yields the normalized switch peak current

$$\frac{I_{SM}}{I_I} = 1 - \frac{\pi(1-D)}{sin\pi D sin(\pi D+\phi)}, \quad \text{for} \quad \omega t_{im} < 2\pi D. \tag{13.20}$$

However, at low values of the duty cycle D, the peak switch current occurs at $\omega t_{im} = 2\pi D$, which gives

$$\frac{I_{SM}}{I_I} = \frac{2\pi(1-D)sin(2\pi D+\phi)}{cos\phi - cos(2\pi D+\phi)} + 1, \quad \text{for} \quad \omega t_{im} = 2\pi D. \tag{13.21}$$

Differentiating the switch voltage waveform in (13.13) gives the value of ωt at which the peak value of the switch voltage occurs

$$\omega t_{vm} = 2\pi - \phi + arcsin\left\{\frac{cos\phi - cos(2\pi D + \phi)}{2\pi(1 - D)}\right\}. \tag{13.22}$$

Substituting this into (13.13) yields the switch peak value V_{SM}/V_I in numerical form. Figure 13.5 depicts plots of ωt_{im}, ωt_{vm}, I_{SM}/I_I, and V_{SM}/V_I versus D.

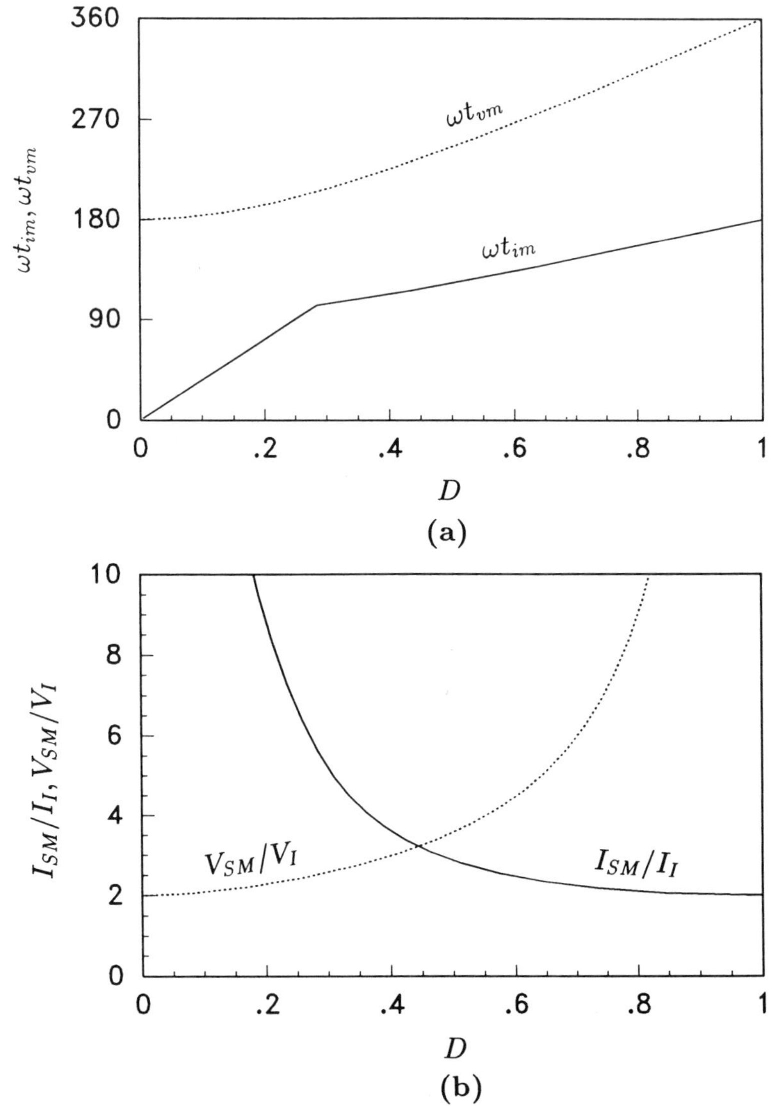

Figure 13.5: Peak values of the switch current and voltage. (a) ωt_{im} and ωt_{vm} versus D. (b) Normalized switch peak current I_{SM}/I_I and normalized switch peak voltage V_{SM}/V_I versus D.

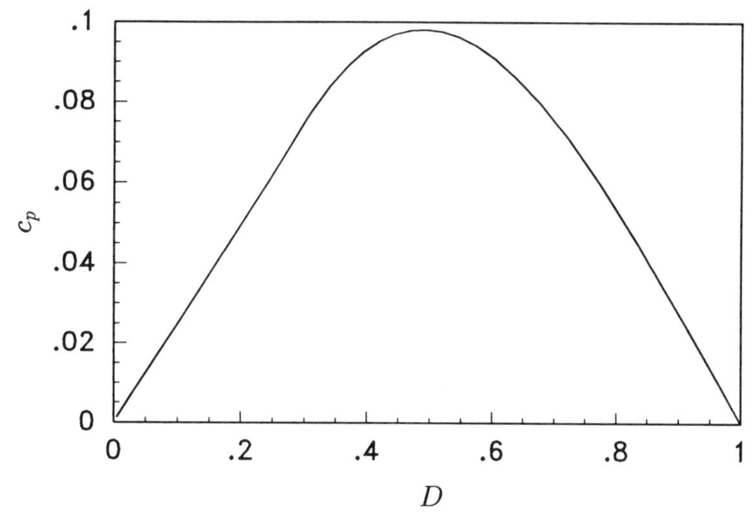

Figure 13.6: Power-output capability c_p versus D.

Neglecting power losses, the ac output power P_{Ri} is equal to the dc input power $P_I = V_I I_I$. Hence, using I_{SM}/I_I and V_{SM}/V_I , one obtains the power-output capability

$$c_p \equiv \frac{P_{R_i}}{I_{SM}V_{SM}} = \frac{I_I V_I}{I_{SM}V_{SM}}. \tag{13.23}$$

A plot of c_p versus D is displayed in Fig. 13.6.

13.3.4 Input Impedance of the Resonant Circuit

The current through the series-resonant circuit is sinusoidal. Consequently, higher harmonics of the input power are zero. Therefore, it is sufficient to consider the input impedance of the series-resonant circuit at the operating frequency f. Figure 13.7 shows an equivalent circuit of the series-resonant circuit above resonance at the operating frequency f. The fundamental component of the input voltage of the series-resonant circuit at the operating frequency is

$$v_{R1} = v_{Ri} + v_{Li} = V_{Rim}sin(\omega t + \phi) + V_{Lim}cos(\omega t + \phi). \tag{13.24}$$

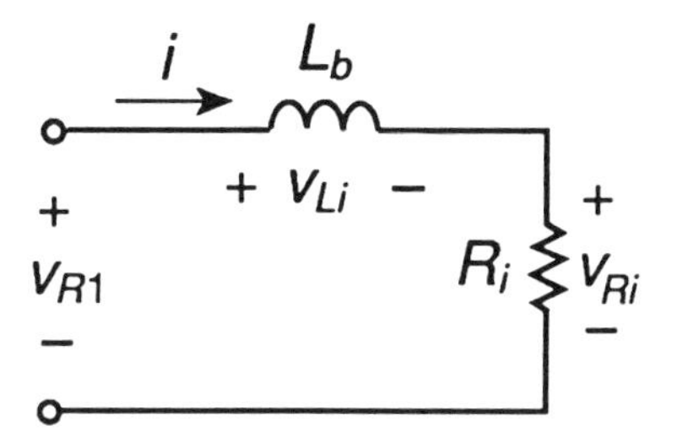

Figure 13.7: Equivalent circuit of the series-resonant circuit above resonance at the operating frequency f.

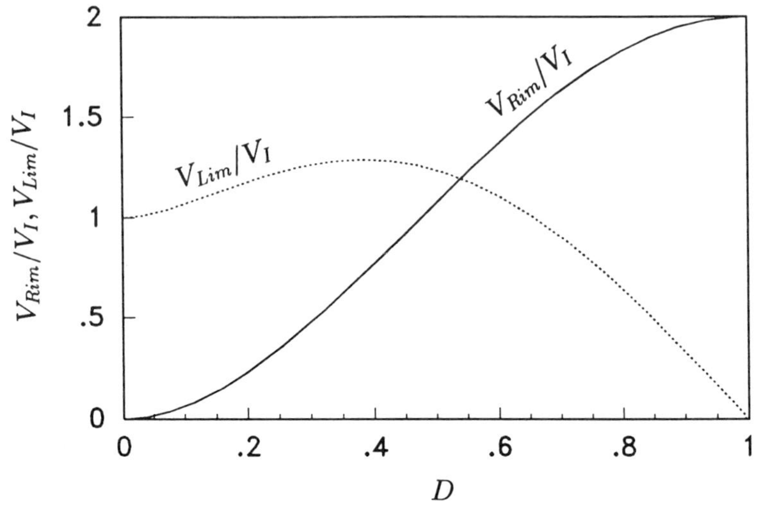

Figure 13.8: Normalized amplitudes of the fundamental components V_{Rim}/V_I and V_{Lim}/V_I versus D.

Using (13.13) and the Fourier formula,

$$V_{Rim} = \frac{1}{\pi}\int_{2\pi D}^{2\pi} v_S sin(\omega t + \phi) d(\omega t) = -\frac{2 sin \pi D sin(\pi D + \phi)}{\pi(1 - D)} V_I. \quad (13.25)$$

Substituting (13.13) into the Fourier formula and using (13.16), the amplitude of the fundamental component of the voltage across the input reactance of the series-resonant circuit (equal to the reactance of the inductance L_b) is obtained as

$$\begin{aligned} V_{Lim} &= \omega L_b I_m = \frac{1}{\pi}\int_{2\pi D}^{2\pi} v_S cos(\omega t + \phi) d(\omega t) \\ &= \frac{1 - 2(1-D)^2\pi^2 - 2cos\phi cos(2\pi D + \phi) + cos 2(\pi D + \phi)[cos 2\pi D - \pi(1-D) sin 2\pi D]}{2(1-D)\pi cos(\pi D + \phi)[(1-D)\pi cos \pi D + sin \pi D]} V_I. \end{aligned} \quad (13.26)$$

Figure 13.8 shows V_{Rim}/V_I and V_{Lim}/V_I as functions of the duty cycle D.

13.3.5 Output Power

From (13.25), one obtains the output power

$$P_{Ri} = \frac{V_{Rim}^2}{2R_i} = \frac{2 sin^2 \pi D sin^2(\pi D + \phi) V_I^2}{\pi^2(1 - D)^2 R_i}. \quad (13.27)$$

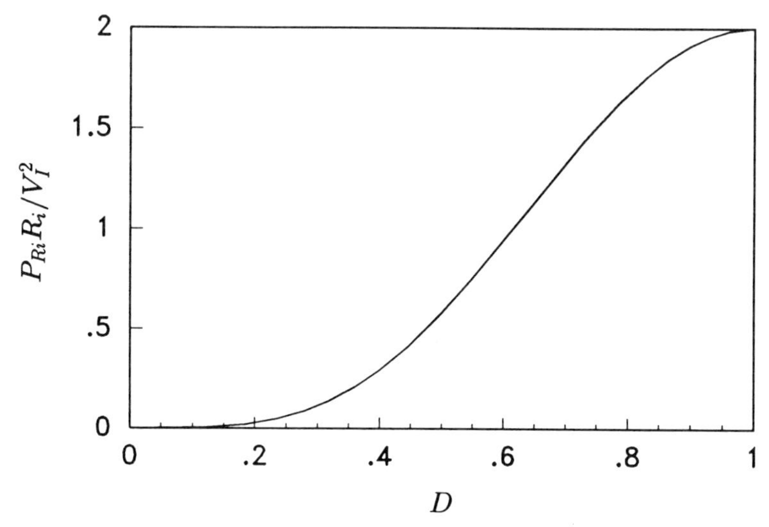

Figure 13.9: Normalized output power $P_{Ri}R_i/V_I^2$ as a function of the duty cycle D.

Figure 13.9 shows a plot of the normalized output power $P_{Ri}R_i/V_I^2$ as a function of the duty cycle D.

13.3.6 Component Values

Combining (13.10), (13.16), and (13.25),

$$\omega C_1 R_i = \frac{2 sin\pi D cos(\pi D + \phi) sin(\pi D + \phi)[(1-D)\pi cos\pi D + sin\pi D]}{\pi^2(1-D)}. \quad (13.28)$$

Similarly, using (13.10), (13.16), and (13.26),

$$\frac{\omega L_b}{R_i} = \frac{2(1-D)^2\pi^2 - 1 + 2cos\phi cos(2\pi D+\phi) - cos2(\pi D+\phi)[cos2\pi D - \pi(1-D)sin2\pi D]}{4 sin\pi D cos(\pi D+\phi) sin(\pi D+\phi)[(1-D)\pi cos\pi D + sin\pi D]}. \quad (13.29)$$

The product of (13.28) and (13.29) yields

$$\omega^2 L_b C_1 = \frac{2(1-D)^2\pi^2 - 1 + 2cos\phi cos(2\pi D+\phi) - cos2(\pi D+\phi)[cos2\pi D - \pi(1-D)sin2\pi D]}{2\pi^2(1-D)}. \quad (13.30)$$

Figure 13.10 shows plots of $\omega C_1 R_i$, $\omega L_b/R_i$, and $\omega^2 L_b C_1$ as functions of the duty cycle D. From (13.1), (13.2), and (13.29), the reactance of the resonant inductor is

$$\omega L = Q_L R_i \quad (13.31)$$

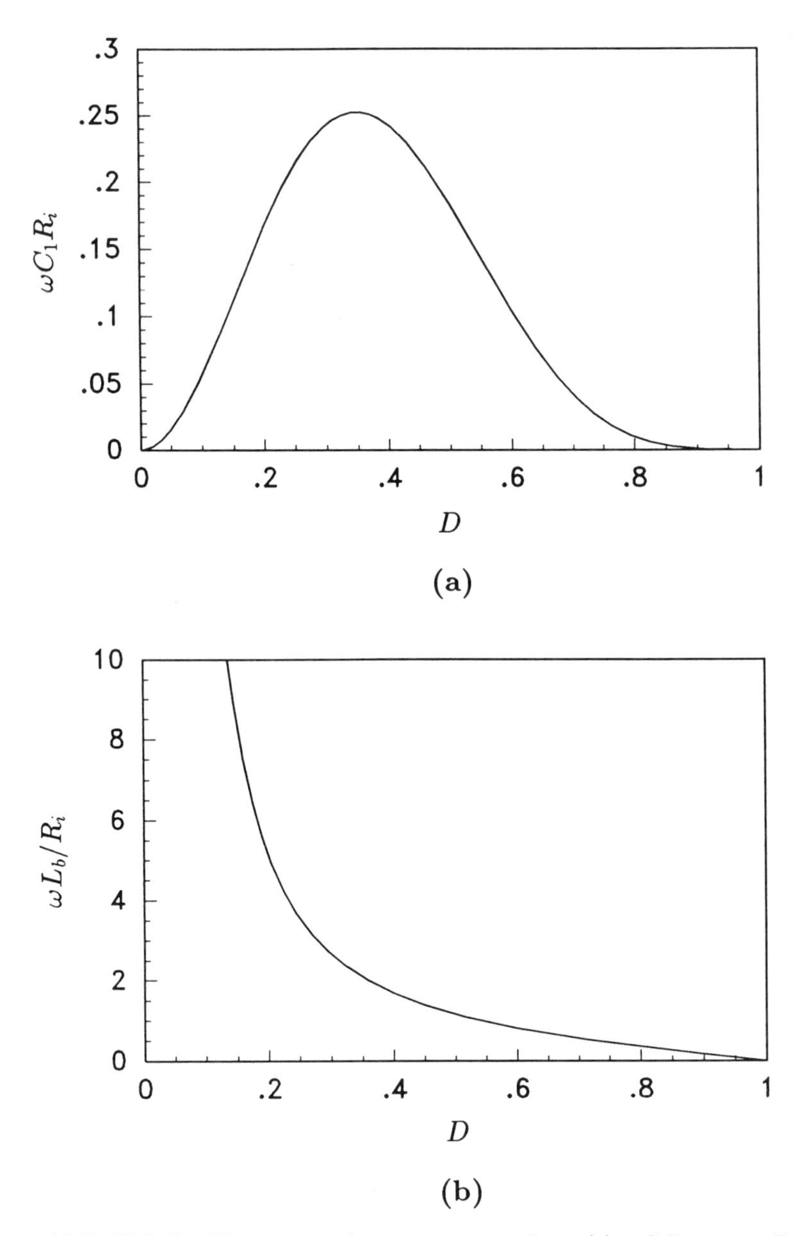

Figure 13.10: Relationships among the component values. (a) $\omega C_1 R_i$ versus D. (b) $\omega L_b / R_i$ versus D. (c) $\omega^2 L_b C_1$ versus D.

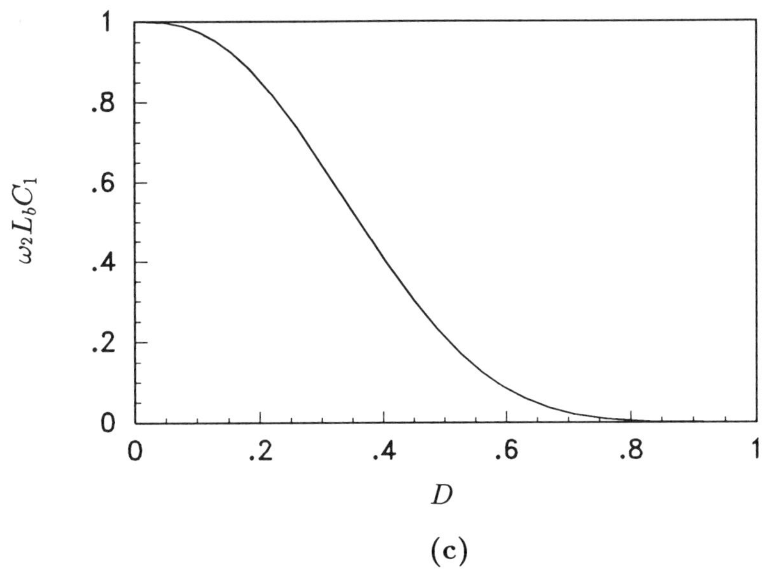

(c)

Figure 13.10 (*Continued*)

and the reactance of the resonant capacitor is

$$\frac{1}{\omega C} = \omega L_a = \omega(L - L_b) = Q_L R_i - \omega L_b. \tag{13.32}$$

The minimum value of the choke inductance L_{fmin} that ensures the current ripple to be less than 10% of the dc current I_I is

$$L_{fmin} = 2\left(\frac{\pi^2}{4} + 1\right)\frac{R_i}{f} = \frac{7R_i}{f}. \tag{13.33}$$

13.4 PARAMETERS AT D = 0.5

The parameters of the Class E ZVS inverter for the duty cycle $D = 0.5$ are as follows:

$$\frac{i_S}{I_I} = \begin{cases} \frac{\pi}{2} sin\omega t - cos\omega t + 1, & \text{for} \quad 0 < \omega t \leq \pi, \\ 0, & \text{for} \quad \pi < \omega t \leq 2\pi \end{cases} \tag{13.34}$$

$$\frac{v_S}{V_I} = \begin{cases} 0, & \text{for } 0 < \omega t \leq \pi, \\ \pi\left(\omega t - \frac{3\pi}{2} - \frac{\pi}{2} cos\omega t - sin\omega t\right), & \text{for } \pi < \omega t \leq 2\pi \end{cases} \tag{13.35}$$

$$\frac{i_{C1}}{I_I} = \begin{cases} 0, & \text{for} \quad 0 < \omega t \leq \pi, \\ \frac{\pi}{2} sin\omega t - cos\omega t + 1, & \text{for} \quad \pi < \omega t \leq 2\pi \end{cases} \tag{13.36}$$

$$tan\phi = -\frac{2}{\pi} \tag{13.37}$$

$$sin\phi = \frac{2}{\sqrt{\pi^2+4}} \tag{13.38}$$

$$cos\phi = -\frac{\pi}{\sqrt{\pi^2+4}} \tag{13.39}$$

$$\phi = \pi - arctan\left(\frac{2}{\pi}\right) = 2.5747 \text{ rad} = 147.52^\circ \tag{13.40}$$

$$R_{DC} \equiv \frac{V_I}{I_I} = \frac{1}{\pi\omega C_1} = \frac{\pi^2+4}{8}R_i = 1.7337R_i \tag{13.41}$$

$$\frac{I_{SM}}{I_I} = \frac{\sqrt{\pi^2+4}}{2} + 1 = 2.862 \tag{13.42}$$

$$\frac{V_{SM}}{V_I} = 2\pi(\pi - \phi) = 3.562 \tag{13.43}$$

$$c_p = \frac{I_I V_I}{I_{SM} V_{SM}} = \frac{1}{\pi(\pi-\phi)(2+\sqrt{\pi^2+4})} = 0.0981 \tag{13.44}$$

$$\frac{I_m}{I_I} = \frac{\sqrt{\pi^2+4}}{2} = 1.8621 \tag{13.45}$$

$$\frac{V_{Rim}}{V_I} = \frac{4}{\sqrt{\pi^2+4}} = 1.074 \tag{13.46}$$

$$\frac{V_{Lim}}{V_I} = \frac{\pi(\pi^2-4)}{4\sqrt{\pi^2+4}} = 1.2378 \tag{13.47}$$

$$P_{Ri} = \frac{V_{Rim}^2}{2R_i} = \frac{8}{\pi^2+4}\frac{V_I^2}{R_i} = 0.5768\frac{V_I^2}{R_i} \tag{13.48}$$

$$\omega C_1 R_i = \frac{8}{\pi(\pi^2+4)} = 0.1836 \tag{13.49}$$

$$\frac{\omega L_b}{R_i} = \frac{\pi(\pi^2-4)}{16} = 1.1525 \tag{13.50}$$

$$\omega^2 L_b C_1 = \frac{\pi^2-4}{2(\pi^2+4)} = 0.2116 \tag{13.51}$$

$$\frac{1}{\omega C R_i} = \left[Q_L - \frac{\omega L_b}{R_i}\right] = \left[Q_L - \frac{\pi(\pi^2-4)}{16}\right] \approx Q_L - 1.1525. \tag{13.52}$$

13.5 EFFICIENCY

The power losses and the efficiency of the Class E inverter will be considered for the duty cycle $D = 0.5$. The current through the input choke inductor I_I is essentially constant. Hence, from (13.45), the rms value of the inductor current is

$$I_{Lfrms} \approx I_I = \frac{2I_m}{\sqrt{\pi^2 + 4}}. \tag{13.53}$$

The inverter efficiency is defined as $\eta_I = P_{Ri}/P_I$ and $P_{Ri} = R_i I_m^2/2$. From (13.48) and (13.53), the power loss in the dc ESR r_{Lf} of the choke inductor L_f is

$$P_{rLf} = r_{Lf} I_{Lfrms}^2 = \frac{4I_m^2 r_{Lf}}{(\pi^2 + 4)} = \frac{8r_{Lf}}{(\pi^2 + 4)R_i} P_{Ri}. \tag{13.54}$$

For the duty cycle $D = 0.5$, the rms value of the switch current is found from (13.34)

$$I_{Srms} = \sqrt{\frac{1}{2\pi}\int_0^{\pi} i_S^2 d(\omega t)} = \frac{I_I\sqrt{\pi^2 + 28}}{4} = \frac{I_m}{2}\sqrt{\frac{\pi^2 + 28}{\pi^2 + 4}}, \tag{13.55}$$

resulting in the switch conduction loss

$$P_{rDS} = r_{DS} I_{Srms}^2 = \frac{r_{DS} I_m^2 (\pi^2 + 28)}{4(\pi^2 + 4)} = \frac{(\pi^2 + 28) r_{DS}}{2(\pi^2 + 4)R_i} P_{Ri}. \tag{13.56}$$

Using (13.36), the rms value of the current through the shunt capacitor C_1 is

$$I_{C1rms} = \sqrt{\frac{1}{2\pi}\int_{\pi}^{2\pi} i_{C1}^2 d(\omega t)} = \frac{I_I\sqrt{\pi^2 - 4}}{4} = \frac{I_m}{2}\sqrt{\frac{\pi^2 - 4}{\pi^2 + 4}}, \tag{13.57}$$

which leads to the power loss in the ESR r_{C1} of shunt capacitor C_1

$$P_{rC1} = r_{C1} I_{C1rms}^2 = \frac{r_{C1} I_m^2 (\pi^2 - 4)}{4(\pi^2 + 4)} = \frac{(\pi^2 - 4) r_{C1}}{2(\pi^2 + 4)R_i} P_{Ri}. \tag{13.58}$$

The power losses in the ESR r_L of the resonant inductor L and in the ESR r_C of resonant capacitor C are

$$P_{rL} = \frac{r_L I_m^2}{2} = \frac{r_L}{R_i} P_{Ri} \tag{13.59}$$

and

$$P_{rC} = \frac{r_C I_m^2}{2} = \frac{r_C}{R_i} P_{Ri}. \tag{13.60}$$

The turn-on switching loss is zero if the ZVS condition is satisfied. The turn-off switching loss can be estimated as follows. Assume that the transistor current during the turn-off time t_f decreases linearly

$$i_S = 2I_I\left(1 - \frac{\omega t - \pi}{\omega t_f}\right), \quad \text{for} \quad \pi < \omega t \le \pi + \omega t_f. \tag{13.61}$$

The sinusoidal current through the resonant circuit does not change significantly during the fall time t_f and is $i \approx 2I_I$. Hence, the current through the shunt capacitor C_1 can be approximated by

$$i_{C1} \approx \frac{2I_I(\omega t - \pi)}{\omega t_f}, \quad \text{for} \quad \pi < \omega t \le \pi + \omega t_f, \tag{13.62}$$

which gives the voltage across the shunt capacitor and the switch

$$\begin{aligned} v_S &= \frac{1}{\omega C_1}\int_{\pi}^{\omega t} i_{C1}\, d(\omega t) = \frac{I_I}{\omega C_1}\frac{(\omega t)^2 - 2\pi\omega t + \pi^2}{\omega t_f} \\ &= \frac{V_I \pi[(\omega t)^2 - 2\pi\omega t + \pi^2]}{\omega t_f}. \end{aligned} \tag{13.63}$$

Thus, the average value of the power loss associated with the fall time t_f

$$P_{tf} = \frac{1}{2\pi}\int_{\pi}^{\pi+\omega t_f} i_S v_S\, d(\omega t) = P_I\frac{(\omega t_f)^2}{12} \approx P_{Ri}\frac{(\omega t_f)^2}{12}. \tag{13.64}$$

From (13.54), (13.56), (13.64), (13.59), and (13.60), one obtains the overall power loss

$$\begin{aligned} P_{LS} &= P_{rLf} + P_{rDS} + P_{C1} + P_{rL} + P_{rC} + P_{tf} \\ &= P_{Ri}\left[\frac{8r_{Lf}}{(\pi^2+4)R_i} + \frac{(\pi^2+28)r_{DS}}{2(\pi^2+4)R_i} + \frac{r_{C1}(\pi^2-4)}{2(\pi^2+4)R_i} + \frac{r_L + r_C}{R_i} + \frac{(\omega t_f)^2}{12}\right]. \end{aligned} \tag{13.65}$$

This leads to the efficiency of the Class E inverter

$$\begin{aligned} \eta_I &\equiv \frac{P_{Ri}}{P_I} = \frac{P_{Ri}}{P_{Ri} + P_{LS}} = \frac{1}{1 + \frac{P_{Ri}}{P_{LS}}} \\ &= \frac{1}{1 + \frac{8r_{Lf}}{(\pi^2+4)R_i} + \frac{(\pi^2+28)r_{DS}}{2(\pi^2+4)R_i} + \frac{(\pi^2-4)r_{C1}}{2(\pi^2+4)R_i} + \frac{r_L + r_C}{R_i} + \frac{(\omega t_f)^2}{12}}. \end{aligned} \tag{13.66}$$

Figure 13.11 shows a plot of the efficiency η_I as a function of R_i for $r_{Lf} = 0.1\ \Omega$, $r_{DS} = 0.5\ \Omega$, $r_L = 0.5\ \Omega$, $r_C = 0.02\ \Omega$, $r_C = 0.02\ \Omega$, $r_{C1} = 0.02\ \Omega$, and $\omega t_f = \pi/18$.

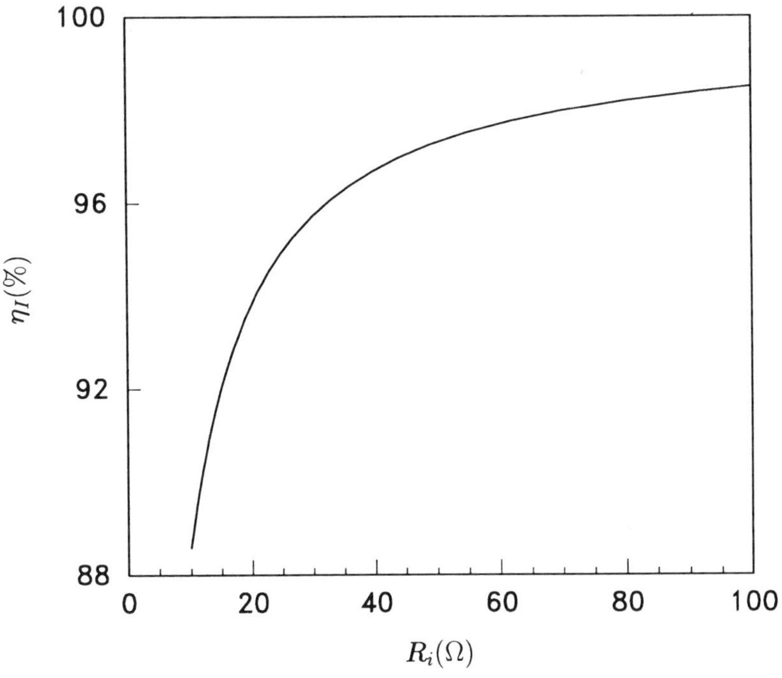

Figure 13.11: Efficiency of the Class E ZVS inverter versus load resistance R_i for $r_{Lf} = 0.1\ \Omega$, $r_{DS} = 0.5\ \Omega$, $r_L = 0.5\ \Omega$, $r_C = 0.02\ \Omega$, $r_{C1} = 0.02\ \Omega$, and $\omega t_f = \pi/18$.

The gate-drive power of each MOSFET that is required to charge and discharge a highly nonlinear MOSFET input capacitance is given by

$$P_G = fV_{GSm}Q_g, \tag{13.67}$$

where V_{GSm} is peak value of the gate-to-source voltage v_{GS} and Q_g is the gate charge at $v_{GS} = V_{GSm}$.

The power gain of the Class E ZVS inverter at $D = 0.5$ is given by

$$k_p \equiv \frac{P_{Ri}}{P_G} = \frac{8}{\pi^2 + 4} \times \frac{V_I^2}{R_i f V_{GSm} Q_g}. \tag{13.68}$$

At high frequencies, the output capacitance of the switch C_{out} becomes higher than that required to achieve zero-voltage-switching operation. For $D = 0.5$, the maximum operating frequency at which Class E operation is achievable is [27]

$$f_{max} = \frac{0.1971}{2\pi R_i C_{out}} = \frac{0.05439 P_{Ri}}{V_I^2 C_{out}}. \tag{13.69}$$

Above this frequency, a Class C-E operation can be obtained, which offers reasonably high efficiency [27].

13.6 MATCHING RESONANT CIRCUITS

13.6.1 Basic Circuit

The component values of the resonant circuit of the basic Class E inverter shown in Fig. 13.1(a) for optimum operation at $D = 0.5$ obtained from (13.2), (13.48), (13.49), and (13.52) are as follows:

$$R_i = \frac{8}{\pi^2 + 4} \frac{V_I^2}{P_{Ri}} \approx 0.5768 \frac{V_I^2}{P_{Ri}} \tag{13.70}$$

$$X_{C1} = \frac{1}{\omega C_1} = \frac{\pi(\pi^2 + 4)R_i}{8} \approx 5.4466 R_i \tag{13.71}$$

$$X_L = \omega L = Q_L R_i \tag{13.72}$$

$$X_C = \frac{1}{\omega C} = \left[Q_L - \frac{\pi(\pi^2 - 4)}{16}\right] R_i \approx (Q_L - 1.1525)R_i. \tag{13.73}$$

Suboptimum operation (i.e., ZVS operation) occurs for load resistance $R_{i(sub)}$ lower than that given in (13.70), that is,

$$0 \le R_{i(sub)} < R_i. \tag{13.74}$$

Notice that the Class E ZVS inverter with the basic resonant circuit shown in Fig. 13.1(a) operates safely under short-circuit conditions.

The basic resonant circuit of Fig. 13.1(a) does not have matching capability. In order to transfer a specified amount of power P_{Ri} at a specified dc voltage V_I, the load resistance R_i must be of the value determined by (13.70).

13.6.2 Resonant Circuit π1a

According to (13.70), V_I, P_{Ri}, and R_i are dependent quantities. In many applications, the load resistance is given and is different from that given in (13.70). Therefore, there is a need for matching circuits that provide impedance transformation. Figure 13.12 shows various matching resonant circuits. In the circuits shown in Fig. 13.12(a) and (c), impedance transformation is accomplished by tapping the resonant capacitance C, and in the circuits shown in Fig. 13.11(b) and (d) by tapping the resonant inductance L.

Figure 13.13 shows an equivalent circuit of the matching circuit shown in Fig. 13.12(a). Let us assume that the load resistance R_i is given. Using (13.70), the series equivalent resistance for optimum operation at $D = 0.5$ is given by

$$R_s = \frac{8}{\pi^2 + 4} \frac{V_I^2}{P_{Ri}} \approx 0.5768 \frac{V_I^2}{P_{Ri}}. \tag{13.75}$$

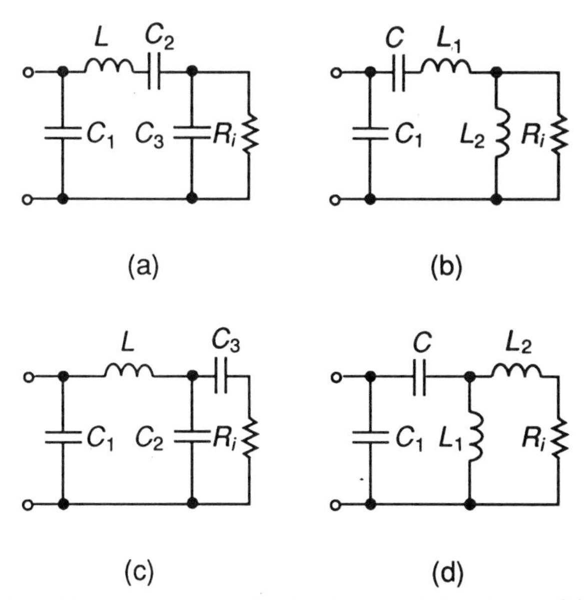

Figure 13.12: Matching resonant circuits. (a) Resonant circuit π1a. (b) Resonant circuit π2a. (c) Resonant circuit π1b. (d) Resonant circuit π4a.

The components C_1 and L are given by (13.71) and (13.72). The reactance factor for the R_i–C_3 and R_s–C_s equivalent two-port networks is

$$q = \frac{R_i}{X_{C_3}} = \frac{X_{Cs}}{R_s}. \tag{13.76}$$

Resistances R_s and R_i as well as the reactances X_{Cs} and X_{C3} are related by

$$R_s = \frac{R_i}{1+q^2} = \frac{R_i}{1+\left(\frac{R_i}{X_{C3}}\right)^2} \tag{13.77}$$

and

$$X_{Cs} = \frac{X_{C3}}{1+\frac{1}{q^2}} = \frac{X_{C3}}{1+(\frac{X_{C3}}{R_i})^2}. \tag{13.78}$$

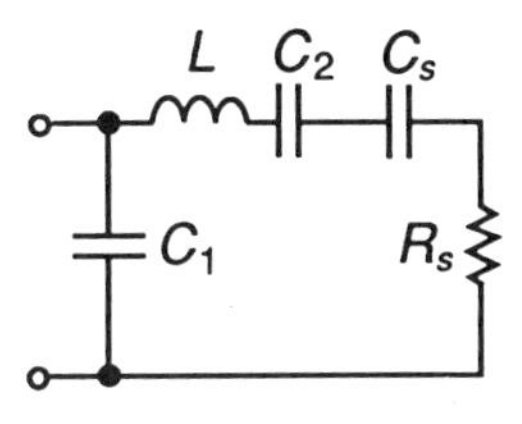

Figure 13.13: Equivalent circuit of the matching circuit π1a.

Rearrangement of (13.77) gives

$$q = \sqrt{\frac{R_i}{R_s} - 1}. \tag{13.79}$$

Substitution of (13.79) into (13.76) yields

$$X_{Cs} = R_s\sqrt{\frac{R_i}{R_s} - 1}. \tag{13.80}$$

Referring to Fig. 13.13 and using (13.73) and (13.80), one arrives at

$$\begin{aligned} X_{C2} &= \frac{1}{\omega C_2} = X_C - X_{Cs} = R_s\left[Q_L - \frac{\pi(\pi^2-4)}{16}\right] - qR_s \\ &= R_s\left[Q_L - \frac{\pi(\pi^2-4)}{16} - \sqrt{\frac{R_i}{R_s} - 1}\right]. \end{aligned} \tag{13.81}$$

From (13.76) and (13.79),

$$X_{C3} = \frac{1}{\omega C_3} = \frac{R_i}{q} = \frac{R_i}{\sqrt{\frac{R_i}{R_s} - 1}}. \tag{13.82}$$

It follows from (13.82) that the circuit shown in Fig. 13.12(a) can match the resistances that satisfy the inequality

$$R_s < R_i. \tag{13.83}$$

Suboptimum operation is obtained for

$$0 \leq R_{s(sub)} < R_s, \tag{13.84}$$

which corresponds to

$$R_i < R_{i(sub)} < \infty. \tag{13.85}$$

Expressions (13.77) and (13.78) are illustrated in Fig. 13.14. As R_i is increased from 0 to X_{C3}, R_s increases to $X_{C3}/2$, R_s reaches the maximum value $R_{smax} = X_{C3}/2$ at $R_i = X_{C3}$, and as R_i is increased from X_{C3} to ∞, R_s decreases from X_{C3} to 0. Thus, the $R_i - C_3$ circuit acts as an *impedance inverter* [20] for $R_i > X_{C3}$. If the optimum operation occurs at $R_i = X_{C3}$, then $R_{smax} = X_{C3}/2$ and the inverter operates under ZVS conditions at any load resistance R_i [28]. This is because $R_s \leq R_{smax} = X_{C3}/2$ at any values of R_i.

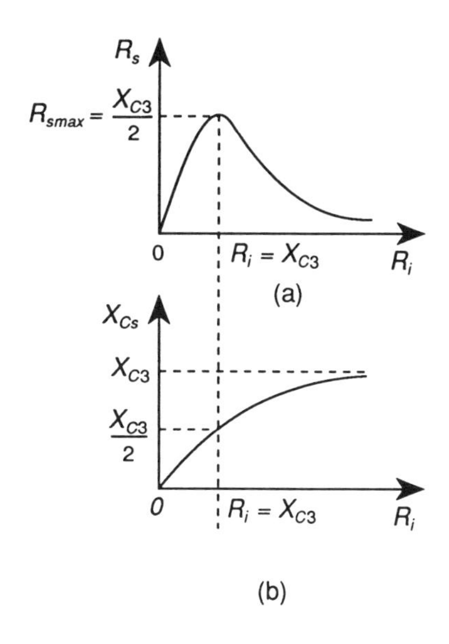

Figure 13.14: Series equivalent resistance R_s and reactance X_{Cs} as functions of load resistance R_i in the circuit π1a. (a) R_s versus R_i. (b) X_{Cs} versus R_i.

13.6.3 Resonant Circuit π2a

The values of R_s, X_{C1}, and X_C for the resonant circuit shown in Fig. 13.11(b) can be calculated for optimum operation at $D = 0.5$ from (13.75), (13.71), and (13.73), respectively. The reactances of L_1 and L_2 are

$$X_{L1} = \omega L_1 = \left(Q_L - \sqrt{\frac{R_i}{R_s} - 1} \right) R_s \tag{13.86}$$

$$X_{L2} = \omega L_2 = \frac{R_i}{\sqrt{\frac{R_i}{R_s} - 1}}. \tag{13.87}$$

The range of resistances that can be matched by the circuit shown in Fig. 13.11(b) is

$$R_s < R_i. \tag{13.88}$$

Suboptimum operation takes place for

$$0 \le R_{s(sub)} < R_s, \tag{13.89}$$

and consequently for

$$R_i < R_{i(sub)} < \infty. \tag{13.90}$$

The relationship between resistances R_s and R_i is

$$R_s = \frac{R_i}{1+q^2} = \frac{R_i}{1+\left(\frac{R_i}{X_{L2}}\right)^2}. \tag{13.91}$$

Thus, as R_i is increased from 0 to X_{L2}, R_s increases from 0 to $R_{smax} = X_{L2}/2$, and as R_i is increased from X_{L2} to ∞, R_i decreases from $X_{L2}/2$ to 0. It is clear that the $R_i - L_2$ circuit behaves as an *impedance inverter* for $R_i > X_{L2}$. If the optimum operation occurs at $R_i = X_{L2}$, then $Rs = X_{L2}/2$ and the ZVS operation occurs at any load resistance [25], [30]. In this case, $R_s \leq R_{smax} = X_{L2}/2$.

13.6.4 Resonant Circuit π1b

The values of R_s, X_{C1}, and X_L for the circuit shown in Fig. 13.12(c) can be calculated for for optimum operation at $D = 0.5$ from (13.75), (13.71), and (13.72), respectively. The values of X_{C2} and X_{C3} can be found from

$$X_{C2} = \frac{1}{\omega C_2} = R_i\sqrt{\frac{R_s[(Q_L - 1.1525)^2 + 1]}{R_i} - 1} \tag{13.92}$$

$$X_{C3} = \frac{1}{\omega C_3} = \frac{R_s[(Q_L - 1.1525)^2 + 1]}{Q_L - 1.1525 - \sqrt{\frac{R_s[(Q_L - 1.1525)^2+1]}{R_i} - 1}}. \tag{13.93}$$

The resistances that can be matched by the aforementioned circuit are

$$\frac{R_i}{(Q_L - 1.1525)^2 + 1} < R_s < R_i. \tag{13.94}$$

Suboptimum operation takes place for

$$R_{s(sub)} > R_s \tag{13.95}$$

and therefore for

$$R_{i(sub)} < R_i. \tag{13.96}$$

13.6.5 Resonant Circuit π4a

The values of R_s, X_{C1}, and X_C for the circuit shown in Fig. 13.12(d) can be calculated for optimum operation at $D = 0.5$ from (13.75), (13.71), and (13.72), respectively. The reactances of L_1 and L_2 are

$$X_{L1} = \omega L_1 = R_s\sqrt{\frac{R_s(Q_L^2+1)}{R_i} - 1} \quad (13.97)$$

$$X_{L2} = \omega L_2 = \frac{R_s(Q_L^2+1)}{Q_L - \sqrt{\frac{R_s(Q_L^2+1)}{R_i} - 1}}. \quad (13.98)$$

This circuit can match resistances

$$\frac{R_i}{Q_L^2+1} < R_s < R_i. \quad (13.99)$$

Suboptimum operation takes place for

$$R_{s(sub)} > R_s \quad (13.100)$$

and therefore for

$$R_{i(sub)} < R_i. \quad (13.101)$$

13.7 DESIGN EXAMPLE

Example 13.1

Design the Class E ZVS inverter of Fig. 13.1(a) to satisfy the following specifications: $V_I = 100$ V, $P_{Rimax} = 80$ W, and $f = 1.2$ MHz. Assume $D = 0.5$.

Solution: It is sufficient to design the inverter for the full power. Using (13.48), the full-load resistance is

$$R_i = \frac{8}{\pi^2+4}\frac{V_I^2}{P_{Ri}} = 0.5768 \times \frac{100^2}{80} = 72.1\ \Omega. \quad (13.102)$$

From (13.41), the dc resistance of the inverter is

$$R_{DC} = \frac{\pi^2+4}{8}R_i = 1.7337 \times 72.1 = 125\ \Omega. \quad (13.103)$$

The amplitude of the output voltage is computed from (13.46)

$$V_{Rim} = \frac{4}{\sqrt{\pi^2+4}} = 1.074 \times 100 = 107.4\ \text{V}. \quad (13.104)$$

The maximum voltage across the switch and the shunt capacitor can be calculated from (13.43) as

$$V_{SM} = 3.562V_I = 3.562 \times 100 = 356.2\ \text{V}. \quad (13.105)$$

From (13.41), the dc input current is

$$I_I = \frac{8}{\pi^2 + 4}\frac{V_I}{R_i} = 0.5768 \times \frac{100}{72.1} = 0.8 \text{ A}. \quad (13.106)$$

The maximum switch current obtained using (13.42) is

$$I_{SM} = \left(\frac{\sqrt{\pi^2 + 4}}{2} + 1\right) I_I = 2.862 \times 0.8 = 2.29 \text{ A}. \quad (13.107)$$

The amplitude of the current through the resonant circuit computed from (13.45) is

$$I_m = \frac{I_I\sqrt{\pi^2 + 4}}{2} = 1.8621 \times 0.8 = 1.49 \text{ A}. \quad (13.108)$$

Assuming $Q_L = 7$ and using (13.31), (13.49), and (13.52), the component values of the load network are

$$L = \frac{Q_L R_i}{\omega} = \frac{7 \times 72.1}{2\pi \times 1.2 \times 10^6} = 66.9 \ \mu\text{H}, \quad (13.109)$$

$$C_1 = \frac{8}{\pi(\pi^2 + 4)\omega R_i} = \frac{8}{2\pi^2(\pi^2 + 4) \times 1.2 \times 10^6 \times 72.1} = 338 \text{ pF}, \quad (13.110)$$

and

$$C = \frac{1}{\omega R_i \left[Q_L - \frac{\pi(\pi^2 - 4)}{16}\right]} = \frac{1}{2\pi \times 1.2 \times 10^6 \times 72.1(7 - 1.1525)} = 315 \text{ pF}. \quad (13.111)$$

It follows from (13.33) that in order to keep the current ripple in the choke inductor below 10% of the full-load dc input current I_I, the value of the choke inductance must be greater than

$$L_f = 2\left(\frac{\pi^2}{4} + 1\right)\frac{R_i}{f} = 2\left(\frac{\pi^2}{4} + 1\right) \times \frac{72.1}{1.2 \times 10^6} = 416.6 \ \mu\text{H}. \quad (13.112)$$

The peak voltages across the resonant capacitor C and inductor L are

$$V_{Cm} = \frac{I_m}{\omega C} = \frac{1.49}{2\pi \times 1.2 \times 10^6 \times 315 \times 10^{12}} = 627.4 \text{ V} \quad (13.113)$$

and

$$V_{Lm} = \omega L I_m = 2\pi \times 10^6 \times 66.9 \times 10^6 \times 1.49 = 751.6 \text{ V}. \quad (13.114)$$

Assume that the dc ESR of the choke L_f is $r_{Lf} = 0.15 \ \Omega$. Hence, from

(13.54) the power loss in r_{Lf} is

$$P_{rLf} = r_{Lf} I_I^2 = 0.15 \times 0.8^2 = 0.096 \text{ W}. \tag{13.115}$$

From (13.55), the rms value of the switch current is

$$I_{Srms} = \frac{I_I \sqrt{\pi^2 + 28}}{4} = 0.8 \times 1.5385 = 1.231 \text{ A}. \tag{13.116}$$

If the MPT5N40 MOSFET is used whose on-resistance $r_{DS} = 1\ \Omega$, the transistor conduction power loss is

$$P_{rDS} = r_{DS} I_{Srms}^2 = 1 \times 1.231^2 = 1.515 \text{ W}. \tag{13.117}$$

Using (13.57), one obtains the rms current through the shunt capacitor

$$I_{C1rms} = \frac{I_I \sqrt{\pi^2 - 4}}{4} = 0.8 \times 0.6057 = 0.485 \text{ A}. \tag{13.118}$$

Assuming the ESR of C_1 to be $r_{C1} = 76 \text{ m}\Omega$, one arrives at the conduction power loss in r_{C1}

$$P_{rC1} = r_{C1} I_{C1rms}^2 = 0.076 \times 0.485^2 = 0.018 \text{ W}. \tag{13.119}$$

Assume the ESRs of the resonant inductor and capacitor to be $r_L = 0.5\ \Omega$ and $r_C = 50 \text{ m}\Omega$. Hence, the power losses in the resonant components are

$$P_{rL} = \frac{r_L I_m^2}{2} = \frac{0.5 \times 1.49^2}{2} = 0.555 \text{ W} \tag{13.120}$$

$$P_{rC} = \frac{r_C I_m^2}{2} = \frac{0.05 \times 1.49^2}{2} = 0.056 \text{ W}. \tag{13.121}$$

The total conduction loss is

$$\begin{aligned} P_r &= P_{rDS} + P_{rLf} + P_{rC1} + P_{rL} + P_{rC} \\ &= 1.515 + 0.096 + 0.018 + 0.555 + 0.056 = 2.24 \text{ W}. \end{aligned} \tag{13.122}$$

The inverter efficiency associated with the conduction loss is

$$\eta_I = \frac{P_{Ri}}{P_{Ri} + P_r} = \frac{80}{80 + 2.24} = 97.28\%. \tag{13.123}$$

To estimate the turn-off switching loss, assume $t_f = 0.05T = 0.05/f = 0.05/(1.2 \times 10^6) = 41.7$ ns. Hence, $\omega t_f = 2\pi \times 1.2 \times 10^6 \times 41.7 \times 10^{-9} = 0.314$ rad. Thus, from (13.64) the turn-off switching loss is

$$P_{tf} = \frac{(\omega t_f)^2 P_{Ri}}{12} = \frac{0.314^2 \times 80}{12} = 0.657 \text{ W}. \tag{13.124}$$

For the MTP5N40 power MOSFET, $Q_g = 27$ nC. Hence, assuming $V_{GSm} = 8$ V, one obtains the gate-drive power

$$P_G = fV_{GSm}Q_g = 1.2 \times 10^6 \times 8 \times 27 \times 10^{-9} = 0.259 \text{ W}. \tag{13.125}$$

Thus, the power loss is

$$P_{LS} = P_r + P_{tf} + P_G = 2.24 + 0.657 + 0.259 = 3.156 \text{ W}. \tag{13.126}$$

The efficiency of the inverter becomes

$$\eta_I = \frac{P_{Ri}}{P_{Ri} + P_{LS}} = \frac{80}{80 + 3.156} = 96.20\%. \tag{13.127}$$

The power gain of the inverter is

$$k_p = \frac{P_{Ri}}{P_G} = \frac{80}{0.259} = 309. \tag{13.128}$$

The equivalent capacitance when the switch is off is $C_{eq} = CC_1/(CC_1) = 163$ pF and the resonant frequencies are $f_{o1} = 1/(2\pi\sqrt{LC}) = 1.096$ MHz and $f_{o2} = 1/(2\pi\sqrt{LC_{eq}}) = 1.524$ MHz. Notice that the operating frequency f is between the resonant frequencies f_{o1} and f_{o2}.

13.8 SUMMARY

- The transistor output capacitance, the choke parasitic capacitance, and the stray capacitance are absorbed into the shunt capacitance C_1 in Class E ZVS inverter.
- The turn-on switching loss is zero.
- The operating frequency f is greater than the resonant frequency $f_o = 1/(2\pi\sqrt{LC})$ of the series resonant circuit. This results in an inductive load for the switch when it is on.
- The antiparallel diode of the switch turns off at low di/dt and zero voltage, reducing reverse-recovery effects. Therefore, the MOSFET body diode can be used and there is no need for a fast diode.
- The zero-voltage-switching operation can be accomplished in the basic topology for load resistances ranging from zero to R_{iopt}. Matching circuits can be used to match any impedance to the desired load resistance.
- The peak voltage across the transistor is about four times higher than

the input dc voltage. Therefore, the circuit is suitable for low input voltage applications.

- The drive circuit is easy to build because the gate-to-source voltage of the transistor is referenced to ground.
- The circuit is very efficient and can be operated at high frequencies.
- The large choke inductance with a low current ripple can be replaced by a low inductance with a large current ripple. In this case, the equations describing the inverter operation will change [24].
- The loaded quality factor of the resonant circuit can be small. In the extreme case, the resonant capacitor becomes a large dc-blocking capacitor. The mathematical description will change [18].
- The maximum operating frequency at which Class E operation is achievable is limited by the output capacitance of the switch and is given by (13.69).

13.9 REFERENCES

1. N. O. Sokal and A. D. Sokal, "Class E—A new class of high–efficiency tuned single-ended switching power amplifiers," *IEEE J. Solid-State Circuits*, vol. SC-10, pp. 168–176, June 1975.
2. N. O. Sokal and A. D. Sokal, "High efficiency tuned switching power amplifier," U.S. Patent no. 3,919,656, November 11, 1975.
3. J. Ebert and M. Kazimierczuk, "High efficiency RF power amplifier," *Bull. Acad. Pol. Sci., Ser. Sci. Tech.*, vol. 25, no. 2, pp. 13–16, February 1977.
4. N. O. Sokal, "Class E can boost the efficiency," *Electronic Design*, vol. 25, no. 20, pp. 96–102, September 27, 1977.
5. F. H. Raab, "Idealized operation of the Class E tuned power amplifier," *IEEE Trans. Circuits Syst.*, vol. CAS-24, pp. 725–735, December 1977.
6. N. O. Sokal and F. H. Raab, "Harmonic output of Class E RF power amplifiers and load coupling network design," *IEEE J. Solid-State Circuits*, vol. SC-12, pp. 86–88, February 1977.
7. F. H. Raab, "Effects of circuit variations on the Class E tuned power amplifier," *IEEE J. Solid-State Circuits*, vol. SC-13, pp. 239–247, April 1978.
8. F. H. Raab and N. O. Sokal, "Transistor power losses in the Class E tuned power amplifier," *IEEE J. Solid-State Circuits*, vol. SC-13, pp. 912–914, December 1978.
9. N. O. Sokal and A. D. Sokal, "Class E switching-mode RF power amplifiers—Low power dissipation, low sensitivity to component values (including transistors) and well–defined operation," *RF Design*, vol. 3, pp. 33–38, no. 41, July/August 1980.
10. J. Ebert and M. K. Kazimierczuk, "Class E high-efficiency tuned oscillator," *IEEE J. Solid-State Circuits*, vol. SC-16, pp. 62–66, April 1981.
11. N. O. Sokal, "Class E high-efficiency switching-mode tuned power amplifier with only one inductor and only one capacitor in load network—Approximate analysis," *IEEE J. Solid-State Circuits*, vol. SC-16, pp. 380–384, August 1981.

12. M. K. Kazimierczuk, "Effects of the collector current fall time on the Class E tuned power amplifier," *IEEE J. Solid-State Circuits*, vol. SC-18, no. 2, pp. 181–193, April 1983.
13. M. K. Kazimierczuk, "Exact analysis of Class E tuned power amplifier with only one inductor and one capacitor in load network," *IEEE J. Solid-State Circuits*, vol. SC-18, no. 2, pp. 214–221, April 1983.
14. M. K. Kazimierczuk, "Parallel operation of power transistors in switching amplifiers," *Proc. IEEE*, vol. 71, no. 12, pp. 1456–1457, December 1983.
15. M. K. Kazimierczuk, "Charge-control analysis of Class E tuned power amplifier," *IEEE Trans. Electron Devices*, vol. ED-31, no. 3, pp. 366–373, March 1984.
16. B. Molnár, "Basic limitations of waveforms achievable in single-ended switching-mode (Class E) power amplifiers," *IEEE J. Solid-State Circuits*, vol. SC-19, no. 1, pp. 144–146, February 1984.
17. M. K. Kazimierczuk, "Collector amplitude modulation of the Class E tuned power amplifier," *IEEE Trans. Circuits Syst.*, vol. CAS-31, no. 6, pp. 543–549, June 1984.
18. M. K. Kazimierczuk, "Class E tuned power amplifier with nonsinusoidal output voltage," *IEEE J. Solid-State Circuits*, vol. SC-21, no. 4, pp. 575–581, August 1986.
19. M. K. Kazimierczuk, "Generalization of conditions for 100-percent efficiency and nonzero output power in power amplifiers and frequency multipliers," *IEEE Trans. Circuits Syst.*, vol. CAS-33, no. 8, pp. 805–506, August 1986.
20. M. K. Kazimierczuk and K. Puczko, "Impedance inverter for Class E dc/dc converters," *29th Midwest Symposium on Circuits and Systems*, Lincoln, NE, August 10–12, 1986, pp. 707–710.
21. G. Lüttke and H. C. Reats, "High voltage high frequency Class-E converter suitable for miniaturization," *IEEE Trans. Power Electronics*, vol. PE-1, pp. 193–199, October 1986.
22. M. K. Kazimierczuk and K. Puczko, "Exact analysis of Class E tuned power amplifier at any Q and switch duty cycle," *IEEE Trans. Circuits Syst.*, vol. CAS-34, no. 2, pp. 149–159, February 1987.
23. G. Lüttke and H. C. Reats, "220 V 500 kHz Class E converter using a BIMOS," *IEEE Trans. Power Electronics*, vol. PE-2, pp. 186–193, July 1987.
24. R. E. Zulinski and J. W. Steadman, "Class E power amplifiers and frequency multipliers with finite dc-feed inductance," *IEEE Trans. Circuits Syst.*, vol. CAS-34, no. 9, pp. 1074–1087, September 1987.
25. M. K. Kazimierczuk and X. T. Bui, "Class E amplifier with an inductive impedance inverter," *IEEE Trans. Industrial Electronics*, vol. IE-37, pp. 160–166, April 1990.
26. M. K. Kazimierczuk and W. A. Tabisz, "Class C-E high-efficiency tuned power amplifier," *IEEE Trans. Circuits Syst.*, vol. CAS-36, no. 3, pp. 421–428, March 1989.
27. M. K. Kazimierczuk and K. Puczko, "Power-output capability of Class E amplifier at any loaded Q and switch duty cycle," *IEEE Trans. Circuits Syst.*, vol. CAS-36, no. 8, pp. 1142–1143, August 1989.
28. M. K. Kazimierczuk and X. T. Bui, "Class E dc/dc converter with a capacitive impedance inverter," *IEEE Trans. Power Electronics*, vol. PE-4, pp. 124–135, January 1989.

29. M. K. Kazimierczuk and K. Puczko, "Class E tuned power amplifier with antiparallel diode or series diode at switch, with any loaded Q and switch duty cycle," *IEEE Trans. Circuits Syst.*, vol. CAS-36, no. 9, pp. 1201–1209, September 1989.
30. M. K. Kazimierczuk and X. T. Bui, "Class E amplifier with an inductive impedance inverter," *IEEE Trans. Industrial Electronics*, vol. IE-37, pp. 160–166, April 1990.
31. R. E. Zulinski and K. J. Grady, "Load-independent Class E power inverters: Part I—Theoretical development," *IEEE Trans. Circuits Syst.*, vol. CAS-37, pp. 1010–1018, August 1990.
32. K. Thomas, S. Hinchliffe, and L. Hobson, "Class E switching-mode power amplifier for high-frequency electric process heating applications," *Electron. Lett.*, vol. 23, no. 2, pp. 80–82, January 1987.
33. D. Collins, S. Hinchliffe, and L. Hobson, "Optimized Class-E amplifier with load variation," *Electron. Lett.*, vol. 23, no. 18, pp. 973–974, August 1987.
34. D. Collins, S. Hinchliffe, and L. Hobson, "Computer control of a Class E amplifier," *Int. J. Electron.*, vol. 64, no. 3, pp. 493–506, 1988.
35. S. Hinchliffe, L. Hobson, and R. W. Houston, "A high-power Class E amplifier for high frequency electric process heating," *Int. J. Electron.*, vol. 64, no. 4, pp. 667–675, 1988.
36. S. Ghandi, R. E. Zulinski, and J. C. Mandojana, "On the feasibility of load-independent output current in Class E amplifiers," *IEEE Trans. Circuits Syst.*, vol. CAS-39, pp. 564–567, July 1992.

13.10 REVIEW QUESTIONS

13.1 What is the ZVS technique?

13.2 Is the transistor output capacitance absorbed into the Class E ZVS inverter topology?

13.3 Is it possible to obtain ZVS operation at any load using the basic topology of the Class E ZVS inverter?

13.4 Is the turn-on switching loss zero in the Class E ZVS inverter?

13.5 Is the turn-off switching loss zero in the Class E ZVS inverter?

13.6 Is it possible to achieve the ZVS condition at any operating frequency?

13.7 Is the basic Class E ZVS inverter short-circuit proof?

13.8 Is the basic Class E ZVS inverter open-circuit proof?

13.9 Is it possible to use a finite dc-feed inductance in series with the dc input voltage source V_I?

13.10 Is it required to use a high value of the loaded quality factor of the resonant circuit in the Class E ZVS inverter?

13.11 PROBLEMS

13.1 Design an optimum Class E ZVS inverter to meet the following specifications: $P_{Ri} = 125$ W, $V_I = 48$ V, and $f = 2$ MHz. Assume $Q_L = 5$.

13.2 The rms value of the U.S. utility voltage is from 92 to 132 V. This voltage is rectified by a bridge peak rectifier to supply a Class E ZVS inverter that is operated at a switch duty cycle of 0.5. What is the required value of the voltage rating of the switch?

13.3 Repeat Problem 13.2 for the European utility line, whose rms voltage is $220 \pm 15\%$.

13.4 Derive the design equations for the component values for the matching resonant circuit π2a shown in Fig. 13.12(b).

13.5 Find the maximum operating frequency at which pure Class E operation is still achievable for $V_I = 200$ V, $P_{Ri} = 75$ W, and $C_{out} = 100$ pF.

CHAPTER 14

CLASS E ZERO-CURRENT-SWITCHING RESONANT INVERTER

14.1 INTRODUCTION

In this chapter, a Class E zero-current-switching (ZCS) inverter [1]–[3] is presented and analyzed. In this inverter, the switch is turned off at zero current, yielding zero turn-off switching loss. A shortcoming of the Class E ZCS inverter is that the switch output capacitance is not included in the basic inverter topology. The switch turns on at a nonzero voltage, and the energy stored in the switch output capacitance is dissipated in the switching device, reducing the efficiency. Therefore, the upper operating frequency of the Class E ZCS inverter is lower than that for the Class E ZVS inverter.

14.2 CIRCUIT DESCRIPTION

A circuit of a Class E ZCS inverter is depicted in Fig. 14.1(a). This circuit was introduced by Kazimierczuk [1]. It consists of a single transistor and a load network. The transistor operates cyclically as a switch at the desired operating frequency $f = \omega/(2\pi)$. The simplest type of load network consists of a resonant inductor L_1 connected in series with the dc source V_I, and an L-C-R_i series-resonant circuit. The resistance R_i is the ac load.

The equivalent circuit of the Class E ZCS inverter is shown in Fig. 14.1(b). The capacitance C is divided into two series capacitances, C_a and C_b, so that capacitance C_a is series resonant with L at the operating frequency $f = \omega/2\pi$

$$\omega = \frac{1}{\sqrt{LC_a}}. \tag{14.1}$$

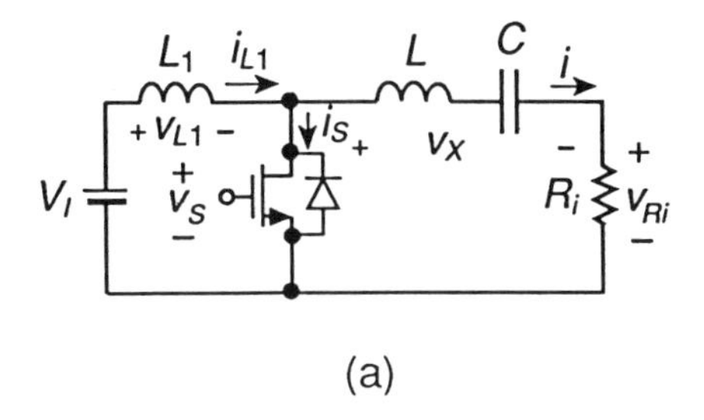

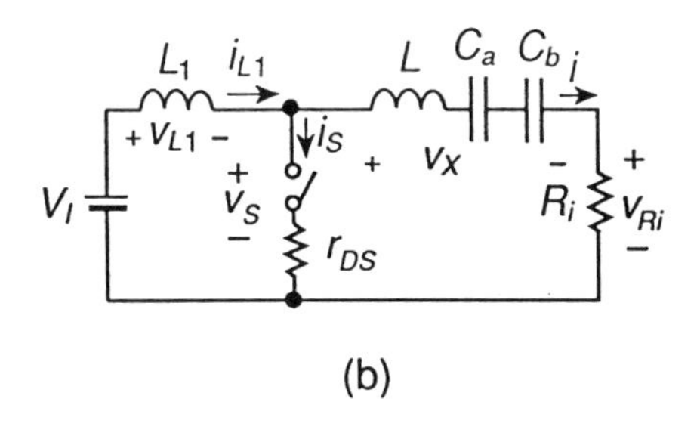

Figure 14.1: Class E zero-current-switching inverter. (a) Circuit. (b) Equivalent circuit.

The additional capacitance C_b signifies the fact that the operating frequency f is lower than the resonant frequency of the series-resonant circuit when the switch is on $f_{o1} = 1/(2\pi\sqrt{LC})$. The loaded quality factor Q_L is defined by the expression

$$Q_L = \frac{X_{Cr}}{R_i} = \frac{C_a + C_b}{\omega R_i C_a C_b}. \tag{14.2}$$

The choice of Q_L involves the usual tradeoff among 1) low harmonic content of the power delivered to R_i (high Q_L), 2) low change of inverter performance with frequency (low Q_L), 3) high efficiency of the load network (low Q_L), and 4)high bandwidth (low Q_L).

14.3 PRINCIPLE OF OPERATION

The equivalent circuit of the inverter is shown in Fig. 14.1(b). It is based on the following assumptions:

1) The elements of the load network are ideal.
2) The loaded quality factor Q_L of the series-resonant circuit is high enough that the output current is essentially a sinusoid at the operating frequency.
3) The switching action of the transistor is instantaneous and lossless; the transistor has zero output capacitance, zero saturation resistance, zero saturation voltage, and infinite "off" resistance.

It is assumed for simplicity that the switch duty ratio is 50%, that is, the switch is ON for half of the ac period and OFF for the remainder of the period; however, the duty ratio can be any arbitrarily chosen value if the circuit component values are chosen to be appropriate for the chosen duty ratio. It will be explained in Section 14.4 that a duty ratio of 50% is one of the conditions for optimum inverter operation.

The inverter operation is determined by the switch when it is closed and by the transient response of the load network when the switch is open. The principle of the inverter operation is explained by the current and voltage waveforms shown in Fig. 14.2. Figure 14.2(a) depicts the waveforms for optimum operation. When the switch is open, its current i_S is zero. Hence, the inductor current i_{L1} is equal to a nearly sinusoidal output current i_{Ri}. The current i_{L1} produces the voltage drop v_{L1} across the inductor L_1. This voltage is approximately a section of a sine wave. The difference between the supply voltage V_I and the voltage v_{L1} is the voltage across the switch v_S. When the switch is closed, the voltage v_S is zero, and voltage v_{L1} equals the supply voltage V_I. This voltage produces the linearly increasing current i_{L1}. The difference between the current i_{L1} and current i_{Ri} flows through the switch.

In the Class E ZCS inverter, it is possible to eliminate power losses due to the on-to-off transition of the transistor, yielding high efficiency. Assuming that the transistor is turned off at $\omega t_{off} = 2\pi$, the ZCS condition at turn-off is

$$i_S(2\pi) = 0. \tag{14.3}$$

For optimum operation, the following condition should also be satisfied

$$\frac{di_S}{d(\omega t)}\Big|_{\omega t=2\pi} = 0. \tag{14.4}$$

If condition (14.3) is not satisfied, the transistor turns off at nonzero current. Consequently, there is a fall time of the drain (or collector) current during which the transistor acts as a current source. During the fall time, the drain current increases and the drain-source voltage increases. Since the transistor current and voltage overlap during the turn-off interval, there is a turn-off power loss. However, if the transistor current is already zero at turn-off, the transistor current fall time is also zero, there is no overlap of the transistor current and voltage, and the turn-off switching loss is zero.

Condition (14.3) eliminates dangerous voltage spikes at the output of the transistor. If this condition is not satisfied, the current i_S changes rapidly during turn-off of the transistor. Hence, the inductor current i_{L1} also changes rapidly during turn-off. Therefore, inductive voltage spikes appear at the output of the transistor and device failure may occur. The rapid change of i_{L1} during turn-off of the transistor causes a change of the energy stored in the inductor L_1. A part of this energy is dissipated in the transistor as

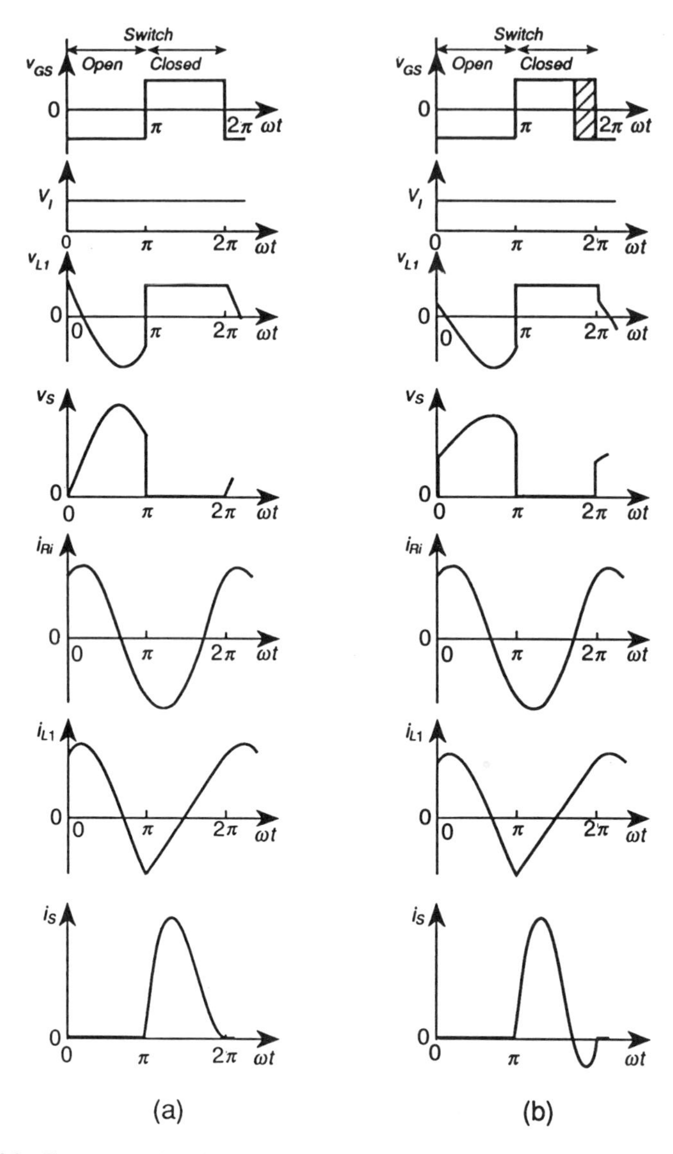

Figure 14.2: Current and voltage waveforms in the Class E ZCS inverter. (a) For optimum operation. (b) For suboptimum operation.

heat, and the remainder is delivered to the series-resonant circuit L, C, and R_i. If condition (14.4) is satisfied, the switch current is always positive and the antiparallel diode never conducts. Furthermore, the voltage across the switch at the turn-off instant will be zero, that is, $v_S(2\pi) = 0$, and during the "off" state the voltage v_S will start to increase from zero only gradually. This zero starting voltage v_S is desirable in the case of the real transistor because the energy stored in the parasitic capacitance across the transistor is zero at the instant the transistor switches off. The parasitic capacitance comprises the transistor capacitances, the winding capacitance of L_1, and stray winding capacitance. The optimum operating conditions can be accomplished by a proper choice of the load-network components. The load resistance at which the ZCS condition is satisfied is $R_i = R_{iopt}$.

Figure 14.2(b) shows the waveforms for suboptimum operation. This operation occurs when only the ZCS condition is satisfied. If the slope of the switch current at the time the switch current reaches zero is positive, the switch current will be negative during a portion of the period. If the transistor is OFF, the antiparallel diode conducts the negative switch current. If the transistor is ON, either only the transistor conducts or both the transistor and the antiparallel diode conduct. The transistor should be turned off during the time interval the switch current is negative. When the switch current reaches zero, the antiparallel diode turns off.

The voltage across the inductor L_1 is described by the expression

$$v_{L1} = \omega L_1 \frac{di_{L1}}{d(\omega t)}. \tag{14.5}$$

At the switch turn-on, the derivative of the inductor current i_{L1} changes rapidly from a negative to a positive value. This causes a step change in the inductor voltage v_{L1} and consequently in the switch voltage v_S.

According to assumption 3), the conduction power loss and the turn-on switching power loss are neglected. The conduction loss dominates at low frequencies, and the turn-on switching loss dominates at high frequencies. The off-to-on switching time is especially important in high-frequency operation. The parasitic capacitance across the transistor is discharged from the voltage $2V_I$ to zero when the transistor switches on. This discharge requires a nonzero length of time. The switch current i_S is increasing during this time. Since the switch voltage v_S and the switch current i_S are simultaneously nonzero, the power is dissipated in the transistor. The off-to-on switching loss becomes comparable to saturation loss at high frequencies. Moreover, the transient response of the load network depends on the parasitic capacitance when the switch is open. This influence is neglected in this analysis. According to assumption 1), the power losses in the parasitic resistances of the load network are also neglected.

14.4 ANALYSIS

14.4.1 Steady-State Current and Voltage Waveforms

The basic equations of the equivalent inverter circuit shown in Fig. 14.1(b) are

$$i_S = i_{L1} - i_{Ri} \tag{14.6}$$

$$v_S = V_I - v_{L1}. \tag{14.7}$$

The series-resonant circuit forces a sinusoidal output current

$$i = I_m sin(\omega t + \varphi). \tag{14.8}$$

The switch is OFF for the interval $0 < \omega t \leq \pi$. Therefore,

$$i_S = 0, \quad \text{for} \quad 0 < \omega \leq \pi. \tag{14.9}$$

From (14.6), (14.8), and (14.9),

$$i_{L1} = i_{Ri} = I_m sin(\omega t + \varphi), \quad \text{for} \quad 0 < \omega t \leq \pi. \tag{14.10}$$

The voltage across the inductor L_1 is

$$v_{L1} = \omega L_1 \frac{di_{L1}}{d(\omega t)} = \omega L_1 I_m cos(\omega t + \varphi), \quad \text{for} \quad 0 < \omega t \leq \pi. \tag{14.11}$$

Hence, (14.7) becomes

$$v_S = V_I - v_{L1} = V_I - \omega L_1 I_m cos(\omega t + \varphi), \quad \text{for} \quad 0 < \omega t \leq \pi. \tag{14.12}$$

Using (14.10) and taking into account the fact that the inductor current i_{L1} is continuous,

$$i_{L1}(\pi+) = i_{L1}(\pi-) = I_m sin(\pi + \varphi) = -I_m sin\varphi. \tag{14.13}$$

The switch is ON for the interval $\pi < \omega t \leq 2\pi$ during which

$$v_S = 0, \quad \text{for} \quad \pi < \omega t \leq 2\pi. \tag{14.14}$$

Substitution of this into (14.7) then produces

$$v_{L1} = V_I, \quad \text{for} \quad \pi < \omega t \leq 2\pi. \tag{14.15}$$

Thus, from (14.13) and (14.15) the current through the inductor L_1 is

$$\begin{aligned} i_{L1} &= \frac{1}{\omega L_1} \int_{\pi}^{\omega t} v_{L1}(u)du + i_{L1}(\pi+) = \frac{1}{\omega L_1} \int_{\pi}^{\omega t} V_I(u)du + i_{L1}(\pi+) \\ &= \frac{V_I}{\omega L_1}(\omega t - \pi) - I_m sin(\omega t + \varphi), \quad \text{for} \quad \pi < \omega t \leq 2\pi. \end{aligned} \tag{14.16}$$

From (14.6) and (14.8),

$$i_S = i_{L1} - i_{Ri} = \frac{V_I}{\omega L_1}(\omega t - \pi) - I_m[sin(\omega t + \varphi) + sin\varphi], \qquad \text{for} \quad \pi < \omega t \leq 2\pi. \tag{14.17}$$

Substituting the ZCS conditon $i_S(2\pi) = 0$ into (14.17),

$$I_m = V_I \frac{\pi}{2\omega L_1 sin\varphi}. \tag{14.18}$$

Because $I_m > 0$,

$$0 < \varphi < \pi. \tag{14.19}$$

From (14.9), (14.17), and (14.18),

$$i_S = \begin{cases} 0, & 0 < \omega t \leq \pi, \\ \frac{V_I}{\omega L_1}\left[\omega t - \frac{3\pi}{2} - \frac{\pi}{2sin\varphi} sin(\omega t + \varphi)\right], & \pi < \omega t \leq 2\pi. \end{cases} \tag{14.20}$$

Substitution of the condition of optimum operation given by (14.4) into (14.20) yields

$$tan\varphi = \frac{\pi}{2}. \tag{14.21}$$

From (14.19) and (14.21),

$$\varphi = arctan\left(\frac{\pi}{2}\right) = 1.0039 \text{ rad} = 57.52^\circ. \tag{14.22}$$

Consideration of trigonometric relationships shows that

$$sin\varphi = \frac{\pi}{\sqrt{\pi^2 + 4}} \tag{14.23}$$

$$cos\varphi = \frac{2}{\sqrt{\pi^2 + 4}}. \tag{14.24}$$

From (14.20) and (14.21),

$$i_S = \begin{cases} 0, & 0 < \omega t \leq \pi, \\ \frac{V_I}{\omega L_1}\left(\omega t - \frac{3\pi}{2} - \frac{\pi}{2} cos\omega t - sin\omega t\right), & \pi < \omega t \leq 2\pi. \end{cases} \tag{14.25}$$

Using the Fourier formula, the supply dc current is

$$I_I = \frac{1}{2\pi}\int_{\pi}^{2\pi} i_S d(\omega t) = \frac{V_I}{\pi \omega L_1}. \tag{14.26}$$

The amplitude of the output current can be found from (14.18), (14.23), and (14.26)

$$I_m = \frac{\sqrt{\pi^2+4}}{2}\frac{V_I}{\omega L_1} = \frac{\pi\sqrt{\pi^2+4}}{2}I_I = 5.8499 I_I. \tag{14.27}$$

Substitution of (14.27) into (14.25) yields the steady-state normalized switch current waveform

$$\frac{i_S}{I_I} = \begin{cases} 0, & 0 < \omega t \leq \pi \\ \pi\left(\omega t - \frac{3\pi}{2} - \frac{\pi}{2}cos\omega t - sin\omega t\right), & \pi < \omega t \leq 2\pi. \end{cases} \tag{14.28}$$

From (14.12), (14.18), and (14.21), the normalized switch voltage waveform is found as

$$\frac{v_S}{V_I} = \begin{cases} \frac{\pi}{2}sin\omega t - cos\omega t + 1, & 0 < \omega t \leq \pi \\ 0, & \pi < \omega t \leq 2\pi. \end{cases} \tag{14.29}$$

14.4.2 Peak Switch Current and Voltage

The peak switch current I_{SM} and voltage V_{SM} can be determined by differentiating waveforms (14.28) and (14.29), and setting the results equal to zero. Finally, we obtain

$$I_{SM} = \pi(\pi - 2\varphi)I_I = 3.562 I_I \tag{14.30}$$

and

$$V_{SM} = \left(\frac{\sqrt{\pi^2+4}}{2} + 1\right)V_I = 2.8621 V_I. \tag{14.31}$$

Neglecting power losses, the output power equals the dc input power $P_I = I_I V_I$. Thus, the power-output capability c_p can be computed from the expression

$$c_p = \frac{P_{Ri}}{I_{SM}V_{SM}} = \frac{I_I V_I}{I_{SM}V_{SM}} = 0.0981. \tag{14.32}$$

It has the same value as the Class E ZVS inverter with a shunt capacitor. It can be proved that the maximum power output capability occurs at a duty ratio of 50% .

14.4.3 Fundamental-Frequency Components

The output voltage is sinusoidal and has the form

$$v_{R1} = V_m sin(\omega t + \varphi), \tag{14.33}$$

where

$$V_m = R_i I_m. \tag{14.34}$$

The voltage v_X across the elements L, C_a, and C_b is not sinusoidal. The fundamental-frequency component v_{X1} of the voltage v_X appears only across the capacitor C_b because the inductance L and the capacitance C_a are resonant at the operating frequency f and their reactance $\omega L - 1/(\omega C_a) = 0$. This component is

$$v_{X1} = V_{X1} cos(\omega t + \varphi), \tag{14.35}$$

where

$$V_{X1} = -\frac{I_m}{\omega C_b}. \tag{14.36}$$

The fundamental-frequency component of the switch voltage is

$$v_{S1} = v_{R1} + v_{X1} = V_m sin(\omega t + \varphi) + V_{X1} cos(\omega t + \varphi). \tag{14.37}$$

The phase shift between the voltages v_{R1} and v_{S1} is determined by the expression

$$tan\psi = \frac{V_{X1}}{V_m} = -\frac{1}{\omega C_b R_i}. \tag{14.38}$$

Using (14.29) and the Fourier formulas, we can obtain

$$V_m = \frac{1}{\pi}\int_0^{2\pi} v_S sin(\omega t + \varphi) d(\omega t) = \frac{4}{\pi\sqrt{\pi^2 + 4}} V_I = 0.3419 V_I \tag{14.39}$$

and

$$V_{X1} = \frac{1}{\pi}\int_0^{2\pi} v_S cos(\omega t + \varphi) d(\omega t) = -\frac{\pi^2 + 12}{4\sqrt{\pi^2 + 4}} V_I = -1.4681 V_I. \tag{14.40}$$

Substituting (14.26) and (14.27) into (14.39) and (14.40),

$$V_m = \frac{8}{\pi(\pi^2 + 4)} \omega L_1 I_m \tag{14.41}$$

$$V_{X1} = -\frac{\pi^2 + 12}{2(\pi^2 + 4)} \omega L_1 I_m. \tag{14.42}$$

The fundamental-frequency components of the switch current $i_{s1} = I_{s1} sin(\omega t + \gamma)$, I_{s1} and γ, and voltage $v_{s1} = V_{s1} sin(\omega t + \vartheta)$, V_{s1} and ϑ, are

$$I_{s1} = I_I \sqrt{\left(\frac{\pi^2}{4} - 2\right)^2 + \frac{\pi^2}{2}} = 1.6389 I_I \tag{14.43}$$

$$\gamma = 180^\circ + arctan\left(\frac{\pi^2 - 8}{2\pi}\right) = 196.571^\circ \tag{14.44}$$

$$V_{s1} = \sqrt{V_m^2 + V_{X1}^2} = V_I \sqrt{\frac{16}{\pi^2(\pi^2+4)} + \frac{(\pi^2+12)^2}{16(\pi^2+4)}} = 1.5074 V_I \tag{14.45}$$

$$\vartheta = \varphi + \psi = -19.372^\circ. \tag{14.46}$$

The phase ϕ of the input impedance of the load network at the operating frequency is

$$\phi = 180^\circ + \vartheta - \gamma = -35.945^\circ. \tag{14.47}$$

This indicates that the input impedance is capacitive.

14.5 POWER RELATIONSHIPS

The dc input power P_I is

$$P_I = I_I V_I, \tag{14.48}$$

and from (14.39) the output power P_{Ri} is

$$P_{Ri} = \frac{V_m^2}{2R_i} = \frac{8}{\pi^2(\pi^2+4)} \frac{V_I^2}{R_i} = 0.05844 \frac{V_I^2}{R_i}. \tag{14.49}$$

14.6 ELEMENT VALUES OF LOAD NETWORK

From (14.34), (14.38), (14.41), and (14.42),

$$\frac{\omega L_1}{R_i} = \frac{\pi(\pi^2+4)}{8} = 5.4466, \tag{14.50}$$

$$\omega C_b R_i = \frac{16}{\pi(\pi^2+12)} = 0.2329, \tag{14.51}$$

and

$$\psi = arctan\left(\frac{V_{X1}}{V_m}\right) = -arctan\left[\frac{\pi(\pi^2+12)}{16}\right] = -76.89^\circ. \tag{14.52}$$

Hence, according to Fig. 14.1(b), the capacitor C_b should be connected in series with C_a, L, and R_i. The values of L and C_a can be found from formulas (14.1) and (14.2).

From (14.27) and (14.50),

$$I_m = \frac{4}{\pi\sqrt{\pi^2+4}}\frac{V_I}{R_i}, \tag{14.53}$$

and from (14.26) and (14.50),

$$I_I = \frac{8}{\pi^2(\pi^2+4)}\frac{V_I}{R_i}. \tag{14.54}$$

The dc input resistance of the inverter is obtained from (14.26) and (14.50)

$$R_{DC} \equiv \frac{V_I}{I_I} = \pi\omega L_1 = \frac{\pi^2(\pi^2+4)}{8}R_i = \frac{2\pi(\pi^2+4)}{(\pi^2+12)}\frac{1}{\omega C_b}. \tag{14.55}$$

The element values of the load network can be computed from the following expressions:

$$R_i = \frac{8}{\pi^2(\pi^2+4)}\frac{V_I^2}{P_{Ri}} = 0.05844\frac{V_I^2}{P_{Ri}} \tag{14.56}$$

$$L_1 = \frac{\pi(\pi^2+4)}{8}\frac{R_i}{\omega} = 5.4466\frac{R_i}{\omega} \tag{14.57}$$

$$C_b = \frac{16}{\pi(\pi^2+12)}\frac{1}{\omega R_i} = \frac{0.2329}{\omega R_i} \tag{14.58}$$

$$C = \frac{1}{\omega R_i Q_L} \tag{14.59}$$

$$L = \left[Q_L - \frac{\pi(\pi^2+12)}{16}\right]\frac{R_i}{\omega} = (Q_L - 4.2941)\frac{R_i}{\omega}. \tag{14.60}$$

The factor Q_L can be chosen freely, according to the considerations discussed at the end of Section 14.1. It is apparent from (14.60) that the factor Q_L must be greater than 4.2941.

14.7 DESIGN EXAMPLE

Example 14.1

Design the Class E inverter of Fig. 14.1(a) to meet the following specifications: $V_I = 100$ V, $P_{Rimax} = 50$ W, and $f = 1$ MHz.

Solution: It is sufficient to design the inverter for the full power. From (14.56), the full-load resistance is

$$R_i = \frac{8}{\pi^2(\pi^2+4)} \frac{V_I^2}{P_{Ri}} = 0.05844 \times \frac{100^2}{50} = 11.7\ \Omega. \tag{14.61}$$

According to Section 14.6, the factor Q_L must be greater than 4.2941. Let $Q_L = 4.5$. Thus, using (14.57), (14.59), and (14.60) the values of the elements of the load network are

$$L_1 = \frac{\pi^2+4}{16} \frac{R_i}{f} = 0.8669 \times \frac{11.7}{10^6} = 10.1\ \mu\text{H}, \tag{14.62}$$

$$C = \frac{1}{\omega R_i Q_L} = \frac{1}{2 \times \pi \times 10^6 \times 11.7 \times 4.5} = 3.02\ \text{nF}, \tag{14.63}$$

and

$$L = \left[Q_L - \frac{\pi(\pi^2+12)}{16}\right] \frac{R_i}{\omega} = \left[4.5 - \frac{\pi(\pi^2+12)}{16}\right] \times \frac{11.7}{2 \times \pi \times 10^6}$$
$$= 0.383\ \mu\text{H}. \tag{14.64}$$

The maximum voltage across the switch can be obtained using (14.31) as

$$V_{SM} = \left(\frac{\sqrt{\pi^2+4}}{2} + 1\right) V_I = 2.8621 \times 100 = 286.2\ \text{V}. \tag{14.65}$$

From (14.54), the dc input current is

$$I_I = \frac{8}{\pi^2(\pi^2+4)} \frac{V_I}{R_i} = 0.0584 \times \frac{100}{11.7} = 0.5\ \text{A}. \tag{14.66}$$

The maximum switch current is calculated using (14.30) as

$$I_{SM} = \pi(\pi - 2\varphi) I_I = 3.562 \times 0.5 = 1.78\ \text{A} \tag{14.67}$$

and from (14.53) the maximum amplitude of the current through the resonant circuit is

$$I_m = \frac{4}{\pi\sqrt{\pi^2+4}} \frac{V_I}{R_i} = 0.3419 \times \frac{100}{11.7} = 2.92\ \text{A}. \tag{14.68}$$

The resonant frequency of the L-C series-resonant circuit when the switch is on is $f_{o1} = 1/(2\pi\sqrt{LC}) = 4.66$ MHz, and the resonant frequency of the L_1-L-C series-resonant circuit when the switch is off is $f_{o2} = 1/(2\pi\sqrt{C(L+L_1)})$ $= 0.911$ MHz.

14.8 SUMMARY

- In the Class E ZCS inverter, the transistor turns off at zero current, reducing turn-off switching loss to zero, even if the transistor switching time is an appreciable fraction of the cycle of the operating frequency.
- The transistor output capacitance is not absorbed into the topology of the Class E ZCS inverter.
- The transistor turns on at nonzero voltage, causing turn-on power loss.
- The efficiency of the Class E ZCS inverter is lower than that of the Class E ZVS inverter at the same frequency and with the same transistor.
- The voltage stress in the Class E ZCS inverter is lower than that of the Class E ZVS inverter.
- The ZCS condition can be satisfied for load resistances ranging from a minimum value R_{iopt} to infinity.
- The load network of the inverter can be modified for impedance transformation and harmonic suppression.

14.9 REFERENCES

1. M. K. Kazimierczuk, "Class E tuned amplifier with shunt inductor," *IEEE J. Solid-State Circuits*, vol. SC-16, no. 1, pp. 2–7, February 1981.
2. N. C. Voulgaris and C. P. Avratoglou, "The use of a switching device in a Class E tuned power amplifier," *IEEE Trans. Circuits Syst.*, vol. CAS-34, pp. 1248–1250, October 1987.
3. C. P. Avratoglou and N. C. Voulgaris, "A Class E tuned amplifier configuration with finite dc-feed inductance and no capacitance," *IEEE Trans. Circuits Syst.*, vol. CAS-35, pp. 416–422, April 1988.

14.10 REVIEW QUESTIONS

14.1 What is the ZCS technique?

14.2 What is the turn-off switching loss in the Class E ZCS inverter?

14.3 What is the turn-on switching loss in the Class E ZCS inverter?

14.4 Is the transistor output capacitance absorbed into the Class E ZCS inverter topology?

14.5 Is the inductance connected in series with the dc input source V_I a large high-frequency choke in the Class E ZCS inverter?

14.6 What are the switch voltage and current stresses for the Class D ZCS inverter at $D = 0.5$?

14.7 Compare the voltage and current stresses for the Class E ZVS and ZCS inverters at $D = 0.5$.

14.11 PROBLEMS

14.1 A Class E ZCS inverter is powered from a 340-V power supply. What is the required voltage rating of the switch if the switch duty cycle is 0.5?

14.2 Design a Class E ZCS inverter to meet the following specifications: $V_I = 180$ V, $P_{Ri} = 250$ W, and $f = 200$ kHz.

14.3 It has been found that a Class E ZCS inverter has the following parameters: $D = 0.5$, $f = 400$ kHz, $L_1 = 20$ μH, and $P_{Ri} = 100$ W. What is the maximum voltage across the switch in this inverter?

PART III

CONVERTERS

CHAPTER 15

CLASS D SERIES RESONANT CONVERTER

15.1 INTRODUCTION

A block diagram of a dc-dc resonant converter is shown in Fig. 15.1. The resonant converter consists of a resonant inverter and a high-frequency rectifier that should be compatible to each other. Current-driven rectifiers should be connected to inverters with a current output and voltage-driven rectifiers should be connected to inverters with a voltage output. The dc voltage transfer function of the converter M_V is a product of the voltage transfer function of the inverter M_{VI} and the voltage transfer function of the rectifier M_{VR}

$$M_V \equiv \frac{V_O}{V_I} = M_{VI} M_{VR} . \tag{15.1}$$

Similarly, the efficiency of the converter is a product of the efficiency of the inverter and the efficiency of the rectifier

$$\eta \equiv \frac{P_O}{P_I} = \eta_I \eta_R . \tag{15.2}$$

In this chapter expressions for the voltage transfer functions and the efficiences of various types of the series resonant converters (SRCs) are derived. An example of a design procedure of the SRC is given, and characteristic waveforms in practical implementation of the SRC are presented. The SRC is obtained by replacing the ac load in the series resonant inverter (SRI) discussed in Chapter 6 by one of the Class D current-driven rectifiers considered in Chapter 2. For the sufficiently high loaded quality factor of the

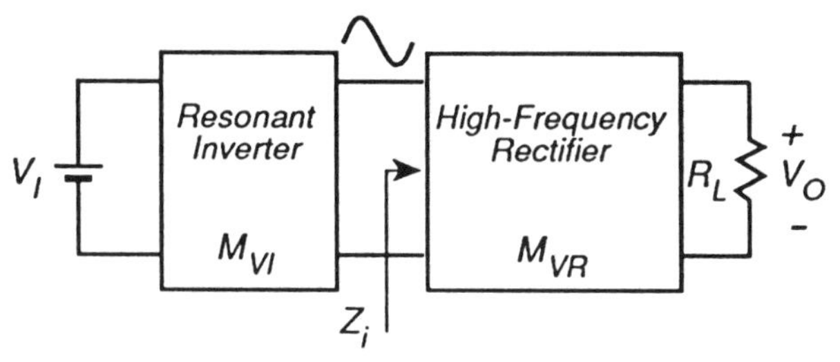

Figure 15.1: Block diagram of a dc-dc resonant converter.

resonant circuit Q_L and the switching frequency close to the resonant frequency, the output of the SRI acts as a sinusoidal current source. For this reason, current-driven rectifiers are compatible with the SRI.

15.2 HALF-BRIDGE SERIES RESONANT CONVERTER

15.2.1 Circuit Description

A half-bridge (HB) series resonant converter [1]–[33] consists of a half-bridge series resonant inverter analyzed in Chapter 6 and one of the current-driven rectifiers analyzed in Chapter 2, as shown in Fig. 15.2. Figure 15.3 depicts idealized current and voltage waveforms in the half-bridge SRC with a Class D current-driven half-wave rectifier for the switching frequency f higher than the resonant frequency f_o. Operation above resonance is preferred because it offers higher efficiency, as discussed in Chapter 6. The transfer function of the resonant circuit depends on the ratio of the switching frequency to the resonant frequency f/f_o (Chapter 6). Therefore, regulation of the dc output voltage V_O against variations in the load resistance R_L (or load current I_O) and the input voltage V_I can be achieved by varying the switching frequency.

The SRC cannot regulate the dc output voltage at no load and light loads. In accordance with the analysis of the transfer function of the resonant circuit in Chapter 6, regulation of the output voltage at light loads would require a very high normalized switching frequency f/f_o. A preload is required to achieve no-load and light-load regulation. A power resistor connected in parallel with the load can be used as a preload, but the resistor consumes power and reduces efficiency. Another solution is to use an additional low-power converter, which is activated as a preload at light loads and no load [32]. The additional converter transfers energy from the output of the main converter to the dc input voltage V_I at light loads and no load. If the main converter is step-down, the additional converter is step-up, and vice versa.

The converter is short-circuit proof if the switching frequency is sufficiently higher or lower than the resonant frequency. In this case, the impedance of the resonant circuit limits the current through the transistors. However, the input impedance of the resonant circuit is very low at the resonant frequency;

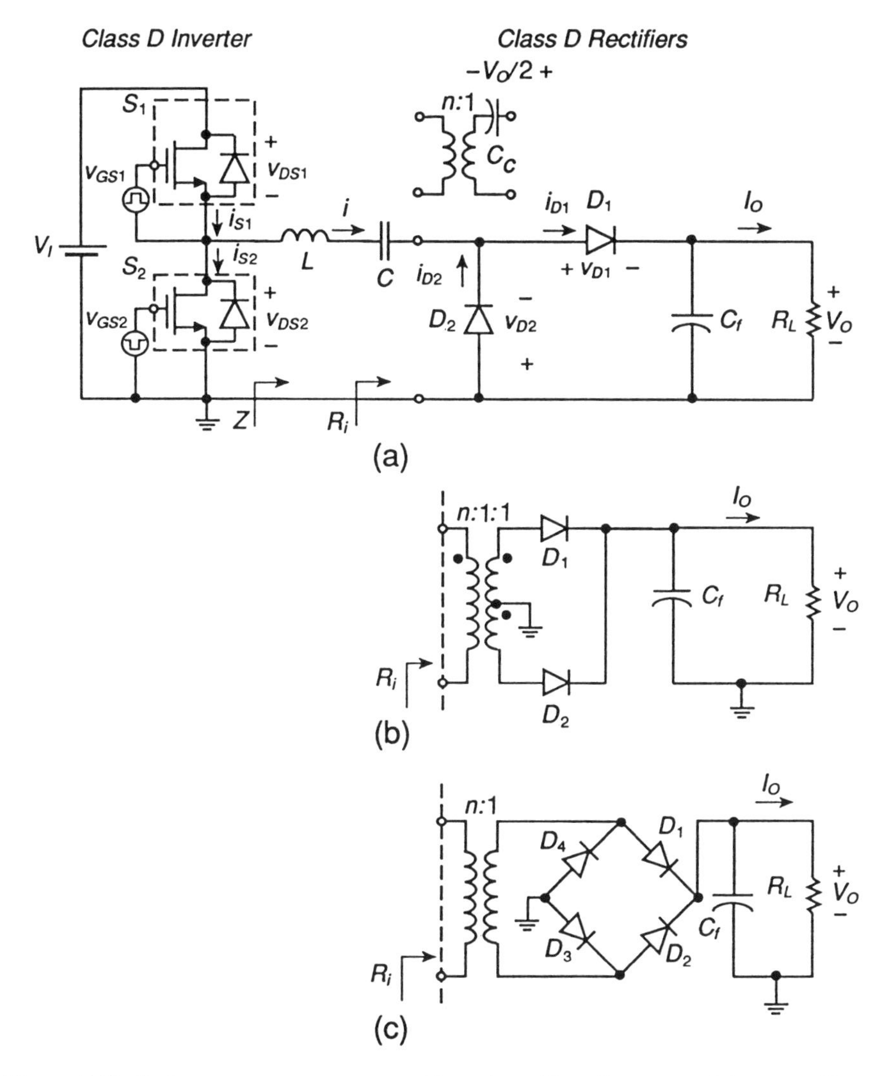

Figure 15.2: Series resonant converter with various Class D current-driven rectifiers. (a) With a half-wave rectifier. (b) With a transformer center-tapped rectifier. (c) With a bridge rectifier.

this impedance is equal to the sum of the parasitic resistances. As a result, a very high current will flow through the resonant circuit and the switches, causing destruction of the converter. The SRC can operate safely at no load because no current flows through the resonant circuit.

Henceforth in this chapter, expressions for the efficiencies the voltage transfer functions are derived under the assumption that switching losses and drive power of the MOSFETs can be neglected.

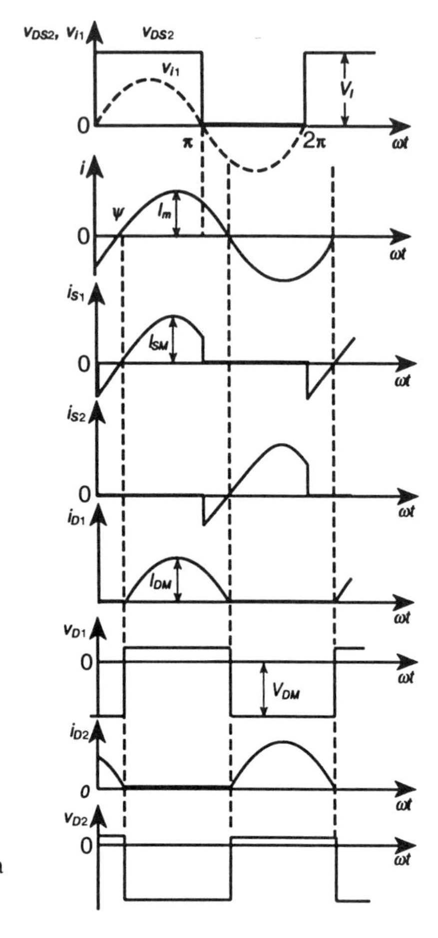

Figure 15.3: Waveforms in the SRC with a half-wave rectifier for $f > f_o$ at high Q_L.

15.2.2 Half-Bridge SRC with Half-Wave Rectifier

The half-bridge SRC with a half-wave rectifier is depicted in Fig. 15.2(a). From (2.27), (2.28), and (6.69), the efficiency of the SRC with a Class D half-wave rectifier is

$$\eta = \eta_{Ir}\eta_R = \frac{\eta_{tr}}{1 + \frac{2V_F}{V_O} + \frac{\pi^2 R_F}{2R_L} + \frac{\pi^2 r \eta_{tr}}{2n^2 R_L} + \frac{r_C}{R_L}\left(\frac{\pi^2}{4} - 1\right)}, \tag{15.3}$$

where $r = r_{DS} + r_L + r_C$.

From (2.31), (6.64), and (15.3), one obtains the dc-to-dc voltage transfer function of the converter

$$M_V = \frac{V_O}{V_I} = M_{VI}M_{VR} = \frac{\sqrt{2}\eta_{Ir}}{\pi\sqrt{1+Q_L^2(\frac{\omega}{\omega_o}-\frac{\omega_o}{\omega})^2}} \times \frac{\pi\eta_R}{n\sqrt{2}} = \frac{\eta}{n\sqrt{1+Q_L^2(\frac{\omega}{\omega_o}-\frac{\omega_o}{\omega})^2}}$$
$$= \frac{\eta_{tr}}{n\sqrt{1+Q_L^2(\frac{\omega}{\omega_o}-\frac{\omega_o}{\omega})^2}[1+\frac{2V_F}{V_O}+\frac{\pi^2 R_F}{2R_L}+\frac{\pi^2 r\eta_{tr}}{2n^2R_L}+\frac{r_C}{R_L}(\frac{\pi^2}{4}-1)]}. \quad (15.4)$$

The range of M_V is from zero to approximately $1/n$.

15.2.3 Half-Bridge SRC with Transformer Center-Tapped Rectifier

The efficiency of the half-bridge SRC with a transformer center-tapped rectifier, shown in Fig. 15.2(b), is obtained from (2.71), (2.72), and (6.69) as

$$\eta = \eta_{Ir}\eta_R = \frac{\eta_{tr}}{1+\frac{V_F}{V_O}+\frac{\pi^2 R_F}{8R_L}+\frac{\pi^2 r\eta_{tr}}{8n^2R_L}+\frac{r_C}{R_L}(\frac{\pi^2}{8}-1)}. \quad (15.5)$$

The product of (6.64) and (2.73) yields, after using (15.5), the dc-to-dc voltage transfer function of the converter

$$M_V = \frac{V_O}{V_I} = M_{VI}M_{VR} = \frac{\eta}{2n\sqrt{1+Q_L^2(\frac{\omega}{\omega_o}-\frac{\omega_o}{\omega})^2}}$$
$$= \frac{\eta_{tr}}{2n\sqrt{1+Q_L^2(\frac{\omega}{\omega_o}-\frac{\omega_o}{\omega})^2}[1+\frac{V_F}{V_O}+\frac{\pi^2 R_F}{8R_L}+\frac{\pi^2 r\eta_{tr}}{8n^2R_L}+\frac{r_C}{R_L}(\frac{\pi^2}{8}-1)]}. \quad (15.6)$$

The range of M_V is from zero to approximately $1/(2n)$.

15.2.4 Half-Bridge SRC with Bridge Rectifier

Using (2.96), (2.97), and (6.69), the efficiency of the SRC with a Class D bridge rectifier depicted in Fig. 15.2(c) can be derived as

$$\eta = \eta_{Ir}\eta_R = \frac{\eta_{tr}}{1+\frac{2V_F}{V_O}+\frac{\pi^2 R_F}{4R_L}+\frac{\pi^2 r\eta_{tr}}{8n^2R_L}+\frac{r_C}{R_L}(\frac{\pi^2}{8}-1)}. \quad (15.7)$$

Combining (6.64), (2.98), and (15.7) produces the dc-to-dc voltage transfer function of the converter

$$M_V = \frac{V_O}{V_I} = M_{VI}M_{VR} = \frac{\eta}{2n\sqrt{1+Q_L^2(\frac{\omega}{\omega_o}-\frac{\omega_o}{\omega})^2}}$$
$$= \frac{\eta_{tr}}{2n\sqrt{1+Q_L^2(\frac{\omega}{\omega_o}-\frac{\omega_o}{\omega})^2}[1+\frac{2V_F}{V_O}+\frac{\pi^2 R_F}{4R_L}+\frac{\pi^2 r\eta_{tr}}{8n^2R_L}+\frac{r_C}{R_L}(\frac{\pi^2}{8}-1)]}. \quad (15.8)$$

The range of M_V is from zero to approximately $1/(2n)$.

Example 15.1

Plot the efficiencies η versus load resistance R_L for Class D half-bridge series resonant converters with transformerless half-wave, center-tapped, and bridge rectifiers at the output voltage $V_O = 100$ V. The full power is $P_O =$ 50 W. Assume the total parasitic resistance of the inverter to be $r = 3\ \Omega$. The parameters of the rectifiers are $V_F = 0.7$ V, $R_F = 0.1\ \Omega$, and $r_C = 25$ mΩ. Discuss the results.

Solution: The full-power load resistance is

$$R_L = \frac{V_O^2}{P_O} = \frac{100^2}{50} = 200\ \Omega. \tag{15.9}$$

The plots of the efficiencies were computed using (15.3), (15.5), and (15.7). Since the converters are to be transformerless, the transformer turns ratio n was set to 1 and the transformer efficiency η_{tr} to 100%. The plots are shown in Fig. 15.4. The highest efficiency is for the converter that employs the center-tapped rectifier. The efficiency of the converter with the bridge rectifier is lower because of higher losses in the rectifier diodes. The converter with the half-wave rectifier has the lowest efficiency because the lowest efficiency of the rectifier (see Example 2.6) and the load resistance reflected to the input of the half-wave rectifier is four times lower than that for the other two rectifiers. Consequently, the efficiency of the inverter for the converter

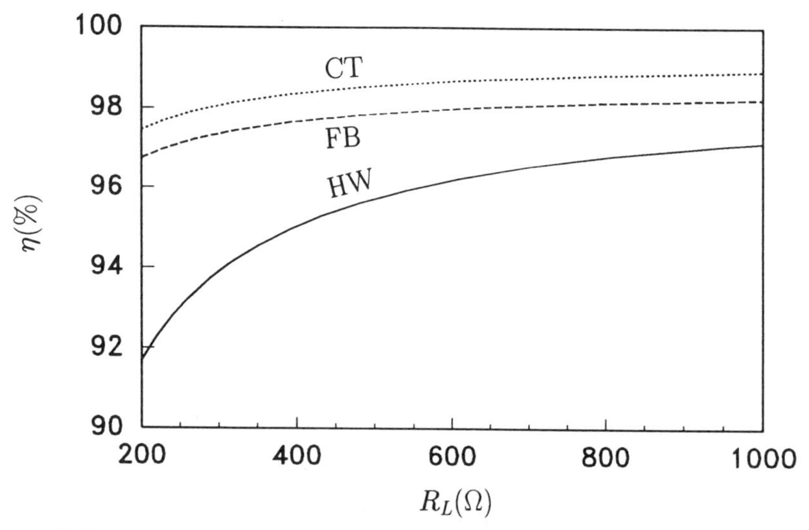

Figure 15.4: Plots of efficiencies for Class D half-bridge series resonant converters with half-wave (HW), center-tapped (CT), and bridge (FB) rectifiers of Example 15.1.

with a half-wave rectifier is the lowest. The efficiencies of all the converters increase with increasing load resistance.

15.3 FULL-BRIDGE SERIES RESONANT CONVERTER

A full-bridge (FB) series resonant converter consists of a full-bridge resonant inverter described in Section 6.9 and one of the current-driven rectifiers of Chapter 2. The expressions for the efficiency remain the same, but the total parasitic resistance of the inverter is now given by

$$r = 2r_{DS} + r_L + r_C. \tag{15.10}$$

Thus, the efficiency of the full-bridge converter is slightly lower than the efficiency of the half-bridge converter with the same component and output voltage values. The voltage transfer function for each full-bridge converter is two times higher than that for the corresponding half-bridge converter.

15.3.1 Full-Bridge SRC with Half-Wave Rectifier

The efficiency of the FB SRC with a Class D half-wave rectifier is given by (15.3). Using (2.31) and (6.166), one arrives at the dc-to–dc voltage transfer function of the FB SRC converter

$$\begin{aligned} M_V &= \frac{V_O}{V_I} = M_{VI}M_{VR} = \frac{2\eta}{n\sqrt{1 + Q_L^2(\frac{\omega}{\omega_o} - \frac{\omega_o}{\omega})^2}} \\ &= \frac{2\eta_{tr}}{n\sqrt{1 + Q_L^2(\frac{\omega}{\omega_o} - \frac{\omega_o}{\omega})^2}[1 + \frac{2V_F}{V_O} + \frac{\pi^2 R_F}{2R_L} + \frac{\pi^2 r\eta_{tr}}{2n^2 R_L} + \frac{r_C}{R_L}(\frac{\pi^2}{4} - 1)]}. \end{aligned} \tag{15.11}$$

The range of M_V is from zero to approximately $2/n$.

15.3.2 Full-Bridge SRC with Transformer Center-Tapped Rectifier

The efficiency of the FB SRC with a transformer center-tapped rectifier is expressed by (15.5). The product of (6.166) and (2.73) leads to the dc-to-dc voltage transfer function of the FB SRC converter

$$\begin{aligned} M_V &= \frac{V_O}{V_I} = M_{VI}M_{VR} = \frac{\eta}{n\sqrt{1 + Q_L^2(\frac{\omega}{\omega_o} - \frac{\omega_o}{\omega})^2}} \\ &= \frac{\eta_{tr}}{n\sqrt{1 + Q_L^2(\frac{\omega}{\omega_o} - \frac{\omega_o}{\omega})^2}[1 + \frac{V_F}{V_O} + \frac{\pi^2 R_F}{8R_L} + \frac{\pi^2 r\eta_{tr}}{8n^2 R_L} + \frac{r_C}{R_L}(\frac{\pi^2}{8} - 1)]}. \end{aligned} \tag{15.12}$$

The range of M_V is from zero to approximately $1/n$.

Table 15.1. Voltage Transfer Function M_V of Lossless Series Resonant Converters

Rectifier	Half-Bridge Converter	Full-Bridge Converter
Half-wave rectifier	$\frac{1}{n\sqrt{1+Q_L^2(\frac{\omega}{\omega_o}-\frac{\omega_o}{\omega})^2}}$	$\frac{2}{n\sqrt{1+Q_L^2(\frac{\omega}{\omega_o}-\frac{\omega_o}{\omega})^2}}$
Center-tapped rectifier	$\frac{1}{2n\sqrt{1+Q_L^2(\frac{\omega}{\omega_o}-\frac{\omega_o}{\omega})^2}}$	$\frac{1}{n\sqrt{1+Q_L^2(\frac{\omega}{\omega_o}-\frac{\omega_o}{\omega})^2}}$
Bridge rectifier	$\frac{1}{2n\sqrt{1+Q_L^2(\frac{\omega}{\omega_o}-\frac{\omega_o}{\omega})^2}}$	$\frac{1}{n\sqrt{1+Q_L^2(\frac{\omega}{\omega_o}-\frac{\omega_o}{\omega})^2}}$

15.3.3 Full-Bridge SRC with Bridge Rectifier

The efficiency of the FB SRC with a Class D bridge rectifier is given by (15.7). From (6.166) and (2.98), one obtains the dc-to-dc voltage transfer function of the FB SRC converter

$$M_V = \frac{V_O}{V_I} = M_{VI}M_{VR} = \frac{\eta}{n\sqrt{1+Q_L^2(\frac{\omega}{\omega_o}-\frac{\omega_o}{\omega})^2}}$$
$$= \frac{\eta_{tr}}{n\sqrt{1+Q_L^2(\frac{\omega}{\omega_o}-\frac{\omega_o}{\omega})^2}[1+\frac{2V_F}{V_O}+\frac{\pi^2 R_F}{4R_L}+\frac{\pi^2 r\eta_{tr}}{8n^2R_L}+\frac{r_C}{R_L}(\frac{\pi^2}{8}-1)]}. \quad (15.13)$$

The range of M_V is from zero to approximately $1/n$. The voltage transfer function M_V of the half-bridge and full-bridge converters with three different rectifiers is given in Table 15.1, neglecting losses.

15.4 DESIGN OF HALF-BRIDGE SRC

To illustrate the significance and facilitate the understanding of the theoretical results obtained in previous chapters and sections, an example of a design of transformerless version of the half-bridge SRC with a half-wave rectifier of Fig. 15.2(a) will be given.

Example 15.2

Design a transformerless series resonant converter with a half-wave rectifier. The following specifications should be satisfied: $V_I = 180$ V, $V_O = 100$ V, and $R_L = 200\ \Omega$ to 1 kΩ. The parameters of the rectifier are $V_F = 0.7$ V, $R_F = 0.1\ \Omega$, and $r_C = 25$ mΩ. Assume that the total inverter efficiency is $\eta_I = 0.92$, the resonant frequency is $f_o = 100$ kHz, and the switching frequency at full load is $f = 110$ kHz. Draw the efficiency η of the designed converter as a function of the load resistance R_L.

Solution: It is sufficient to design the converter for the full load resistance $R_{Lmin} = 200\ \Omega$. The maximum value of the dc output current is $I_{Omax} = V_O/R_{Lmin} = 100/200 = 0.5$ A, and the maximum value of the dc output power is $P_{Omax} = V_O I_{Omax} = 100 \times 0.5 = 50$ W. Using (2.28), (2.27), (2.31), (2.14), and (2.15), the design procedure for the rectifier is as follows:

$$R_i = \frac{2R_L}{\pi^2 \eta_{tr}}\left[1 + \frac{2V_F}{V_O} + \frac{\pi^2 R_F}{2R_L} + \frac{r_C}{R_L}\left(\frac{\pi^2}{4} - 1\right)\right]$$
$$= \frac{2 \times 200}{\pi^2}\left[1 + \frac{2 \times 0.7}{100} + \frac{\pi^2 \times 0.1}{2 \times 200} + \frac{0.025}{200}\left(\frac{\pi^2}{4} - 1\right)\right] = 41.2\ \Omega \quad (15.14)$$

$$\eta_R = \frac{\eta_{tr}}{1 + \frac{2V_F}{V_O} + \frac{\pi^2 R_F}{2R_L} + \frac{r_C}{R_L}(\frac{\pi^2}{4} - 1)}$$
$$= \frac{1}{1 + \frac{2\times 0.7}{100} + \frac{\pi^2 \times 0.1}{2\times 200} + \frac{0.025}{200}(\frac{\pi^2}{4} - 1)} = 98.4\ \% \quad (15.15)$$

$$M_{VR} = \frac{\pi \eta_{tr}}{\sqrt{2}[1 + \frac{2V_F}{V_O} + \frac{\pi^2 R_F}{2R_L} + \frac{r_C}{R_L}(\frac{\pi^2}{4} - 1)]}$$
$$= \frac{\pi}{\sqrt{2}[1 + \frac{2\times 0.7}{100} + \frac{\pi^2 \times 0.1}{2\times 200} + \frac{0.025}{200}(\frac{\pi^2}{4} - 1)]} = 2.184 \quad (15.16)$$

$$I_{DM} = \pi I_O = \pi \times 0.5 = 1.57 \text{ A} \quad (15.17)$$

$$V_{DM} = V_O = 100 \text{ V}. \quad (15.18)$$

The value $\eta_{tr} = 1$ is used in calculations because the transformerless version of the converter is being designed.

The voltage transfer function of the converter is $M_V = V_O/V_I = 100/180 = 0.5556$. Equations (15.4), (6.57), and (6.64) yield the magnitude of the voltage transfer function of the resonant circuit

$$M_{Vr} = \frac{M_V}{M_{Vs} M_{VR}} = \frac{0.5556}{0.45 \times 2.184} = 0.563. \quad (15.19)$$

Since $f/f_o = 1.1$, (6.59) gives the loaded quality factor

$$Q_L = \frac{\sqrt{\frac{\eta_I^2}{M_{Vr}^2} - 1}}{\mid \frac{\omega}{\omega_o} - \frac{\omega_o}{\omega} \mid} = \frac{\sqrt{\frac{0.92^2}{0.563^2} - 1}}{1.1 - \frac{1}{1.1}} = 6.78. \quad (15.20)$$

The overall resistance of the inverter is

$$R = \frac{R_i}{\eta_I} = \frac{41.2}{0.92} = 44.8\ \Omega. \quad (15.21)$$

From (6.10), the component values of the resonant circuit are

$$L = \frac{Q_L R}{\omega_o} = \frac{6.78 \times 44.8}{2 \times \pi \times 100 \times 10^3} = 483.4 \ \mu\text{H} \tag{15.22}$$

$$C = \frac{1}{\omega_o Q_L R} = \frac{1}{2 \times \pi \times 100 \times 10^3 \times 6.78 \times 44.8} = 5.24 \text{ nF}, \tag{15.23}$$

which lead to $Z_o = \sqrt{L/C} = 303.7\ \Omega$. The worst case for the voltage stresses across the resonant components occurs at the corner frequency f_o. Hence, from (6.49),

$$V_{Cm} = V_{Lm} = Z_o I_{mr} = \frac{2V_I Q_L}{\pi} = \frac{2 \times 180 \times 6.78}{\pi} = 777 \text{ V}. \tag{15.24}$$

The overall converter efficiency is $\eta = \eta_I \eta_R = 0.92 \times 0.984 = 0.9$. Hence, the dc input power at full load is

$$P_{Imax} = \frac{P_{Omax}}{\eta} = \frac{50}{0.9} = 55.56 \text{ W}. \tag{15.25}$$

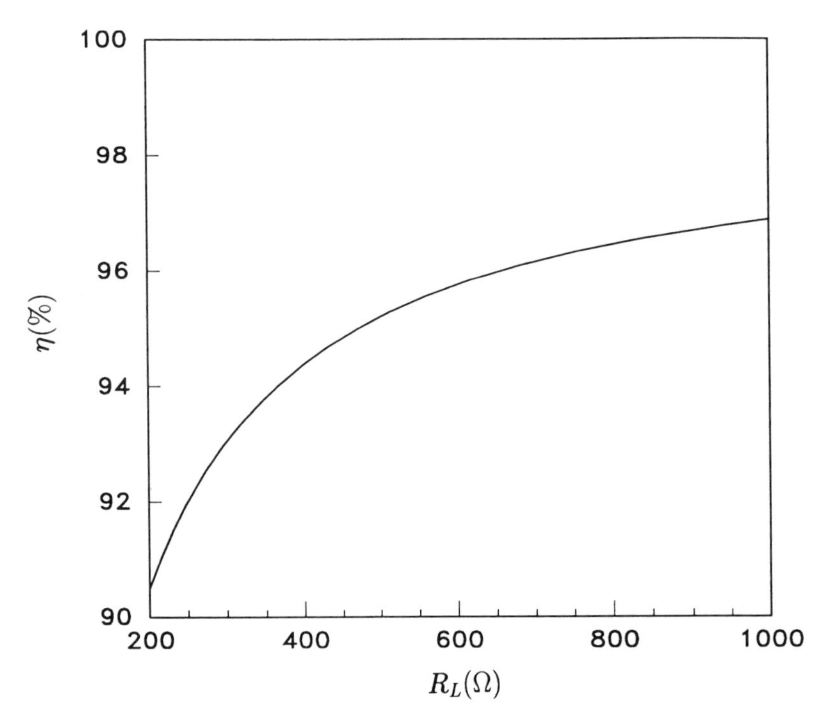

Figure 15.5: Calculated total converter efficiency η versus load resistance R_L at $V_I = 180$ V and $V_O = 100$ V.

The peak value of the switch current is

$$I_{SM(max)} = I_{m(max)} = \sqrt{\frac{2P_{Imax}\eta_I}{R_i}} = \sqrt{\frac{2 \times 55.56 \times 0.92}{41.2}} = 1.58 \text{ A}. \quad (15.26)$$

Figure 15.5 shows a plot of the calculated converter efficiency η as a function of load resistance R_L at $V_I = 180$ V and $V_O = 100$ V.

15.5 SUMMARY

- The transformerless SRC is a step-down converter, except for the full-bridge SRC with the half-wave rectifier.
- The converter can operate safely with an open circuit at the output, but cannot regulate the output voltage.
- The converter can regulate the output voltage V_O over a limited load range from a full load to a reduced load, but it cannot regulate V_O at no load and light loads because its voltage transfer function is not sensitive to frequency. A preload is required to obtain no-load and light-load regulation capability. In the case of a step-down converter, a small step-up converter can be used as preload that is activated at light loads and no load and transfers power from the output to the input voltage source V_I [32]. Likewise, in the case of a step-up converter, a step-down converter can be used as a preload.
- The frequency range required to regulate V_O against load variations depends on the magnitude of the voltage transfer function M_{Vr} of the resonant circuit: (a) If M_{Vr} is close to 0, the frequency range is wide, and (b) if M_{Vr} is close to 1, the frequency range is narrow.
- The efficiency of the converter is higher at part loads than at full load because the rectifier input resistance R_i and the ratio R_i/r increase with increasing R_L.
- The converter is inherently short-circuit protected by the resonant-circuit impedance at all frequencies that are sufficiently higher or lower than the resonant frequency. At the resonant frequency, the input impedance of the resonant circuit is very low and equal to the sum of the parasitic resistances. Therefore, operation at the resonant frequency and a short circuit at the output can cause an excessive current and a catastrophic failure of the converter.
- The converter contains a capacitive output filter, and a large ac current flows through the filter capacitor, causing large conduction losses in the ESR and ripple voltage.
- Since the rectifier diode current is a half-sine wave, the rectifier diodes turn off at low di/dt, reducing noise. In the case of *pn* junction diodes, the reverse recovery current is not a spike, but a portion of a sinusoid whose magnitude is low.

15.6 REFERENCES

1. F. C. Schwarz, "A method of resonant current pulse modulation for power converters," *IEEE Trans. Industrial Electronics and Control Instrumentation*, vol. IECI-17, pp. 209–221, no. 3, May 1970.
2. F. C. Schwarz, "An improved method of resonant current pulse modulation for power converters," *IEEE Trans. Industrial Electronics and Control Instrumentation*, vol. IECI-23, pp. 133–141, no. 2, May 1976.
3. R. J. King and T. A. Stuart, "A normalized model for the half-bridge series resonant converter," *IEEE Trans. Aerospace and Electronic Systems*, vol. AES-17, pp. 190–198, Mar. 1981.
4. V. T. Ranganthan, P. D. Ziogas, and V. R. Stefanovic, "A regulated dc-dc voltage source converter using a high frequency link," *IEEE Trans. Industry Applications*, vol. IA-18, pp. 279–287, no. 3, May/June 1982.
5. R. J. King and T. A. Stuart, "Modeling the full-bridge series resonant converter," *IEEE Trans. Aerospace and Electronic Systems*, vol. AES-18, pp. 449–458, July 1982.
6. V. Vorpérian and S. Ćuk, "A complete dc analysis of the series resonant converter," *IEEE Power Electronics Specialists Conf. Rec.*, 1982, pp. 85–100.
7. R. J. King and T. A. Stuart, "Input overload protection for the series resonant converter," *IEEE Trans. Aerospace and Electronic Systems*, vol. AES-19, pp. 820–830, Nov. 1983.
8. R. J. King and T. A. Stuart, "A large-signal dynamic simulation for the series resonant converter," *IEEE Trans. Aerospace and Electronic Systems*, vol. AES-19, pp. 859–870, Nov. 1983.
9. R. L. Steigerwald, "High-frequency resonant transistor dc-dc converter," *IEEE Trans. Ind. Electron.*, vol. IE-31, pp. 181–191, May 1984.
10. R. L. Steigerwald, "Analysis of a resonant transistor dc-dc converter with capacitive filter," *IEEE Trans. Ind. Electron.*, vol. IE-31, pp. 439–458, July 1984.
11. S. W. H. De Haan and H. Huisman, "Novel operation and control modes for series resonant converters," *IEEE Trans. Ind. Electron.*, vol. IE-32, pp. 150–157, May 1985.
12. R. Oruganti and F. C. Lee, "Resonant power processors, Part I–State plane analysis," *IEEE Trans. Ind. Appl.*, vol. IA-21, pp. 1453–1960, Nov./Dec. 1985.
13. R. Oruganti and F. C. Lee, "Resonant power processors, Part II–Methods of control," *IEEE Trans. Ind. Appl.*, vol. IA-21, pp. 1461–1971, Nov./Dec. 1985.
14. A. F. Witulski and R. W. Erickson, "Design of the series resonant converter for minimum stress," *IEEE Trans. Aerospace and Electronic Systems*, vol. AES-22, pp. 356–363, July 1986.
15. R. Siri and C. Q. Lee, "Analysis and design of series resonant converter by state-space diagram," *IEEE Trans. Aerospace and Electronic Systems*, vol. AES-22, pp. 757–763, no. 6, Nov. 1986.
16. R. J. King and R. L. Laubacher, "A modified series resonant converter," *IEEE Power Electronics Specialists Conf. Rec.*, 1986, pp. 343–350.
17. K. K. Sum, *Recent Developments in Resonant Power Conversion,* Ventura, CA: Intertec Communication Press, 1988.

18. K. D. T. Ngo, "Analysis of a series resonant converter pulsewidth-modulated or current-controlled for low switching losses," *IEEE Trans. Power Electronics*, vol. PE-3, pp. 55–63, no. 1, Jan. 1988.
19. R. L. Steigerwald, "A comparison of half-bridge resonant converter topologies," *IEEE Trans. Power Electron.*, vol. PE-3, pp. 174–182, Apr. 1988.
20. J. P. Vandelac and P. D. Ziogas, "A dc to dc PWM series resonant converter operated at resonant frequency," *IEEE Trans. Ind. Electron.*, vol. IE-35, pp. 451–460, Aug. 1988.
21. R. Oruganti, J. J. Yang, and F. C. Lee, "Implementation of optimal trajectory control of series resonant converters," *IEEE Trans. Power Electronics*, vol. PE-3, pp. 318–327, no. 3, Oct. 1988.
22. N. Mohan, T. M. Undeland, and W. P. Robbins, *Power Electronics: Converters, Applications and Design,* New York: John Wiley & Sons, 1989, ch. 7, pp. 154–203.
23. P. Jain, "Performance comparison of pulse width modulated resonant mode dc/dc converter in space applications," in *IEEE Industry Applications Society Annual Meeting*, Oct. 1989, pp. 1106–1114.
24. T. Sloane, "Design of high-efficiency series-resonant converters above resonance," *IEEE Trans. Aerospace and Electronic Systems*, vol. AES-26, pp. 393–402, Mar. 1990.
25. P. Jain, "A novel frequency domain modelling of a series resonant dc/dc converter," *12th International Telecommunications Energy Conf.*, 1990, pp. 343–350.
26. S. S. Valtchev and J. B. Klaassens, "Efficient resonant power conversion," *IEEE Trans. Industrial Electronics*, vol. IE-37, pp. 490–495, no. 6, Dec. 1990.
27. J. G. Kassakian, M. F. Schlecht, and G. C. Verghese, *Principles of Power Electronics*, Reading, MA: Addison-Wesley, 1991, ch. 9, pp. 197–234.
28. M. K. Kazimierczuk, "Class D current-driven rectifiers for dc/dc converter applications," *IEEE Trans. Ind. Electron.*, vol. IE-38, pp. 1165–1172, Nov. 1991.
29. B. K. Bose (Ed.), *Modern Power Electronics: Evolution, Technology, and Applications*, Piscataway, NJ: IEEE Press, 1992 ch. 5, pp. 321–367.
30. M. K. Kazimierczuk and S. Wang, "Frequency-domain analysis of series resonant converter for continuous conduction mode," *IEEE Trans. Power Electron.*, vol. PE-7, pp. 270–279, April 1992.
31. P. R. K. Chetty, "Resonant power supplies: their history and status," *IEEE Aerospace and Electronic Systems Magazine*, vol. 7, pp. 23–29, Apr. 1992.
32. A. Dmowski, R. Bugyj, and P. Szewczyk, "A novel series-resonant dc/dc converter with full control of output voltage at no-load conditions. Computer simulation based design aspect," *IEEE Industry Applications Society Meeting*, Houston, TX, Oct. 4–9, 1992, pp. 924–928.
33. P. A. Thollot (Ed.) *Power Electronics Technology and Applications 1993*, Piscataway, NJ: IEEE Press, 1993, ch. 3, pp. 118–127.

15.7 REVIEW QUESTIONS

15.1 Which series resonant converter would you choose for high-input low-output voltage applications? Give your reasons.

15.2 Assuming identical components and input voltage, which transformerless version of the series resonant converter has the highest and which has the lowest efficiency?

15.3 Assuming identical components and input voltage, which transformerless version of the series resonant converter has the highest and which has the lowest dc-to-dc voltage transfer function?

15.4 Does the amplitude of the current through the resonant circuit in the SRC depend on the load resistance?

15.5 Is the part load efficiency of the SRC high or low?

15.6 Is the SRC capable of regulating the dc output voltage for light loads and no load?

15.7 Is the SRC safe under open-circuit conditions?

15.8 Is the SRC safe under short-circuit conditions?

15.9 Does the SRC require a wide range of the switching frequency to regulate dc-dc output voltage against load and line variations?

15.10 Is the ripple current through the filter capacitor low or high in the SRC?

15.11 What is the order of the output filter in the SRC?

15.12 Is the corner frequency of the output filter load dependent in the SRC?

15.8 PROBLEMS

15.1 A full-bridge series resonant converter with a transformer center-tapped rectifier has the following parameters: input voltage $V_I = 200$ V, output voltage $V_O = 24$ V, full-load resistance $R_L = 5\ \Omega$, inverter efficiency $\eta_I = 94\%$, rectifier efficiency $\eta_R = 90\%$, resonant inductance $L = 400\ \mu$H, resonant capacitance $C = 1.5$ nF, and transformer turns ratio $n = 5$. Find the full-load operating frequency above the resonant frequency for this converter.

15.2 A half-wave rectifier with a diode threshold voltage $V_F = 0.4$ V, diode forward resistance $R_F = 0.025\ \Omega$, ESR of the filter capacitor $r_C = 20$ mΩ, transformer turns ratio $n = 5$, and efficiency of the transformer $\eta_{tr} = 96\%$ is used in a half-bridge series resonant converter. The efficiency of the inverter is $\eta_I = 92\%$. The converter output voltage is $V_O = 5\ V$, and the maximum output current is $I_O = 10$ A. Calculate the overall efficiency of the converter and its input voltage if $Q_L = 4$ and $f/f_o = 1.1$ at full load. Neglect switching losses and the drive power.

15.3 Design a half-bridge SRC with a bridge rectifier to meet the following specifications: input voltage $V_I = 110$ V, output voltage $V_O = 270$ V, and minimum load resistance $R_L = 500\ \Omega$. The parameters of the rectifier are $V_F = 0.7$ V, $R_F = 0.1\ \Omega$, and $r_C = 25$ mΩ. Assume that the total inverter efficiency is $\eta_I = 0.9$, the resonant frequency is $f_o = 200$ kHz, and the switching frequency at full load is $f = 208$ kHz.

CHAPTER 16

CLASS D PARALLEL RESONANT CONVERTER

16.1 INTRODUCTION

The parallel resonant converter (PRC) [1]–[14] is obtained by cascading either a half-bridge or a full-bridge Class D parallel resonant inverter (PRI) studied in Chapter 7 and one of the Class D voltage-driven rectifiers studied in Chapter 3. The ac load in the PRI is connected in parallel with a resonant capacitor. If the loaded quality factor of the resonant circuit Q_L is high enough ($Q_L > 2.5$) and the switching frequency is close to the resonant frequency, the output of the PRI behaves almost like a sinusoidal voltage source. Therefore, the voltage-driven rectifiers are compatible with the PRI. The purpose of this chapter is to derive analytical expressions for the steady-state characteristics of the PRC using a high-Q_L assumption.

16.2 HALF-BRIDGE PARALLEL RESONANT CONVERTER

16.2.1 Principle of Operation

A circuit of a half-bridge PRC with three rectifiers is shown in Fig. 16.1. It consists of a Class D parallel resonant inverter discussed in Chapter 7 and one of the three Class D voltage-driven rectifiers discussed in Chapter 3: a half-wave rectifier [Fig. 16.1(a)], a transformer center-tapped rectifier [Fig. 16.1(b)], or a bridge rectifier [Fig. 16.1(c)].

The inverter is composed of two bidirectional two-quadrant switches S_1 and S_2, an L-C resonant circuit, and a dc-blocking capacitor C_c. The switches consist of transistors and antiparallel diodes (usually, the MOSFET's body

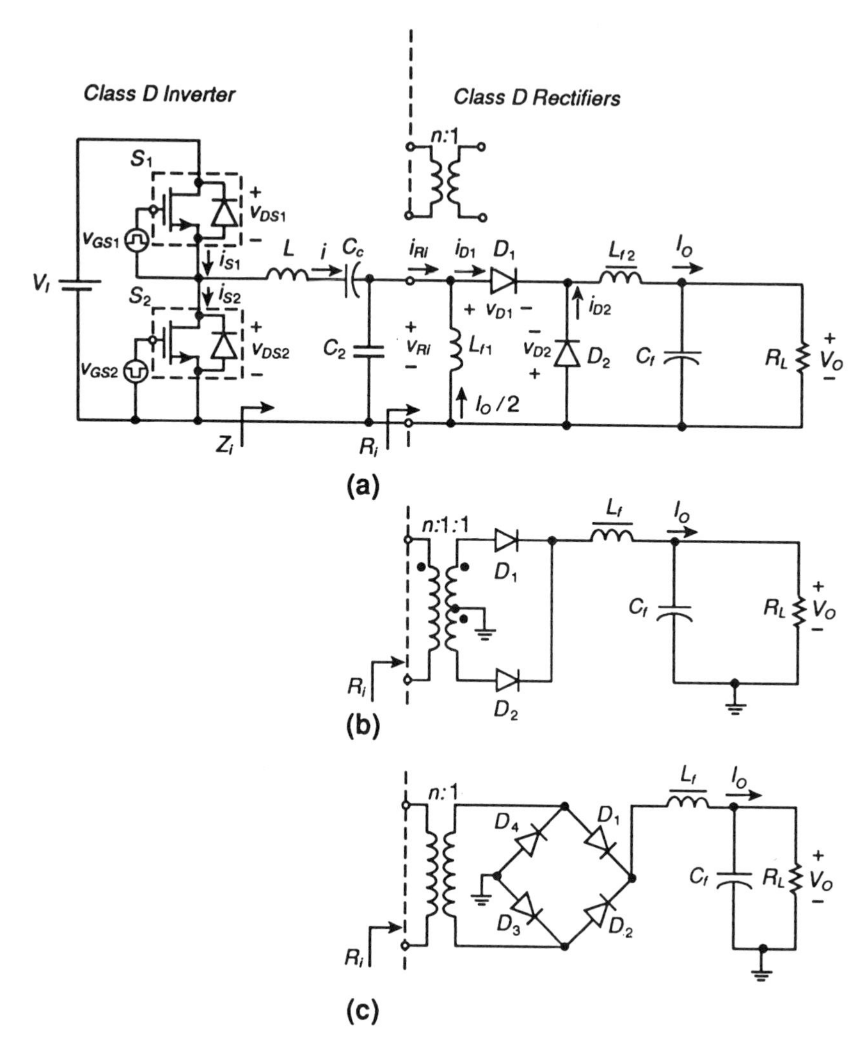

Figure 16.1: Parallel resonant dc-dc converter. (a) With a half-wave rectifier. (b) With a transformer center-tapped rectifier. (c) With a bridge rectifier.

diodes for inductive loads), which can conduct current in both directions. The transistors are driven by rectangular-wave voltage sources v_{GS1} and v_{GS2} at the switching frequency $f = \omega/2\pi$ with a duty cycle of 50%. The L-C resonant circuit converts the square-wave voltage v_{DS2} into a sinusoidal output voltage v_{Ri} if the loaded quality factor Q_L is high. The dc-blocking capacitor C_c protects the inverter from the dc short-circuit through the choke L_{f1} or the primary of the transformer.

The half-wave rectifier of Fig. 16.1(a) consists of two diodes, a second-order low-pass filter L_{f2}-C_f, and a choke L_{f1}. This choke can be replaced by a transformer. The dc component of the diode current i_{D1} flows through the choke L_{f1} or the secondary of the transformer. The transformer center-tapped rectifier of Fig. 16.1(b) comprises two diodes and an output filter. The bridge rectifier of Fig. 16.1(c) is composed of four diodes and an output filter. In this circuit, the transformer may be replaced by a choke; however, the ground of the Class D inverter cannot be connected with the ground of the load.

Equivalent circuits of the converter with a half-wave rectifier are shown in Fig. 16.2. In Fig. 16.2(a), the rectifier is replaced by a square-wave current sink. If the loaded quality factor of the resonant circuit Q_L is sufficiently high (i.e., $Q_L \geq 2.5$), the output voltage of the inverter v_{Ri} is nearly sinusoidal. In this case, the input power of the rectifier contains only the fundamental component. In all three rectifiers, the fundamental component of the square-wave input current is in phase with the input voltage. As a result, each rectifier can be replaced by its input resistance R_i, as shown in Fig. 16.2(b). The parallel R_i-C circuit of Fig. 16.2(b) is converted into a series R_s-C_s circuit of Fig. 16.2(c). In Fig. 16.2(d), the dc voltage source V_I and the switches S_1 and S_2 are replaced by a square-wave voltage source with a low level of zero and a high level of V_I.

The principle of operation of the converter for $f > f_r$ is explained by the current and voltage waveforms shown in Fig. 16.3, where $f_r = 1/(2\pi\sqrt{LC_s})$ is the resonant frequency of the L-C_s-R_s circuit. The input voltage v_{DS2} of the resonant circuit is a square wave whose low level is nearly zero and high level is approximately V_I. If the loaded quality factor Q_L is high, the current i through inductor L is nearly a sine wave and the converter operates in the continuous conduction mode. This current flows through switch S_1 during the first half of the cycle when S_1 is ON and through switch S_2 during the second half of the cycle when S_2 is ON. The operation of the converter above resonance is preferred because the reverse recovery of the MOSFET's diodes is not detrimental.

The output voltage of the inverter, which is the input voltage of the rectifier v_{Ri}, is sinusoidal under the high Q_L assumption. When the instantaneous values of v_{Ri} are positive, diode D_1 is ON and diode D_2 is OFF. The voltage across D_1 is approximately equal to zero and the voltage across D_2 is $v_{D2} \approx -v_{Ri}$. When v_{Ri} is negative, D_2 is ON and D_1 is OFF and $v_{D1} \approx v_{Ri}$. Assuming that the choke inductance L_{f2} is high enough, the ripples of the choke current are negligible and the choke current approximately equals the dc load current I_O. The choke current flows through D_1 when D_1 is ON and through D_2 when D_2 is ON. This means that the current flowing through each diode is a square wave with the magnitude I_O.

The choke inductance L_{f1} is assumed to be high enough so that it conducts only a dc current, equal to $I_O/2$. The input current of the half-wave rectifier $i_{Ri} = (i_{D1} - I_O/2)/n$ is a square wave whose peak values are $I_O/2n$ and $-I_O/2n$. In the transformer center-tapped rectifier and the bridge rectifier, the input current is a square wave with peak values I_O/n and $-I_O/n$. A

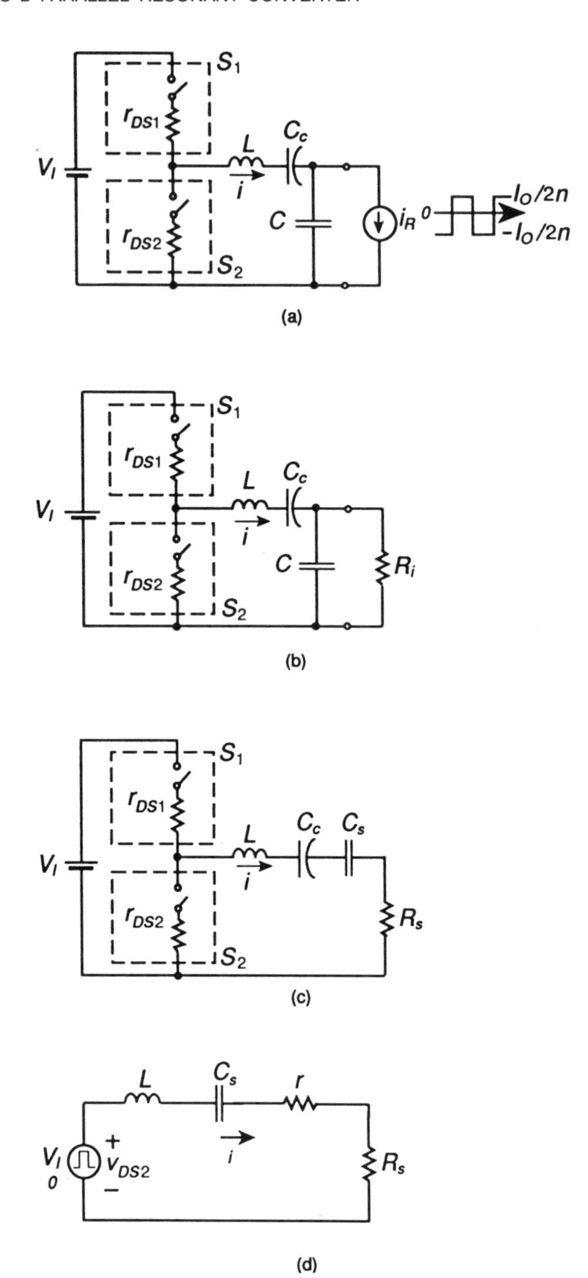

Figure 16.2: Equivalent circuits of the converter. (a) The rectifier is represented by a square-wave current sink. (b) The rectifier is represented by its input resistance R_i. (c) The parallel R_i-C circuit is converted into the R_s-C_s circuit. (d) The dc input source V_I and the transistors are replaced by a square-wave voltage source.

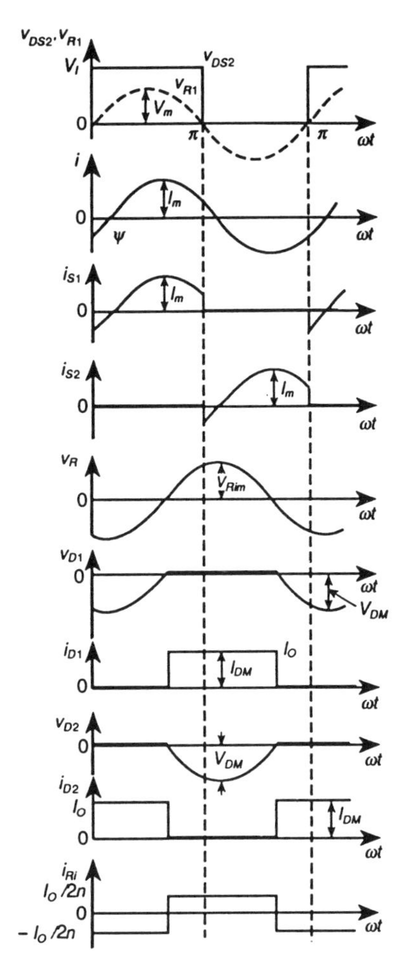

Figure 16.3: Waveforms in the parallel resonant converter with a Class D half-wave rectifier for $f > f_r$.

variable frequency control of the dc output voltage V_O can be exercised because the voltage transfer function of the L-C-R_i resonant circuit depends on the switching frequency f. In this chapter expressions for the efficiencies and the voltage transfer functions of the PRCs arc given neglecting switching losses and drive power of the MOSFETs.

16.2.2 Half-Bridge PRC with Half-Wave Rectifier

From (3.30) and (7.60), one obtains the converter efficiency

$$\eta = \eta_I \eta_R = \frac{\eta_{tr}}{\{1 + \frac{r}{Z_o Q_L}[1 + (\frac{\omega}{\omega_o})^2 Q_L^2]\}(1 + \frac{V_F}{V_O} + \frac{R_F + r_{LF}}{R_L} + a_{hw}^2 \frac{r_{ac} R_L}{f^2 L_f^2})}. \quad (16.1)$$

Combining (7.45), (7.46), and (3.34) produces the dc-to-dc voltage transfer function of the converter with a half-wave rectifier

$$M_V \equiv \frac{V_O}{V_I} = \eta_I M_{VI} M_{VR}$$
$$= \frac{2\eta_I \eta_{tr}}{n\pi^2 \sqrt{[1 - (\frac{\omega}{\omega_o})^2]^2 + [\frac{1}{Q_L}(\frac{\omega}{\omega_o})]^2} (1 + \frac{V_F}{V_O} + \frac{R_F + r_{LF}}{R_L} + a_{hw}^2 \frac{r_{ac} R_L}{f^2 L_f^2})}. \tag{16.2}$$

16.2.3 Half-Bridge PRC with Transformer Center-Tapped Rectifier

The converter efficiency is obtained using (7.60) and (3.65)

$$\eta = \eta_I \eta_R = \frac{\eta_{tr}}{\{1 + \frac{r}{Z_o Q_L}[1 + (\frac{\omega}{\omega_o})^2 Q_L^2]\}(1 + \frac{V_F}{V_O} + \frac{R_F + r_{LF}}{R_L} + a_{ct}^2 \frac{r_{ac} R_L}{f^2 L_f^2})}. \tag{16.3}$$

Using (7.45), (7.46), and (3.67), one arrives at the dc-to-dc voltage transfer function of the converter with a transformer center-tapped rectifier

$$M_V \equiv \frac{V_O}{V_I} = \eta_I M_{VI} M_{VR}$$
$$= \frac{4\eta_I \eta_{tr}}{n\pi^2 \sqrt{[1 - (\frac{\omega}{\omega_o})^2]^2 + [\frac{1}{Q_L}(\frac{\omega}{\omega_o})]^2} (1 + \frac{V_F}{V_O} + \frac{R_F + r_{LF}}{R_L} + a_{ct}^2 \frac{r_{ac} R_L}{f^2 L_f^2})}. \tag{16.4}$$

16.2.4 Half-Bridge PRC with Bridge Rectifier

Combining (7.60) and (3.80) gives the converter efficiency

$$\eta = \eta_I \eta_R = \frac{\eta_{tr}}{\{1 + \frac{r}{Z_o Q_L}[1 + (\frac{\omega}{\omega_o})^2 Q_L^2]\}(1 + \frac{2V_F}{V_O} + \frac{2R_F + r_{LF}}{R_L} + a_b^2 \frac{r_{ac} R_L}{f^2 L_f^2})}. \tag{16.5}$$

From (7.45), (7.46), and (3.82), one obtains the dc-to-dc voltage transfer function of the converter with a bridge rectifier

$$M_V \equiv \frac{V_O}{V_I} = \eta_I M_{VI} M_{VR}$$
$$= \frac{4\eta_I \eta_{tr}}{n\pi^2 \sqrt{[1 - (\frac{\omega}{\omega_o})^2]^2 + [\frac{1}{Q_L}(\frac{\omega}{\omega_o})]^2} (1 + \frac{2V_F}{V_O} + \frac{2R_F + r_{LF}}{R_L} + a_b^2 \frac{r_{ac} R_L}{f^2 L_f^2})}. \tag{16.6}$$

16.3 DESIGN OF HALF-BRIDGE PRC

Example 16.1

Design a transformerless parallel resonant converter of Fig. 16.1(a) to satisfy the following specifications: $V_I = 200$ V, $V_O = 100$ V, $P_{Omax} = 50$ W, and the switching frequency $f = 120$ kHz. Assume the total efficiency of the converter at full load $\eta = 90\%$, the rectifier efficiency $\eta_R = 97\%$, and the corner frequency of the inverter $f_o = 115$ kHz. Plot the designed converter efficiency η as a function of load resistance R_L for $V_F = 0.9$ V, $R_F = 0.1$ Ω, $r_{LF} = 0.1$ Ω, $r_{ac} = 1.9$ Ω, and $L_f = 1$ mH.

Solution: The full-load resistance of the converter is

$$R_{Lmin} = \frac{V_O^2}{P_{Omax}} = \frac{100^2}{50} = 200\ \Omega, \tag{16.7}$$

and the maximum value of the dc load current is

$$I_{Omax} = \frac{V_O}{R_{Lmin}} = \frac{100}{200} = 0.5\ \text{A}. \tag{16.8}$$

The maximum dc input power can be calculated as

$$P_{Imax} = \frac{P_{Omax}}{\eta} = \frac{50}{0.9} = 55.6\ \text{W}, \tag{16.9}$$

and the maximum value of the dc input current

$$I_{Imax} = \frac{P_{Imax}}{V_I} = \frac{55.6}{200} = 0.278\ \text{A}. \tag{16.10}$$

Using (3.31), one obtains the equivalent input resistance of the rectifier

$$R_i = \frac{\pi^2 R_L}{2\eta_R} = \frac{\pi^2 \times 200}{2 \times 0.97} = 1017.5\ \Omega. \tag{16.11}$$

The voltage transfer function of the rectifier can be found from (3.34)

$$M_{VR} = \frac{\sqrt{2}\eta_R}{\pi} = \frac{\sqrt{2} \times 0.97}{\pi} = 0.4367. \tag{16.12}$$

According to (3.14) and (3.14), the diode peak current I_{DM} and voltage V_{DM} are

$$I_{DM} = I_O = 0.5\ \text{A} \tag{16.13}$$

$$V_{DM} = \pi V_O = \pi \times 100 = 314 \text{ V}. \tag{16.14}$$

The dc-to-dc voltage transfer function of the converter is $M_V = V_O/V_I = 100/200 = 0.5$. The required efficiency of the inverter is $\eta_I = \eta/\eta_R = 0.9/0.97 = 0.9278$. Combining (7.33), (7.35), (7.46), and (3.34), one obtains

$$M_V = \eta_I M_{Vs} M_{Vr} M_{VR}, \tag{16.15}$$

from which

$$M_{Vr} = \frac{M_V}{\eta_I M_{Vs} M_{VR}} = \frac{0.5}{0.9278 \times 0.4502 \times 0.4367} = 2.741. \tag{16.16}$$

Rearranging (7.35), the loaded quality factor Q_L can be calculated

$$Q_L = \frac{\omega/\omega_o}{\sqrt{\frac{1}{M_{Vr}^2} - [1 - (\frac{\omega}{\omega_o})^2]^2}} = \frac{1.043}{\sqrt{\frac{1}{2.741^2} - (1 - 1.043^2)^2}} = 2.95. \tag{16.17}$$

From (7.13),

$$\begin{aligned} cos\psi &= \frac{1}{\sqrt{1 + \{Q_L(\frac{\omega}{\omega_o})[(\frac{\omega}{\omega_o})^2 + \frac{1}{Q_L^2} - 1]\}^2}} \\ &= \frac{1}{\sqrt{1 + [2.95 \times 1.043 \times (1.043^2 + \frac{1}{2.95^2} - 1)]^2}} = 0.8484, \end{aligned} \tag{16.18}$$

which gives $\psi = 31.96^\circ$. Using (7.57), the equivalent series resistance of the resonant circuit is

$$R_s = \frac{2V_I^2 cos^2\psi}{\pi^2 P_O} = \frac{2 \times 200^2 \times 0.8484^2}{\pi^2 \times 50} = 116.7 \ \Omega. \tag{16.19}$$

From (7.15),

$$f_r = f_o\sqrt{1 - 1/Q_L^2} = 115 \times \sqrt{1 - 1/2.95^2} = 108.2 \text{ kHz}, \tag{16.20}$$

and from (7.17),

$$Q_r = \sqrt{Q_L^2 - 1} = \sqrt{2.95^2 - 1} = 2.78. \tag{16.21}$$

The component values of the resonant circuit are

$$L = \frac{R_i}{\omega_o Q_L} = \frac{1017.5}{2 \times \pi \times 115 \times 10^3 \times 2.95} = 477.4 \ \mu\text{H} \tag{16.22}$$

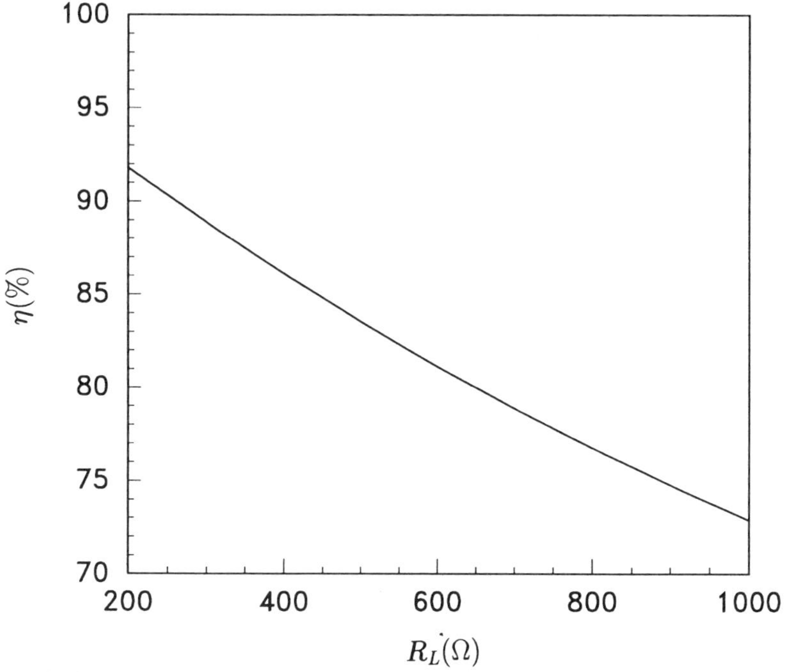

Figure 16.4: Total converter efficiency η as a function of load resistance R_L at $V_I = 200$ V and $V_O = 100$ V.

$$C = \frac{Q_L}{\omega_o R_i} = \frac{2.95}{2 \times \pi \times 115 \times 10^3 \times 1017.5} = 4.01 \text{ nF}. \tag{16.23}$$

Because $Z_o = \sqrt{L/C} = 345\ \Omega$, (7.48) gives

$$I_m = I_{SM} = \frac{2V_I M_{Vr}\sqrt{1 + (Q_L \frac{\omega}{\omega_o})^2}}{\pi Z_o Q_L}$$
$$= \frac{2 \times 200 \times 2.741\sqrt{1 + (2.95 \times 1.043)^2}}{\pi \times 345 \times 2.95} = 1.11 \text{ A}. \tag{16.24}$$

The calculated converter efficiency as a function of load resistance R_L at $V_I = 200$ V and $V_O = 100$ V is depicted in Fig. 16.4. The calculated efficiency of the rectifier turns out to be greater than that assumed for the design. That is why the total calculated efficiency is greater than 90% at full load.

16.4 FULL-BRIDGE PARALLEL RESONANT CONVERTER

A full-bridge parallel resonant converter consists of a full-bridge resonant inverter described in Section 7.6 and one of the voltage-driven rectifiers of

Chapter 3. The expressions for the efficiency remain the same as for the half-bridge PRC, but the total parasitic resistance of the inverter is $r = 2r_{DS} + r_L + r_C$. The efficiency of the full-bridge converter is slightly lower than that of the half-bridge converter with the same components and output voltages. The voltage transfer function for each full-bridge converter is two times higher than that for the corresponding half-bridge converter.

16.4.1 Full-Bridge PRC with Half-Wave Rectifier

The efficiency of the full-bridge PRC with a Class D half-wave rectifier is given by (16.1). Using (3.34), (7.46), and (7.98), one arrives at the dc-to-dc voltage transfer function of the full-bridge converter

$$M_V \equiv \frac{V_O}{V_I} = \eta_I M_{VI} M_{VR}$$
$$= \frac{4\eta_I \eta_{tr}}{n\pi^2 \sqrt{[1-(\frac{\omega}{\omega_o})^2]^2 + [\frac{1}{Q_L}(\frac{\omega}{\omega_o})]^2}(1 + \frac{V_F}{V_O} + \frac{R_F + r_{LF}}{R_L} + a_{hw}^2 \frac{r_{ac} R_L}{f^2 L_f^2})}. \quad (16.25)$$

The range of M_V is from zero to approximately $2/n$.

16.4.2 Full-Bridge PRC with Transformer Center-Tapped Rectifier

The efficiency of the full-bridge PRC with a transformer center-tapped rectifier is expressed by (16.3). The product of (7.98), (7.46), and (3.67) yields the dc-to-dc voltage transfer function of the converter

$$M_V \equiv \frac{V_O}{V_I} = \eta_I M_{VI} M_{VR}$$
$$= \frac{8\eta_I \eta_{tr}}{n\pi^2 \sqrt{[1-(\frac{\omega}{\omega_o})^2]^2 + [\frac{1}{Q_L}(\frac{\omega}{\omega_o})]^2}(1 + \frac{V_F}{V_O} + \frac{R_F + r_{LF}}{R_L} + a_{ct}^2 \frac{r_{ac} R_L}{f^2 L_f^2})}. \quad (16.26)$$

The range of M_V is from zero to approximately $1/n$.

16.4.3 Full-Bridge PRC with Bridge Rectifier

The efficiency of the full-bridge PRC with a Class D bridge rectifier is given by (16.5). From (7.98), (7.46), and (3.82), one obtains the dc-to-dc voltage transfer function of the converter

Table 16.1. Voltage Transfer Function M_V of Lossless Parallel Resonant Converters

Rectifier	Half-Bridge Converter	Full-Bridge Converter
HW rectifier	$\frac{2}{n\pi^2\sqrt{[1-(\frac{\omega}{\omega_o})^2]^2+[\frac{1}{Q_L}(\frac{\omega}{\omega_o})]^2}}$	$\frac{4}{n\pi^2\sqrt{[1-(\frac{\omega}{\omega_o})^2]^2+[\frac{1}{Q_L}(\frac{\omega}{\omega_o})]^2}}$
CT rectifier	$\frac{4}{n\pi^2\sqrt{[1-(\frac{\omega}{\omega_o})^2]^2+[\frac{1}{Q_L}(\frac{\omega}{\omega_o})]^2}}$	$\frac{8}{n\pi^2\sqrt{[1-(\frac{\omega}{\omega_o})^2]^2+[\frac{1}{Q_L}(\frac{\omega}{\omega_o})]^2}}$
BR rectifier	$\frac{4}{n\pi^2\sqrt{[1-(\frac{\omega}{\omega_o})^2]^2+[\frac{1}{Q_L}(\frac{\omega}{\omega_o})]^2}}$	$\frac{8}{n\pi^2\sqrt{[1-(\frac{\omega}{\omega_o})^2]^2+[\frac{1}{Q_L}(\frac{\omega}{\omega_o})]^2}}$

$$M_V \equiv \frac{V_O}{V_I} = \eta_I M_{VI} M_{VR}$$

$$= \frac{8\eta_I\eta_{tr}}{n\pi^2\sqrt{[1-(\frac{\omega}{\omega_o})^2]^2+[\frac{1}{Q_L}(\frac{\omega}{\omega_o})]^2}(1+\frac{2V_F}{V_O}+\frac{2R_F+r_{LF}}{R_L}+a_b^2\frac{r_{ac}R_L}{f^2L_f^2})}. \quad (16.27)$$

The range of M_V is from zero to approximately $1/n$.

The voltage transfer function M_V for the half-bridge and full-bridge PRCs with a half-wave (HW), a transformer center-tapped (CT) rectifier, and a bridge (BR) rectifier is given in Table 16.1.

16.5 SUMMARY

- The parallel resonant converter can be used as both a step-down and step-up converter at the transformer turns ratio $n = 1$.
- It can regulate the output voltage from full load to no load, using a narrow frequency range.
- The converter efficiency decreases with increasing load resistance R_L because the inverter series equivalent load resistance R_s and the ratio R_s/r decrease with increasing R_L.
- The PRC is inherently short-circuit protected.
- It is prone to catastrophic failure in open circuit operation if the switching frequency f is close to the corner frequency f_o of the resonant circuit. The control circuit should prevent such an operation.
- The rectifier diodes turn off at a high di/dt. Therefore, if pn junction diodes are used, the reverse-recovery spikes are generated, causing switching noise.
- The PRC contains an inductive output filter and, thereby, the current through the filter capacitor is low, reducing conduction loss in the ESR and the ripple voltage. Therefore, the converter is suitable for low-voltage/high-current applications. The corner frequency of the second-

order output filter is independent of load resistance R_L, maintaining a wide closed loop bandwidth and a fast response to rapid load and line changes at any load (including light loads).

- The boundary between an inductive and capacitive load to the transistors depends on the load resistance R_L for $Q_L > 1$. However, the load is inductive at any load resistance R_L for $Q_L \leq 1$.
- The rectifiers have two modes of operation. They behave like Class D rectifiers at low load resistances R_L (i.e., high $\omega L_{f2}/R_L$) and like Class E rectifiers at high load resistances R_L (i.e., low $\omega L_{f2}/R_L$).
- For an inductive load, the turn-off switching loss can be reduced to nearly zero by adding a capacitor in parallel with one of the transistors and using the gate driver with a dead time. Since the turn-on switching loss is inherently zero for an inductive load, the total switching loss can be reduced to a negligible level. Because of low switch voltages, low on-resistance MOSFETs can be used, reducing conduction loss.

16.6 REFERENCES

1. V. T. Ranganthan, P. D. Ziogas, and V. R. Stefanovic, "A regulated dc-dc voltage source converter using a high frequency link," *IEEE Trans. Industrial Electronics*, vol. IA-18, no. 3, pp. 279–287, May/June 1982.
2. R. L. Steigerwald, "High-frequency resonant transistor dc-dc converters," *IEEE Trans. Industrial Electronics*, vol. IE-31, pp. 181–191, May 1984.
3. R. Oruganti and F. C. Lee, "Resonant power processors," Parts I and II, *IEEE Trans. Ind. Appl.*, vol. IA-21, pp. 1453–1971, Nov./Dec. 1985.
4. R. L. Steigerwald, "A comparison of half-bridge resonant converter topologies," *IEEE Trans. Power Electronics*, vol. PE-3, pp. 174–182, Apr. 1988.
5. S. D. Johnson and R. W. Erickson, "Steady-state analysis and design of the parallel resonant converter," *IEEE Power Electronics Specialists Conference Record*, 1986, pp. 154–165; reprinted in *IEEE Trans. Power Electronics*, vol. PE-3, pp. 93–104, Jan. 1988.
6. S. Deb, A. Joshi, and S. R. Doradla, "A novel frequency-domain model for a parallel resonant converter," *IEEE Trans. Power Electronics*, vol. PE-3, no. 2, pp. 208–215, April 1988.
7. Y. Kang and A. K. Upadhyay, "Analysis and design of a half-bridge parallel resonant converter," *IEEE Trans. Power Electronics*, vol. PE-3, pp. 231–243, July 1988.
8. Y. Kang and A. K. Upadhyay, "Analysis and design of a half-bridge parallel resonant converter operating above resonance," *IEEE Industry Applications Society Annual Meeting*, Oct. 1988, pp. 827–836.
9. S. D. Johnson, A. F. Witulski, and R. W. Erickson, "Comparison of resonant topologies in high-voltage dc applications, *IEEE Trans. Aerospace and Electronic Systems*, vol. AES-24, pp. 263–273, May 1988.
10. A. K. S. Bhat and M. M. Swamy, "Analysis of parallel resonant converter op-

erating above resonance," *IEEE Trans. Aerospace and Electronic Systems*, vol. AES-25, pp. 449–458, July 1989.

11. A. K. S. Bhat and M. M. Swamy, "Loss calculation in transistorized parallel resonant converter operating above resonance," *IEEE Trans. Power Electronics*, vol. PE-4, pp. 391–401, October 1989.
12. M. K. Kazimierczuk and J. Jóźwik, "Class E zero-voltage-switching and zero-current-switching rectifiers," *IEEE Trans. Circuits Syst.*, vol. CAS-37, pp. 436–444, Mar. 1990.
13. A. K. S. Bhat and M. M. Swamy, "Analysis and design of a parallel resonant converter including the effect of a high-frequency transformer," *IEEE Trans. Industrial Electronics*, vol. IE-37, pp. 297–306, August 1990.
14. M. K. Kazimierczuk, W. Szaraniec, and S. Wang, "Analysis and design of parallel resonant converter at high Q_L," *IEEE Trans. Aerospace and Electronic Systems*, vol. AES-28, pp. 35–50, January 1992.

16.7 REVIEW QUESTIONS

16.1 Is the part-load efficiency of the PRC high?

16.2 Does the PRC require a wide range of the switching frequency to regulate the dc output voltage against load and line variations?

16.3 Is the PRC short-circuit proof?

16.4 Is the PRC open-circuit proof?

16.5 Does the resonant circuit in the PRC behave as a voltage or a current source?

16.6 Does the resonant circuit in the PRC exhibit impedance transformation capability?

16.7 Is the amplitude of the current through the resonant inductor and the switches load dependent in the PRC?

16.8 Is the peak-to-peak value of the current through the filter capacitor high in the PRC?

16.9 What is the order of the output filter in the PRC?

16.8 PROBLEMS

16.1 A full-bridge PRC with a bridge rectifier has the following parameters: $M_V = 5$, $\eta_I = 95\%$, $\eta_R = 92\%$, $n = 1/3$, and $f/f_o = 0.9$. The input resistance of the rectifier is $R_i = 500\ \Omega$. What is the value of the resonant inductance L if the resonant capacitance is $C = 4.7$ nF?

16.2 A transformerless half-bridge PRC with a half-wave rectifier supplies 100 W power to a resistance of 25 Ω. The input voltage is $V_I = 200$ V, the normalized switching frequency is $f/f_o = 0.9$, and the loaded quality factor is $Q_L = 3$. What is the efficiency of the converter?

16.3 Design a full-bridge parallel resonant converter with a transformer center-tapped rectifier to meet the following specifications: $V_I = 400$ V, $V_O = 180$ V, $R_{Lmin} = 125\ \Omega$, and the switching frequency $f = 180$ kHz. Assume the total efficiency of the converter at full load $\eta = 90\%$, the rectifier efficiency $\eta_R = 97\%$, the transformer turns ratio $n = 4$, and the corner frequency of the inverter $f_o = 200$ kHz.

CHAPTER 17

CLASS D SERIES-PARALLEL RESONANT CONVERTER

17.1 INTRODUCTION

The series-parallel resonant converter (SPRC) [1]–[12] combines the advantageous properties of both the series resonant converter and the parallel resonant converter. Its topologies are obtained by cascading a half-bridge or a full-bridge series-parallel resonant inverter (SPRI) covered in Chapter 8 and one of the Class D voltage-driven rectifiers discussed in Chapter 3. The load in the SPRI is connected in parallel with one of the resonant capacitors. If the reactance of this capacitor is lower than the input impedance of the rectifier, the output voltage of the inverter is approximately sinusoidal and the inverter acts as a sinusoidal voltage source. Accordingly, the SPRI is compatible with voltage-driven rectifiers. If the aforementioned assumption is not satisfied, the output voltage of the inverter differs somewhat from a sinusoid. The objectives of this chapter are 1) to present various topologies of the SPRC, 2) to derive analytical equations for the SPRC operating in the continuous conduction mode (CCM), 3) to determine operating conditions for achieving high efficiency at part loads, 4) to find the boundary frequency between the capacitive and inductive loads, and 5) to determine the frequency range for safe operation under short-circuit and open-circuit conditions.

17.2 CIRCUIT DESCRIPTION

A circuit of the SPRC with three rectifiers is shown in Fig. 17.1. It consists of a Class D inverter and one of the three Class D voltage-driven rectifiers: a half-wave rectifier [Fig. 17.1(a)], a transformer center-tapped rectifier [Fig. 17.1(b)], or a bridge rectifier [Fig. 17.1(c)]. The inverter is composed of two bidirectional two-quadrant switches S_1 and S_2 and a resonant circuit L-C_1-C_2. Capacitor C_1 is connected in series with resonant inductor L as in the SRC, and capacitor C_2 is connected in parallel with the load as in the

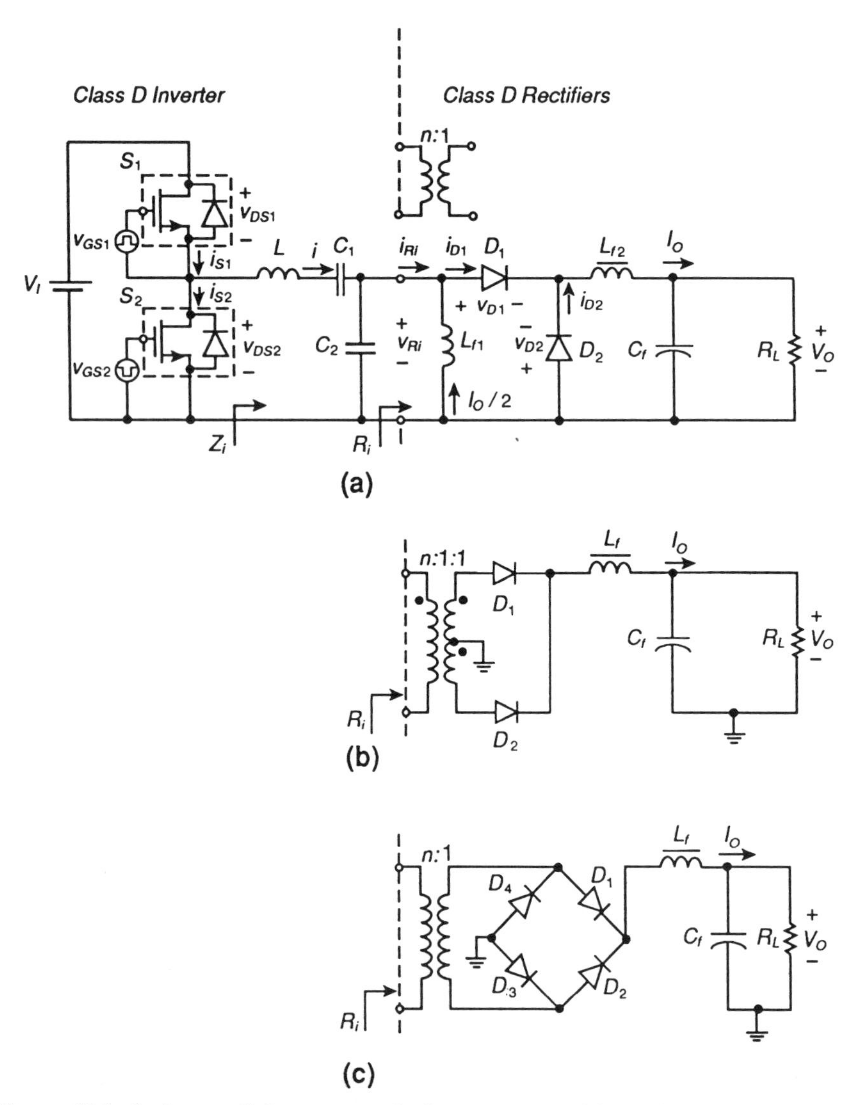

Figure 17.1: Series-parallel resonant dc-dc converter with various Class D voltage-driven rectifiers. (a) With a half-wave rectifier. (b) With a transformer center-tapped rectifier. (c) With a bridge rectifier.

PRC. The switches consist of MOSFETs and their body diodes. Each switch can conduct a positive or a negative current. The transistors are driven by rectangular-wave voltage sources v_{GS1} and v_{GS2}. Switches S_1 and S_2 are alternately turned ON and OFF at the switching frequency $f = \omega/2\pi$ with a duty cycle of 50%. If capacitance C_1 becomes very large (i.e., capacitor C_1 is replaced by a dc-blocking capacitor), the SPRC becomes the PRC. If capacitance C_2 becomes zero (i.e., capacitor C_2 is removed from the circuit) and the Class D voltage-driven rectifiers are replaced by the Class D current-driven rectifiers, the SPRC becomes the SRC. In fact, the transformer version of the Class D inverter of the SRC is the same as that of the SPRC because of the transformer stray capacitance. For the reasons given above, the SPRC exhibits intermediate characteristics between those of the SRC and the PRC.

The output voltage of the inverter is rectified by one of the rectifiers shown in Fig. 17.1. Because this voltage is sensitive to the switching frequency f, the dc output voltage V_O can be controlled by varying the switching frequency. The choke inductor L_{f1} can be replaced by a transformer to achieve an isolation and/or a desired voltage transfer function. The capacitance C_2 can be moved to the transformer secondary. In this case, the winding capacitance is absorbed into the capacitance C_2, and the leakage and magnetizing inductances are absorbed into the resonant inductance L. Thus, the parasitic transformer components are incorporated into the converter topology.

Figure 17.2 shows equivalent circuits of the converter with a Class D voltage-driven half-wave rectifier. The MOSFETs are modeled by switches whose on-resistances are r_{DS1} and r_{DS2}. The rectifier can be replaced by a square-wave current sink, as shown in Fig. 17.2(a). The peak values $-I_O/(2n)$ and $I_O/(2n)$ of the current waveform i_R are valid for the half-wave rectifier. For the transformer center-tapped rectifier and the bridge rectifier, the current waveform i_R is a square wave with peak values $-I_O/n$ and I_O/n. If the output voltage of the inverter v_{Ri} is sinusoidal, the input power of the rectifier contains only the fundamental component. In all three rectifiers, the fundamental component of the square-wave input current is in phase with the input voltage. Therefore, each rectifier can be replaced by its input resistance R_i, as shown in Fig. 17.2(b). The parallel R_i-C_2 circuit of Fig. 17.2(b) is converted into a series R_s-C_s of Fig. 17.2(c). In Fig. 17.2(d), the dc voltage source V_I and the switches S_1 and S_2 are replaced by a square-wave voltage source with a low level of zero and a high level of V_I. The equivalent averaged on-resistance of the MOSFETs is $r_{DS} = (r_{DS1} + r_{DS2})/2$. Neglecting the ESRs of resonant capacitors C_1 and C_2, the total parasitic resistance is $r = r_{DS} + r_L$ and the total resistance of the inverter is $R = R_s + r$.

Current and voltage waveforms of the converter with a half-wave rectifier for $f > f_r = 1/(2\pi\sqrt{LC_{eq}})$ are similar to those for the PRC shown in Fig. 16.3. Note that $C_{eq} = C_1C_s/(C_1 + C_s)$. Operation of the converter above resonance is preferred because the reverse recovery of the MOSFET's diodes does not affect adversely the circuit performance. The input voltage v_{DS2} of

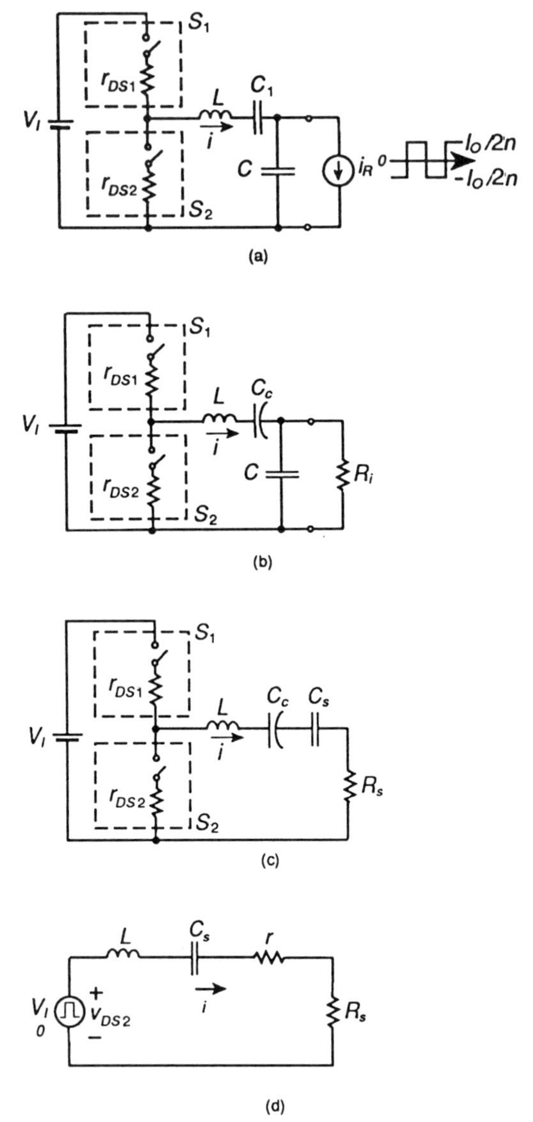

Figure 17.2: Equivalent circuits of the converter with a Class D half-wave rectifier. (a) The rectifier is represented by a square-wave current sink. (b) The rectifier is represented by its input resistance R_i. (c) The parallel R_i-C_2 circuit is converted into the R_s-C_s circuit. (d) The dc input source V_I and the transistors are replaced by a square-wave voltage source.

the resonant circuit is a square wave. If the loaded quality factor is high, the inductor current i is nearly sinusoidal and flows alternately through switches S_1 or S_2 .

In the SRC, the amplitude I_m of the current through the series-resonant circuit is inversely proportional to the load resistance, reducing the conduction loss at light loads and yielding high part-load efficiency. In the PRC, the amplitude I_m of the current through the resonant inductor is almost independent of the load resistance because most of the inductor current flows through the resonant capacitor connected in parallel with R_i, resulting in a constant conduction loss and poor part-load efficiency. In the SPRC, if $R_i \ll X_{C2} = 1/\omega C_2$, most of the inductor current flows through the load resistance, and therefore I_m is inversely proportional to the load resistance, resulting in high part-load efficiency like in the SRC. When the rectifier input resistance R_i becomes greater than X_{C2}, most of the inductor current flows through the resonant capacitor C_2, making I_m independent of R_i like in the PRC.

Switching losses and drive-power of the MOSFETs are neglected in this chapter.

17.3 HALF-BRIDGE SERIES-PARALLEL RESONANT CONVERTER

17.3.1 Half-Bridge SPRC with Half-Wave Rectifier

From (8.40) and (3.30), one obtains the converter efficiency

$$\eta = \eta_I \eta_R = \frac{\eta_{tr}}{\left\{1 + \frac{r}{R_i}\{1 + [\frac{R_i}{Z_o}(\frac{\omega}{\omega_o})(1 + A)]^2\}\right\}(1 + \frac{V_F}{V_O} + \frac{R_F + r_{LF}}{R_L} + a_{hw}^2 \frac{r_{ac} R_L}{f^2 L_f^2})}. \tag{17.1}$$

Combining (8.27), (8.28), and (3.34) produces the dc-to-dc voltage transfer function of the converter with a half-wave rectifier

$$M_V \equiv \frac{V_O}{V_I} = \eta_I M_{VI} M_{VR}$$

$$= \frac{2\eta_I \eta_{tr}}{n\pi^2 \sqrt{(1 + A)^2[1 - (\frac{\omega}{\omega_o})^2]^2 + [\frac{1}{Q_L}(\frac{\omega}{\omega_o} - \frac{\omega_o}{\omega}\frac{A}{A+1})]^2}(1 + \frac{V_F}{V_O} + \frac{R_F + r_{LF}}{R_L} + a_{hw}^2 \frac{r_{ac} R_L}{f^2 L_f^2})}. \tag{17.2}$$

17.3.2 Half-Wave SPRC with Transformer Center-Tapped Rectifier

Using (8.40) and (3.65), the converter efficiency is

$$\eta = \eta_I \eta_R = \frac{\eta_{tr}}{\left\{1 + \frac{r}{R_i}\{1 + [\frac{R_i}{Z_o}(\frac{\omega}{\omega_o})(1 + A)]^2\}\right\}(1 + \frac{V_F}{V_O} + \frac{R_F + r_{LF}}{R_L} + a_{ct}^2 \frac{r_{ac} R_L}{f^2 L_f^2})}. \tag{17.3}$$

The dc-to-dc voltage transfer function of the converter with a center-tapped rectifier is obtained from (8.27), (8.28), and (3.67)

$$M_V \equiv \frac{V_O}{V_I} = \eta_I M_{VI} M_{VR}$$

$$= \frac{4\eta_I \eta_{tr}}{n\pi^2\sqrt{(1+A)^2[1-(\frac{\omega}{\omega_o})^2]^2+[\frac{1}{Q_L}(\frac{\omega}{\omega_o}-\frac{\omega_o}{\omega}\frac{A}{A+1})]^2}(1+\frac{V_F}{V_O}+\frac{R_F+r_{LF}}{R_L}+a_{ct}^2\frac{r_{ac}R_L}{f^2L_f^2})}. \tag{17.4}$$

17.3.3 Half-Bridge SPRC with Bridge Rectifier

Combining (8.40) and (3.80), one arrives at the converter efficiency

$$\eta = \eta_I \eta_R = \frac{\eta_{tr}}{\left\{1+\frac{r}{R_i}\{1+[\frac{R_i}{Z_o}(\frac{\omega}{\omega_o})(1+A)]^2\}\right\}(1+\frac{2V_F}{V_O}+\frac{2R_F+r_{LF}}{R_L}+a_b^2\frac{r_{ac}R_L}{f^2L_f^2})}. \tag{17.5}$$

Using (8.27), (8.28), and (3.82), the dc-to-dc voltage transfer function of the converter with a bridge rectifier is

$$M_V \equiv \frac{V_O}{V_I} = \eta_I M_{VI} M_{VR}$$

$$= \frac{4\eta_I \eta_{tr}}{n\pi^2\sqrt{(1+A)^2[1-(\frac{\omega}{\omega_o})^2]^2+[\frac{1}{Q_L}(\frac{\omega}{\omega_o}-\frac{\omega_o}{\omega}\frac{A}{A+1})]^2}(1+\frac{2V_F}{V_O}+\frac{2R_F+r_{LF}}{R_L}+a_b^2\frac{r_{ac}R_L}{f^2L_f^2})}. \tag{17.6}$$

17.4 DESIGN OF HALF-BRIDGE SPRC

Example 17.1

Design a transformerless SPRC with a half-wave rectifier shown in Fig. 17.1(a) to meet the following specifications: $V_I = 250$ V, $V_O = 40$ V, and $I_O = 0$ to 2 A. Neglect losses due to the ripple current in the rectifier. Plot the efficiency of the designed converter η versus load resistance R_L.

Solution: The maximum output power is

$$P_{Omax} = V_O I_{Omax} = 40 \times 2 = 80 \text{ W}. \tag{17.7}$$

The full-load resistance of the converter is

$$R_{Lmin} = \frac{V_O}{I_{Omax}} = \frac{40}{2} = 20\ \Omega. \tag{17.8}$$

Assuming that the total efficiency of the converter is $\eta = 90$ %, one obtains the maximum dc input power

$$P_{Imax} = \frac{P_{Omax}}{\eta} = \frac{80}{0.9} = 88.9 \text{ W} \tag{17.9}$$

and the maximum value of the dc input current

$$I_{Imax} = \frac{P_{Imax}}{V_I} = \frac{88.9}{250} = 0.36 \text{ A}. \tag{17.10}$$

The pn-junction silicon diodes have usually the threshold voltage $V_F = 0.7$ V and the forward resistance $R_F = 0.1\ \Omega$. Let us assume that the dc ESR of the filter inductor is $r_{LF} = 0.1\ \Omega$. Hence, (3.31) gives the equivalent input resistance of the rectifier at full load

$$R_{imin} = \frac{\pi^2}{2} R_{Lmin}\left(1 + \frac{V_F}{V_O} + \frac{R_F + r_{LF}}{R_{Lmin}}\right)$$

$$= \frac{\pi^2}{2} \times 20\left(1 + \frac{0.7}{40} + \frac{0.1 + 0.1}{20}\right) = 101.4\ \Omega. \tag{17.11}$$

From (3.30), the efficiency of rectifier η_R is

$$\eta_R = \frac{P_O}{P_i} = \frac{1}{1 + \frac{V_F}{V_O} + \frac{R_F + r_{LF}}{R_{Lmin}}} = \frac{1}{1 + \frac{0.7}{40} + \frac{0.1+0.1}{20}} = 97.3\ \%. \tag{17.12}$$

The voltage transfer function of the rectifier can be found from (3.34)

$$M_{VR} = \frac{\sqrt{2}}{\pi\left(1 + \frac{V_F}{V_O} + \frac{R_F + r_{LF}}{R_{Lmin}}\right)} = \frac{\sqrt{2}}{\pi\left(1 + \frac{0.7}{40} + \frac{0.1+0.1}{20}\right)} = 0.44. \tag{17.13}$$

The voltage transfer function of the converter is

$$M_V = \frac{V_O}{V_I} = \frac{40}{250} = 0.16. \tag{17.14}$$

Hence, using (8.23) and (17.13), the required transfer function of the resonant circuit is

$$M_{Vr} = \frac{M_V}{\eta M_{Vs} M_{VR}} = \frac{0.16}{0.9 \times 0.45 \times 0.44} = 0.9. \tag{17.15}$$

Assuming that $A = 1$, $Q_L = 0.2$, and $f_o = 100$ kHz and solving numerically (8.27), one obtains the normalized switching frequency $f/f_o = 0.7953$ and $f = 79.53$ kHz.

The component values of the resonant circuit are

$$L = \frac{R_{imin}}{\omega_o Q_L} = \frac{101.4}{2 \times \pi \times 100 \times 10^3 \times 0.2} = 807\ \mu\text{H} \tag{17.16}$$

$$C = \frac{Q_L}{\omega_o R_{imin}} = \frac{0.2}{2 \times \pi \times 100 \times 10^3 \times 101.4} = 3.1\ \text{nF} \tag{17.17}$$

$$C_1 = C(1 + \frac{1}{A}) = 6.2\ \text{nF} \tag{17.18}$$

$$C_2 = C(1 + A) = 6.2\ \text{nF}. \tag{17.19}$$

The characteristic impedance of the resonant circuit is $Z_o = R_{imin}/Q_L = 507\ \Omega$.

From (8.30), the maximum switch current equal to the amplitude of the current through the resonant circuit is

$$I_{SM} = I_m = \frac{2V_I}{\pi R_{imin}} \sqrt{\frac{1 + [Q_L(\frac{\omega}{\omega_o})(1 + A)]^2}{(1 + A)^2[1 - (\frac{\omega}{\omega_o})^2]^2 + \frac{1}{Q_L^2}(\frac{\omega}{\omega_o} - \frac{\omega_o}{\omega}\frac{A}{A+1})^2}}$$

$$= \frac{2 \times 250}{\pi \times 101.4} \sqrt{\frac{1 + [0.2 \times 0.7953(1 + 1)]^2}{(1 + 1)^2(1 - 0.7953^2)^2 + \frac{1}{0.2^2}(0.7953 - \frac{1}{0.7953} \times \frac{1}{1+1})^2}} = 1.48\ \text{A}. \tag{17.20}$$

The peak values of the voltages across the reactive components can be obtained from (8.32), (8.33), and (8.34) as

$$V_{Lm} = \omega L I_m = 2\pi \times 79.53 \times 10^3 \times 807 \times 10^{-6} \times 1.48 = 596.8\ \text{V} \tag{17.21}$$

$$V_{C1m} = \frac{I_m}{\omega C_1} = \frac{1.48}{2\pi \times 79.53 \times 10^3 \times 6.2 \times 10^{-9}} = 477.7\ \text{V} \tag{17.22}$$

and

$$V_{C2m} = \frac{2V_I}{\pi\sqrt{(1 + A)^2[1 - (\frac{\omega}{\omega_o})^2]^2 + [\frac{1}{Q_L}(\frac{\omega}{\omega_o} - \frac{\omega_o}{\omega}\frac{A}{A+1})]^2}}$$

$$= \frac{2 \times 250}{\pi\sqrt{(1 + 1)^2(1 - 0.7953^2)^2 + [\frac{1}{0.2}(0.7953 - \frac{1}{0.7953} \times \frac{1}{1+1})]^2}} = 143.3\ \text{V}. \tag{17.23}$$

A plot of the calculated overall converter efficiency η as a function of the load resistance R_L at $V_I = 250$ V and $V_O = 40$ V is displayed in Fig. 17.3.

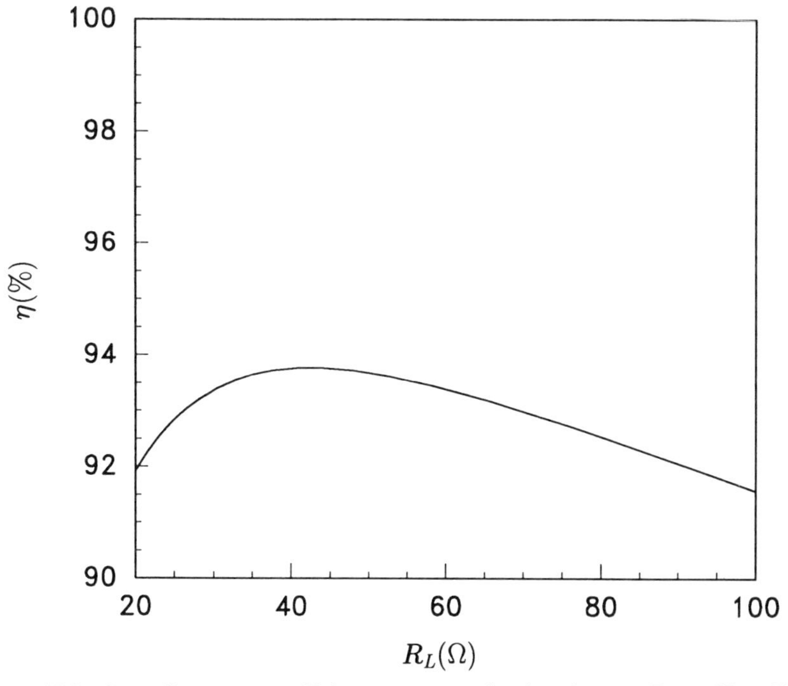

Figure 17.3: Overall converter efficiency η versus load resistance R_L at $V_I = 250\ V$ and $V_O = 40\ V$.

17.5 FULL-BRIDGE SERIES-PARALLEL RESONANT CONVERTER

A full-bridge series-parallel resonant converter consists of a full-bridge resonant inverter described in Section 8.5 and one of the voltage-driven rectifiers of Chapter 3. The expressions for the efficiency remain the same, but the total parasitic resistance of the inverter is now given by (8.76) and is higher by r_{DS} than that for the half-bridge inverter given by (8.39). Thus, the efficiency of the full-bridge converter is slightly lower than the efficiency of the half-bridge converter with the same component and output voltage values. The voltage transfer function for each full-bridge converter is two times higher than that for the corresponding half-bridge converter.

17.5.1 Full-Bridge SPRC with Half-Wave Rectifier

The efficiency of the SPRC with a Class D half-wave rectifier is given by (17.1). Using (3.34), (8.28), and (8.67), one arrives at the dc-to-dc voltage transfer function of the converter

$$M_V \equiv \frac{V_O}{V_I} = \eta_I M_{VI} M_{VR}$$
$$= \frac{4\eta_I \eta_{tr}}{n\pi^2\sqrt{(1+A)^2[1-(\frac{\omega}{\omega_o})^2]^2 + [\frac{1}{Q_L}(\frac{\omega}{\omega_o} - \frac{\omega_o}{\omega}\frac{A}{A+1})]^2}(1+\frac{V_F}{V_O}+\frac{R_F+r_{LF}}{R_L}+a_{hw}^2\frac{r_{ac}R_L}{f^2L_f^2})}. \quad (17.24)$$

17.5.2 Full-Bridge SPRC with Transformer Center-Tapped Rectifier

The efficiency of the SPRC with a transformer center-tapped rectifier is expressed by (17.3). The product of (8.67), (8.28), and (3.67) leads to the dc-to-dc voltage transfer function of the converter

$$M_V \equiv \frac{V_O}{V_I} = \eta_I M_{VI} M_{VR}$$
$$= \frac{8\eta_I \eta_{tr}}{n\pi^2\sqrt{(1+A)^2[1-(\frac{\omega}{\omega_o})^2]^2 + [\frac{1}{Q_L}(\frac{\omega}{\omega_o} - \frac{\omega_o}{\omega}\frac{A}{A+1})]^2}(1+\frac{V_F}{V_O}+\frac{R_F+r_{LF}}{R_L}+a_{ct}^2\frac{r_{ac}R_L}{f^2L_f^2})}. \quad (17.25)$$

17.5.3 Full-Bridge SPRC with Bridge Rectifier

The efficiency of the SPRC with a Class D bridge rectifier is given by (17.5). From (8.67), (8.28), and (3.82), one obtains the dc-to-dc voltage transfer function of the converter

$$M_V \equiv \frac{V_O}{V_I} = \eta_I M_{VI} M_{VR}$$
$$= \frac{8\eta_I \eta_{tr}}{n\pi^2\sqrt{(1+A)^2[1-(\frac{\omega}{\omega_o})^2]^2 + [\frac{1}{Q_L}(\frac{\omega}{\omega_o} - \frac{\omega_o}{\omega}\frac{A}{A+1})]^2}(1+\frac{2V_F}{V_O}+\frac{2R_F+r_{LF}}{R_L}+a_b^2\frac{r_{ac}R_L}{f^2L_f^2})}. \quad (17.26)$$

Table 17.1 gives the dc-to-dc voltage transfer function M_V for the half-bridge and full-bridge lossless converters with a half-wave (HW) rectifier, a transformer center-tapped (CT) rectifier, and a bridge (BR) rectifier.

Table 17.1. Voltage Transfer Function M_V of Lossless Series-Parallel Resonant Converters

Rectifier	Half-Bridge Converter	Full-Bridge Converter
HW rectifier	$\frac{2}{n\pi^2\sqrt{(1+A)^2[1-(\frac{\omega}{\omega_O})^2]^2+[\frac{1}{Q_L}(\frac{\omega}{\omega_O}-\frac{\omega_O}{\omega}\frac{A}{A+1})]^2}}$	$\frac{4}{n\pi^2\sqrt{(1+A)^2[1-(\frac{\omega}{\omega_O})^2]^2+[\frac{1}{Q_L}(\frac{\omega}{\omega_O}-\frac{\omega_O}{\omega}\frac{A}{A+1})]^2}}$
CT rectifier	$\frac{4}{n\pi^2\sqrt{(1+A)^2[1-(\frac{\omega}{\omega_O})^2]^2+[\frac{1}{Q_L}(\frac{\omega}{\omega_O}-\frac{\omega_O}{\omega}\frac{A}{A+1})]^2}}$	$\frac{8}{n\pi^2\sqrt{(1+A)^2[1-(\frac{\omega}{\omega_O})^2]^2+[\frac{1}{Q_L}(\frac{\omega}{\omega_O}-\frac{\omega_O}{\omega}\frac{A}{A+1})]^2}}$
BR rectifier	$\frac{4}{n\pi^2\sqrt{(1+A)^2[1-(\frac{\omega}{\omega_O})^2]^2+[\frac{1}{Q_L}(\frac{\omega}{\omega_O}-\frac{\omega_O}{\omega}\frac{A}{A+1})]^2}}$	$\frac{8}{n\pi^2\sqrt{(1+A)^2[1-(\frac{\omega}{\omega_O})^2]^2+[\frac{1}{Q_L}(\frac{\omega}{\omega_O}-\frac{\omega_O}{\omega}\frac{A}{A+1})]^2}}$

17.6 SUMMARY

- The series-parallel resonant converter is able to regulate the output voltage from full load to no load.
- The voltage transfer function of the resonant circuit is independent of the load at the resonant frequency $f_{rs} = 1/(2\pi\sqrt{LC_1})$ of the L-C_1 resonant circuit. The whole resonant circuit represents a capacitive load to the transistors at f_{rs} because $f_{rs} < f_r$.
- If full load occurs at a low value of Q_L, the magnitude of the current through the switches and the resonant inductor decreases with increasing load resistance, reducing the conduction loss and maintaining high part-load efficiency. However, beyond a certain value of Q_L, the amplitude of the current becomes essentially constant, reducing the efficiency at light loads.
- If full load occurs at a high value of Q_L, the magnitude of the switch current is almost independent of the load, keeping a constant conduction loss and reducing efficiency at part load (as for the PRC).
- The resonant frequency f_r, which forms the boundary between a capacitive and an inductive load, is dependent on the load.
- The converter cannot operate safely with an open circuit at frequencies close to the corner frequency f_o.
- The converter cannot operate safely with a short circuit at frequencies close to the resonant frequency f_r.
- The dc voltage transfer function M_V at switching frequencies f, different from corner frequency f_o, is lower for the SPRC than that for the PRC; it decreases with increasing $A = C_1/C_2$.
- The sensitivity of the dc voltage transfer function to the load decreases with increasing C_1/C_2 for high values of Q_L.
- For regulating V_O against load variations, the normalized frequency range $\Delta f/f_o$ decreases with increasing C_1/C_2. Therefore, low values of M_V are easier to achieve.
- Neither the filter capacitor nor the filter inductor carries a high ripple current, reducing the conduction losses in ESRs and making the converter suitable for applications in low-voltage and high-current power supplies.

17.7 REFERENCES

1. A. K. S. Bhat and S. B. Dewan, "Analysis and design of a high frequency resonant converter using LCC-type commutation," *Proc. IEEE Industry Applications Society Annual Meeting*, 1986, pp. 657–663; reprinted in *IEEE Trans. Power Electronics*, vol. PE-2, pp. 291–301, Oct. 1987.
2. A. K. S. Bhat and S. B. Dewan, "Steady-state analysis of a LCC-type commutated

high-frequency inverter," *IEEE Power Electronics Specialists Conference Record*, Kyoto, Japan, Apr. 11–14, 1988, pp. 1220–1227.

3. R. L. Steigerwald, "A comparison of half-bridge resonant converter topologies," *IEEE Trans. Power Electronics*, vol. PE-3, pp. 174–182, Apr. 1988.
4. I. Batarseh, R. Liu, and C. Q. Lee, "Design of parallel resonant converter with LCC-type commutation," *Electronics Letters*, vol. 24, no. 3, pp. 177–179, Feb. 1988.
5. I. Batarseh, R. Liu, C. Q. Lee, and A. K. Upadhyay, "150 watts and 140 kHz multi-output LCC-type parallel resonant converter," in *IEEE Applied Power Electronics Conference*, 1989, pp. 221–230.
6. I. Batarseh and C. Q. Lee, "High-frequency high-order parallel resonant converter," *IEEE Trans. Industrial Electronics*, vol. IE-36, pp. 485–498, Nov. 1989.
7. A. K. S. Bhat, "Analysis, optimization and design of a series-parallel resonant converter," in *IEEE Applied Power Electronics Conference*, Los Angeles, CA, Mar. 11–19, 1990, pp. 155–164.
8. S. Shah and A. K. Upadhyay, "Analysis and design of a half-bridge series-parallel resonant converter operating in the discontinuous conduction mode," in *IEEE Applied Power Electronics Conference*, Los Angeles, CA, Mar. 11–19, 1990, pp. 165–174.
9. I. Batarseh, R. Liu, C. Q. Lee, and A. K. Upadhyay, "Theoretical and experimental studies of the LCC-type parallel resonant converter," *IEEE Trans. Power Electronics*, vol. PE-5, pp. 140–150, Apr. 1990.
10. C. Q. Lee, I. Batarseh, and R. Liu, "Design of capacitively coupled LCC-type parallel resonant converter," *IECON Conf. Rec.*, Oct. 1988.
11. A. K. S. Bhat, "Analysis and design of series-parallel resonant power supply," *IEEE Trans. Aerospace and Electronic Systems*, vol. AES-28, pp. 249–258, January 1992.
12. M. K. Kazimierczuk, N. Thirunarayan, and S. Shan, "Analysis of series-parallel resonant converter," *IEEE Trans. Aerospace and Electronic Systems*, vol. AES-29, pp. 88–99, January 1993.

17.8 REVIEW QUESTIONS

17.1 Is the magnitude of the current through the resonant capacitor and the switches dependent on the load resistance in the SPRC?

17.2 Is the part-load efficiency of the SPRC high?

17.3 Is it possible to regulate the dc output voltage over the load range from full load to no load in the SPRC?

17.4 When is the dc voltage transfer function of the SPRC independent of the load? Is the load of the switching lags capacitive or inductive in this case?

17.5 Is the boundary between capacitive and inductive loads dependent on the frequency in the SPRC?

17.6 Can the SPRC operate safely with a short circuit at the output?

17.7 Can the SPRC operate safely with an open circuit at the output?

17.8 How does the sensitivity of the dc voltage transfer function change with the ratio of the resonant capacitances C_1/C_2?

17.9 Is the operation of the SPRC safe with a short circuit at the output?

17.10 Is the operation of the SPRC safe with an open circuit at the output?

17.9 PROBLEMS

17.1 A transformerless half-bridge SPRC with a half-wave rectifier has the following parameters: the output power $P_O = 50$ W, the load resistance $R_L = 50\ \Omega$, the input voltage $V_I = 280$ V, the normalized switching frequency $f/f_o = 0.9$, the loaded quality factor $Q_L = 0.3$, and the ratio of capacitances $A = 1.2$. What is the efficiency of the converter?

17.2 A full-bridge SPRC with a transformer center-tapped rectifier supplies power to a 40-Ω load resistance. The converter efficiency is $\eta = 91\%$, the efficiency of the inverter is $\eta_I = 96\%$, and the transformer turns ratio is $n = 2$. What is the resonant inductance of the inverter if the equivalent capacitance is $C = 1$ nF and the loaded quality factor is $Q_L = 0.4$?

17.3 Design a full-bridge SPRC with a transformer center-tapped rectifier. The following specifications should be met: input voltage V_I=270 V, output voltage $V_O = 48$ V, output current $I_O = 0$ to 4 A, and operating frequency $f = 200$ kHz. Assume that the efficiency of the converter is $\eta = 90\%$, the efficiency of the inverter is $\eta_I = 93\%$, the ratio of capacitances is $A = 1$, the normalized switching frequency is $f/f_o = 0.95$, and the transformer turns ratio is $n = 4$.

CHAPTER 18

CLASS D CLL RESONANT CONVERTER

18.1 INTRODUCTION

The CLL resonant dc-dc converter [1]–[7] is obtained by replacing the ac load in the CLL resonant inverter studied in Chapter 9 with one of the Class D voltage-driven rectifiers covered in Chapter 3. The output voltage ot the CLL resonant inverter is approximately sinusoidal. This assumption is satisfied if the reactance of the inductor in parallel with the load is lower than the input resistance of the rectifier. In this case, the inverter acts as sinusoidal voltage source and therefore is suitable to drive voltage-driven rectifiers. The purpose of this chapter is 1) to introduce different circuits of the CLL resonant converters, 2) to derive analytical equations for the CLL converter operating in the CCM, 3) to find the boundary frequency between capacitive and inductive loads, and 4) to determine the frequency range for safe operation under short-circuit and open-circuit conditions. The analysis is carried out in the frequency domain using Fourier series techniques. Design equations describing the steady-state operation are derived. It is found that the dc voltage transfer function of the converter is almost *insensitive* to the load variations. In addition, the circuit has high efficiency over a wide range of load resistance.

18.2 CIRCUIT DESCRIPTION

A circuit of the CLL converter [1]–[7] is shown in Fig. 18.1. It consists of a Class D inverter and one of the three Class D voltage-driven rectifiers: a half-wave rectifier [Fig. 18.1(a)], a transformer center-tapped rectifier

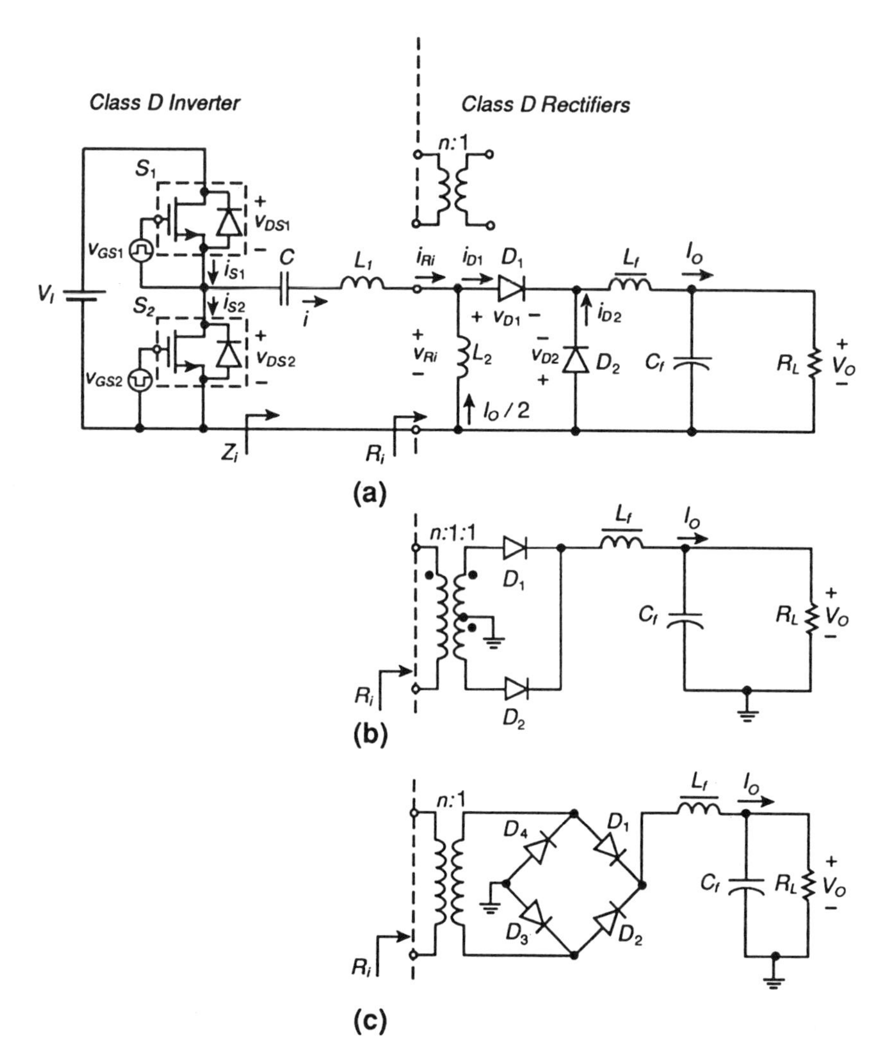

Figure 18.1: CLL resonant dc-dc converter with various Class D voltage-driven rectifiers. (a) With a half-wave rectifier. (b) With a transformer center-tapped rectifier. (c) With a bridge rectifier.

[Fig. 18.1(b)], or a bridge rectifier [Fig. 18.1(c)]. The inverter is composed of two bidirectional two-quadrant switches S_1 and S_2 and a resonant circuit C-L_1-L_2. The resonant capacitor C is connected in series with the tapped inductor L_1-L_2. The load is connected in parallel with a part of the inductor. The switches consist of MOSFETs and their body diodes. Each switch can conduct a positive or a negative current. The transistors are driven by rectangular-wave voltage sources v_{GS1} and v_{GS2}. Switches S_1 and S_2 are alternately turned ON and OFF at the switching frequency $f = \omega/2\pi$ with a duty cycle of 50%. The output voltage of the inverter is rectified by one of the rectifiers. Because this voltage is sensitive to the switching frequency f, the dc output voltage V_O can be controlled by varying the switching frequency. The resonant inductor L_2 can be replaced by a transformer to achieve an isolation and/or a desired voltage transfer function.

Equivalent circuits of the converter with the Class D voltage-driven half-wave rectifier are shown in Fig. 18.2. In Fig. 18.2(a), the rectifier is replaced by a square-wave current sink and the MOSFETs are replaced by switches with on-resistances r_{DS1} and r_{DS2}. Assuming that the output voltage of the inverter v_{Ri} is sinusoidal, the input power of the rectifier contains only the fundamental component. In all three rectifiers, the fundamental component of the square-wave input current is in phase with the input voltage. Therefore, the input impedance of each rectifier at the switching frequency is purely resistive and the rectifiers can be modeled by an input resistance R_i, as shown in Fig. 18.2(b). The parallel R_i-L_2 circuit of Fig. 18.2(b) can be converted into a series R_s-L_s circuit of Fig. 18.2(c) at a given frequency. The dc voltage source V_I and the switches S_1 and S_2 in Fig. 18.2(d) are modeled by a square-wave voltage source, where the low level of the square wave is zero and the high level is V_I. The equivalent on-resistance of the MOSFETs is $r_{DS} \approx (r_{DS1} + r_{DS2})/2$. The parasitic resistance r of the inverter is associated with the resistance of the switch r_{DS}, the equivalent series resistance (ESR) of the capacitor r_{Cr}, and the ESRs of the inductors r_{L1} and r_{L2}.

Waveforms of the converter with a half-wave rectifier for $f > f_r = 1/(2\pi\sqrt{CL_{eq}}) = 1/(2\pi\sqrt{C(L_1 + L_s)}$ are similar to those for the PRC depicted in Fig. 16.3. Operation of the converter above resonance is preferred because the MOSFET body diodes turn off gradually and do not generate current spikes associated with the reverse recovery. The input voltage v_{DS2} of the resonant circuit is a square wave. Assuming that loaded quality factor Q_r at the resonant frequency f_r is high, the capacitor current i is nearly sinusoidal and flows alternately through switches S_1 or S_2 .

In the series resonant converter (SRC), the amplitude I_m of the current through the series-resonant circuit is inversely proportional to the load resistance, reducing the conduction loss at light loads and yielding high part-load efficiency. In the parallel resonant converter (PRC), the amplitude I_m of the current through the resonant inductor is almost independent of the load resistance because most of the inductor current flows through the resonant capacitor connected in parallel with R_i, resulting in a constant conduction loss and poor part-load efficiency. In the CLL converter, if $R_i \ll X_{L2} = \omega L_2$, most of the capacitor current flows through the load resistance and therefore I_m is inversely proportional to the load resistance, resulting in high part-load efficiency like in the SRC. When the rectifier input resistance R_i becomes greater than X_{L2}, most of the capacitor current i flows through the resonant inductor L_2, making I_m independent of R_i like in the PRC.

The converter is not safe under short-circuit and open-circuit conditions. At $R_L = 0$, the inductor L_2 is shorted circuited and the resonant circuit

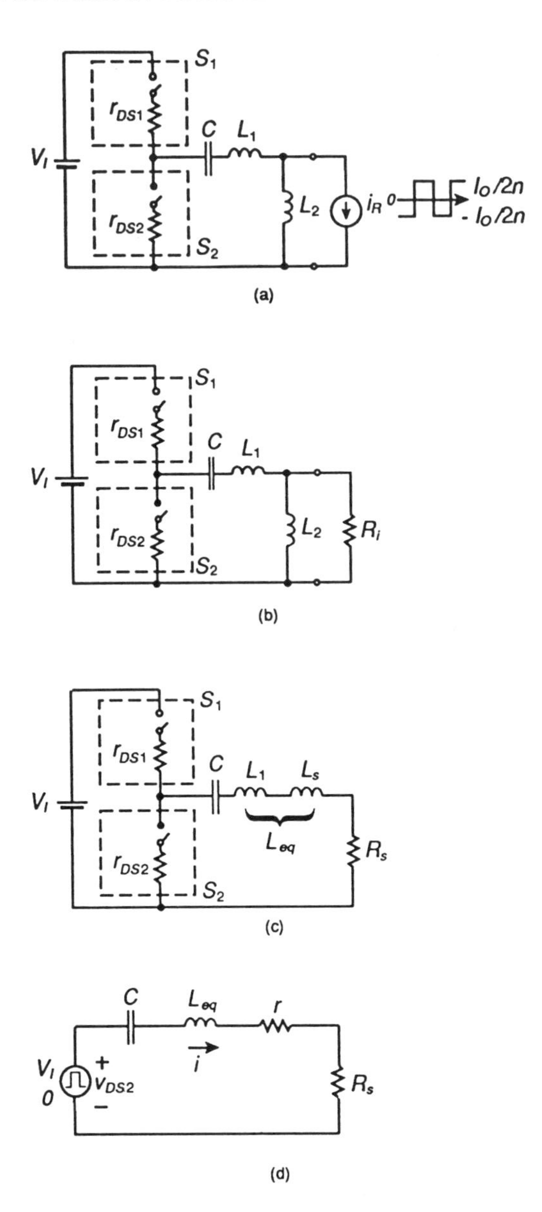

Figure 18.2: Equivalent circuits of the converter with a Class D half-wave rectifier. (a) The rectifier is represented by a square-wave current sink. (b) The rectifier is represented by its input resistance R_i. (c) The parallel R_i-L_2 circuit is converted into the R_s-L_s circuit. (d) The dc input source V_I and the transistors are replaced by a square-wave voltage source.

consists of L_1 and C. If the switching frequency f is equal to the resonant frequency of the C-L_1 circuit $f_{rs} = 1/(2\pi\sqrt{L_1 C})$, the magnitude of the current through the switches and the C-L_1 resonant circuit is $I_m \approx 2V_I/(\pi r)$. This current may become excessive and may destroy the circuit. If f is far from f_{rs}, I_m is limited by the reactance of the resonant circuit. At $R_L = \infty$, the resonant circuit comprises of C and the series combination of L_1 and L_2. Consequently, its resonant frequency is equal to f_o, $I_m \approx 2V_I/(\pi r)$, and the converter is not safe at or close to this frequency. In this chapter, switching losses and drive power of the MOSFETs are neglected.

18.3 HALF-BRIDGE CLL RESONANT CONVERTER

18.3.1 Half-Bridge CLL RC with Half-Wave Rectifier

From (9.34) and (3.30), one obtains the converter efficiency

$$\eta = \eta_I \eta_R = \frac{\eta_{tr}}{\left\{1 + \frac{r}{R_i}\{1 + [\frac{R_i}{Z_o}(\frac{\omega_o}{\omega})(1+A)]^2\}\right\}(1 + \frac{V_F}{V_O} + \frac{R_F + r_{LF}}{R_L} + a_{hw}^2 \frac{r_{ac} R_L}{f^2 L_f^2})}. \tag{18.1}$$

Combining (9.26), (9.27), and (3.34) produces the dc-to-dc voltage transfer function of the converter with a half-wave rectifier

$$M_V \equiv \frac{V_O}{V_I} = \eta_I M_{VI} M_{VR}$$

$$= \frac{2\eta_I \eta_{tr}}{n\pi^2\sqrt{(1+A)^2[1-(\frac{\omega_o}{\omega})^2]^2 + [\frac{1}{Q_L}(\frac{\omega}{\omega_o}\frac{A}{A+1} - \frac{\omega_o}{\omega})]^2}(1 + \frac{V_F}{V_O} + \frac{R_F + r_{LF}}{R_L} + a_{hw}^2 \frac{r_{ac} R_L}{f^2 L_f^2})}. \tag{18.2}$$

18.3.2 Half-Bridge CLL RC with Transformer Center-Tapped Rectifier

Using (9.34) and (3.65), the converter efficiency is

$$\eta = \eta_I \eta_R = \frac{\eta_{tr}}{\left\{1 + \frac{r}{R_i}\{1 + [\frac{R_i}{Z_o}(\frac{\omega_o}{\omega})(1+A)]^2\}\right\}(1 + \frac{V_F}{V_O} + \frac{R_F + r_{LF}}{R_L} + a_{ct}^2 \frac{r_{ac} R_L}{f^2 L_f^2})}. \tag{18.3}$$

The dc-to-dc voltage transfer function of the converter with a center-tapped rectifier is obtained from (9.26), (9.27), and (3.67)

$$M_V \equiv \frac{V_O}{V_I} = \eta_I M_{VI} M_{VR}$$

$$= \frac{4\eta_I \eta_{tr}}{n\pi^2 \sqrt{(1+A)^2[1-(\frac{\omega_o}{\omega})^2]^2 + [\frac{1}{Q_L}(\frac{\omega}{\omega_o}\frac{A}{A+1} - \frac{\omega_o}{\omega})]^2}(1 + \frac{V_F}{V_O} + \frac{R_F + r_{LF}}{R_L} + a_{ct}^2 \frac{r_{ac} R_L}{f^2 L_f^2})}. \tag{18.4}$$

18.3.3 Half-Bridge CLL RC with Bridge Rectifier

Combining (9.34) and (3.80), one arrives at the converter efficiency

$$\eta = \eta_I \eta_R = \frac{\eta_{tr}}{\left\{1 + \frac{r}{R_i}\{1 + [\frac{R_i}{Z_o}(\frac{\omega_o}{\omega})(1+A)]^2\}\right\}(1 + \frac{2V_F}{V_O} + \frac{2R_F + r_{LF}}{R_L} + a_b^2 \frac{r_{ac} R_L}{f^2 L_f^2})}. \tag{18.5}$$

Using (9.26), (9.27), and (3.82), the dc-to-dc voltage transfer function of the converter with a bridge rectifier is

$$M_V \equiv \frac{V_O}{V_I} = \eta_I M_{VI} M_{VR}$$

$$= \frac{4\eta_I \eta_{tr}}{n\pi^2 \sqrt{(1+A)^2[1-(\frac{\omega_o}{\omega})^2]^2 + [\frac{1}{Q_L}(\frac{\omega}{\omega_o}\frac{A}{A+1} - \frac{\omega_o}{\omega})]^2}(1 + \frac{2V_F}{V_O} + \frac{2R_F + r_{LF}}{R_L} + a_b^2 \frac{r_{ac} R_L}{f^2 L_f^2})}. \tag{18.6}$$

18.4 DESIGN OF HALF-BRIDGE CLL RC

Example 18.1

Design a transformerless half-bridge CLL RC with a half-wave rectifier of Fig. 18.1(a) to meet the following specifications: $V_I = 250$ V, $V_O = 40$ V, and $I_O = 04$ to 2 A. Neglect losses due to the ripple current in the rectifier. Sketch the designed converter efficiency η as a function of load resistance R_L.

Solution: The maximum output power is

$$P_{Omax} = V_O I_{Omax} = 40 \times 2 = 80 \text{ W}. \tag{18.7}$$

The full-load resistance of the converter is

$$R_{Lmin} = \frac{V_O}{I_{Omax}} = \frac{40}{2} = 20\ \Omega. \tag{18.8}$$

Assuming that the total efficiency of the converter is $\eta = 90\%$, one obtains the maximum dc input power

$$P_{Imax} = \frac{P_{Omax}}{\eta} = \frac{80}{0.9} = 88.9 \text{ W} \tag{18.9}$$

and the maximum value of the dc input current

$$I_{Imax} = \frac{P_{Imax}}{V_I} = \frac{88.9}{250} = 0.36 \text{ A}. \tag{18.10}$$

The *pn*-junction silicon diodes have usually the threshold voltage $V_F = 0.7$ V and the forward resistance $R_F = 0.1\ \Omega$. The dc ESR of the filter inductor r_L is 0.1 Ω. Hence, (3.31) gives the equivalent input resistance of the rectifier at full load,

$$\begin{aligned} R_{imin} &= \frac{\pi^2}{2} R_{Lmin} \left(1 + \frac{V_F}{V_O} + \frac{R_F + r_L}{R_{Lmin}}\right) \\ &= \frac{\pi^2}{2} \times 20 \left(1 + \frac{0.7}{40} + \frac{0.1 + 0.1}{20}\right) = 101.4\ \Omega. \end{aligned} \tag{18.11}$$

From (3.30), the efficiency of rectifier η_R is

$$\eta_R = \frac{P_O}{P_i} = \frac{1}{1 + \frac{V_F}{V_O} + \frac{R_F + r_{LF}}{R_{Lmin}}} = \frac{1}{1 + \frac{0.7}{40} + \frac{0.1+0.1}{20}} = 97.3\ \%. \tag{18.12}$$

The voltage transfer function of the rectifier can be found from (3.34)

$$M_{VR} = \frac{\sqrt{2}}{\pi\left(1 + \frac{V_F}{V_O} + \frac{R_F + r_{LF}}{R_{Lmin}}\right)} = \frac{\sqrt{2}}{\pi\left(1 + \frac{0.7}{40} + \frac{0.1+0.1}{20}\right)} = 0.44. \tag{18.13}$$

The voltage transfer function of the converter is

$$M_V = \frac{V_O}{V_I} = \frac{40}{250} = 0.16. \tag{18.14}$$

From (9.21) and (18.13), the required transfer function of the resonant circuit is

$$M_{Vr} = \frac{M_V}{\eta M_{Vs} M_{VR}} = \frac{0.16}{0.9 \times 0.45 \times 0.44} = 0.9. \tag{18.15}$$

Assuming that $A = 1$, $Q_{Lmin} = 0.2$, and $f_o = 100$ kHz and solving numerically (9.23), one obtains the normalized switching frequency $f/f_o = 1.471$ and $f = 147.1$ kHz. From (9.35), the maximum efficiency occurs at $Q_L = (f/f_o)/2$. However, a lower value of Q_L at full load was selected to ensure good part-load efficiency.

The component values of the resonant circuit are

$$C = \frac{Q_L}{\omega_o R_{imin}} = \frac{0.2}{2 \times \pi \times 100 \times 10^3 \times 101.4} = 3.1 \text{ nF} \tag{18.16}$$

$$L = \frac{R_{imin}}{\omega_o Q_L} = \frac{101.4}{2 \times \pi \times 100 \times 10^3 \times 0.2} = 807 \ \mu\text{H} \tag{18.17}$$

$$L_1 = \frac{L}{1 + \frac{1}{A}} = 403.5 \ \mu\text{H} \tag{18.18}$$

$$L_2 = \frac{L}{1 + A} = 403.5 \ \mu\text{H}. \tag{18.19}$$

The characteristic impedance of the resonant circuit is $Z_o = R_{imin}/Q_L = 507 \ \Omega$.

From (9.29), the maximum switch current equal to the amplitude of the current through the resonant circuit is

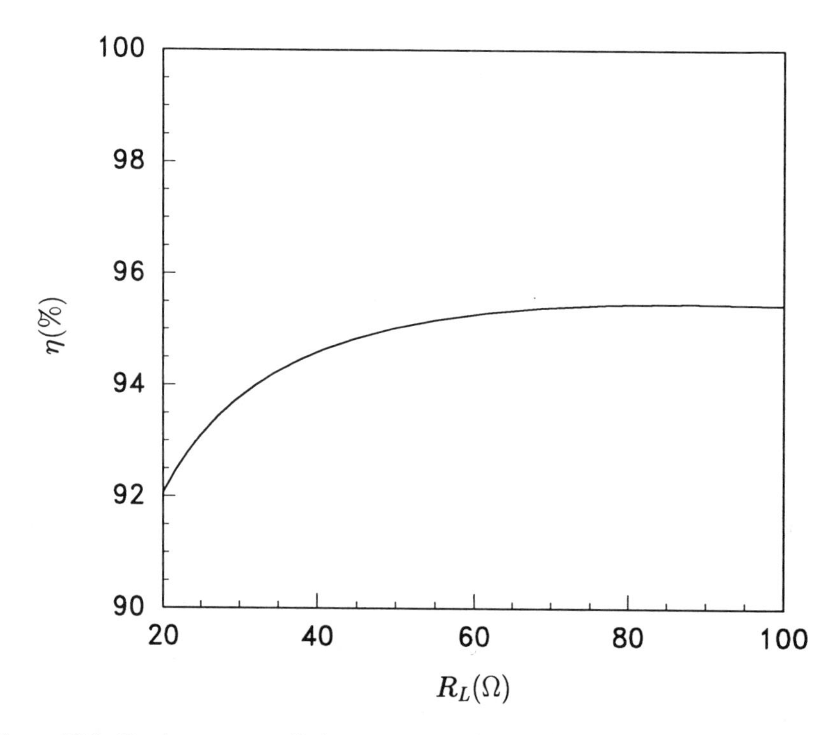

Figure 18.3: Total converter efficiency η versus load resistance R_L at $V_I = 250 \ V$ and $V_O = 40 \ V$.

$$I_{SM} = I_m = \frac{2V_I M_{Vr}}{\pi R_{imin}} \sqrt{1 + \left[Q_L \left(\frac{\omega}{\omega_o}\right)(1+A)\right]^2}$$
$$= \frac{2 \times 250 \times 0.9}{\pi \times 101.4} \sqrt{1 + [0.2 \times 1.471(1+1)]^2} = 1.64 \text{ A.} \quad (18.20)$$

The maximum switch voltage is $V_{SM} = V_I = 250$ V.

The peak values of the voltages across the reactive components can be obtained from (9.36), (9.37), and (9.38) as

$$V_{L1m} = \omega L_1 I_m = 2\pi \times 147.1 \times 10^3 \times 403.5 \times 10^{-6} \times 1.64 = 612 \text{ V}, \quad (18.21)$$

$$V_{L2m} = \frac{2V_I}{\pi\sqrt{(1+A)^2[1-(\frac{\omega_o}{\omega})^2]^2 + [\frac{1}{Q_L}(\frac{\omega}{\omega_o}\frac{A}{A+1} - \frac{\omega_o}{\omega})]^2}},$$
$$= \frac{2 \times 250}{\pi\sqrt{(1+1)^2(1-\frac{1}{1.471^2})^2 + [\frac{1}{0.2}(1.471 \times \frac{1}{1+1} - \frac{1}{1.471})]^2}} = 143.3 \text{ V}, \quad (18.22)$$

and

$$V_{Cm} = \frac{I_m}{\omega C} = \frac{1.64}{2\pi \times 147.1 \times 10^3 \times 3.1 \times 10^{-9}} = 572.4 \text{ V.} \quad (18.23)$$

The diode peak current is $I_{DM} = I_O = 2$ A. The diode peak voltage is $V_{DM} = \pi V_O = 125.7$ V. Figure 18.3 shows the calculated overall converter efficiency η versus load resistance R_L at $V_I = 250$ V and $V_O = 40$ V.

18.5 FULL-BRIDGE CLL RESONANT CONVERTER

A full-bridge CLL resonant converter consists of a full-bridge resonant inverter described in Section 9.5 and one of the voltage-driven rectifiers of Chapter 3. The expressions for the efficiency remain the same, but the total parasitic resistance of the inverter is now given by (9.72) and is higher by r_{DS} than that for the half-bridge inverter given by (9.33). Thus, the efficiency of the full-bridge converter is slightly lower than the efficiency of the half-bridge converter with the same component and output voltage values. The voltage transfer function for each full-bridge converter is two times higher than that for the corresponding half-bridge converter.

18.5.1 Full-Bridge CLL RC with Half-Wave Rectifier

The efficiency of the CLL RC with a Class D half-wave rectifier is given by (18.1). Using (9.27), (9.63), and (3.34), one arrives at the dc-to-dc voltage transfer function of the converter

$$M_V \equiv \frac{V_O}{V_I} = \eta_I M_{VI} M_{VR}$$

$$= \frac{4\eta_I \eta_{tr}}{n\pi^2 \sqrt{(1+A)^2[1-(\frac{\omega_o}{\omega})^2]^2 + [\frac{1}{Q_L}(\frac{\omega}{\omega_o}\frac{A}{A+1} - \frac{\omega_o}{\omega})]^2}(1 + \frac{V_F}{V_O} + \frac{R_F + r_{LF}}{R_L} + a_{hw}^2 \frac{r_{ac} R_L}{f^2 L_f^2})}. \quad (18.24)$$

18.5.2 Full-Bridge CLL RC with Transformer Center-Tapped Rectifier

The efficiency of the CLL RC with a transformer center-tapped rectifier is expressed by (18.3). The product of (9.27), (9.63), and (3.67) gives the dc-to-dc voltage transfer function of the converter

$$M_V \equiv \frac{V_O}{V_I} = \eta_I M_{VI} M_{VR}$$

$$= \frac{8\eta_I \eta_{tr}}{n\pi^2 \sqrt{(1+A)^2[1-(\frac{\omega_o}{\omega})^2]^2 + [\frac{1}{Q_L}(\frac{\omega}{\omega_o}\frac{A}{A+1} - \frac{\omega_o}{\omega})]^2}(1 + \frac{V_F}{V_O} + \frac{R_F + r_{LF}}{R_L} + a_{ct}^2 \frac{r_{ac} R_L}{f^2 L_f^2})}. \quad (18.25)$$

18.5.3 Full-Bridge CLL RC with Bridge Rectifier

The efficiency of the CLL RC with a Class D bridge rectifier is given by (18.5). From (9.27), (9.63), and (3.82), one obtains the dc-to-dc voltage transfer function of the converter

$$M_V \equiv \frac{V_O}{V_I} = \eta_I M_{VI} M_{VR}$$

$$= \frac{8\eta_I \eta_{tr}}{n\pi^2 \sqrt{(1+A)^2[1-(\frac{\omega_o}{\omega})^2]^2 + [\frac{1}{Q_L}(\frac{\omega}{\omega_o}\frac{A}{A+1} - \frac{\omega_o}{\omega})]^2}(1 + \frac{2V_F}{V_O} + \frac{2R_F + r_{LF}}{R_L} + a_b^2 \frac{r_{ac} R_L}{f^2 L_f^2})}. \quad (18.26)$$

Table 18.1. Voltage Transfer Function M_V of Lossless CLL Resonant Converters

Rectifier	Half-Bridge Converter	Full-Bridge Converter
HW rect.	$\frac{2}{n\pi^2\sqrt{(1+A)^2[1-(\frac{\omega_O}{\omega})^2]^2+[\frac{1}{Q_L}(\frac{\omega}{\omega_O}\frac{A}{A+1}-\frac{\omega_O}{\omega})]^2}}$	$\frac{4}{n\pi^2\sqrt{(1+A)^2[1-(\frac{\omega_O}{\omega})^2]^2+[\frac{1}{Q_L}(\frac{\omega}{\omega_O}\frac{A}{A+1}-\frac{\omega_O}{\omega})]^2}}$
CT rect.	$\frac{4}{n\pi^2\sqrt{(1+A)^2[1-(\frac{\omega_O}{\omega})^2]^2+[\frac{1}{Q_L}(\frac{\omega}{\omega_O}\frac{A}{A+1}-\frac{\omega_O}{\omega})]^2}}$	$\frac{8}{n\pi^2\sqrt{(1+A)^2[1-(\frac{\omega_O}{\omega})^2]^2+[\frac{1}{Q_L}(\frac{\omega}{\omega_O}\frac{A}{A+1}-\frac{\omega_O}{\omega})]^2}}$
BR rect.	$\frac{4}{n\pi^2\sqrt{(1+A)^2[1-(\frac{\omega_O}{\omega})^2]^2+[\frac{1}{Q_L}(\frac{\omega}{\omega_O}\frac{A}{A+1}-\frac{\omega_O}{\omega})]^2}}$	$\frac{8}{n\pi^2\sqrt{(1+A)^2[1-(\frac{\omega_O}{\omega})^2]^2+[\frac{1}{Q_L}(\frac{\omega}{\omega_O}\frac{A}{A+1}-\frac{\omega_O}{\omega})]^2}}$

Voltage transfer functions for lossless half-bridge and full-bridge CLL converters with various rectifiers are given in Table 18.1.

18.6 SUMMARY

- The dc voltage transfer function of the CLL converter is *independent* of the load resistance for the normalized switching frequency $f/f_o = \sqrt{1 + L_2/L_1}$. This occurs at inductive loads of the switches, which is a very desirable feature if the power MOSFETs are used as switches.
- The efficiency decreases with increasing load resistance for light loads.
- The resonant frequency f_r, which forms the boundary between a capacitive and an inductive load, is dependent on load.
- The converter cannot operate safely with a short circuit at frequencies close to the resonant frequency f_r because of the excessive peak value of the current through the resonant capacitor and switches.
- The converter cannot operate safely with an open circuit at frequencies close to the corner frequency f_o.

18.7 REFERENCES

1. H. A. Kojori, J. D. Lavers, and S. B. Dewan, "Steady-state analysis and design of an inductor-transformer dc-dc resonant converter," *IEEE Industry Applications Society Annual Meeting*, 1987, pp. 984–989.
2. H. A. Kojori, J. D. Lavers, and S. B. Dewan, "State-plane analysis of resonant dc-dc converter incorporating integrated magnetics," *IEEE Trans. Magnetics*, vol. 24, pp. 2998–2900, Nov. 1988.
3. E. G. Schmidtner, "A new high frequency resonant converter topology," *High Frequency Power Conversion Conf.*, 1988, pp. 390–403.
4. M. K. Kazimierczuk and J. Jóźwik, "Class E zero-voltage-switching and zero-current-switching rectifiers," *IEEE Trans. Circuits Syst.*, vol. CAS-37, pp. 436–444, Mar. 1990.
5. M. K. Kazimierczuk, "Class D voltage-switching MOSFET power amplifier," *Proc. IEE, Pt. B, Electric Power Appl.*, vol. 138, pp. 285–296, Nov. 1991.
6. R. P. Severns, "Topologies of three-element resonant converters," *IEEE Trans. Power Electronics*, vol. PE-7, pp. 89–98, Jan. 1992.
7. A. K. S. Bhat, "Analysis and design of a fixed frequency LCL-type series resonant converter," *Proc. of the IEEE Applied Power Electronics Conference*, Boston, Feb. 23–27, 1992.

18.8 REVIEW QUESTIONS

18.1 Is the part-load efficiency of the CLL resonant converter high?

18.2 Is the boundary between inductive and capacitive loads frequency dependent?

18.3 Is the dc voltage transfer function always load dependent?

18.4 Is the load capacitive or inductive when the dc voltage transfer function is independent of the load?

18.5 Can the converter regulate the dc output voltage from full load to no load?

18.6 Is the converter short-circuit proof?

18.7 Is the converter open-circuit proof?

18.8 How does the sensitivity of the dc voltage transfer function change with increasing ratio of the resonant inductance L_1/L_2?

18.9 PROBLEMS

18.1 A transformerless full-bridge CLL RC with a half-wave rectifier delivers 121 W power to 25 Ω load resistance with an efficiency of 89%. The inverter operates at the normalized operating frequency $f/f_o = 1.15$. What is the characteristic impedance of the inverter if the input voltage of the converter is $V_I = 280$ V, the inductance ratio is $A = 1$, and the efficiency of the rectifier is $\eta_R = 96\%$?

18.2 A half-bridge CLL RC with a bridge rectifier has the following parameters: input voltage $V_I = 100$ V, output voltage $V_O = 360$ V, and operating frequency $f = 150$ kHz. The loaded quality factor of the inverter is $Q_L = 0.5$, the resonant capacitance is $C = 4.7$ nF, and the resonant inductances are $L_1 = 250$ μH and $L_2 = 200$ μH. The transformer turns ratio is $n = 1/5$ and the efficiency of the rectifier is $\eta_R = 96\%$. Calculate the input power of the converter.

18.3 Design a transformerless full-bridge CLL RC with a half-wave rectifier that meets the following specifications: $V_I = 200$ V, $V_O = 50$ V, and $I_O =$ 0 to 4 A. Neglect losses due to the ripple current in the rectifier. Assume the ratio of inductances $A = 0.5$, the resonant frequency $f = 100$ kHz, the normalized operating frequency $f/f_o = 1.41$, the efficiency of the rectifier $\eta_R = 96\%$, and the efficiency of the inverter $\eta_I = 94\%$.

CHAPTER 19

CLASS D CURRENT-SOURCE RESONANT CONVERTER

19.1 INTRODUCTION

The Class D current-source resonant converter (CSRC) is obtained by replacing the ac load in the Class D current-source resonant inverter analyzed in Chapter 11 by one of the Class D voltage-driven rectifiers studied in Chapter 3. The inverter contains a parallel-resonant circuit, which acts as a sinusoidal voltage source and is suitable to drive voltage-driven rectifiers. The objectives of this chapter are to 1) introduce the current-source parallel-resonant converter, 2) present a comprehensive frequency-domain analysis of this current-source converter for steady-state operation, and 3) give a design example.

19.2 CIRCUIT DESCRIPTION

Class D current-source dc-dc resonant converters with three different Class D voltage-driven rectifiers are shown in Fig. 19.1. The characteristics of these converters are derived below.

19.2.1 CSRC with Half-Wave Rectifier

Combining (11.24) and (3.34) produces the dc-to-dc voltage transfer function of the converter with a half-wave rectifier

$$M_V \equiv \frac{V_O}{V_I} = M_{VI} M_{VR} = \frac{\eta_I \eta_{tr} R_i \sqrt{1 + [Q_L(\frac{\omega}{\omega_o} - \frac{\omega_o}{\omega})]^2}}{nR(1 + \frac{V_F}{V_O} + \frac{R_F + r_L}{R_L} + a_{hw}^2 \frac{r_{ac} R_L}{f^2 L_f^2})}. \quad (19.1)$$

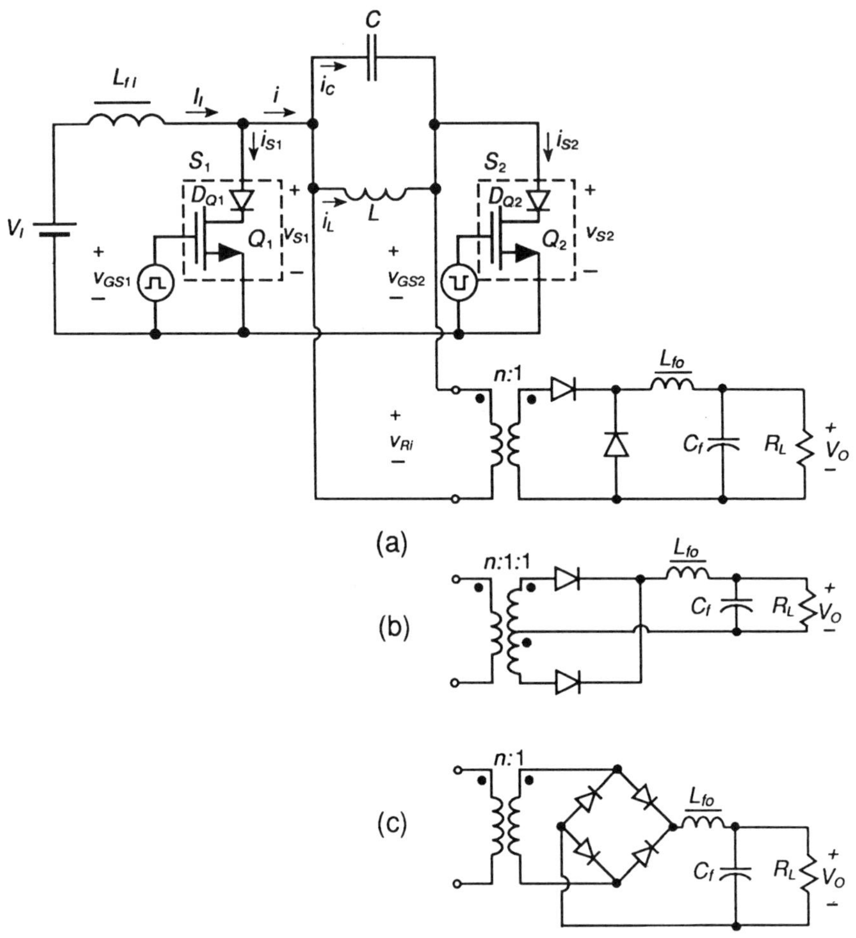

Figure 19.1: Class D current-source converter with a parallel-resonant circuit. (a) With a half-wave rectifier. (b) With a transformer center-tapped rectifier. (c) With a bridge rectifier.

From (11.47) and (3.30), one obtains the converter efficiency

$$\eta = \eta_I \eta_R = \frac{\eta_{tr}}{\{1 + \frac{\pi^2 r R_i [1 + Q_L^2 (\frac{\omega}{\omega_o} - \frac{\omega_o}{\omega})^2]}{2R^2} + \frac{V_{Fi}}{V_I} + \frac{R_i}{R_d}\}(1 + \frac{V_F}{V_O} + \frac{R_F + r_{LF}}{R_L} + a_{hw}^2 \frac{r_{ac} R_L}{f^2 L_f^2})}. \tag{19.2}$$

19.2.2 CSRC with Transformer Center-Tapped Rectifier

Using (11.24) and (3.67), one arrives at the dc-to-dc voltage transfer function of the converter with a transformer center-tapped rectifier

$$M_V \equiv \frac{V_O}{V_I} = M_{VI}M_{VR} = \frac{2\eta_I\eta_{tr}R_i\sqrt{1+[Q_L(\frac{\omega}{\omega_o}-\frac{\omega_o}{\omega})]^2}}{nR(1+\frac{V_F}{V_O}+\frac{R_F+r_L}{R_L}+a_{ct}^2\frac{r_{ac}R_L}{f^2L_f^2})}. \tag{19.3}$$

The efficiency of the inverter is obtained from (11.47) and (3.65)

$$\eta = \eta_I\eta_R = \frac{\eta_{tr}}{\left\{1+\frac{\pi^2 rR_i[1+Q_L^2(\frac{\omega}{\omega_o}-\frac{\omega_o}{\omega})^2]}{2R^2}+\frac{V_{Fi}}{V_I}+\frac{R_i}{R_d}\right\}(1+\frac{V_F}{V_O}+\frac{R_F+r_{LF}}{R_L}+a_{ct}^2\frac{r_{ac}R_L}{f^2L_f^2})}. \tag{19.4}$$

19.2.3 CSRC with Class D Bridge Rectifier

From (11.24) and (3.82), one obtains the dc-to-dc voltage transfer function of the converter with a bridge rectifier

$$M_V \equiv \frac{V_O}{V_I} = M_{VI}M_{VR} = \frac{2\eta_I\eta_{tr}R_i\sqrt{1+[Q_L(\frac{\omega}{\omega_o}-\frac{\omega_o}{\omega})]^2}}{nR(1+\frac{2V_F}{V_O}+\frac{2R_F+r_L}{R_L}+a_b^2\frac{r_{ac}R_L}{f^2L_f^2})}. \tag{19.5}$$

The converter efficiency is calculated from (11.47) and (3.80)

$$\eta = \eta_I\eta_R = \frac{\eta_{tr}}{\{1+\frac{\pi^2 rR_i[1+Q_L^2(\frac{\omega}{\omega_o}-\frac{\omega_o}{\omega})^2]}{2R^2}+\frac{V_{Fi}}{V_I}+\frac{R_i}{R_d}\}(1+\frac{2V_F}{V_O}+\frac{2R_F+r_{LF}}{R_L}+a_b^2\frac{r_{ac}R_L}{f^2L_f^2})}. \tag{19.6}$$

19.3 DESIGN OF CSRC

Example 19.1

Design a current-source converter with a transformer center-tapped rectifier. The following specifications should be satisfied: $V_I = 120$ V, $V_O = 48$ V, and $P_{Omax} = 25$ W. Assume the total converter efficiency $\eta = 91\%$ and the resonant frequency $f_o = 100$ kHz. Neglect losses due to the ripple current in the rectifier. Plot the efficiency of the designed converter versus load resistance R_L for $r = 3.4\ \Omega$ and $R_d = 200$ kΩ.

Solution: The maximum dc input power is

$$P_{Imax} = \frac{P_{Omax}}{\eta} = \frac{25}{0.91} = 27.5 \text{ W}, \tag{19.7}$$

and the maximum value of the dc input current is

$$I_{Imax} = I_{SMmax} = \frac{P_{Imax}}{V_I} = \frac{27.5}{120} = 0.23 \text{ A}. \tag{19.8}$$

The full-load resistance of the converter can be calculated as

$$R_{Lmin} = \frac{V_O^2}{P_{Omax}} = \frac{48^2}{25} = 92 \ \Omega. \tag{19.9}$$

The maximum value of the dc load current is

$$I_{Omax} = \frac{V_O}{R_{Lmin}} = \frac{48}{92} = 0.52 \text{ A}. \tag{19.10}$$

An MUR860 *pn*-silicon diode (Motorola) has the threshold voltage $V_F = 0.7$ V and the forward resistance $R_F = 0.1 \ \Omega$. The dc ESR of the filter inductor can be expected to be $r_L = 0.1 \ \Omega$, Assuming the transformer turns ratio $n = 5$, the transformer efficiency $\eta_{tr} = 97\%$, and using (3.66), one obtains the equivalent input resistance of the rectifier

$$\begin{aligned} R_i &= \frac{\pi^2 n^2}{8\eta_{tr}} R_L \left(1 + \frac{V_F}{V_O} + \frac{R_F + r_L}{R_L}\right) \\ &= \frac{\pi^2 \times 5^2}{8 \times 0.97} \times 92 \times \left(1 + \frac{0.7}{48} + \frac{0.2}{92}\right) = 2974 \ \Omega. \end{aligned} \tag{19.11}$$

From (3.65), the efficiency of rectifier η_R is

$$\eta_R = \frac{P_O}{P_i} = \frac{\eta_{tr}}{1 + \frac{V_F}{V_O} + \frac{R_F + r_L}{R_L}} = \frac{0.97}{1 + \frac{0.7}{48} + \frac{0.2}{92}} = 95.4 \ \%. \tag{19.12}$$

The voltage transfer function of the rectifier can be found from (3.67)

$$\begin{aligned} M_{VR} &= \frac{2\sqrt{2}\eta_{tr}}{\pi n \left(1 + \frac{V_F}{V_O} + \frac{R_F + r_L}{R_L}\right)} \\ &= \frac{2\sqrt{2} \times 0.97}{\pi \times 5 \left(1 + \frac{0.7}{48} + \frac{0.2}{92}\right)} = 0.172. \end{aligned} \tag{19.13}$$

The diode peak current I_{DM} and voltage V_{DM} are

$$I_{DM} = I_O = 0.52 \text{ A} \tag{19.14}$$

$$V_{DM} = \pi V_O = 3.14 \times 48 = 150.8 \text{ V}. \tag{19.15}$$

The dc-to-dc voltage transfer function of the converter is

$$M_V = \frac{V_O}{V_I} = \frac{48}{120} = 0.4. \tag{19.16}$$

The required efficiency of the inverter is

$$\eta_I = \frac{\eta}{\eta_R} = \frac{0.91}{0.954} = 0.954. \tag{19.17}$$

Assuming the ratio $R/R_i = 0.97$, $\omega/\omega_o = 0.95$, and rearranging (19.3), the loaded quality factor is

$$Q_L = \frac{\sqrt{\left(\frac{nM_V R}{2\eta R_i}\right)^2 - 1}}{\left|\frac{\omega}{\omega_o} - \frac{\omega_o}{\omega}\right|} = \frac{\sqrt{\left(\frac{5\times0.4\times0.97}{2\times0.91}\right)^2 - 1}}{\left|0.95 - \frac{1}{0.95}\right|} = 3.6. \tag{19.18}$$

The total resistance of the inverter is $R = R_i(R/R_i) = 2974 \times 0.97 = 2885\ \Omega$. The component values of the resonant circuit are

$$L = \frac{R}{\omega_o Q_L} = \frac{2885}{2 \times \pi \times 100 \times 10^3 \times 3.6} = 1.276\ \text{mH} \tag{19.19}$$

$$C = \frac{Q_L}{\omega_o R} = \frac{3.6}{2 \times \pi \times 100 \times 10^3 \times 2885} = 1.99\ \text{nF}. \tag{19.20}$$

The characteristics impedance is $Z_o = \sqrt{L/C} = 801\ \Omega$. Equation (11.17) gives the maximum amplitude of the input current to the resonant circuit $I_m = (2/\pi)I_{Imax} = 0.146$ A. The maximum switch voltage V_{SM} is equal to the peak value of the voltage across the resonant circuit. Therefore, from (11.24),

$$\begin{aligned} V_{SM} &= \sqrt{2}V_{Ri} = \pi V_I \eta_I \frac{R_i}{R}\sqrt{1 + \left[Q_L\left(\frac{\omega}{\omega_o} - \frac{\omega_o}{\omega}\right)\right]^2} \\ &= \pi \times 120 \times 0.954 \times \frac{1}{0.97}\sqrt{1 + \left[3.6\left(0.95 - \frac{1}{0.95}\right)\right]^2} = 395.3\ \text{V}. \end{aligned} \tag{19.21}$$

The total converter efficiency η versus R_L at $V_I = 120$ V and $V_O = 48$ V is shown in Fig. 19.2. It can be seen that it is slightly higher than that assumed for the design. This is because the switching losses and the gate-power loss are not included in (19.4).

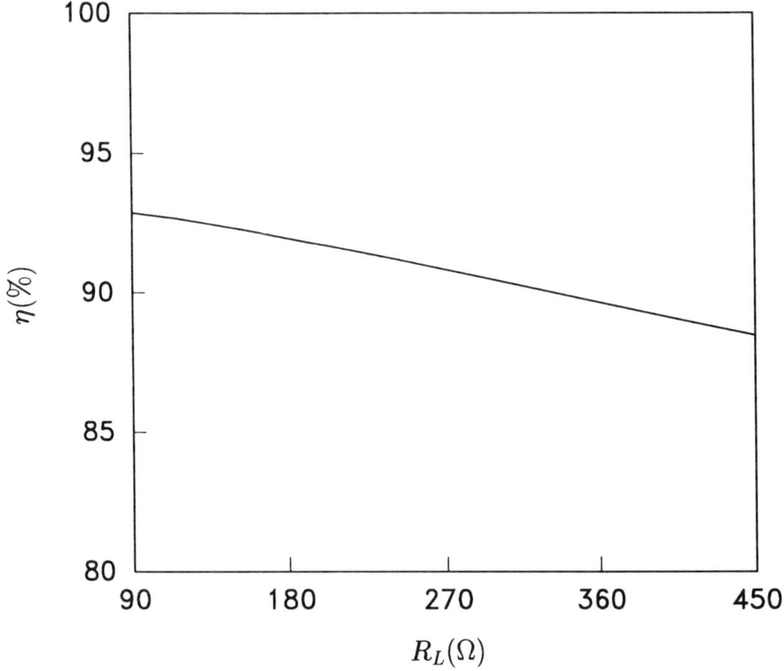

Figure 19.2: Total converter efficiency η as a function of load resistance R_L at $V_I =$ 120 V and $V_O = 48$ V.

19.4 SUMMARY

- The current-source resonant converter can regulate the output voltage from full load to 20% of full load, using a narrow frequency range on the order of 6%.
- It is an easy circuit to build and drive. Since the gates of the MOSFETs are referenced to ground, a pulse transformer or an optical coupler is not needed to drive the MOSFETs.
- The converter efficiency decreases with increasing load resistance R_L because the ratio R_i/R_d increases with increasing R_L.
- Operation closer to the resonant frequency is desirable because the converter does not draw high current.
- The rectifier diodes turn off at a high di/dt. Therefore, if *pn* junction diodes are used, the reverse-recovery spikes are generated, causing switching noise.
- The input current of the converter is nonpulsating.

19.5 REFERENCES

1. P. J. Baxandall, "Transistor sine-wave *LC* oscillators, some general considerations and new developments," *Proc. IEE*, vol. 106, pt. B, pp. 748–758, May 1959.
2. M. R. Osborne, "Design of tuned transistor power inverters," *Electron. Eng.*, vol. 40, no. 486, pp. 436–443, 1968.
3. W. J. Chudobiak and D. F. Page, "Frequency and power limitations of Class-D transistor inverter," *IEEE J. Solid-State Circuits*, vol. SC-4, pp. 25–37, Feb. 1969.
4. H. L. Krauss, C. W. Bostian, and F. H. Raab, *Solid State Radio Engineering*, New York: John Wiley & Sons, ch. 14.1, pp. 439–441, 1980.
5. J. G. Kassakian, "A new current mode sine wave inverter," *IEEE Trans. Industry Applications*, vol. IA-18, pp. 273–278, May/June 1982.
6. R. L. Steigerwald, "High-frequency resonant transistor dc-dc converters," *IEEE Trans. Industrial Electronics*, vol. IE-31, pp. 181–191, May 1984.
7. J. G. Kassakian, M. F. Schlecht, and G. C. Verghese, *Principles of Power Electronics*, Reading, MA: Addison-Wesley, 1991, ch. 9.3, pp. 212–217.
8. M. K. Kazimierczuk and A. Abdulkarim, "Current-source parallel-resonant dc-dc converter," *IEEE Trans. Industrial Electronics,* vol. 42, 1995.

19.6 REVIEW QUESTIONS

19.1 Is is difficult to drive the transistors in the CSRC?

19.2 How does the efficiency of the CSRC change versus load resistance?

19.3 What is the current stress of the switches in the CSRC?

19.4 What is the voltage stress of the switches in the CSRC?

19.5 What is the current stress of the diodes in the CSRC?

19.6 What is the voltage stress of the diodes in the CSRC?

19.7 Is the CSRS short-circuit proof?

19.8 Is the CSRS open-circuit proof?

19.7 PROBLEMS

19.1 A current-source resonant converter with a bridge rectifier supplies power of 150 W to a load resistance. The parameters of the converter are input voltage $V_I = 48$ V, output voltage $V_O = 200$ V, loaded quality factor $Q_L = 3.5$, normalized resonant frequency $\omega/\omega_o = 0.93$, efficiency of the resonant circuit $\eta_{rc} = 98\%$, and transformer turns ratio $n = 1/2$. What is the input power of the converter?

19.2 A current-source resonant converter with a half-wave rectifier has the following parameters: input voltage $V_I = 180$ V, output voltage $V_O =$ 100 V, load resistance $R_L = 100\ \Omega$, resonant frequency $f_o = 100$ kHz, efficiency of the inverter $\eta_I = 95\%$, efficiency of the resonant circuit $\eta_{rc} = 99\%$, efficiency of the rectifier $\eta_R = 95\%$, resonant inductance $L = 1$ mH, transformer turns ratio $n = 2$, and operation below the resonant frequency. Calculate the operating frequency of the converter.

19.3 Design a current-source converter with a bridge rectifier to meet the following specifications: $V_I = 48$ V, $V_O = 280$ V, and $P_{Omax} = 100$ W. Assume the total converter efficiency $\eta = 90\%$, the efficiency of the rectifier $\eta_R = 94\%$, the efficiency of the resonant circuit $\eta_{rc} = 99\%$, the transformer turns ratio $n = 1/3$, the normalized switching frequency $\omega/\omega_o = 0.95$, and the resonant frequency $f_o = 100$ kHz. Neglect losses due to the ripple current in the rectifier.

CHAPTER 20

CLASS D-E RESONANT CONVERTER

20.1 INTRODUCTION

Class D-E resonant dc-dc converters [1]–[5] may be synthesized by cascading one of the Class D resonant inverters considered in Chapters 6 through 11 and one of the Class E rectifiers studied in Chapters 4 and 5. The purpose of this chapter is to present one example of Class D-E resonant converters. This example demonstrates that any current-output inverter can be cascaded with any current-driven rectifier and any voltage-output inverter can be cascaded with any voltage-driven rectifier to synthesize a dc-dc converter.

20.2 CIRCUIT DESCRIPTION

Figure 20.1(a) shows the circuit of a Class D-E resonant dc-dc converter [1], [2]. It consists of a Class D series resonant inverter and a Class E current-driven low dv/dt rectifier. The Class E rectifier is composed of a diode, a capacitor C_2 connected in parallel with the diode, a large filter capacitor C_3, and an inductor L_2. Resistance R_L represents a dc load. The circuit is called a Class E rectifier because the diode current and voltage waveforms are time-inversed images of the corresponding transistor waveforms in a Class E inverter. In order to design the Class D inverter, its load impedance is needed. Since the series-resonant circuit forces a sinusoidal current driving the rectifier, only the power at the fundamental frequency is transferred from the inverter to the rectifier. For this reason, it is sufficient to determine the rectifier input impedance at the switching frequency. This impedance is defined as the ratio of the phasor of the fundamental component of the rectifier in-

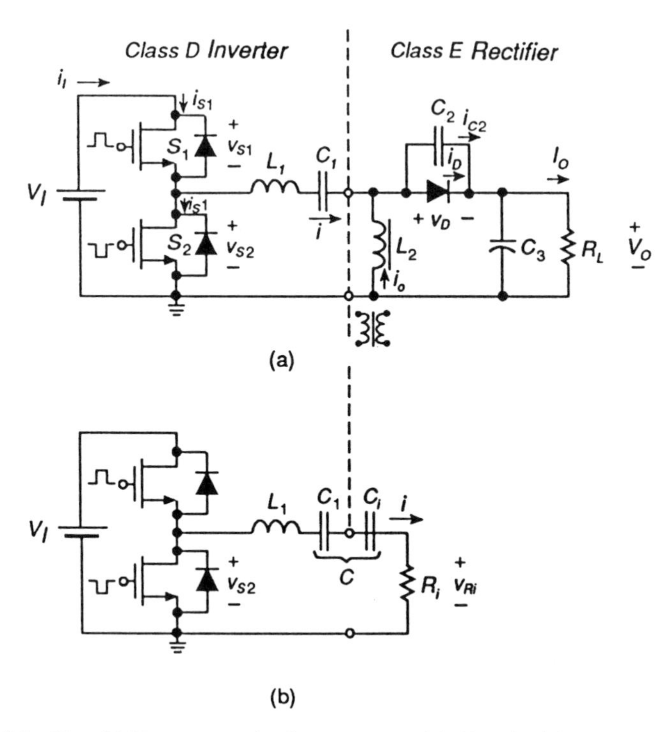

Figure 20.1: Class D-E resonant dc-dc converter. (a) Circuit. (b) Basic circuit of Class D series resonant inverter.

put voltage to the phasor of the rectifier input current. The input impedance may be represented by a series combination of an input resistor R_i and an input capacitor C_i, as shown in Fig. 20.1(b). The inductor L_2 closes the path for the dc component of the diode current, which is equal to the dc output current I_O. The inductance L_2 is assumed to be large enough so that the ac ripple of the output current may be neglected, but a finite dc-feed inductance can be used as well. The inductor L_2 may be replaced by a transformer to provide a dc isolation and a proper value of the dc-to-dc transfer function $M_V = V_O/V_I$. The rectifier diode may be inverted to obtain a negative output voltage V_O in the transformerless converter. The advantage of the rectifier topology is that the diode parasitic capacitance and the parasitic capacitance of L_2 are absorbed into C_2 and, therefore, do not adversely affect the circuit operation.

The Class D inverter employs a pair of bidirectional switches S_1 and S_2 and a series-resonant circuit L_1-C_1-C_i-R_i, as shown in Fig. 20.1(b). Each switch is comprised of a transistor and an antiparallel diode. The MOSFET's body diodes may be used as antiparallel diodes for the operation above resonance. A very important advantage of the Class D inverter is that the peak voltage of the transistors has the lowest possible value, equal to the dc input voltage

V_I. Therefore, this inverter is especially suitable for high voltage applications, such as off-line power supplies.

20.3 PRINCIPLE OF OPERATION

The principle of the converter operation is explained by the current and voltage waveforms depicted in Fig. 20.2. Switches S_1 and S_2 are ON and OFF alternately with a duty cycle of 0.5, applying a nearly square-wave voltage to the series-resonant circuit. The resonant frequency is $\omega_o = 1/\sqrt{L_1 C_1 C_i/(C_1 + C_i)}$. The total resistance R is the sum of the rectifier input resistance R_i and a parasitic resistance r associated with the MOSFET on-resistances and equivalent series resistances of the resonant components. If the loaded quality

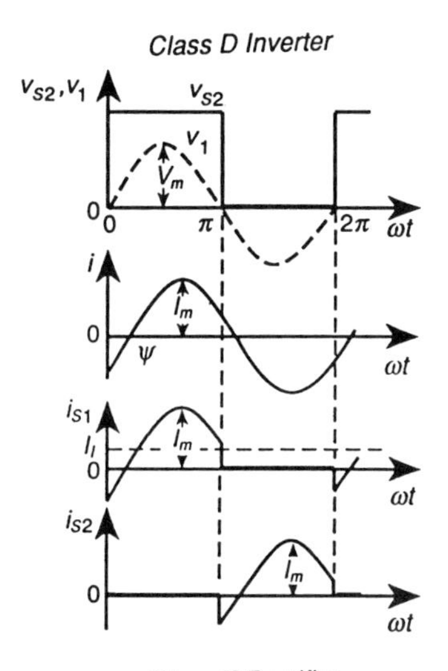

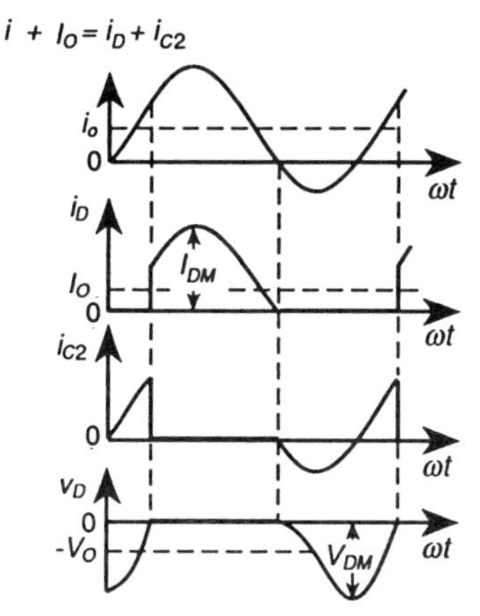

Figure 20.2: Waveforms in Class D-E converter for $f > f_o$.

factor $Q_L = \omega_o L_1/R$ is sufficiently high (e.g., $Q_L \geq 5$), the current through the resonant circuit is nearly sinusoidal. When S_1 is ON and S_2 is OFF, the current flows through S_1, and *vice versa*. In general, the operating frequency f differs from the resonant frequency f_o in dc-dc converters. Two cases can be distinguished: 1) $f < f_o$ and 2) $f > f_o$. For both cases, the phase shift ψ between the current i and the fundamental component v_1 of voltage v_{DS2} is nonzero. The positive switch current flows through the transistor, but the negative switch current can flow through the antiparallel diode and/or the transistor.

The Class E rectifier is driven by a nearly sinusoidal current i, produced by the series-resonant circuit of the Class D inverter. The current through the choke inductor L_2 is approximately equal to I_O. Consequently, the current through the parallel combination of the diode and capacitor C_2 is a shifted sinusoid $i + I_O$. This current flows through the diode when the diode is ON and through capacitor C_2 when the diode is OFF. The diode begins to turn off when its current reaches zero. The current through the capacitor shapes the voltage across the capacitor and the diode, in accordance with the equation $i_{C2} = C_2 d(v_{C2})/dt$. Because i_{C2} is zero at turn-off, the diode turns off at zero dv/dt. The diode voltage then decreases gradually when i_{C2} is negative, reaches its minimum value V_{DM} when i_{C2} crosses zero, and rises when i_{C2} is positive. Once the diode voltage reaches its threshold value, the diode turns on. Since the capacitor current at turn-on is limited by the series-resonant circuit and the choke L_2, the diode turns on at low dv/dt, reducing turn-on switching loss and noise. The diode current, however, has a step change at turn-on, generating noise. The diode turns off at zero dv/dt and turns on at low dv/dt. Therefore, switching losses associated with charging and discharging the diode capacitance are considerably reduced. In addition, the diode turns off at low di/dt, reducing the reverse recovery effect. Due to the smooth diode voltage and current waveforms and thereby "soft" diode turn-on and turn-off transitions, the noise level is significantly reduced. Another property of the Class E rectifier is a large diode ON duty cycle D, which is independent of the output voltage ripple. The duty cycle may assume any value between zero and one. The diode current waveform composed of wide and smooth pulses contains low higher harmonics, reducing the level of conducted and radiated electromagnetic interference (EMI/RFI). The Class E rectifier should be driven by a sinusoidal current source. Thus, the Class E rectifier is compatible with the Class D inverter with a series-resonant circuit because this circuit acts as a sinusoidal current source. A detailed analysis of the Class E rectifier can be found in Chapter 4. The final results are summarized in Table 4.1.

The dc output voltage of the converter can be regulated against load and line variations by varying the switching frequency f. This is possible because the amplitude of the current through the series-resonant circuit is dependent on the switching frequency in accordance with the resonance curve. In turn, the dc output voltage depends on the amplitude of the current in the series-resonant circuit.

20.4 RECTIFIER PARAMETERS FOR D = 0.5

A detail analysis of the Class E rectifier at any value of the diode on-duty cycle D is given in Chapter 4. The maximum power-output capability occurs at $D = 0.5$. The rectifier parameters at a duty cycle of 0.5 are as follows:

$$M_{VR} \equiv \frac{V_O}{V_{Ri}} = \sqrt{\frac{\pi^2 + 4}{8}} \approx 1.3167 \tag{20.1}$$

$$\omega C_2 R_L = \frac{1}{\pi} \approx 0.3183 \tag{20.2}$$

$$\frac{R_i}{R_L} = \frac{8}{\pi^2 + 4} \approx 0.5768 \tag{20.3}$$

$$\frac{C_i}{C_2} = \frac{2(\pi^2 + 4)}{\pi^2 - 4} \approx 4.726 \tag{20.4}$$

$$\omega C_2 R_i = \frac{8}{\pi(\pi^2 + 4)} \approx 0.1836 \tag{20.5}$$

$$\frac{I_{DM}}{I_O} = \frac{\sqrt{\pi^2 + 4}}{2} + 1 \approx 2.862 \tag{20.6}$$

$$\frac{V_{DM}}{V_O} = 2\pi arctan(\frac{2}{\pi}) \approx 3.562. \tag{20.7}$$

From (6.64) and (20.1), one obtains the dc-to-dc voltage transfer function of the converter

$$M_V \equiv \frac{V_O}{V_I} = M_{VI} M_{VR} = \frac{\eta_I \sqrt{\pi^2 + 4}}{2\pi \sqrt{1 + Q_L^2 (\frac{\omega}{\omega_o} - \frac{\omega_o}{\omega})^2}}. \tag{20.8}$$

At the resonant frequency, $M_V \approx 0.6$. Expressions (20.1) to (20.8) may be used to design the converter for full load. Unfortunately, equations describing the Class E rectifier are transcendental. Therefore, it is not possible to derive closed-form relationship among V_I, V_O, and f at any R_L. A qualitative description of the converter behavior at light loads is as follows. As the load resistance R_L is increased from its minimum value R_{Lmin} to ∞, the diode on-duty cycle D decreases from its maximum value D_{max} to 0, the voltage transfer function M_{VR} increases from its minimum value M_{Rmin} to ∞, the input capacitance C_i decreases from its maximum values C_{imax} to C_2, and the input resistance initially increases and then decreases. Consequently, the overall resonant capacitance C decreases from its maximum value

$$C_{max} = \frac{C_1 C_{imax}}{C_1 + C_{imax}} \tag{20.9}$$

to its minimum value

$$C_{min} = \frac{C_1 C_{imin}}{C_1 + C_{imin}} \tag{20.10}$$

Thus, the resonant frequency f_o increases from its minimum value

$$f_{omin} = \frac{1}{2\pi\sqrt{L_1 C_{max}}} \tag{20.11}$$

to its maximum value

$$f_{omax} = \frac{1}{2\pi\sqrt{L_1 C_{min}}}. \tag{20.12}$$

The loaded quality factor Q_L is a complicated function of R_L. Since the rectifier input resistance R_i decreases with increasing load R_L, the inverter voltage transfer function M_{Vr} decreases [Fig. 6.17(b)]. On the other hand, the rectifier voltage transfer function M_{VR} increases with increasing R_L. Therefore, only a relatively small increase in the switching frequency is required to regulate the dc output voltage V_O.

The voltage transfer function of the Class E rectifier can be approximated by

$$M_{VR} \approx 0.8\omega C_2 R_L + 1. \tag{20.13}$$

Hence, using (6.64) the voltage transfer function of the converter at any load is

$$M_V \approx \frac{\sqrt{2}\eta_I(0.8\omega C_2 R_L + 1)}{\pi\sqrt{1 + Q_L^2(\frac{\omega}{\omega_o} - \frac{\omega_o}{\omega})^2}}. \tag{20.14}$$

20.5 DESIGN OF CLASS D-E RESONANT CONVERTER

Example 19.1

Design the Class D-E converter of Fig. 20.1(a) to meet the following specifications: $V_I = 100$ V, $V_O = 50$ V, and $R_L = 50\ \Omega$ to ∞. Assume the resonant frequency $f_o = 100$ kHz, the converter efficiency $\eta = 90\%$, and the inverter efficiency $\eta_I = 94\%$.

Solution: It is sufficient to design the converter for the full-load resistance $R_{Lmin} = 50\ \Omega$. The maximum value of the dc output current is $I_{Omax} = V_O/R_{Lmin} = 50/50 = 1$ A, and the maximum value of the dc output power is $P_{Omax} = V_O I_{Omax} = 50\times1 = 50$ W. Let us design the rectifier for $D = 0.5$ that gives the best utilization of the rectifier's diode (maximum power-output capability). The voltage transfer function of the converter is $M_V = V_O/V_I = 50/100 = 0.5$. Assuming $f/f_o = 1.05$ and using (20.2) to (20.7), the parameters of the Class E rectifier are as follows:

$$C_2 = \frac{1}{2\pi^2 f R_{Lmin}} = \frac{1}{2\pi^2 \times 105 \times 10^3 \times 50} = 9.65 \text{ nF}, \tag{20.15}$$

$$R_i = \frac{8R_L}{\pi^2 + 4} = 0.5769 \times 50 = 28.84 \ \Omega, \tag{20.16}$$

$$C_i = \frac{2(\pi^2 + 4)C_2}{\pi^2 - 4} = 4.7259 \times 9.65 \times 10^{-9} = 45.6 \text{ nF}, \tag{20.17}$$

$$I_{DM} = (\frac{\sqrt{\pi^2 + 4}}{2} + 1)I_{Omax} = 2.862 \times 1 = 2.862 \text{ A}, \tag{20.18}$$

and

$$V_{DM} = 2\pi arctan\left(\frac{2}{\pi}\right) V_O = 3.562 \times 50 = 178.1 \text{ V}. \tag{20.19}$$

From (20.8) and (20.1), one obtains the loaded quality factor as

$$Q_L = \frac{\sqrt{\frac{\eta_I^2(\pi^2+4)}{4\pi^2 M_V^2} - 1}}{\mid \frac{\omega}{\omega_o} - \frac{\omega_o}{\omega} \mid} = \frac{\sqrt{\frac{0.94^2(\pi^2+4)}{4\pi^2 \times 0.5^2} - 1}}{\mid 1.05 - \frac{1}{1.05} \mid} = 5.04. \tag{20.20}$$

Hence, the parameters of the inverter are

$$L_1 = \frac{Q_L R_i}{\omega_o} = \frac{5.04 \times 28.84}{2\pi \times 100 \times 10^3} = 231.3 \ \mu\text{H}, \tag{20.21}$$

$$C = \frac{1}{\omega_o Q_L R_i} = \frac{1}{2\pi \times 100 \times 10^3 \times 5.04 \times 28.84} = 10.95 \text{ nF}, \tag{20.22}$$

$$C_1 = \frac{CC_i}{C_i - C} = \frac{10.95 \times 45.6}{45.6 - 10.95} = 14.41 \text{ nF}, \tag{20.23}$$

and

$$P_{Imax} = \frac{P_{Omax}}{\eta} = \frac{50}{0.9} = 55.56 \text{ W}. \tag{20.24}$$

Since the assumed value of the rectifier efficiency is $\eta_R = \eta/\eta_I = 0.9/0.94 = 0.9575$, the amplitude of the current through the resonant circuit equal to the maximum switch current is

$$I_m = I_{SM} = \sqrt{\frac{2P_{Omax}}{\eta_R R_i}} = \sqrt{\frac{2 \times 50}{0.9575 \times 28.84}} = 1.9 \text{ A}. \tag{20.25}$$

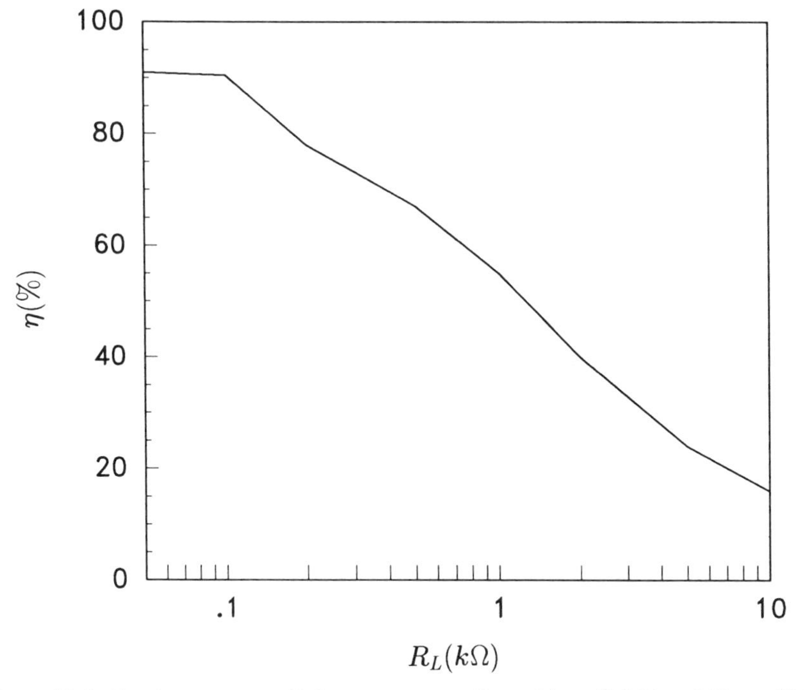

Figure 20.3: Total converter efficiency η versus R_L at $V_I = 100$ V and $V_O = 50$ V.

At $R_L = \infty$, $C_i = C_2$, the total resonant capacitance is $C_{min} = C_1C_2/(C_1 + C_2) = 5.78$ nF, and the resonant frequency is $f_{omax} = 1/(2\pi\sqrt{L_1C_{min}}) = 137.7$ kHz.

The amplitude of the voltage across the resonant capacitor is

$$V_{C1m} = \frac{I_m}{\omega C_1} = \frac{1.9}{2\pi \times 105 \times 10^3 \times 14.41 \times 10^{-9}} = 199.9 \text{ V}, \qquad (20.26)$$

and the amplitude of the voltage across the resonant inductor is

$$V_{L1m} = \omega L_1 I_m = 2\pi \times 105 \times 10^3 \times 231.3 \times 10^{-6} \times 1.9 = 289.9 \text{ V}. \quad (20.27)$$

The designed converter was built using MTP5N40 MOSFETs (Motorola) as switches S_1 and S_2, an MR826 fast recovery diode (Motorola), $L_1 = 226$ μH, $C_1 = 15$ nF, $C_2 = 10$ nF, $L_2 = 20$ mH, and $C_3 = 100$ μF. Figure 20.3 shows the plot of the measured converter efficiency η as a function of load resistance R_L.

20.6 SUMMARY

- Class D-E resonant converters can regulate the dc output voltage for load resistances ranging from full load to no load.
- The range of the switching frequency required to regulate the dc output voltage against load and line variations is narrow.
- The maximum values of the peak voltages and currents occur at full load.
- The input capacitance of the rectifier decreases with increasing load resistance, reducing the total resonant capacitance and increasing the resonant frequency of the inverter.
- Any current-output inverter can be cascaded with any current-driven rectifier and any voltage-output inverter can be be cascaded with any voltage-driven rectifier to form a dc-dc converter. Therefore, Class D or Class E inverters can be cascaded with compatible Class D or Class E rectifiers.

20.7 REFERENCES

1. M. K. Kazimierczuk and W. Szaraniec, "Class D-E resonant dc/dc converter," *IEEE Trans. Aerospace and Electronic Systems*, vol. AES-29, pp. 963–976, Jan. 1993.
2. M. K. Kazimierczuk and J. Jóźwik, "Resonant dc/dc converter with Class-E inverter and Class-E rectifier," *IEEE Trans. Ind. Electron.*, vol. 36, pp. 568–578, Nov. 1989.
3. M. K. Kazimierczuk and J. Jóźwik, "Class E^2 narrow-band resonant dc/dc converter, *IEEE Trans. Instrum. and Meas.*, vol. 38, pp. 1064–1068, Dec. 1989.
4. J. Jóźwik and M. K. Kazimierczuk, "Analysis and design of Class E^2 dc/dc converter," *IEEE Trans. Ind. Electron.*, vol. 37, pp. 173–183, Apr. 1990.
5. M. K. Kazimierczuk and J. Jóźwik, "Class E zero-voltage-switching and zero-current-switching rectifiers," *IEEE Trans. Circuits Syst.*, vol. CAS-37, pp. 436–444, Mar. 1990.

20.8 REVIEW QUESTIONS

20.1 Is it possible to regulate the dc output voltage from full load to no load in the Class D-E resonant converter?

20.2 When do the maximum current and voltage stresses occur?

20.3 Is the noise level in the Class D-E resonant converter high?

20.4 Is the capacitance of the rectifier diode included in the rectifier topology?

20.5 Is the resonant frequency of the inverter dependent on the load resistance?

20.9 PROBLEMS

20.1 A rectifier in a Class D-E resonant converter of Fig. 20.1(a) operates with a duty ratio $D = 0.5$. The parameters of the inverter are input voltage $V_I = 200$ V, efficiency $\eta_I = 96\%$, loaded quality factor $Q_L = 2.85$, and normalized switching frequency $f/f_o = 1.1$. What is the output voltage of the converter?

20.2 A Class D-E resonant converter operating with a switching frequency 200 kHz supplies 100 W power at a 50 V output voltage. The inverter has the following parameters: input voltage $V_I = 100$ V, efficiency $\eta_I = 94\%$, loaded quality factor $Q_L = 3$, and resonant frequency $f_o = 180$ kHz. Calculated the approximate value of the shunt capacitor C_2 in the rectifier.

20.3 Design the Class D-E converter of Fig. 20.1(a) to meet the following specifications: input voltage $V_I = 200$ V, output voltage $V_O = 100$ V, and load resistance $R_L = 50\ \Omega$ to ∞. Assume the resonant frequency $f_o = 100$ kHz, the normalized switching frequency $f/f_o = 1.07$, the converter efficiency $\eta = 90\%$, and the inverter efficiency $\eta_I = 95\%$.

CHAPTER 21

PHASE-CONTROLLED RESONANT CONVERTERS

21.1 INTRODUCTION

Fixed-frequency phase-controlled dc-dc converters [1]–[22] are obtained by cascading phase-controlled full-bridge resonant inverters considered in Chapter 12 and rectifiers that are compatible with these inverters. The dc output voltage can be regulated against load current and line voltage variations by varying the phase shift between the gate-drive voltages of the switching legs while maintaining a fixed operating frequency. The objective of this chapter is to present detailed characteristics of one of the many possible phase-controlled converters, namely, a single-capacitor phase-controlled resonant converter (SC PC SRC) [16].

21.2 CIRCUIT DESCRIPTION OF SC PC SRC

A single-capacitor phase-controlled full-bridge Class D series resonant converter is shown in Fig. 21.1. It is composed of a phase-controlled single-capacitor resonant inverter analyzed in Chapter 12 and one of the Class D current-driven rectifiers studied in Chapter 2. The phase-controlled Class D inverter consists of a dc input voltage source V_I, two switching legs, two resonant inductors L, one resonant capacitor C, and a coupling capacitor C_C. Each switching leg comprises two switches with antiparallel diodes. If one of the Class D current driven rectifiers is connected in series with capacitor C as an ac load, a single-capacitor phase-controlled series resonant converter is obtained. Regulation of the dc output voltage V_O against load and line variations can be accomplished by varying the phase shift between the voltages,

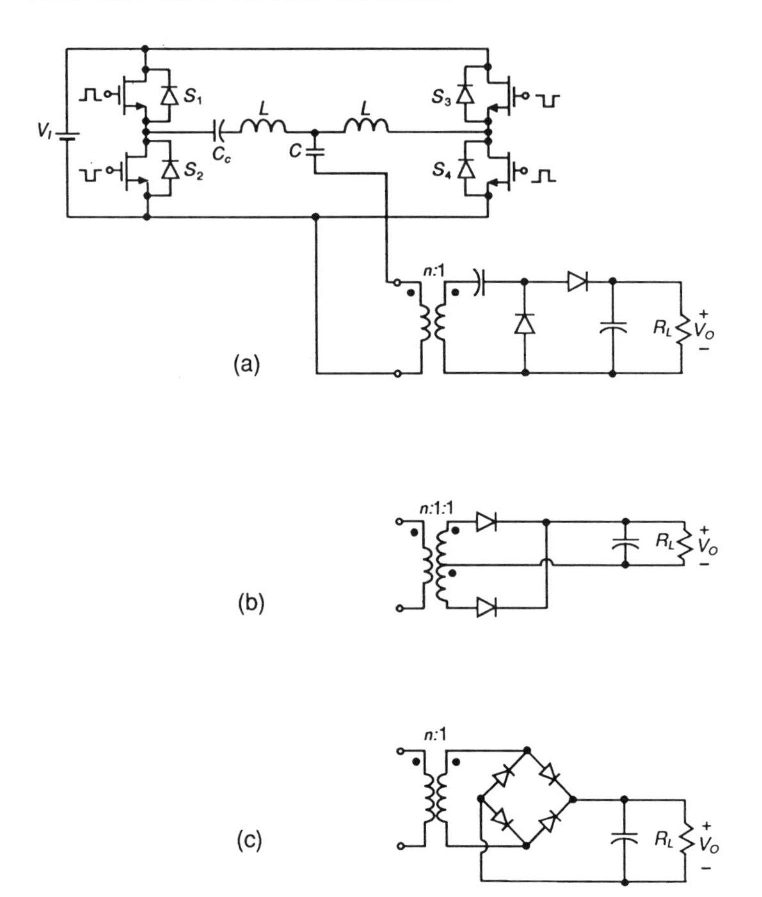

Figure 21.1: Single-capacitor phase-controlled Class D series resonant converter. (a) With a half-wave rectifier. (b) With a transformer center-tapped rectifier. (c) With a bridge rectifier.

which drive the switching legs while the operating frequency is maintained constant.

21.2.1 SC PC SRC with Half-Wave Rectifier

The single-capacitor phase-controlled series resonant converter with a half-wave rectifier is depicted in Fig. 21.1(a). From (12.14) and (2.31), one obtains the dc-to-dc voltage transfer function of the converter

$$M_V = \frac{V_O}{V_I} = \eta_I M_{VI} M_{VR}$$

$$= \frac{\eta_I \eta_{tr} cos(\frac{\phi}{2})}{n\sqrt{1 + Q_L^2(\frac{\omega}{\omega_o} - \frac{\omega_o}{\omega})^2}[1 + \frac{2V_F}{V_O} + \frac{\pi^2 R_F}{2R_L} + \frac{r_{ESR}}{R_L}(\frac{\pi^2}{4} - 1)]}. \quad (21.1)$$

From (12.29) and (2.27), the converter efficiency is

$$\eta = \eta_I \eta_R$$

$$= \frac{1}{1 + \frac{1}{Z_o Q_L cos^2(\frac{\phi}{2})}\{r[Q_L^2 + sin^2(\frac{\phi}{2})(\frac{\omega_o}{\omega})^2(Q_L^2\frac{\omega_o^2}{\omega^2} - 2Q_L^2 + 1)] + 2r_C Q_L^2 cos^2(\frac{\phi}{2})\}}$$

$$\times \frac{\eta_{tr}}{1 + \frac{2V_F}{V_O} + \frac{\pi^2 R_F}{2R_L} + \frac{r_{ESR}}{R_L}(\frac{\pi^2}{4} - 1)}. \quad (21.2)$$

21.2.2 SC PC SRC with Transformer Center-Tapped Rectifier

Figure 21.1(b) shows a single-capacitor phase-controlled series resonant converter with a transformer center-tapped rectifier. From (12.14) and (2.73), the voltage transfer function of the converter is

$$M_V = \frac{V_O}{V_I} = \eta_I M_{VI} M_{VR}$$

$$= \frac{\eta_I \eta_{tr} cos(\frac{\phi}{2})}{2n\sqrt{1 + Q_L^2(\frac{\omega}{\omega_o} - \frac{\omega_o}{\omega})^2}[1 + \frac{V_F}{V_O} + \frac{\pi^2 R_F}{8R_L} + \frac{r_{ESR}}{R_L}(\frac{\pi^2}{8} - 1)]}. \quad (21.3)$$

Using (12.29) and (2.71), one obtains the converter efficiency

$$\eta = \eta_I \eta_R$$

$$= \frac{1}{1 + \frac{1}{Z_o Q_L cos^2(\frac{\phi}{2})}\{r[Q_L^2 + sin^2(\frac{\phi}{2})(\frac{\omega_o}{\omega})^2(Q_L^2\frac{\omega_o^2}{\omega^2} - 2Q_L^2 + 1)] + 2r_C Q_L^2 cos^2(\frac{\phi}{2})\}}$$

$$\times \frac{\eta_{tr}}{1 + \frac{V_F}{V_O} + \frac{\pi^2 R_F}{8R_L} + \frac{r_{ESR}}{R_L}(\frac{\pi^2}{8} - 1)}. \quad (21.4)$$

21.2.3 SC PC SRC with Bridge Rectifier

A single-capacitor phase-controlled series resonant converter with a bridge rectifier is shown in Fig. 21.1(c). The voltage transfer function of the converter can be obtained from (12.14) and (2.98) as

$$M_V = \frac{V_O}{V_I} = \eta_I M_{VI} M_{VR}$$

$$= \frac{\eta_I \eta_{tr} cos(\frac{\phi}{2})}{2n\sqrt{1 + Q_L^2(\frac{\omega}{\omega_o} - \frac{\omega_o}{\omega})^2}[1 + \frac{2V_F}{V_O} + \frac{\pi^2 R_F}{4R_L} + \frac{r_{ESR}}{R_L}(\frac{\pi^2}{8} - 1)]}. \quad (21.5)$$

From (12.29) and (2.96), the converter efficiency is

$$\eta = \eta_I \eta_R$$
$$= \frac{1}{1 + \frac{1}{Z_o Q_L cos^2(\frac{\phi}{2})}\{r[Q_L^2 + sin^2(\frac{\phi}{2})(\frac{\omega_o}{\omega})^2(Q_L^2 \frac{\omega_o^2}{\omega^2} - 2Q_L^2 + 1)] + 2r_C Q_L^2 cos^2(\frac{\phi}{2})\}}$$
$$\times \frac{\eta_{tr}}{1 + \frac{2V_F}{V_O} + \frac{\pi^2 R_F}{4R_L} + \frac{r_{ESR}}{R_L}(\frac{\pi^2}{8} - 1)]}. \tag{21.6}$$

21.3 DESIGN EXAMPLE

Example 21.1

Design a transformer single-capacitor phase-controlled series resonant converter with a transformer center-tapped rectifier of Fig. 21.1(b). The specifications are $V_I = 270$ to 300 V, $V_O = 28$ V, and $R_{Lmin} = 10\ \Omega$. Assume the resonant frequency $f_o = 150$ kHz, the inverter efficiency $\eta_I = 94\%$, and the rectifier efficiency $\eta_R = 95\%$. Draw the efficiency of the designed converter η as a function of load resistance R_L.

Solution: To assure that switches are loaded inductively, the ratio f/f_o must be greater than 1.15 (see Section 12.2.5). Let $f/f_o = 1.33$. Hence,

$$f = \left(\frac{f}{f_o}\right) f_o = 1.33 \times 150 \times 10^3 = 200 \text{ kHz}. \tag{21.7}$$

Consider the case for full power, which is

$$P_{Omax} = \frac{V_O^2}{R_{Lmin}} = \frac{28^2}{10} = 78.4 \text{ W}. \tag{21.8}$$

The maximum output current is

$$I_{Omax} = \frac{V_O}{R_{Lmin}} = \frac{28}{10} = 2.8 \text{ A}. \tag{21.9}$$

Let us pick the transformer turns ratio to be $n = 2.5$. From (2.72), the input resistance of the rectifier is found to be

$$R_i = \frac{8n^2 R_L}{\pi^2 \eta_R} = \frac{8 \times 2.5^2 \times 10}{\pi^2 \times 0.95} = 53.3\ \Omega. \tag{21.10}$$

Using (2.73), one can calculate the voltage transfer function of the rectifier as

$$M_{VR} = \frac{\pi \eta_R}{2\sqrt{2}n} = \frac{\pi \times 0.95}{2\sqrt{2} \times 2.5} = 0.422. \quad (21.11)$$

It follows from (21.3) that the maximum required voltage transfer function of the inverter is

$$M_{VI} = \frac{V_O}{\eta_I V_{Imin} M_{VR}} = \frac{28}{0.94 \times 270 \times 0.422} = 0.261. \quad (21.12)$$

Assume $cos(\phi/2) = 0.9$. From (12.14), one obtains

$$Q_L = \frac{\sqrt{\frac{2cos^2(\frac{\phi}{2})}{M_{VI}^2 \pi^2} - 1}}{\mid \frac{\omega}{\omega_o} - \frac{\omega_o}{\omega} \mid} = \frac{\sqrt{\frac{2 \times 0.9^2}{0.261^2 \pi^2} - 1}}{\mid 1.33 - \frac{1}{1.33} \mid} = 2.03. \quad (21.13)$$

Hence, using (12.10) and (12.11),

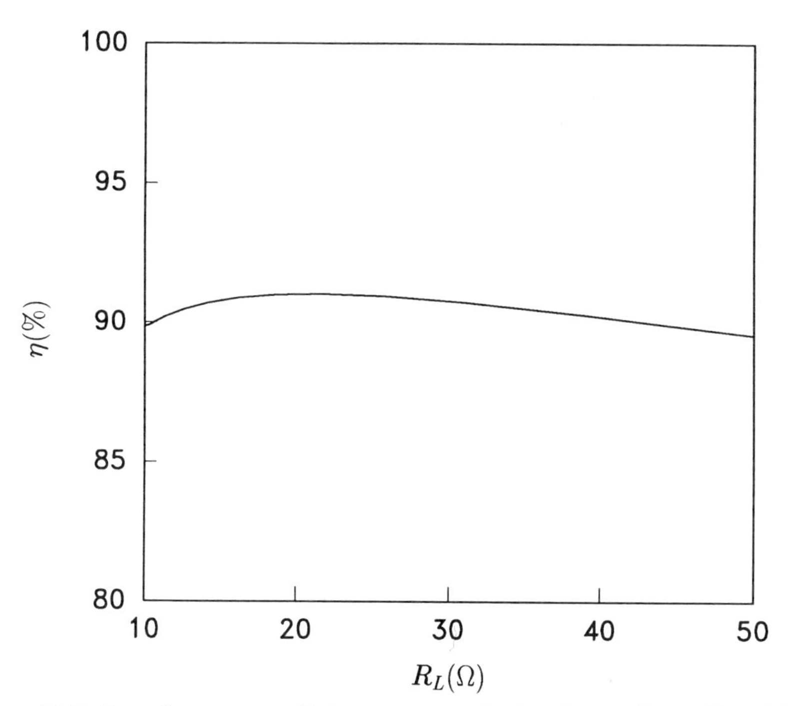

Figure 21.2: Overall converter efficiency η versus load resistance R_L at $V_I = 270$ V and $V_O = 28$ V at $r = 2\ \Omega$, $r_C = 0.2\ \Omega$, $V_F = 0.4$ V, $R_F = 0.075\ \Omega$, and $r_{ESR} = 0.05\ \Omega$.

$$L = \frac{2R_iQ_L}{\omega_o} = \frac{2 \times 53.3 \times 2.03}{2 \times \pi \times 150 \times 10^3} = 229.7\ \mu\text{H} \tag{21.14}$$

and

$$C = \frac{2}{\omega_o^2 L} = \frac{2}{(2 \times \pi \times 150 \times 10^3)^2 \times 229.7 \times 10^{-6}} = 9.8\ \text{nF}. \tag{21.15}$$

A plot of the calculated total converter efficiency η (excluding the drive power) versus R_L is shown in Fig. 21.2 for $V_I = 270$ V and $V_O = 28$ V.

21.4 SUMMARY

- The single-capacitor phase-controlled series resonant converter can regulate the dc output voltage V_O from full load to no load and over a wide range of the line voltage by varying the phase shift between the drive voltages of the two inverters while maintaining a fixed operating frequency.
- Fixed operating frequency of the converter allows for optimization of the magnetic and filter components and easiest reduction of EMI levels.
- The switching legs of both inverters are loaded by inductive loads for $f/f_o > 1.15$.
- The efficiency of the converter is high at light loads because the rectifier input resistance R_i and, thereby, the ratio R_i/r increases with increasing R_L.
- The converter is inherently short-circuit and open-circuit protected by the impedances of the resonant circuits.
- The converter with the transformer turns ratio $n = 1$ is a step-down converter.
- Unlike in most other converters, very low values of the dc-to-dc voltage transfer function are achievable in a single stage.
- The operation at a constant frequency and the inductive loads for both switching legs is achieved in the single-capacitor phase-controlled converter at the expense of a second resonant inductor.
- Fixed-frequency phase-controlled Class E ZVS and ZCS converters can also be obtained.

21.5 REFERENCES

1. H. Chireix, "High power outphasing modulation," *Proc. IRE*, vol. 23, pp. 1370–1392, Nov. 1935.
2. L. F. Gaudernack, "A phase-opposition system of amplitude modulation," *Proc. IRE*, vol. 26, pp. 983–1008, Aug. 1938.
3. F. H. Raab, "Efficiency of outphasing RF power-amplifier systems," *IEEE Trans. Commun.*, vol. COM-33, pp. 1094–1099, Oct. 1985.

4. P. Savary, M. Nakaoka, and T. Maruhashi, "Resonant vector control base high frequency inverter," *IEEE Power Electronics Specialists Conference Record*, 1985, pp. 204–213.
5. I. J. Pitel, "Phase-modulated resonant power conversion techniques for high-frequency link inverters," *in Proc. IEEE Industrial Applications Society Annual Meeting*, 1985; reprinted in *IEEE Trans. Ind. Appl.*, vol. IA-22, pp. 1044–1051, Nov./Dec. 1986.
6. F. S. Tsai and F. C. Y. Lee, "Constant-frequency phase-controlled resonant power processors," *Proceedings of IEEE Industry Applications Society Annual Meeting*, Denver, CO, 1986, Sept. 28–Oct. 3, pp. 617–622
7. F. S. Tsai, P. Materu, and F. C. Y. Lee, "Constant-frequency clamped-mode resonant converter," *IEEE Power Electronics Specialists Conference Record*, 1987, pp. 557–566.
8. F. S. Tsai and F. C. Y. Lee, "A complete DC characterization of a constant-frequency, clamped-mode, series-resonant converter," *IEEE Power Electronics Specialists Conference Record*, Kyoto, Japan, 1988, April 11–14, pp. 987–996.
9. F. S. Tsai, Y. Chin, and F. C. Y. Lee, "State-plane analysis of a constant-frequency clamped-mode parallel-resonant converter," *IEEE Trans. Power Electron.*, vol. PE-3, pp. 364–378, July 1988.
10. Y. Chin and F. C. Y. Lee, "Constant-frequency parallel resonant converter," *IEEE Trans. Ind. Appl.*, vol. IA-25, pp. 133–142, Jan./Feb. 1989.
11. P. Jain, "Performance comparison of pulse width modulated resonant mode dc/dc converter in space applications," in *IEEE Industry Applications Society Annual Meeting*, Oct. 1989, pp. 1106–1114.
12. C. Q. Hu, X. Z. Zhang, and S. P. Huang, "Class-E combined-converter by phase-shift control," in *IEEE Power Electronics Specialists Conf. Rec.*, 1989, pp. 229–234.
13. M. K. Kazimierczuk, "Synthesis of phase-modulated resonant dc/ac inverters and dc/dc convertors," *IEE Proc., Pt. B, Electric Power Appl.*, vol. 139, pp. 387–394, July 1992.
14. D. Czarkowski and M. K. Kazimierczuk, "Phase-controlled CLL resonant converter," *IEEE Applied Power Electronics Conference (APEC'93)*, San Diego, CA, March 7–11, 1993, pp. 432–438.
15. M. K. Kazimierczuk and D. Czarkowski, "Phase-controlled series resonant converter," *IEEE Power Electronics Specialists Conf. (PESC'93)*, Seattle, WA, June 20–24, 1993, pp. 1002–1008.
16. D. Czarkowski and M. K. Kazimierczuk, "Single-capacitor phase-controlled CLL resonant converter," *IEEE Trans. Circuits Syst.*, vol. CAS-40, pp. 383–391, June 1993.
17. D. Czarkowski and M. K. Kazimierczuk, "Phase-controlled series-parallel resonant converter," *IEEE Trans. Power Electron.*, vol. PE-8, pp. 309–319, July 1993.
18. M. K. Kazimierczuk and M. K. Jutty, "Phase-modulated series-parallel converter with series load," *IEE Proc., Pt. B, Electric Power Appl.*, vol. 140, pp. 297–306, Sept. 1993.
19. M. K. Kazimierczuk, D. Czarkowski, and N. Thirunarayan, "A new phase-controlled parallel resonant converter," *IEEE Trans. Ind. Electron.*, vol. IE-40, pp. 542–552, Dec. 1993.
20. R. L. Steigerwald and K. D. T. Ngo, "Half-bridge lossless switching converter," U.S. Patent no. 4,864,479, Sept. 5, 1989.

21. R. Redl, N. O. Sokal, and L. Balogh, "A novel soft-switching full-bridge dc/dc converter: analysis, design considerations, and experimental results at 1.5 kW, 100 kHz," *IEEE Trans. Power Electron.*, vol. 6, pp. 408–418, July 1991.
22. F. S. Tsai, J. Sabate, and F. C. Y. Lee, "Constant-frequency zero-voltage-switched parallel-resonant converter," *IEEE INTELEC Conf.*, Florence, Italy, Oct. 15–18, 1991, Paper 16.4, pp 1–7.

21.6 REVIEW QUESTIONS

21.1 Is it possible to regulate the dc output voltage in phase-controlled converters at a constant switching frequency?

21.2 Is it possible to regulate the dc output voltage in phase-controlled converters from full load to no load?

21.3 When are the switching legs in the single-capacitor phase-controlled series resonant converter loaded by inductive loads?

21.4 Is it possible to achieve zero-voltage-switching turn-on for all the transistors in the single-capacitor phase-controlled converter?

21.5 Is the single-capacitor phase-controlled series resonant converter short-circuit proof?

21.6 Is the single-capacitor phase-controlled series resonant converter open-circuit proof?

21.7 PROBLEMS

21.1 A transformerless single-capacitor phase-controlled series resonant converter with a half-wave rectifier supplies power of 200 W to a 50-Ω load resistance. The parameters of the inverter are input voltage $V_I = 220$ V, normalized switching frequency $\omega/\omega_o = 1.3$, loaded quality factor $Q_L = 3$, efficiency $\eta_I = 94\%$, and $\phi = 30°$. Calculate the efficiency of the converter.

21.2 A single-capacitor phase-controlled series resonant converter with a transformer center-tapped rectifier converts 280 V to 48 V with 91% efficiency. The inverter has the following parameters: switching frequency $f = 200$ kHz, resonant inductance $L = 542.2$ μH, resonant capacitance $C = 4.7$ nF, and $\phi = 25°$. The efficiency of the rectifier is 95% and the transformer turns ratio is $n = 1$. What is the output power of the converter?

21.3 Design a single-capacitor phase-controlled series resonant converter with a half-wave rectifier. The following specifications should be met: $V_I = 200$ V, $V_O = 28$ V, and $P_{Omax} = 50$ W. Assume the resonant frequency $f_o = 120$ kHz, the normalized switching frequency $f/f_o = 1.25$, the inverter efficiency $\eta_I = 94\%$, the rectifier efficiency $\eta_R = 95\%$, the transformer turns ratio $n = 4$, and $cos(\phi/2) = 0.9$ at full load.

ANSWERS TO SELECTED PROBLEMS

2.1 With *pn* junction diodes, $\eta_R = 70.88\%$.
With Schottky diodes, $\eta_R = 83.4\%$.

2.3 $\eta_R = 67.35\%$.
$M_{VR} = 0.1496$.
$R_i = 7.52\ \Omega$.

2.5 $r_C = 10.4\ \text{m}\Omega$.

2.6 $r_C = 38.2\ \text{m}\Omega$.

2.7 $f_{z2} = 311.3$ kHz.

3.1 $\eta_R = 95.9\%$.
$M_{VR} = 0.216$.
$R_i = 2058.3\ \Omega$.

3.2 $\eta_R = 96\%$.
$M_{VR} = 0.4322$.
$R_i = 514\ \Omega$.

3.4 $\eta_R = 54.5\%$.
$M_{VR} = 0.098$.
$R_i = 14.15\ \Omega$.

6.1 $f_o = 1$ MHz.
$Z_o = 529.2\ \Omega$.
$Q_L = 2.627$.
$Q_o = 365$.
$Q_{Lo} = 378$.
$Q_{Co} = 10584$.

6.2 $Q = 65.2$ VA.
$P_R = 24.8$ W.

6.3 $I_m = 0.4964$ A.
$V_{Cm} = V_{Lm} = 262.7$ V.
$Q = 65$ VA.

6.4 $\eta_r = 99.28\%$.

6.6 $V_{SM} = 400$ V for both inverters.

6.7 $V_{Cm} = V_{Lm} = 585.65$ V.

7.1 $f_r = 149.5$ kHz.

7.2 $V_{Cm} = 304.26$ V.
$V_{Lm} = 319.9$ V.

8.2 $V_{Rim} = 226.4$ V.

8.3 $V_{Lm} = 407.1$ V.
$V_{C1m} = 245.8$ V.

9.3 $V_{Rim} = 80.15$ V.

9.4 $V_{Cm} = 111.4$ V.

10.1 $t_1 = 0.12$ μs.

10.2 $V_{GSm} = 26.55$ V.

11.1 $R_{Lp} = 98.02$ kΩ.

11.2 $P_R = 559.2$ W.

11.3 $\omega_o = 10^6$ rad/s.
$I_{Lm} = I_{Cm} = 1.2$ A.
$V_{Rim} = V_{Lm} = V_{Cm} = 600$ V.

12.1 $\phi = 89.97°$.

12.2 $P_{Ri} = 33.98$ W.

12.3 $\mathbf{M_{VI}} = \frac{cos(\frac{\phi}{2})}{\sqrt{2}[1+jQ_L(\frac{\omega}{\omega_o} - \frac{\omega_o}{\omega})]}$.

12.4 $\mathbf{M_{VI}} = \frac{cos(\frac{\phi}{2})}{\sqrt{2}[1-(\frac{\omega}{\omega_o})^2+j\frac{1}{Q_L}\frac{\omega}{\omega_o}]}$.

13.2 $V_{SM} = 682$ V.

13.3 $V_{SM} = 1307$ V.

13.5 $f_{max} = 1.069$ MHz.

14.1 $V_{SM} = 973.1$ V.

14.3 $V_{SM} = 359.7$ V.

15.1 $f = 230.5$ kHz.

15.2 $\eta = 60.3\%$.
$V_I = 52.2$ V.

16.1 $L = 209.7$ μH.

16.2 $\eta = 43.81\%$.

17.1 $\eta = 93.87\%$.

17.2 $L = 270.9$ μH.

18.1 $Z_o = 772.2$ Ω.

18.2 $P_I = 46$ W.

19.1 $P_I = 164.8$ W.

19.2 $f = 89.9$ kHz.

20.1 $V_O = 100$ V.

20.2 $C_2 = 15.9$ nF.

21.1 $\eta = 88.5\%$.

21.2 $P_O = 27.42$ W.

INDEX